Advances in Finite Geometries and Designs

Advances in Finite Geometries and Designs

Proceedings of the Third Isle of Thorns Conference 1990

Edited by

J. W. P. Hirschfeld
Reader in Mathematics, University of Sussex

D. R. Hughes
Professor of Mathematics, Queen Mary and Westfield College, London

J. A. Thas
Professor of Mathematics, State University of Ghent

OXFORD   NEW YORK   TOKYO
OXFORD UNIVERSITY PRESS
1991

Oxford University Press, Walton Street, Oxford OX2 6DP
Oxford New York Toronto
Delhi Bombay Calcutta Madras Karachi
Petaling Jaya Singapore Hong Kong Tokyo
Nairobi Dar es Salaam Cape Town
Melbourne Auckland
and associated companies in
Berlin Ibadan

Oxford is a trade mark of Oxford University Press

Published in the United States
by Oxford University Press, New York

© Oxford University Press, 1991

A catalogue record for this book is available from the British Library

Library of Congress Cataloging in Publication Data
Isle of Thorns Conference (3rd: 1990: University of Sussex)
Advances in finite geometries and designs: proceedings of the
Third Isle of Thorns Conference 1990/edited by J. W. P. Hirschfeld,
D. R. Hughes, J. A. Thas.
p. cm.
1. Finite geometries—Congresses. 2. Combinatorial designs and
configurations—Congresses. I. Hirschfeld, J. W. P. (James William
Peter), 1940– . II. Hughes, D. R. (Daniel R.) III. Thas, J. A.
(Joseph Adolf) IV. Title.
QA167.2.174 1990 511'.6—dc20 91–16277
ISBN 0–19–853592–9

Printed in Great Britain by
Courier International Ltd
Tiptree, Essex

Preface

This book contains articles based on talks at the Third Isle of Thorns Conference on Finite Geometries and Designs, which took place from 15 to 21 July, 1990, at the White House, Isle of Thorns, at Chelwood Gate in East Sussex, a conference centre run by the University of Sussex.

There are 34 articles in all. There were 67 participants and 45 talks; all these are listed at the end of the book. Several good articles were rejected because of insufficient space.

The editors, who were also the conference organisers, are grateful for the financial support of the Science and Engineering Research Council, the London Mathematical Society and the School of Mathematical and Physical Sciences of the University of Sussex. Both David Tranah of Cambridge University Press and Martin Gilchrist of Oxford University Press presented book exhibitions.

Above all we are grateful to Jery Holliss for the production of the camera-ready copy, using Microsoft Word 4 on an Apple Macintosh computer.

> J.W.P.H.
> D.R.H.
> J.A.T.

> March 1991

The death of Jery Holliss is a sad conclusion to the conference. He was an unrelenting perfectionist to whom I am glad to dedicate this volume.

> J.W.P.H.
> April 1991

Table of Contents

Introduction

As in the introduction of Finite Geometries and Designs (Cambridge University Press, 1981), the proceedings of the previous Isle of Thorns conference, finite projective spaces (Galois geometries) may be considered as the central concept.

All the papers in this collection either consider particular substructures of Galois geometries or related incidence structures. The latter structures usually have important examples embedded in some sense in a finite projective space: non-Desarguesian planes, strongly regular graphs, partial geometries, generalized n-gons, generalized quadrangles, maximum distance separable linear codes, nets, diagram geometries, linear spaces, partial linear spaces, difference sets. Some of the substructures considered are, in a Galois geometry, k-arcs, (hyper)ovals, cubic curves, (k, t)-arcs, nuclei, flocks. A continual theme, either more or less prominent, in both types of structure is the action of an automorphism group.

In this Introduction we will not be quite as ambitious as in the previous proceedings, where the purpose was "to attempt to trace some of the main threads of finite geometry, and to locate the papers of this collection in the warp and woof of its fabric." Here we place the papers within a topic and summarise each of them. The placing of the papers under a topic is somewhat arbitrary both because some papers lie within more than one topic and because some pairs of topics have an intersection. Of necessity some papers are described more precisely than others.

It is convenient to collect initially the axioms for various incidence structures. Let S be a connected incidence structure of points and lines such that the lines cover the points and such that each line contains at least two points; assume that there is at least one line. Consider the following further axioms:

(1)   there are $s + 1$ points on a line, $s > 0$;

(2)   there are $t + 1$ lines through a point, $t > 0$;

(3)   two points are incident with at most one line;

(4)   two lines are incident with at most one point;

(5)   any two flags lie in a circuit of n points and n lines, $n > 2$;

(6)   every proper circuit contains at least n points and n lines;

(7)   given a non-incident point-line pair, there are α lines through the point meeting the line, $\alpha > 0$;

(8)   through any two points there are λ lines.

If S satisfies (1) to (6), it is a generalized n-gon; when $n = 4$ it is a generalized quadrangle. If S satisfies (1) to (4), it is a point-line geometry. If S satisfies (1) to (4) and (7) it is a partial geometry. If S

satisfies (3), it is a partial linear space. If S satisfies (8) with $\lambda = 1$, then it is a linear space. If S satisfies (1), (2) and (8), it is a 2-design; in this case it is usual to replace $s + 1$ by k and $t + 1$ by r.

Generalized quadrangles, n-gons

Kantor examines the method of constructing generalized quadrangles from a family of subgroups of a finite group, which was due to him and was in fact a much more general construction than might have been thought at first sight. Here he examines general properties of the automorphism group of generalized quadrangles constructed in this way. Generalized quadrangles can also be derived from flocks of cones, which are partitions by conics of the points of a quadric cone other than the vertex in $PG(3, q)$; this is a two-way process. Kantor has also shown how a certain generalized hexagon gives rise to a generalized quadrangle. So this means that one can pass from the hexagon to a flock. Bader shows how to reverse this process and pass from a flock to a hexagon. On a connected theme, Lunardon also considers the relation between flocks and translation planes. A flock F_0 determines q others $F_1, ..., F_q$. It has been shown that if the plane from F_0 is a semifield plane or likeable and all F_i are isomorphic then F_0 is a known type of flock. Here the proof is given without relying on the classification of finite simple groups.

Van Maldeghem studies transitivity in generalized n-gons and establishes that much simpler transitivity properties than are in the definition of "Moufang" n-gons are in fact sufficient to characterize the "classical" ones.

Brouwer considers the question of the finiteness of a generalized quadrangle when one of the parameters is finite; that is, when does s finite imply that t is also finite? Cameron proved this for $s = 2$ and Kantor for $s = 3$ by a complicated argument. Here Brouwer gives a simple proof for $s = 3$.

Groups acting on geometries

Batten and Brown examine criteria for a group G to act as the full translation group of an affine space or an affine plane. They make no assumption that G is abelian; the abelian case is an old result of Baer. Brown looks at an arbitrary abelian group acting on a finite plane with point orbit P and line orbit L and considers divisibility properties of $|P|$ and $|L|$. Siemons and Webb look at a group G acting on two sets Ω_1 and Ω_2 with the same number of fixed points in Ω_1 and Ω_2 for every element of G. It is reasonable to take faithful, transitive representations.   They consider Wielandt's problem: if G acts primitively on Ω_1, does it also act primitively on Ω_2?

Cohen looks at the quaternionic reflection groups and decides how to deduce presentations of these groups from Shepherd's diagrams.

$PG(2, q)$

A k-arc in $PG(2, q)$ is a set of k points with at most two on any line. The maximum value of k is $q + 1$ or $q + 2$ according as q is odd or even: a $(q + 2)$-arc is called a hyperoval.   If a k-arc is maximal with respect to inclusion it is called complete.   O'Keefe, Penttila and Praeger calculate, without a computer, the collineation groups of two recently discovered 34-arcs in $PG(2, 32)$, the Cherowitzo and Payne hyperovals: it is shown that they both have small groups, of respective orders five and ten.   Korchmáros looks at a generalization of the hyperoval in $PG(2, q)$, namely a $(q + t)$-set in $PG(2, q)$ such that every line meets it in 0, 2 or t points: the hyperoval is the case $t = 2$. He examines examples with a large group of collineations and classifies those with doubly transitive groups.   A (k, n)-arc is a set of k points with at most n on any line of the plane.   For a fixed n, not much is known about the maximum value of k.   Hirschfeld and Szönyi look at constructions of large (k, n)-arcs via a problem of squares in a finite field.

Blokhuis and Mazzocca look at an essential property of a $(q + 1)$-arc in a plane of even order and extend it to a plane of any order as follows: if $\mathcal{B}$ is a set of size $q + 1$ in $PG(2, q)$, then the point A is a nucleus of $\mathcal{B}$ if every line through A meets $\mathcal{B}$ precisely once.   If $\mathcal{B}$ is not a line then Blokhuis and Wilbrink showed that the size of $N(\mathcal{B})$, the set of nuclei of $\mathcal{B}$, is at most $q - 1$.   Here a new proof of this result is given.   This leads to the study of quasi-odd sets $\mathcal{A}$ in $AG(2, q)$, q even, where every line meets $\mathcal{A}$ in zero or an odd number of points;   then $\mathcal{A}$ has at most $q - 1$ points.

Voloch obtains improvements in the size $m'(2, q)$ of the second largest complete arc for $q = p^{2h+1}$ with $h > 0$.

A plane cubic has intuitively one absolutely invariant, since a cubic is determined by nine conditions and a projective linear transformation by eight.   Kaneta and Maruta examine what happens to invariants of a cubic in the anomalous cases of characteristic two and three, in which some classical results do not apply.

Linear spaces

Blokhuis, Calderbank, Metsch and Seress look at a problem which connects coding theory with partial linear linear spaces. The size of the largest binary code of constant weight with a fixed length and minimum distance is achieved precisely when a partial linear space with particular parameters exists. This leads to an embedding problem of such spaces.

Partial geometries

De Clerck, Del Fra and Ghinelli examine restrictions on the parameters of substructures of a partial geometry. Hughes looks at the properties of many different examples of rank three partial geometries and obtains general results for the bounds on their diameter.

Diagram geometries

Del Fra, Ghinelli and Pasini look at diagram geometries with the Intersection Property; that is, if there is some element incident with both x and y, then there is an element z such that the elements incident with z are precisely the elements incident with both x and y. In particular, all geometries with diagram $pG.A_m.pG$ are considered.

A subspace in a point-line geometry is a subset of the points containing any line any line of which there are two points in the subset. A geometric hyperplane is a proper subspace meeting every line. Cooperstein and Shult investigate geometric hyperplanes in a wide class of geometries.

Buekenhout revisits geometries first constructed by Hering, which begin from the fact that there is a subgroup of $GL(6, 3)$ isomorphic to $SL(2, 13)$ and, more particularly, that this subgroup lies in $Sp(6, 3)$. So it is natural to study $PG(5, 3)$, with a fixed symplectic polarity, and a group G isomorphic to $PSL(2, 13)$ acting transitively on the 364 points of $PG(5, 3)$.

Non-Desarguesian planes

de Resmini analyses in considerable detail the combinatorial structure of a particular non-Desarguesian plane of order 16, and finds the different types of complete arcs, hyperovals and subplanes of orders two and four.

Ho and Gonçalves look at a group G acting on a projective plane when G is totally irregular; that is, the stabilizer of a point is always non-trivial.   With G non-abelian and simple, they examine the cases that there is either an involutory homology or involutory elation in G.   More particular results are obtained when G is $PSU(3, q)$.

Mason is led from the idea of a Galois group of a field extension K over k to consider a two-dimensional vector space V over K and so $2N$-dimensional over k with G a group of k-linear transformations preserving a spread; when k is finite, is V a free kG-module?   This is the basis reduction problem for projective planes.

Strongly regular graphs

Motivated by the example of a unitary polarity in $PG(n, 4)$, Haemers, Higman and Hobart consider the decomposition of a strongly regular graph into two such graphs; they determine the possible parameters of the subgraphs.   Löwe considers conditions for a strongly regular graph to have a sharply 1-transitive group of automorphisms.   Spence shows that the (40, 27, 18) strongly regular graphs can be used to construct at least 389 non-isomorphic 2-(40, 13, 4) designs.

Designs

An s-fold quasimultiple of a 2-(v, k, λ) design is a 2-$(v, k, s\lambda)$ design. Jungnickel and Tonchev look at the possible block intersection numbers and in particular study, in the case that the design is not actually multiple, when it has only two such numbers.

Moorhouse looks at the rank of the code of a k-net of order n and shows, for a translation net N_k with an abelian translation group, that

$$\text{rank}_p N_k - \text{rank}_p N_{k-1} \geq n - k + 1,$$

where N_{k-1} is any $(k-1)$-subnet.

Wild shows how to construct an infinite family of a particular class of difference sets.

$PG(n, q)$

If S is an $n \times n$ non-singular, non-alternating, symmetric matrix over $GF(q)$, q even, it determines a polarity of $PG(n-1, q)$.   That S is non-alternating means that S has non-zero diagonal and so the self-polar points form a hyperplane.   Liu and Wan look at the geometry of the vector space with the corresponding bilinear form and linear group G. They determine the size of the orbits of G acting on the subspaces of m dimensions.

The notion of a plane k-arc can be generalized to $PG(n, q)$; it is a set of k points, $k \geq n + 1$, with at most n in any $(n - 1)$-space. Storme and Thas consider a k-arc on a normal rational curve Γ and decide when it can be extended to a $(k + 1)$-arc only by a point of Γ. This leads to the result that a normal rational curve in $PG((q + 3)/2, q)$ is complete. More generally, when n is sufficiently small compared to q, then a normal rational curve is complete; for an infinity of primes p, the only restriction on the dimension is $2 \leq n \leq p - 1$. This result is connected to maximum distance separable (MDS) codes. A linear code C over $GF(q)$ is semi-cyclic if when $(x, ..., y, z) \in C$, then so is $(tz, x, ..., y)$ for some non-zero t. Maruta studies general properties of these codes and determines when they are MDS.

Wettl considers a different version of the idea of nucleus to the above. A point P of a set K in $PG(n, q)$ is an m-nucleus if every m-space through P meets $K \backslash \{P\}$ in at most m points. The 1-nuclei of $PG(3, q)$ are particularly studied and the results applied to complete k-arcs in $PG(n, q)$.

Flocks of cones and generalized hexagons

L. Bader

University of Rome II

1. Introduction

It is well known (see for instance Thas (1987)) that to every flock of the quadratic cone in $PG(3, q)$ there corresponds a generalized quadrangle of order (q^2, q), defined as a group coset geometry according to Kantor (1986), and conversely.

Moreover, the generalized quadrangle introduced in Kantor (1980) and associated with the flock of Fisher, Thas and Walker (Thas (1987) or Gevaert and Johnson (1988)) is easily constructible from the dual $G_2(q)$ hexagon as described in Kantor (1980) or Kantor (1984, p.101).

The aim of this paper is to discuss the possibility of defining generalized hexagons from flocks of cones by reversing the previous construction of Kantor.

2. Preliminaries

Here we recall only some definitions. For more details, the reader is referred to Gevaert and Johnson (1988), Kantor (1984, 1985), Payne (1980), Payne and Thas (1984).

Let $F = GF(q)$. Denote by G the group whose elements are those of F^5 and whose product is the following:

$$(a,b,c,d,e)(a'b',c',d',e') = (a+a',b+b', c+c'+ da'+eb', d+d', e+e')$$

with $a,b,c,d,e, a',b',c',d',e' \in F$. Let $\{B_t \mid t \in GF(q)\}$ be a given set of upper triangular 2×2 matrices over $GF(q)$ and put $M_t = B_t + B_t^T$. Let ∞ be a symbol, let $t \in GF(q)$ and define $A_2(\infty) = \{(0, 0, 0, d, e) \mid d, e \in F\}$, $A_3(\infty) = \{(0, 0, c, d, e) \mid c, d, e \in F\}$, $A_2(t) = \{(a, b, (a, b)B_t (a, b)^T, (a, b)M_t) \mid a, b \in F\}$, $A_3(t) = \{(a, b, c + (a, b)B_t (a, b)^T, (a, b)M_t) \mid a, b, c \in F\}$. We remark that the maps $t:G \to G$ defined by $(a, b, c, d, e)^t = (a, b, c + (a, b)B_t(a, b)^T, (d, e) + (a, b)M_t)$, for all $t \in GF(q)$, are automophisms of G and we have $A_2(t) = \{(a, b, 0, 0, 0)^t \mid a, b \in F\}$, $A_3(t) = \{(a, b, c, 0, 0)^t \mid a, b, c \in F\}$. As in Kantor (1986) and Payne (1980, 1985) these subgroups of G define an incidence structure $(\mathcal{P}_1, \mathcal{L}_1, I_1)$ which is a generalized quadrangle with parameters (q^2, q), provided either q is odd and $-\det(M_t - M_v)$ is a nonsquare in F or q is even and $X^2 + X + (x_t + x_u)(z_t +$

$z_u)(y_t + y_u)^{-2}$ is irreducible over F, for all $t, v \in F$ with $t \neq v$, in the following way: let $F^+ = GF(q) \cup \{\infty\}$ and put

$$\mathcal{P}_1 = \{\mathcal{E}, A_3(t) g, g \mid g \in G, t \in F^+\},$$
$$\mathcal{L}_1 = \{[t], A_2(t)g \mid g \in G, t \in F^+\},$$

where $\mathcal{E}$ and $[t]$ are symbols, and the line $[t]$ is incident with $\mathcal{E}$ and $A_3(t)g$ for all $g \in G$, while all other incidences are given by inclusion. We recall that for $t, v, u \in F^+$ with $t \neq v \neq u \neq t$, a necessary and sufficient condition for $(\mathcal{P}_1, \mathcal{L}_1, \mathcal{E}_1)$ to be a generalized quadrangle is that the following properties hold:

$$A_3(t) \cap A_2(v) = 1, \quad A_2(t)A_2(v) \cap A_2(u) = 1.$$

Moreover (see Gevaert and Johnson (1988) or Thas (1987)), the above generalized quadrangle is associated with a flock of the quadratic cone of $PG(3, q)$, which is the intersection of the cone with equation $X_0X_1 - X_2^2 = 0$, with the q planes π_t's with equations $x_tX_0 + z_tX_1 + y_tX_2 + X_3 = 0$ for $t \in GF(q)$ and $B_t = \begin{bmatrix} x_t & y_t \\ 0 & z_t \end{bmatrix}$. For the known examples of flocks of cones, see the list in Gevaert and Johnson (1988) and Bader et al. (1990), Johnson (1990), Payne and Thas (1990).

On the other hand, the dual $G_2(q)$ hexagon is the incidence structure $(\mathcal{P}_2, \mathcal{L}_2, \mathcal{I}_2)$ defined as follows (Kantor (1984, pp.100-101) and Tits (1976)). Let Q be the group whose elements are those of F^5 $(F = GF(q))$ and whose product is

$$(\alpha, \beta, \gamma, \delta, \varepsilon)(\alpha', \beta', \gamma', \delta', \varepsilon')$$
$$= (\alpha + \alpha', \beta + \beta', \gamma + \gamma' + \alpha'\varepsilon - 3\beta'\delta, \delta + \delta', \varepsilon + \varepsilon')$$

with $\alpha, \beta, \gamma, \delta, \varepsilon, \alpha', \beta', \gamma', \delta', \varepsilon' \in F$. For each $t \in GF(q)$, define the automorphism t of Q by $(\alpha, \beta, \gamma, \delta, \varepsilon)^t = (\alpha, \beta + \alpha t, \gamma - \alpha^2 t^3 - 3\beta^2 t - 3\alpha\beta t^2, \delta + \alpha t^2 + 2\beta t, \varepsilon + \alpha t^3 + 3\beta t^2 + 3\delta t)$. Let ∞ be a symbol, let $t \in GF(q)$ and define $A'_1(\infty) = \{(0, 0, 0, 0, \varepsilon) \mid \varepsilon \in F\}$, $A'_2(\infty) = \{(0, 0, 0, \delta, \varepsilon) \mid \delta, \varepsilon \in F\}$, $A'_3(\infty) = \{(0, 0, \gamma, \delta, \varepsilon) \mid \gamma, \delta, \varepsilon \in F\}$, $A'_4(\infty) = \{(0, \beta, \gamma, \delta, \varepsilon) \mid \beta, \gamma, \delta, \varepsilon \in F\}$, $A'_1(t) = \{(\alpha, 0, 0, 0, 0)^t \mid \alpha \in F\}$, $A'_2(t) = \{(\alpha, \beta, 0, 0, 0)^t \mid \alpha, \beta \in F\}$, $A'_3(t) = \{(\alpha, \beta, \gamma, 0, 0)^t \mid \alpha, \beta, \gamma \in F\}$, $A'_4(t) = \{(\alpha, \beta, \gamma, \delta, 0)^t \mid \alpha, \beta, \gamma, \delta \in F\}$. Let $F^+ = GF(q) \cup \{\infty\}$ and put

$$\mathcal{P}_2 = \{\mathcal{E}', A'_4(t)g, A'_2(t)g, g \mid g \in Q, t \in F^+\}$$

$$L_2 = \{[t]', A'_3(t)g, A'_1(t)g \mid g \in Q, t \in F^+\}$$

where $\mathcal{E}'$ and $[t]'$ are symbols, and the line $[t]'$ is incident with $\mathcal{E}'$ and $A'_4(t)g$ for all $g \in Q$, while all other incidences are given by inclusion. If $q > 2$ and $q \equiv 2 \pmod{3}$, the incidence structure whose point set and line set are respectively $\mathcal{P}' = \{\mathcal{E}', A'_3(t)g, g \mid g \in Q, t \in F^+\}$ and $L' = \{[t]', A'_2(t)g \mid g \in Q, t \in F^+\}$ and the incidence is "inherited" by the dual $G_2(q)$ hexagon (for more details, see Kantor (1984, p.101)) is a generalized quadrangle with parameters (q^2, q) which is isomorphic to the one introduced in Kantor (1980) and is associated with the flock of Fisher, Thas and Walker (Thas (1987)).

3. Theorem

Let $\mathcal{F}$ be any flock of the quadratic cone of $PG(3, q)$ whose associated generalized quadrangle is defined by the group G and the set of matrices $\{B_t \mid t \in GF(q)\}$ as in Section 2. For each $t \in F^+$, fix the subgroups $A_1(t)$ and $A_4(t)$ of G, having orders resp. q and q^4, such that $A_1(t) \leq A_2(t) \leq A_3(t) \leq A_4(t)$. Define the incidence structure $(\mathcal{P}, L, \mathrm{I})$ as follows:

$$\mathcal{P} = \{\mathcal{E}, A_4(t)g, A_2(t)g, g \mid g \in G, t \in F^+\}$$
$$L = \{[t], A_3(t)g, A_1(t)g \mid g \in G, t \in F^+\}$$

where $\mathcal{E}$ and $[t]$ are symbols, and the incidence is $\mathcal{E}\mathrm{I}[t]$, $A_4(t)g \mathrm{I}[t]$, $g\mathrm{I}A_1(t)g$ for all $g \in G$, $t \in F^+$ while $A_i(t)g\mathrm{I}A_{i+1}(v)h$ if and only if $v = t$ and $g \in A_{i+1}(v)h$ with $i = 1, 2, 3$.

Lemma 1. $(\mathcal{P}, L, \mathrm{I})$ *has a connected incidence graph G with distance at most 8, not containing circuits of length less than 8.*

Proof. Let $t, v \in F^+$, $g, h \in G$. The following properties hold:

6.5 $d(\mathcal{E}, [t]) = 1$;

6.4 $d(\mathcal{E}, A_4(t)g) = 2$ and the unique path of length < 10 is $\mathcal{E}\mathrm{I}[t]\mathrm{I}A_4(t)g$;

6.3 $d(\mathcal{E}, A_3(t)g) = 3$ and the unique path of length < 9 is $\mathcal{E}\mathrm{I}[t] \mathrm{I} A_4(t)g\mathrm{I} A_3(t)g$;

6.2 $d(\mathcal{E}, A_2(t)g) = 4$ and the unique path of length < 8 is $\mathcal{E}\mathrm{I}[t]\mathrm{I}A_4(t)g\mathrm{I}A_3(t)g\mathrm{I} A_2(t)g$;

6.1 $d(\mathcal{E}, A_1(t)g) = 5$ and the unique path of length < 7 is $\mathcal{E}I[t]IA_4(t)gIA_3(t)gI\,A_2(t)gIA_1(t)g$;

6.0 $d(\mathcal{E}, g) = 6$, a path being the one through $[t]$ for all $t \in F^+$;

5.5 $d([t], [v]) = 2$ if $v \neq t$, and the unique path of length < 10 is $[t]\,I\mathcal{E}I[v]$;

5.4.i $d([t], A_4(t)g) = 1$;

5.4.ii $d([t], A_4(v)h) = 3$ if $v \neq t$, and the unique path of length < 9 is the one through $\mathcal{E}$;

5.3.i $d([t], A_3(t)g) = 2$ and the unique path of length < 10 is $[t]IA_4(t)gI\,A_3(t)g$;

5.3.ii $d([t], A_3(v)g) = 4$ if $v \neq t$, and the unique path of length < 8 is the one through $\mathcal{E}$;

5.2.i $d([t], A_2(t)g) = 3$ and the unique path of length < 9 is $[t]IA_4(t)g\,IA_3(t)g\,IA_2(t)g$;

5.2.ii $d([t], A_2(v)g) = 5$ if $v \neq t$ and the unique path of length < 7 is the one through $\mathcal{E}$;

5.1.i $d([t], A_1(t)g) = 4$ and the unique path of length < 8 is $[t]IA_4(t)g\,IA_3(t)gI\,A_2(t)gIA_1(t)g$;

5.1.ii $d([t], A_1(v)g) = 6$ if $v \neq t$, a path being either the one through $\mathcal{E}$ or $[t]IA_4(t)hIA_3(t)hIA_2(t)hIA_1(t)hIhIhIA_1(v)g$ with $h \in A_1(v)g$;

5.0 $d([t], g) = 5$ and the unique path of length < 7 is $[t]IA_4(t)g\,IA_3(t)g\,IA_2(t)g\,IA_1(t)g\,Ig$;

4.4.i $d(A_4(t)g, A_4(t)h) = 2$ if $h \notin A_4(t)g$, and the unique path of length < 10 is $A_4(t)g\,I[t]IA_4(t)h$;

4.4.ii $d(A_4(t)g, A_4(v)h) = 4$ if $v \neq t$, and the unique path of length < 8 is the one through $\mathcal{E}$;

4.3.i $d(A_4(t)g, A_3(t)h) = 1$ if and only if $h \in A_4(t)g$;

4.3.ii $d(A_4(t)g, A_3(t)h) = 3$ if and only if $h \notin A_4(t)g$, and the unique path of length < 9 is the one through $[t]$;

4.3.iii $d(A_4(t)g, A_3(v)h) = 5$ if $v \neq t$, and the unique path of length < 7 is the one through $\mathcal{E}$;

4.2.i $d(A_4(t)g, A_2(t)h) = 2$ if and only if $h \in A_4(t)g$, a path being $A_4(t)g\,IA_3(t)\,hIA_2(t)h$;

4.2.ii $d(A_4(t)g, A_2(t)h) = 4$ if and only if $h \notin A_4(t)g$, and the unique path of length < 8 is the one through $[t]$;

4.2.iii $d(A_4(t)g, A_2(v)h) = 6$ if $v \neq t$, a path being either the one through $\mathcal{E}$ or $A_4(t)g\,IA_3(t)k'_iIA_2(t)k_i\,IA_1(t)k_i\,Ik_i\,IA_1(v)k_i\,IA_2(v)h$ for $k_i \in A_4(t)g \cap A_2(v)h$ (in the generalized quadrangle, each of the q points $A_3(t)k'_i$ $(i = 1, ..., q; k'_i \in A_4(t)g)$ is non incident with the line $A_2(v)h$, hence there exists a unique line $A_2(t)k_i$ through $A_3(t)k'_i$ which is incident with $A_2(v)h$, the common point being k_i);

4.1.i $d(A_4(t)g, A_1(t)h) = 3$ if and only if $h \in A_4(t)g$, a path being $A_4(t)gIA_3(t)h\ IA_2(t)hIA_1(t)h$;

4.1.ii $d(A_4(t)g, A_1(t)h) = 5$ if and only if $h \notin A_4(t)g$, and the unique path of length < 7 is the one through $[t]$;

(4.1.iii)' $d(A_4(t)g, A_1(v)h) \leq 7$ if $v \neq t$ (see, for instance, 4.2.iii);

4.0.i $d(A_4(t)g, h) = 4$ if and only if $h \in A_4(t)g$, a path being $A_4(t)gIA_3(t)hI\ A_2(t)hIA_1(t)hIh$;

4.0.ii $d(A_4(t)g, h) = 6$ if and only if $h \notin A_4(t)g$, a path being the one through $[t]$;

3.3.i $d(A_3(t)g, A_3(t)h) = 2$ if and only if $h \in A_4(t)g \setminus A_3(t)g$, a path being $A_3(t)gIA_4(t)g\ IA_3(t)h$;

3.3.ii $d(A_3(t)g, A_3(t)h) = 4$ if and only if $h \notin A_4(t)g$, a path being the one through $[t]$;

3.3.iii $d(A_3(t)g, A_3(v)h) = 6$ if $v \neq t$, a path being either the one through $\mathcal{E}$ or $A_3(t)g\ IA_2(t)k'_i\ IA_1(t)k_i\ Ik_i\ IA_1(v)k_i\ IA_2(v)k_i\mathcal{E}A_3(v)h$ for $k_i \in A_3(t)g \cap A_3(v)h, i = 1, ..., q$ (in the generalized quadrangle, for each of the q lines $A_2(t)k'_i, i = 1, ..., q$ through the point $A_3(t)g$, there exists a unique line $A_2(v)k_i$ through the point $A_3(v)h$ which is incident with $A_2(t)k'_i$, the common point being k_i);

3.2.i $d(A_3(t)g, A_2(t)h) = 1$ if and only if $h \in A_3(t)g$;

3.2.ii $d(A_3(t)g, A_2(t)h) = 3$ if and only if $h \in A_4(t)g \setminus A_3(t)g$, a path being the one through $A_4(t)g$;

3.2.iii $d(A_3(t)g, A_2(t)h) = 5$ if and only if $h \notin A_4(t)g$, and the unique path of length < 7 is the one through $[t]$;

3.2.iv $d(A_3(t)g, A_2(v)h) = 5$ if $v \neq t$ and the unique path of length < 7 is $A_3(t)gI\ A_2(t)kIA_1(t)kIkIA_1(v)kIA_2(v)h$ with $k = A_3(t)g \cap A_2(v)h$ (in the generalized quadrangle, k is the unique point of the line $A_2(v)h$ which is collinear with the point $A_3(t)g$);

3.1.i $d(A_3(t)g, A_1(t)h) = 2$ if and only if $h \in A_3(t)g$, a path being $A_3(t)g IA_2(t)h\ IA_1(t)h$;

3.1.ii $d(A_3(t)g, A_1(t)h) = 4$ if and only if $h \in A_4(t)g \setminus A_3(t)g$, a path being the one through $A_4(t)g$;

3.1.iii $d(A_3(t)g, A_1(t)h) = 6$ if and only if $h \notin A_4(t)g$, a path being the one through $[t]$;

3.1.iv $d(A_3(t)g, A_1(v)h) = 4$ if and only if $v \neq t$ and $k \in A_1(v)h$, where $\{k\} = A_3(t)g \cap A_2(v)h$ (see 3.2.iv), a path being the one through k and $A_2(v)h$;

3.1.v $d(A_3(t)g, A_1(v)h) = 6$ if and only if $v \neq t$ and $k \notin A_1(v)h$, where $\{k\} = A_3(t)g \cap A_2(v)h$, a path being the one through k;

3.0.i $d(A_3(t)g, h) = 3$ if and only if $h \in A_3(t)g$, a path being $A_3(t)g$ $IA_2(t)h$ $IA_1(t)hIh$;

3.0.ii $d(A_3(t)g, h) = 5$ if $h \in A_4(t)g \setminus A_3(t)g$, a path being the one through $A_4(t)g$;

(3.0.iii)' $d(A_3(t)g, h) \leq 7$ if $h \notin A_4(t)g$ (see, for instance, 3.1.iii);

2.2.i $d(A_2(t)g, A_2(t)h) = 2$ if and only if $h \in A_3(t)g \setminus A_2(t)g$, a path being $A_2(t)gIA_3(t)g$ $IA_2(t)h$;

2.2.ii $d(A_2(t)g, A_2(t)h) = 4$ if and only if $h \in A_4(t)g \setminus A_3(t)g$, a path being the one through $A_4(t)g$;

2.2.iii $d(A_2(t)g, A_2(t)h) = 6$ if and only if $h \notin A_4(t)g$, a path being the one through $[t]$;

2.2.iv $d(A_2(t)g, A_2(v)h) = 4$ if and only if $v \neq t$ and $A_2(t)g \cap A_2(v)h = k$ for some $k \in G$ ($A_2(t)g$ and $A_2(v)h$ are incident lines of the generalized quadrangle), and the unique path of length < 8 is the one through k;

2.2.v $d(A_2(t)g, A_2(v)h) = 6$ if and only if $v \neq t$ and $A_2(t)g \cap A_2(v)h = \varnothing$ ($A_2(t)g$ and $A_2(v)h$ are non incident lines of the generalized quadrangle), a path being $A_2(t)g$ $IA_3(t)gIA_2(t)kIA_1(t)kIkIkIA_1(v)kIA_2(v)h$ with $k = A_3(t)g \cap A_2(v)h$ (see 3.2.iv);

2.1.i $d(A_2(t)g, A_1(t)h) = 1$ if and only if $h \in A_2(t)g$;

2.1.ii $d(A_2(t)g, A_1(t)h) = 3$ if and only if $h \in A_3(t)g \setminus A_2(t)g$, a path being the one through $A_3(t)g$;

2.1.iii $d(A_2(t)g, A_1(t)h) = 5$ if and only if $h \in A_4(t)g \setminus A_3(t)g$, a path being the one through $A_4(t)g$;

2.1.iv $d(A_2(t)g, A_1(v)h) = 3$ if and only if $v \neq t$ and in the generalized quadrangle the lines $A_2(t)g$ and $A_2(v)h$ are incident in a point k such that $k \in A_1(v)h$, a path being the one through k;

2.1.v $d(A_2(t)g, A_1(v)h) = 5$ if $v \neq t$ and if in the generalized quadrangle the lines $A_2(t)g$ and $A_2(v)h$ are incident in a point k such that $k \in A_2(v)h \setminus A_1(v)h$, a path being the one through k;

(2.1.vi)' $d(A_2(t)g, A_1(v)h) \leq 7$ if either $v = t$ and $h \notin A_4(t)g$ or $v \neq t$ and $A_2(t)g, A_2(v)h$ are non incident lines of the generalized quadrangle (see, for instance, 2.2.iii and 2.2.v);

2.0.i $d(A_2(t)g, h) = 2$ if and only if $h \in A_2(t)g$, a path being $A_2(t)g$ $IA_1(t)hIh$;

2.0.ii $d(A_2(t)g, h) = 4$ if $h \in A_3(t)g \setminus A_2(t)g$, a path being the one through $A_3(t)g$;

2.0.iii $d(A_2(t)g, h) = 4$ if $h \notin A_3(t)g$ and if there exists a (unique) pair $(v, k) \in F^+ \times G$ such that $k \in A_1(v)h \cap A_2(t)g$ (in the generalized quadrangle, $A_2(v)h$ is the unique line through h which is incident with $A_2(t)g$, the common point being k, and $k \in A_1(v)h$ holds), a path being the one through k;

2.0.iv $\quad d(A_2(t)g, h) = 6$ if $h \notin A_3(t)g$ and if $A_2(t)g \cap A_1(v)h = \emptyset$ for all $v \in F^+$, a path being the one through k (where in the generalized quadrangle $A_2(v)h$ is the unique line through h which is incident with $A_2(t)g$, the common point being k);

1.1.i $\quad d(A_1(t)g, A_1(t)h) = 2$ if and only if $h \in A_2(t)g \setminus A_1(t)g$, a path being the one through $A_2(t)g$;

1.1.ii $\quad d(A_1(t)g, A_1(t)h) = 4$ if $h \in A_3(t)g \setminus A_2(t)g$, a path being the one through $A_3(t)g$;

1.1.iii $\quad d(A_1(t)g, A_1(t)h) = 4$ if $h \notin A_3(t)g$ and if there exists a point $h_0 \in A_1(t)h$ such that, if $A_2(v)h_0$ is the unique line of the generalized quadrangle through h_0 which is incident with $A_2(t)g$, then there exists $k_0 \in A_1(v)h_0 \cap A_1(t)g$, a path being $A_1(t)g$ I k_0 I$A_1(v)h_0$ Ih_0 I $A_1(t)h$;

(1.1.iv)' $\quad d(A_1(t)g, A_1(t)h) \leq 6$ if $h \in A_4(t)g \setminus A_3(t)g$, a path being the one through $A_4(t)g$;

1.1.v $\quad d(A_1(t)g, A_1(t)h) = 6$ if $h \notin A_4(t)g$ and there exists a point $h_0 \in A_1(t)h$ (resp. $g_0 \in A_1(t)g$) such that the unique line $A_2(v)h_0$ (resp. $A_2(v)g_0$) of the generalized quadrangle through h_0 (resp.g_0) which is incident with $A_2(t)g$ (resp. $A_2(t)h$), the common point being k_0, satisfies either $k_0 \in A_1(v)h_0$ (resp. $k_0 \in A_1(v)g_0$) or $k_0 \in A_1(t)g$ (resp. $k_0 \in A_1(t)h$), a path being either $A_1(t)g$I$A_2(t)g$I $A_1(t)k_0$ Ik_0 I$A_1(v)h_0$ Ih_0 I$A_1(t)h$ (resp. $A_1(t)g$Ig_0I$A_1(v)g_0$Ik_0I$A_1(t)k_0$I$A_2(t)h$I$A_1(t)h$) or $A_1(t)g$Ik_0 I$A_1(v)k_0$ I$A_2(v)h_0$ I$A_1(v)h_0$ I h_0 I $A_1(t)h$ (resp. $A_1(t)g$Ig_0I$A_1(v)g_0$ I$A_2(v)$ g_0 I$A_1(v)$ k_0 Ik_0 I$A_1(t)h$);

(1.1.vi)' $\quad d(A_1(t)g, A_1(t)h) \leq 8$ if $h \notin A_4(t)g$ and for all $h_0 \in A_1(t)h$, $h_1 \in A_2(t)h$, $g_0 \in A_1(t)g$ and $g_1 \in A_2(t)g$ we have $h_0 g_1^{-1}, h_1 g_0^{-1} \notin A_2(v)$ for all $v \in F^+$ (see, for instance, 2.2.iii);

1.1.vii $\quad d(A_1(t)g, A_1(v)h) = 2$ if and only if $v \neq t$ and in the generalized quadrangle the lines $A_2(t)g$ and $A_2(v)h$ are incident in a point k such that $k \in A_1(t)g \cap A_1(v)h$, a path being the one through k;

1.1.viii $\quad d(A_1(t)g, A_1(v)h) = 4$ if $v \neq t$ and in the generalized quadrangle the lines $A_2(t)g$ and $A_2(v)h$ are incident in a point k such that either $k \in A_1(t)g \setminus A_1(v)h$ or $k \in A_1(v)h \setminus A_1(t)g$, a path being the one through k;

1.1.ix $\quad d(A_1(t)g, A_1(v)h) = 6$ if $v \neq t$ and in the generalized quadrangle the lines $A_2(t)g$ and $A_2(v)h$ are incident in a point k such that $k \notin A_1(t)g \cup A_1(v)h$, a path being the one through k;

(1.1.x)' $\quad d(A_1(t)g, A_1(v)h) \leq 8$ if $v \neq t$ and $A_2(t)g$ and $A_2(v)h$ are non incident lines of the generalized quadrangle (see, for instance, 2.2.v);

1.0.i $\quad d(A_1(t)g, h) = 1$ if and only if $h \in A_1(t)g$;

1.0.ii $d(A_1(t)g, h) = 3$ if $h \in A_2(t)g \setminus A_1(t)g$, a path being the one through $A_2(t)g$;

1.0.iii $d(A_1(t)g, h) = 5$ if $h \in A_3(t)g \setminus A_2(t)g$, a path being the one through $A_3(t)g$;

(1.0.iv)' $d(A_1(t)g, h) \leq 7$ if $h \notin A_3(t)g$ (see, for instance, 2.0.iii and 2.0.iv);

0.0.i $d(g, h) = 2$ if and only if $gh^{-1} \in A_1(t)$ for some $t \in F^+$;

0.0.ii $d(g, h) = 4$ if $gh^{-1} \in A_2(t) \setminus A_1(t)$ for some $t \in F^+$, a path being the one through $A_2(t)g$;

(0.0.iii)' $d(g, h) \leq 6$ if $gh^{-1} \in A_3(t)$ for some $t \in F^+$, a path being the one through $A_3(t)g$;

0.0.iv $d(g, h) = 4$ if $gh^{-1} \notin A_2(u)$ for all $u \in F^+$ but there exist $t, v \in F^+$ and $k \in G$ such that $k \in A_1(t)g \cap A_1(v)h$, a path being g $IA_1(t)gIkIA_1(v)hIh$;

(0.0.v)' $d(g, h) \leq 8$ if $gh^{-1} \notin A_3(t)$ for all $t \in F^+$ (see, for instance, 2.0.iv).

Remark 2.   By the proof of the Lemma, the graph G does not contain any circuit of length < 12 through $\mathcal{E}$ or $[t]$, for all $t \in F^+$.

Let us now state the following properties:

(P1) $A_1(t) \cap A_4(v) = 1$ for all $t, v \in F^+$

(P2)   Let $g, h \in G$ and $t \in F^+$ such that $h \notin A_4(t)g$. Let $A_2(t)g_1, ...,$ $A_2(t)g_q$ be the q transversals of $A_2(t)$ contained into $A_3(t)g$ (that is, $[t]$, $A_2(t)g_1, ..., A_2(t)g_q$ are the $q + 1$ lines of the generalized quadrangle through $A_3(t)g$). For all $i = 1, ..., q$, let k_i (resp. $A_2(v_i)k_i$) be the unique point of $A_2(t)g_i$ (resp. line through h) such that $h \in A_2(v_i)k_i$. Then, there exists exactly one $j \in \{1, ..., q\}$ such that $h \in A_1(v_j)h_j$, or, equivalently, exactly one of the lines $A_2(v_i)k_i$ of the generalized quadrangle can be "shortened" to a line of the generalized hexagon (see below)

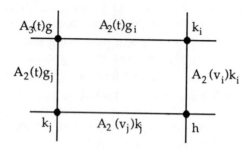

(P3) Let $g, h \in G$, $t, v \in F^+$ and suppose $A_2(t)g$ and $A_2(v)h$ are non incident lines of the generalized quadrangle. If $v = t$, suppose also $h \notin A_4(t)g$. Let $\{h_1, \ldots, h_q\} = A_1(v)h$. For all $i = 1, \ldots, q$, let k_i (resp. $A_2(u_i)k_i$) be the unique point of $A_2(t)g$ (resp. line through h_i) such that $h_j \in A_2(u_i)k_i$. Then, there exists exactly one $j \in \{1, \ldots, q\}$ such that $h_j \in A_1(u_j)k_j$, or, equivalently, exactly one of the lines $A_2(u_i)k_i$ of the generalized quadrangle can be "shortened" to a line of the generalized hexagon (see below)

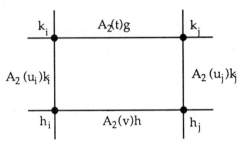

(P4) Let $g, h \in G$ and $t \in F^+$ such that $h \notin A_3(t)g$. Let $A_2(u')h$ (resp. k') be the unique line (resp. point) of the generalized quadrangle such that $k' \in A_2(t)g \cap A_2(u')h$. Suppose that $k' \notin A_1(t)g \cup A_1(u')h$. Then there exist and are uniquely determined $k, f \in G$ and $u, v \in F^+$ such that $k \in A_1(t)g, f \in A_1(u)h \cap A_1(v)k$, or, equivalently, if none of the lines (of the generalized quadrangle) of the path from g to h through k' can be "shortened" then there exists another path from a point k in $A_1(t)g$ to h (the one through f in the picture below) that can be shortened at both the steps

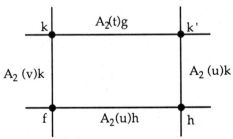

Remark 3. The previous properties can also be formulated in terms of groups. For instance, (P2) is equivalent to $G = \bigcup\limits_{v \in F^+ \setminus \{t\}} A_1(v)A_3(t)$.

Remark 4. If (P1) holds, in (P2) it is sufficient to request $h \notin A_3(t)g$, instead of $h \notin A_4(t)g$ because $h \in A_1(v_j)k_j$ forces $h \notin A_4(t)h_j = A_4(t)g$. If (P1) holds, in (P3) the condition $h \notin A_4(t)g$ if $v = t$, follows from (P1), because $h \in A_1(u_j)k_j$ forces $h \notin A_4(t)h_j = A_4(t)g$. By (P3), if $k' \in A_1(t)g \cup A_1(u')h$, there exist no $k, f \in G$ and $u, v \in F^+$ such that $k \in A_1(t)g$ and $f \in A_1(u)h \cap A_1(v)k$. If (P3) holds, in (P4) it is sufficient to state the existence of k, f, u, v because, by (P3), if such a path exists, it is uniquely determined.

Theorem 5. $(\mathcal{P}, \mathcal{L}, I)$ *is a generalized hexagon if and only if* (P1), (P2), (P3) *and* (P4) *hold.*

Proof. Let $(\mathcal{P}, \mathcal{L}, I)$ be a generalized hexagon. Suppose (P1) is false, i.e. there exist $v, t \in F^+$ such that $g, h \in A_1(v) \cap A_4(t)$ with $g \neq h$: then $A_4(t)gIA_3(t)gI\,A_2(t)gIA_1(t)gIgI\,A_1(v)gIhIA_1(t)hIA_2(t)hIA_3(t)hIA_4(t)g$ is a circuit of length 10. Let $h, g \in G$ and $t \in F^+$ such that $h \notin A_4(t)g$: there is a unique path of length < 6 between h and $A_3(t)g$ and such a path must be of type $A_3(t)gIA_2(t)kIA_1(t)kIkI\,A_1(v)kIh$, so (P2) is proved. Let $g, h \in G$ and $t, v \in F^+$; let $A_2(t)g$ and $A_2(v)h$ be non incident lines of the generalized quadrangle and, for $t=v$, also suppose $h \notin A_4(t)g$: there is a unique path of length < 6 between $A_1(v)h$ and $A_2(t)g$, and such a path must be of type $A_2(t)gIA_1(t)kIkIA_1(u)kIh'IA_1(v)h$, so (P3) is proved. Finally, let $k' \in G$ and $t, u' \in F^+$ such that $t \neq u'$. Then fix $g \in A_2(t)k' \backslash A_1(t)k'$, $h \in A_2(u')k' \backslash A_1(u')k'$; clearly $h \notin A_3(t)g$ because no triangles exist in the generalized quadrangle. There is a unique path of length < 6 between $A_1(t)g$ and h, and such a path must be of type $A_1(t)gIkIA_1(v)kIfIA_1(u)hIh$, and (P4) is proved.

Conversely, suppose (P1), (P2), (P3) and (P4) hold. Then

4.1.iii $d(A_4(t)g, A_1(v)h) = 5$ if $v \neq t$, and the unique path of length < 7 is $A_4(t)gI\,A_3(t)kIA_2(t)kIA_1(t)kIkIA_1(v)h$ with$\{k\} = A_1(v)h \cap A_4(t)g$ (by (P1), the cosets $A_4(t)k_1, ..\ ..., A_4(t)k_q$, for $\{k_1, ..., k_q\} = A_1(v)h$, are pairwise distinct);

3.0.iii $d(A_3(t)g, h) = 5$ if $h \notin A_4(t)g$ by (P2);

2.1.vi $d(A_2(t)g, A_1(v)h) = 5$ if either $v = t$ and $h \notin A_4(t)g$ or $v \neq t$ and in the generalized quadrangle $A_2(t)g$ and $A_1(v)h$ are non incident lines, by (P3);

1.1.iv $d(A_1(t)g, A_1(t)h) = 6$ if $h \in A_4(t)g$ by (P1);

1.0.iv $d(A_1(t)g, h) = 5$ if $h \notin A_2(t)\,g$ by (P4);

1.1.vi $d(A_1(t)g, A_1(t)h) = 6$ if $h \notin A_4(t)g$ by 1.0.iv;

1.1.x $d(A_1(t)g, A_1(t)h) = 6$ if $v \neq t$ and if $A_2(t)g$ and $A_2(v)h$ are non incident lines of the generalized quadrangle, by 2.1.vi;

0.0.v $d(g, h) = 6$ if $gh^{-1} \notin A_3(t)$ for all $t \in F^+$ by 1.0.iv;

To complete the proof, we will show that there are no circuits of length less than 12. By Remark 1, such a circuit does not contain $\mathcal{E}$ or $[t]$, for all $t \in F^+$. Moreover:

(a)   any circuit not containing $\mathcal{E}$ nor any of the lines $[t]$'s, for all $t \in F^+$, but containing some $A_4(t)g$ has length ≥ 12 by (P1);

(b)   any circuit not containing $\mathcal{E}$, nor any of the lines $[t]$'s for all $t \in F^+$, nor any of the points $A_4(t)g$ for all $t \in F^+, g \in G$, but containing some $A_3(t)g$ has length ≥ 12 by (P2);

(c)   any circuit not containing any of the points $\mathcal{E}$ or $A_4(t)g$, nor any of the lines $[t]$'s or $A_3(t)g$ for all $t \in F^+, g \in G$, but containing some $A_2(t)g$ has length ≥ 12 by (P3);

(d)   any circuit containing only points of type $g \in G$ and lines of type $A_1(t)h$ has length ≥ 12 by (P3).

Remark 6.   We explicitly remark that the subgroups $\mathcal{A}_1(t) = \{(\alpha, 0, 0, 0, 0)^t \mid \in F\}$ and $\mathcal{A}_4(t) = \{(\alpha, \beta, \gamma, \delta, 0)^t \mid \alpha, \beta, \gamma, \delta \in F)\}$ for $t \in F$, $\mathcal{A}_1(\infty) = \{(0, 0, 0, 0, \varepsilon) \mid \varepsilon \in F\}$ and $\mathcal{A}_4(\infty) = \{(0, \beta, \gamma, \delta, \varepsilon) \mid \beta, \gamma, \delta, \varepsilon \in F\}$ do not satisfy property (P1), while the subgroups $\overline{\mathcal{A}}_1(t) = \{(0, \beta, 0, 0, 0) \mid \beta \in F\}$ and $\overline{\mathcal{A}}_4(t) = \mathcal{A}_4(t)$ for $t \in F$, $\overline{\mathcal{A}}_1(\infty) = \mathcal{A}_1(\infty)$, $\overline{\mathcal{A}}_4(\infty) = \{(\alpha, 0, \gamma, \delta, \varepsilon) \mid \alpha, \gamma, \delta, \varepsilon \in F\}$ satisfy (P1) but not (P4).

Acknowledgement.   The author is a member of G.N.S.A.G.A. of C.N.R. and has partial financial support by Italian Ministero dell'Università e della Ricerca Scientifica e Tecnologica.

References

L. Bader, G. Lunardon and J.A. Thas (1990), Derivation of flocks of quadratic cones, *Forum Math.* **2**, 163-174.

J.C. Fisher and J.A. Thas (1979), Flocks in $PG(3, q)$, *Math. Z.* **169**, 1-11.

H. Gevaert and N.L. Johnson (1988), Flocks of quadratic cones, generalized quadrangles and translation planes, *Geom. Dedicata* **27**, 301-317.

N.L. Johnson (1990), Derivation of partial flocks of quadratic cones, *Rend. Mat.*, to appear.

W.M. Kantor (1980), Generalized quadrangles associated with $G_2(q)$, *J. Combin. Theory Ser.* **A 29**, 212-219.

W.M. Kantor (1984), Generalized polygons, SCABs and GABs, *Buildings and the Geometry of Diagrams* (ed. L. A. Rosati), Springer-Verlag, Berlin, 79-158.

W.M. Kantor (1985), Generalized quadrangles and translation planes, *Algebras Groups Geom.* **3**, 313-322.

W.M. Kantor (1986), Some generalized quadrangles with parameters (q^2, q), *Math. Z.* **192**, 45-50.

S.E. Payne (1980), Generalized quadrangles as group coset geometries, *Congr. Numer.* **29**, 717-734.

S.E. Payne (1985), A new infinite family of generalized quadrangles, *Congr. Numer.* **49**, 115-128.

S.E. Payne and J.A. Thas (1984), *Finite Generalized Quadrangles*, Pitman, London.

S.E. Payne and J.A. Thas (1990), Conical flocks, partial flocks, derivation, and generalized quadrangles, *Geom. Dedicata*, to appear.

J.A. Thas (1987), Generalized quadrangles and flocks of cones, *European J. Combin.* **8**, 441-452.

J. Tits (1976), Classification of buildings of spherical type and Moufang polygons: a survey, *Teorie Combinatorie*, Volume I, 229-256, Accademia Nazionale dei Lincei, Rome.

Group partitions, affine spaces and translation groups

L.M. Batten and J.M.N Brown
University of Manitoba and York University

1. Introduction

Affine and projective geometries developed 'classically' as the study of invariants of certain groups of projective collineations. The equivalence of desarguesian projective and affine geometries to the well-known constructions from vector spaces over skew-fields was one of the first major results by way of a characterization. For more information on this, the reader is referred to Hilbert (1899), Veblen and Young (1910) and Dembowski (1968).

In the last few decades, much important work has been done to characterize (finite) affine and projective geometries by means of assumptions on subspace structure or on the collineation group. Axioms on points, lines and/or hyperplanes of designs were used by Kantor (1969a) to characterize projective space, whereas Lüneburg (1961) used collineation group conditions; Dembowski and Wagner (1960) used both types of axiom. Dembowski (1968, p.67) gave an extension of the Dembowski-Wagner result.

For affine spaces, point, line and/or hyperplane conditions have been considered by Lenz (1954, 1959), Dembowski (1964, 1967) (Jungnickel and Lenz gave a new proof of this in 1985), Kantor (1969a) and Buekenhout (1969). Characterizations have also been given using collineation group conditions by André (1954), Baer (1963) (see also Lüneburg (1980)), Schulz (1967), Kantor (1969b), Buekenhout (1970) (for order 2 cases), Grochowska (1986) and Sörenson (1988).

In addition, characterizations of the sets of h-dimensional and $(h+1)$-dimensional subspaces of projective and affine spaces have been given by Ray-Chaudhuri and Sprague (1976), Sprague (1981) Bichara and Tallini (1982, 1983) and Biondi and Melone (1985).

In this article, we characterize translation affine spaces of dimension ≥ 2 in terms of group action. Our result is an improvement of a theorem of Baer (1963) which treats the dimension ≥ 3 case using a stronger group hypothesis. The result for dimension 2 is already known and due to André (1954) Our main result is as follows.

Theorem 1. *Let G be a group with non-trivial partition Π such that*

$$G_i \cap G_j G_k = \{1\} \text{ or } G_i \text{ for all } G_i, G_j, G_k \in \Pi. \qquad (*)$$

Then G is the full translation group of an affine space or translation affine plane, the point set of which is G and the line set of which is the set of all left cosets xG_i, $x \in G$, $G_i \in \Pi$.

Baer (1963) proved the above under the additional condition that G be abelian. We present two different proofs of Theorem 1. The first is in the spirit of Baer, with a more careful use of the group properties. The second, suggested to us by Peter Johnson, shows that each product $G_i G_j$ of distinct elements G_i, G_j of the partition is a group, and then applies André's result to conclude that each $G_i G_j$, and therefore also G, is abelian.

As an intermediate step in our first proof we have the following result.

Theorem 2. *Let $S = (P, L)$ be a linear space such that $G \subseteq \text{Aut}(S)$ acts sharply transitively on P and G_l is transitive on the points of l for all lines l in L. If there is a point $c \in P$ such that $G_h \cap G_k G_l = \{1\}$ or G_h for all line stabilizers of lines h, k, l on c, then S is an affine space or translation affine plane and G is its full translation group.*

2. Preliminary results

In this section we state an axiomatic characterization of affine space due to Lenz which will be used in the proof presented in §3. We also present three lemmas to be used in later sections.

Affine space can be defined as projective space less a hyperplane. However, the following axiom system of Lenz characterizes affine space (1954, 1959).

An *affine space* consists of a family of objects called points and a family of objects called lines, together with an incidence relation between points and lines. The set of lines falls into pairwise disjoint subsets called *parallel classes*. Two lines are said to be *parallel* when they belong to the same parallel class. Incidence and parallelism satisfy the following axioms.

AS1. Two distinct points p and q are incident with a unique line pq.

AS2. For every line l and point p, there is precisely one line on p parallel to l.

AS3. (Trapezoid axiom) Let pq and rs be distinct parallel lines, p, q, r and s distinct points. Let x be a point on pr, x distinct from p and r. Then there is a point incident with both lines xq and rs.

AS4. (Parallelogram axiom) If no line has more than two points, then for every three distinct points p, q, r the line on r parallel to pq has a point in common with the line through q parallel to pr.

AS5. Every line is incident with at least two points.

AS6. There are two disjoint lines that are not parallel.

A family of points and lines with an incidence relation is called an *affine plane* if it satisfies AS1, AS5 and

AS2'. For every line l and point p, $p \notin l$, there is precisely one line on p missing l.

Note that if a family of points and lines with an incidence relation and a parallel relation as above satisfies AS1, AS2 and AS5 while AS6 fails, then AS2' holds automatically, and so S is an affine plane.

Define a *linear space* to be a pair $S = (P, L)$ where P is a set of elements called *points* and L a set of subsets of P called *lines*, such that AS1 and AS5 hold. We shall call S *non-trivial* if $|L| > 1$. A *partition* of a group G is a set Π of non-identity subgroups G_i of G such that

$$\bigcup_i G_i = G \text{ and } G_i \cap G_j = \{1\} \text{ for all distinct } G_i, G_j \in \Pi. \text{ We say } \Pi \text{ is}$$

trivial if $|\Pi| = 1$.

Lemma 3. *Let G be a group with non-trivial partition Π. Then $S = (P, L)$ where $P = G$ and $L = \{xG_i \mid x \in G, G_i \in \Pi\}$ is a linear space and G, acting on S by $g(x) = gx$ for all $g \in G$, $x \in P$, is a subgroup of $\mathrm{Aut}(S)$. Moreover, G acts sharply transitively on P, G_l is transitive on the points of l for all $l \in L$ and Π is the set $\{G_l \mid 1 \in l \in L\}$.*

Proof. AS5 follows by definition of line. To show that AS1 holds, let p and q be distinct points of P. Then there is a unique $G_i \in \Pi$ such that $1 \ne q^{-1}p \in G_i$. Thus qG_i is the unique line on p and on q.

To show that G acts sharply transitively on P, let $x, y \in P$ and let $g \in G$. Then $g(x) = y$ if and only if $g = yx^{-1}$.

We finally show that for all $l \in L$, G_l is transitive on the points of l. If the line l has the form $1G_i = G_i \in \Pi$ then $G_l = G_i$ since G_i is a group, and therefore it acts transitively on itself. If the line l has the form

xG_i, then $G_l = xG_ix^{-1}$ is transitive on xG_i since G_i is transitive on G_i. That Π is as claimed, also follows now.

Lemma 4. *Let Π be a partition of the group G such that condition (*) of Theorem 1 holds. Let G_i, G_j, G_k be distinct elements of Π. Then $G_k \subseteq G_iG_j$ implies $G_j \subseteq G_iG_k$ and $G_i \subseteq G_kG_j$.*

Proof. Suppose $G_j \not\subseteq G_iG_k$. Then $G_j \cap G_iG_k = \{1\}$ by (*). Let $\alpha \in G_k \backslash G_i \subseteq G_k \subseteq G_iG_j$. Then $\alpha = \beta\gamma$ for some $\beta \in G_i, \gamma \in G_j, \gamma \neq 1$. Then $1 \neq \gamma = \beta^{-1}\alpha \in G_j \cap G_iG_k = \{1\}$. Contradiction. This proves the first implication. The proof of the second implication is analagous.

Lemma 5. *Let Π be a partition of the group G such that condition (*) of Theorem 1 holds. Then $G_iG_j = G_jG_i$ for all $G_i, G_j \in \Pi$.*

Proof. We may assume $G_i \neq G_j$. By (*) it is sufficient to prove (for $G_k \in \Pi$) $G_k \subseteq G_iG_j$, if and only if $G_k \subseteq G_jG_i$. This is trivial for $G_k = G_i$ or G_j. So we may assume $G_k \neq G_i, G_j$. Applying Lemma 4 repeatedly gives $G_k \subseteq G_iG_j$ if and only if $G_j \subseteq G_iG_k$ if and only if $G_i \subseteq G_jG_k$ if and only if $G_k \subseteq G_jG_i$.

3. The first proof

Assume the hypotheses of Theorem 1. Then by Lemma 3, the hypotheses of Theorem 2 hold for S as defined in Lemma 3 and with c defined to be 1. It therefore suffices to prove Theorem 2.

The following remarks will be used from time to time without necessarily referring to them.

Remark (a). For $x \in P$ and $1 \neq g \in G$, g fixes the line $xg(x)$

Remark (b). If $1 \neq g \in G$ and g fixes the line l, then $l = xg(x)$ for each x in l.

Remark (c). If $l \neq h$ are lines for which $G_l \cap G_h \neq \{1\}$, then $l \cap h = \emptyset$.

We must define the parallel classes, prove AS2, AS3 and AS4, and show that the group G is the group of all translations. Define the parallel classes to be the line orbits under G and write $l \parallel h$ if l and h are in the same line orbit.

The parallel class containing the line l will be denoted by $G(l)$.

To prove AS2 we must show that given any line l and point p, then p is contained in precisely one line of $G(l)$. By point transitivity and the definition of parallel class, p is on at least one line of $G(l)$.

The fact that p is on at most one line of $G(l)$ follows from Remarks (a), (b) and (c).

We now prove AS3. Let p, q, r be non-collinear points and let x be a point on pr distinct from p and r. We can let $p = g_p(c), q = g_q(c), r = g_r(c), x = g_x(c)$ with $g_p, g_q, g_r, g_x \in G$. Then there are lines k, l, m on c with $pq = g_q(l) = g_q G_l(c), pr = g_r(m) = g_r G_m(c), qx = g_x(k) = g_x G_k(c)$. Note that the line on r parallel to pq is $g_r(l) = g_r G_l(c)$. We need to prove $g_r(l) \cap g_x(k) \neq \varnothing$.

Then $p \in pq, pr$ implies $g_p(c) = p = g_r \lambda(c) = g_r \mu(c)$ for some $\lambda \in G_l$, $\mu \in G_m$. Also $q \in qx$ implies $g_q(c) = q = g_x \kappa(c)$ for some $\kappa \in G_k$ and $x \in pr$ implies $g_x(c) = x \in g_r \mu'(c)$ for some $\mu' \in G_m$.

Thus by sharp transitivity $g_p = g_q \lambda = g_r \mu, g_q = g_x \kappa, g_x = g_r \mu'$. Also $\mu' \neq \mu, 1$ because $x \neq p, r$.

Now $G_m \setminus \{1\} \ni \mu^{-1} \mu' = (\lambda^{-1} g_q^{-1} g_r)(g_r^{-1} g_x) = \lambda^{-1}(g_q^{-1} g_x) = \lambda^{-1} \kappa^{-1} \in G_l G_k$. Thus $G_m \subseteq G_l G_k$ by (*). But $g_r^{-1} g_x = \mu' \in G_m \subseteq G_l G_k$. Thus $g_r^{-1} g_x = \lambda' \kappa'$ for some $\lambda' \in G_l, \kappa' \in G_k$. And so $g_r \lambda'(c) = g_x \kappa'^{-1}(c) \in g_r G_l(c) \cap g_x G_k(c) = g_r(l) \cap g_x(k)$, proving AS3.

Finally, we prove AS4. Let $p, q,$ and r be non-collinear points. We can let $p = g_p(c), q = g_q(c), r = g_r(c), g_p, g_r \in G$. Then there are lines l and h on c such that $pq = g_q(l) = g_q G_l(c)$, and so $p = g_p(c) = g_q \lambda(c)$ for some $\lambda \in G_l$; and also $pr = g_r(h) = g_r G_h(c)$, and so $p = g_p(c) = g_r \chi(c)$ for some $\chi \in G_h$. Thus $g_q \lambda = g_r \chi$, or $g_q^{-1} g_r = \lambda \chi^{-1} = \chi' \lambda'$ for $\chi' \in G_h, \lambda' \in G_l$ by Lemma 5. Then $g_r \lambda'^{-1}(c) = g_q \chi'(c) \in g_r(l) \cap g_q(h)$ proving AS4.

It remains only to show that G is the full translation group.

Because of our definition of parallelism of lines $g(l) \| l$ for every line l and every $g \in G$. It is an elementary exercise (using for example AS3 and AS4) to show that if g is a fixed-point-free automorphism of an affine space with the property that $g(l) \| l$ for all lines l, then g is a translation. So every $g \in G$ is a translation. By transitivity of G on the set of all points, G is the full translation group.

4. The second proof

Again assume the hypotheses of Theorem 1. By Lemma 5, $G_i G_j = G_j G_i$ for all $G_i, G_j \in \Pi$ and thus the product of any two distinct elements of Π is a group. We shall apply the theorem of André, below, to such a product $\hat{G} = G_i G_j$ and to $\hat{\Pi} = \{G_k \mid G_k \in \Pi, G_k \subseteq G_i G_j\}$ (note that (*) implies $\hat{\Pi}$ is a partition of $\hat{G}$).

In order to apply André we must show that $G_k G_h = G_i G_j$ for all G_k, $G_h \subseteq G_i G_j$ with G_i, G_j, G_h, $G_k \in \Pi$, $G_i \neq G_j$, $G_h \neq G_k$. Obviously $G_k G_h \subseteq G_i G_j G_i G_j =$ (by Lemma 5) $G_i G_j$. It remains only to show $G_i G_j \subseteq G_k G_h$.

First we show $G_i \subseteq G_h G_k$. This is obvious if $G_i = G_k$ or G_h. If $G_i \neq G_k$, G_h then, by Lemma 4, $G_k \subseteq G_i G_j$ implies $G_j \subseteq G_i G_k$. Thus $G_h \subseteq G_i G_j \subseteq G_i G_i G_k \subseteq G_i G_k$. Using Lemma 4 again, we deduce $G_i \subseteq G_h G_k = G_k G_h$. Analogously, $G_j \subseteq G_k G_h$. Thus $G_i G_j \subseteq G_k G_h$.

Theorem (André 1954). *Let Π be a non-trivial partition of the group G such that $G = AB$ for all A, $B \in \Pi$, $A \neq B$. Then G is the translation group of a translation affine plane, the point set of which is G and the line set of which is the set of all left cosets xG_i, $x \in G$, $G_i \in \Pi$. Consequently G is abelian.*

By André, each product $G_i G_j$, of distinct elements of our partition Π is abelian and therefore our group is abelian. We can now apply Baer's Theorem (1963) to complete our proof.

Theorem (Baer 1963). *Let Π be a non-trivial partition of the abelian group G such that for some A, $B \in \Pi$, $G \neq AB$. Suppose also that condition (*) of Theorem 1 holds. Then G is the translation group of an affine space of dimension ≥ 3, the point set of which is G and the line set of which is the set of all cosets xG_i, $x \in G$, $G_i \in \Pi$.*

Acknowledgement. This research was supported by N.S.E.R.C. grants A 3486 and A 8027

References

J. André (1954), Über nicht-Desarguessche Ebenen mit transitiver Translationgruppe, *Math. Z.* **60**, 156-186.
R. Baer (1963), Partitionen abelscher Gruppen, *Arch. Math.* **14**, 73-83.
L.M. Batten (1986), *Combinatorics of Finite Geometries*, Cambridge University Press, Cambridge.
A. Bichara and G. Tallini (1982), On the characterization of the Grassmann manifold representing the planes in a projective space, *Ann. Discrete Math.* **14**, 129-150.
A. Bichara and G. Tallini (1983), On a characterization of Grassmann space representing the h-dimensional subspaces in a projective space, *Ann. Discrete Math.* **18**, 113-132.

P. Biondi and N. Melone (1985), Incidence structures of affine and projective types, *J. Geom.* **25**, 178-191.

F. Buekenhout (1969), Une caractèrisation des espaces affins basée sur la notion de droite, *Math. Z.* **111**, 367-371.

F. Buekenhout (1970), A characterization of the affine spaces of order two as 3-designs, *Math. Z.* **118**, 83-85.

P. Dembowski (1964), Eine Kennzeichnung der endlichen affinen Räume, *Arch. Math.* **15**, 146-154.

P. Dembowski (1967), Berichtigung und Ergänzung zu "Eine Kennzeichnung der endlichen affinen Räume", *Arch. Math.* **18**, 111-112.

P. Dembowski (1968), *Finite Geometries*, Springer-Verlag, Berlin.

P. Dembowski and A. Wagner (1960), Some characterizations of finite projective spaces, *Arch. Math.* **11**, 465-469.

M. Grochowska (1986), Ordered affine geometry based on the notion of positive dilation groups, *J. Geom.* **27**, 103-111.

D. Hilbert (1899), *Grundlagen der Geometrie*, B.G. Teubner, Leipzig, Berlin; English translation: *Foundations of Geometry*, Open Court, Illinois, third printing 1987.

D.R. Hughes and F.C. Piper (1985), *Design Theory*, Cambridge University Press, Cambridge.

D. Jungnickel and H. Lenz (1985), Two remarks on affine designs with classical parameters, *J. Combin. Theory Ser. A.* **38**, 105-109.

W.M. Kantor (1969a), Characterizations of finite projective and affine spaces, *Canad. J. Math.* **21**, 64-75.

W.M. Kantor (1969b), Automorphism groups of designs, *Math. Z.* **109**, 246-252.

H. Lenz (1954), Zur Begründung der analytischen Geometrie, *Bayer. Akad. Wiss.* **1**, 17-72,

H. Lenz (1959), Ein kurzer Weg zur analytischen Geometrie, *Math.-Phys. Semesterber.* **6**, 57-67.

H. Lüneburg (1961), Zentrale Automorphismen von λ-Räumen, *Arch. Math.* **12**, 134-145.

H. Lüneburg (1980), *Translation Planes*, Springer-Verlag, Berlin.

D.K. Ray-Chaudhuri and A.P. Sprague (1976), Characterization of projective incidence structures, *Geom. Dedicata* **5**, 361-376.

R.H. Schulz (1967), Über blockpläne mit transitiver Dilatationsgruppe, *Math. Z.* **98**, 60-82.

K. Sörensen (1988), Über pseudoaffine Räume, *J. Geom.* **31**, 159-171.

A.P. Sprague (1981), Pasch's axiom and projective spaces, *Discrete Math.* **33**, 79-87.

O. Veblen and J.W. Young (1910), *Projective Geometry* 2 vols., Ginn and Co., Boston: (1938), Blaisdell, New York.

On maximal sets of nuclei in $PG(2, q)$ and quasi-odd sets in $AG(2, q)$

A. Blokhuis and F. Mazzocca

Eindhoven University of Technology
and University of Naples

1. Introduction

Let $\mathcal{B}$ be a set of $q + 1$ points in a projective plane Π of order q. A point $A \in \Pi$ is called a *nucleus* of $\mathcal{B}$, if every line through A meets the set $\mathcal{B}$ exactly once. The set of nuclei of $\mathcal{B}$ is denoted by $N(\mathcal{B})$ or $\mathcal{A}$.

There are two fundamental results on the structure of $\mathcal{A}$; both require the plane Π to be Desarguesian.

Result 1 (Blokhuis and Wilbrink 1987). *If $\mathcal{B}$ is not a line, then $|N(\mathcal{B})| \leq q - 1$.*

Result 2 (Segre and Korchmáros 1977). *Let A_1, A_2, A_3 be three non-collinear nuclei of the set $\mathcal{B}$. Then the points of $\mathcal{B}$ on the lines A_1A_2, A_2A_3, A_3A_1 are collinear.*

The proof of Result 1 surprisingly does not use Result 2. In this note we start with an elementary derivation of 1 from 2. This is done by a process we call "lifting".

The following paragraph discusses "quasi-odd" sets and we proceed by establishing the connection between sets with the maximal number of nuclei and quasi-odd sets. We finish with odds and ends.

2. Alternative proof of Result 1, using the Segre-Korchmáros Lemma

Let $|\mathcal{B}| = q + 1$ and $\mathcal{A} = N(\mathcal{B})$ be sets in $PG(2, q)$ and suppose $\mathcal{B}$ is not a line. Then there exists a line l, disjoint from $\mathcal{B}$ and hence from $\mathcal{A}$, and we consider $\mathcal{A}$ and $\mathcal{B}$ as subsets of $AG(2, q) = PG(2, q) \backslash l$ coordinatized in some way. Define a map $f : \mathcal{A} \to GF(q)^*$ as follows: Choose $A \in \mathcal{A}$ arbitrarily and put $f(A) = 1$. For $A_i \in \mathcal{A}$ let B be the unique point of $\mathcal{B}$ on the line AA_i. For some $\lambda \in GF(q) \backslash \{0,1\}$ we have $B = \lambda A + (1 - \lambda)A_i$. Put $f(A_i) = \lambda / (\lambda - 1)$. The idea is that now $(A, f(A))$, $(A_i, f(A_i))$ and $(B, 0)$ are collinear in $AG(3, q) \simeq AG(2, q) \times GF(q)$. It follows immediately from the Segre and Korchmáros result that if $B' \in A_iA_j$ then also

$(A_i, f(A_i))$, $(A_j, f(A_j))$ and $(B', 0)$ are collinear.   In particular, if $A_i \neq A_j$ then $f(A_i) \neq f(A_j)$, so that f is injective, and hence $|A| \leq q - 1$, finishing the proof.   Note that the lifting approach also yields new proofs for the generalizations of Result 1 on nuclei of sets in $AG(n, q)$ by Bruen and Mazzocca (to appear).

3. Quasi-odd sets in $AG(2, q)$, q even

In this section we discuss a seemingly unrelated problem.   The connection with the problems on nuclei will become clear in the next paragraph.

Let A be a set of points in $PG(2, q)$, q even. A is called *odd* if every line of the plane intersects the set A in an odd number of points; A is *even* if every line meets it in an even number of points.   Even sets are equivalent to words in the dual of the binary code spanned by the rows of the incidence matrix of the plane (where rows are indexed by lines).   Odd subsets form a coset of this dual code.   In the affine plane $AG(2, q)$, there are no odd sets.   Here we introduce the notion of a *quasi-odd* set (or perhaps better, *zero-odd* set).   This is a set of points in $AG(2, q)$ with the property that every line intersects it in an odd number of points, or misses the set completely.   Our main result is the following:

Theorem 1. *Let A be a quasi-odd set in the Desarguesian $AG(2, q)$, q even.   Then $|A| \leq q - 1$.*

Proof. Identify $AG(2, q)$ with $GF(q^2)$ and consider the polynomial

$$f(x) = \sum_{a \in A} (x - a)^{q-1}.$$

Suppose A is not empty.   Consider the lines through some fixed $a \in A$. Besides a they contain an even number of points from the set A; so adding these numbers gives that $|A|$ is odd.   It follows that $f(x)$ is not identically zero, for the coefficient of x^{q-1} equals one.   Note that for $x \in GF(q^2)$ the value of $(x - a)^{q-1}$ depends only on the direction of the line joining x and a.   If $x \in A$ every direction will occur an even number of times, so the points of A are all zeroes of $f(x)$.   It follows that $|A| \leq q - 1$. □

We know of only a few examples of quasi-odd sets of maximal size, allthough there are probably many more:
(1) $PG(2, 2)$ in $AG(2, 8)$;
(2) $PG(2, \sqrt{q})$ minus a hyperoval in $AG(2, q)$;
(3) $PG(n, 2)$ embedded as a linear space in $AG(2, 2^{n+1})$ (this example, containing the first one as a special case, is due to M.J. de Resmini).
Finally we mention that the same proof works in arbitrary dimension, giving

Theorem 2. Let $\mathcal{A}$ be a quasi-odd set in $AG(n, q)$, q even. Then $|\mathcal{A}| \leq q - 1$. □

4. The structure of sets with the maximal number of nuclei

The notion of quasi-odd set arose in the investigation of sets of size $q + 1$ with the maximal number, $q - 1$ of nuclei. The precise relation is formulated in the following theorem.

Theorem 3. Let $\mathcal{B}$ be a set of $q + 1$ points in $PG(2, q)$, $q = p^r$, p prime, and $\mathcal{B}$ not a line, such that $\mathcal{A} = N(\mathcal{B})$ has cardinality $q - 1$. Then no line intersects $\mathcal{A} \cup \mathcal{B}$ in 1 (mod p) points. In particular $\mathcal{A} \cup \mathcal{B}$ has no tangents (1-secants), and if $p = 2$, $\mathcal{A}$ is a quasi-odd set (of maximal size) and $\mathcal{A} \cup \mathcal{B}$ is an even set.

Proof. Let l be a line in $PG(2, q)$. We consider two cases: (1) l intersects $\mathcal{B}$ in $k > 1$ points; (2) l intersects $\mathcal{B}$ in one point and $\mathcal{A}$ in $k \geq 0$ points.

(1). *The line l intersects $\mathcal{B}$ in $k > 1$ points*
We want to show that $k \not\equiv 1$ (mod p); so suppose that in fact $k \equiv 1$ (mod p). We identify as usual $AG(2, q) = PG(2, q) \backslash l$ in some way with $GF(q^2)$ and denote by $\mathcal{B}'$ the set of $q + 1 - k$ point of $\mathcal{B}$ not on the line l.
Consider

$$f(x) = \sum_{b \in \mathcal{B}'} (x - b)^{q-1}.$$

We notice two things. First of all, the degree of f is less than $q - 1$ since $|\mathcal{B}'| \equiv 0$ (mod p). Second, each $a \in \mathcal{A}$ sees some $b \in \mathcal{B}'$ in the same set of directions, namely the directions not corresponding to the points in $\mathcal{B} \cap l$. Since the value of $(x - b)^{q-1}$ only depends on the direction of the line joining x and $\mathcal{B}$ we see that there is a constant c such that $f(a) = c$ for all $a \in \mathcal{A}$.

Since $|\mathcal{A}| = q - 1 > \deg(f)$ we get $f(x) \equiv c$. Let

$$\pi_k = \sum_{b \in \mathcal{B}} b.$$

Then

$$f(x) = \sum_{m=0}^{q-1} \pi_m x^{q-1-m} \ ,$$

and we may deduce

$$\pi_0 = \pi_1 = \ldots = \pi_{q-2} = 0, \ \pi_{q-1} = c.$$

In contrast to $f(x)$ we consider the polynomial

$$g(x) = \prod_{b \in \mathcal{B}} (x - b) = \sum_{m=0}^{q+1-k} \sigma_m x^{q+1-k-m} \ .$$

Here $\sigma_m = (-1)^m \sum b_{i_1} \ldots b_{i_m}$ where the sum runs over all subsets of size m of $\mathcal{B}$. The π_m and σ_m are related by the Newton formulas:

$$\sum_{m=0}^{n-1} \pi_{n-m} \sigma_m + n \sigma_n = 0.$$

Now we assume $0 \in \mathcal{B}$: this we may do without loss of generality, since we may choose the origin in $AG(2, q)$ arbitrarily.   It follows that

$$\sigma_{q+1-k} = (-1)^{q+1-k} \prod_{b \in \mathcal{B}} b = 0,$$

but $\sigma_{q-k} \neq 0$.   On the other hand $q - k \equiv -1 \pmod{p}$, so from the Newton formula with $n = q - k < q - 1$ we obtain:

$$\sigma_{q-k} = \sum_{m=0}^{q-k-1} \pi_{q-k-1-m} \sigma_m = 0.$$

This contradiction finishes (1).

In order to handle (2) we must use a trick.

(2). *l intersects $\mathcal{B}$ in one point and $\mathcal{A}$ in $k \geq 0$ points.*
We must show that $k \not\equiv 0 \pmod{p}$, so assume $k \equiv 0 \pmod{p}$.   We start again by identifying $PG(2, q) \backslash l$ with $GF(q^2)$, as follows: $PG(2, q)$ has

homogeneous coordinates of the form (x, y, z), where we take l to be the line with equation $z = 0$. Furthermore we take $(1, 0, 0)$ to be the point of $\mathcal{B}$ on l, and we assume that the line with equation $x = 0$ contains no point of the set $\mathcal{B}$. Let $\beta \in GF(q^2)\setminus GF(q)$. Then we identify the point $(b_1, b_2, ,1) \in PG(2, q)\setminus l$ with $b_1 + \beta b_2 \in GF(q^2)$. Finally we assume (again without loss of generality) that $-1 = (-1, 0, 1) \notin \mathcal{A}$. Let $\mathcal{B}'$ again denote the set of points of $\mathcal{B}$ not on the line l, and $\pi_k = \pi_k(\mathcal{B}') = \sum_{b \in \mathcal{B}'} b^k$ is the coefficient of x^{q-1-k} in $\sum_{b \in \mathcal{B}'} (x - b)^{q-1}$.

First we show that $\pi_j(\mathcal{B}') = 0$ for all $j \leq k$. If $k = 0$ there is nothing to prove. Denote the points of $\mathcal{B}'$ by $x_i + \beta y_i$, $i = 1, ..., ,q$, $x_i, y_i \in GF(q)$. There are k directions such that $\mathcal{B}'$ has exactly one point on each line with that direction. Call those directions $\lambda_1, ..., \lambda_k \in GF(q)$, so that $y_i - \lambda x_i$, $i = 1, ..., q$ takes all different values in $GF(q)$ ($\lambda \in \{\lambda_1, ..., \lambda_k\}$). It follows that for any $j < q - 1$ and $\lambda \in \{\lambda_1, ..., \lambda_k\}$,

$$\sum_{i=1}^{q} (y_i - \lambda x_i)^j = 0.$$

Considering this as a polynomial of degree j in λ we see at least k distinct zeros, so for $j < k$ it is identically zero. It can be deduced from this that, for $j < k$,

$$\pi_j = \sum_{i=1}^{q} (x_i + \beta y_i)^j = 0.$$

Since $k = p \cdot k/p$ we have

$$\pi_k = (\pi_{k/p})^p = 0.$$

This proves the assertion.

We consider again $f(x) = \sum_{b \in \mathcal{B}'} (x - b)^{q-1}$. It follows from $\pi_0 = \pi_1 = ... = \pi_k = 0$ that $\deg(f) < q - 1 - k$. On the other hand we have $f(a) = -1$ for all $a \in \mathcal{A}$ (by our special choice of the point of $\mathcal{B}$ on l). So $f(x) + 1$ vanishes identically, and we have

$$1 + \sum_{b \in \mathcal{B}} (x - b_1 - \beta b_2)^{q-1} \equiv 0. \tag{1}$$

We recall at this point the original proof of Result 1: With $\tilde{\mathcal{A}}, \tilde{\mathcal{B}} \subset GF(q^2)$, $|\tilde{\mathcal{A}}| = q - 1$, $|\tilde{\mathcal{B}}| = q + 1$, $\tilde{\mathcal{A}} = N(\tilde{\mathcal{B}})$ we have

$$(h(x) =) \sum_{\tilde{b} \in \tilde{\mathcal{B}}} (x - \tilde{b})^{q-1} = \prod_{\tilde{a} \in \tilde{\mathcal{A}}} (x - \tilde{a}).$$

Consider the map $(x, y, z) \rightarrow (z, y, x)$; it interchanges the lines with equations $x = 0$ and $z = 0$, and has the property that it maps $\mathcal{B}$ into $AG(2, q) = PG(2, q) \backslash (z = 0)$, since it was assumed that the line with equation $x = 0$ contains no points of $\mathcal{B}$. Let $\tilde{\mathcal{B}}$ be the image of $\mathcal{B}$.

The point $b = (b_1, b_2, 1)$ is mapped to $(1, b_2, b_1) = (1/b_1, \ b_2/b_1, 1) = \tilde{b}$ and $(1, 0, 0) \rightarrow (0, 0, 1)$. Since $b_1 \neq 0$ for $b \in \mathcal{B}'$, so the line $x = 0$ is a passing line. Hence

$$h(x) = \sum_{\tilde{b} \in \tilde{\mathcal{B}}} (x - \tilde{b})^{q-1} = x^{q-1} + \sum_{b \in \mathcal{B}'} (x - \frac{1}{b_1} - \frac{b_2}{b_1} \beta^{q-1})$$

$$= x^{q-1} + \sum_{b \in \mathcal{B}'} (xb_1 - 1 - \beta b_2)^{q-1} = \prod_{\tilde{a} \in \tilde{\mathcal{A}}} (x - \tilde{a}).$$

Now we use a trick: consider

$$h(-1) = (-1)^{q-1} + \sum_{b \in \mathcal{B}'} (-b_1 - 1 - \beta b_2)^{q-1} = 0,$$

because of (1). So $-1 = (-1, 0, 1) \in \tilde{\mathcal{A}}$, or equivalently $-1 \in \mathcal{A}$. We started however with the assumption $-1 \notin \mathcal{A}$, so we arrive again at a contradiction.

5. Another proof of Result 1

The original proof of Result 1 reduced the problem to one in the affine plane. Since this can be done in many ways, one seems to loose information in the proces. Also it seems somewhat artificial. Here we present what we think is the proper (projective) version of this proof. Again $\mathcal{B}$ is a set of $q + 1$ points in $PG(2, q)$, not all on a line, and $\mathcal{A} = N(\mathcal{B})$.

Let (b_1, b_2, b_3) denote homogeneous coordinates for the point $b \in \mathcal{B}$ and consider the polynomial

$$\beta(x, y, z) = \sum_{b \in \mathcal{B}} (b_1 x + b_2 y + b_3 z)^{q-1}.$$

Lemma. *If $a = (a_1, a_2, a_3)$ is a nucleus of $\mathcal{B}$ then $a_1x + a_2y + a_3z$ divides $\beta(x, y, z)$.*

Proof. Let $[u, v, w] = \{(a, b, c) \in PG(2, q) \mid au + bv + cw = 0\}$ be a line intersecting the set $\mathcal{B}$ in precisely one point. Then $\beta(u, v, w) = 0$, since of the $q + 1$ terms in the defining sum exactly one equals zero, and the remaining q are 1. We may assume $a_1 \neq 0$ and then write

$$\beta(x, y, z) = (a_1x + a_2y + a_3z) \cdot \gamma(x, y, z) + \delta(y, z),$$

with δ homogeneous of degree $q - 1$ (or identically zero). Since

$$[u, y, z], \text{ where } u = \frac{-a_2y - a_3z}{a_1},$$

is a line through a and hence contains a unique point of $\mathcal{B}$ we have that $\beta(u, y, z) = 0$ and of course $a_1u + a_2y + a_3z = 0$. It follows that $\delta(y, z) = 0$ for all $y, z \in GF(q)$. Since, if δ is nonzero, it has degree $q - 1$ it follows that δ is identically zero.

As a consequence of the claim we see that

$$\prod_{a \in \mathcal{A}}(a_1x + a_2y + a_3z) \left| \sum_{b \in \mathcal{B}}(b_1x + b_2y + b_3z)^{q-1} \right. .$$

In particular $\mathcal{A}$ has size at most $q - 1$ and in case of equality it should be possible to get strong information from this identity involving the coordinates of the points in $\mathcal{A}$ and $\mathcal{B}$.

6. Final remarks

The lifting process, as well as the identity in section 5, provide strong geometrical and algebraic information concerning the structure of sets with the maximal number of nuclei. For small q it is sufficient to characterize the pairs $(N(\mathcal{B}), \mathcal{B})$ with $|N(\mathcal{B})| = q - 1$. In fact we conjecture that the only examples of such pairs are the following:

(1) $N(\mathcal{B})$ consists of $q - 1$ collinear points, $\mathcal{B}$ of one of the two other points on this line, together with the q remaining points on a line through the non-chosen point.

(2) $N(\mathcal{B}) \cup \mathcal{B}$ is a Desargues configuration in $PG(2, 5)$. The set of nuclei consists of one of the triangles plus the center of the perspectivity.

The result on quasi-odd sets might be of interest for the problem of embedding designs in projective spaces.

We would like to thank Marialuisa de Resmini for stimulating conversations.

Acknowledgement. This research was done whilst the first author was a visiting professor at the University of Naples, sponsored by the Consiglio Nazionale di Ricerca.

References

A. Blokhuis and H.A. Wilbrink (1987), A characterization of exterior lines of certain sets of points in $PG(2, q)$, *Geom. Dedicata* **23**, 253-254.

A.A. Bruen and F. Mazzocca (to appear), Nuclei of sets in finite projective and affine spaces, *Combinatorica*.

B. Segre and G. Korchmáros (1977), Una proprietà degli insiemi di punti di un piano di Galois caratterizzante quelli formati dai punti delle singole rette esterne ad una conica, *Atti Accad. Naz. Lincei Rend.* **62**, 1-7.

An embedding theorem for partial linear spaces

A. Blokhuis[1], A.R. Calderbank[2], K. Metsch[3] and A. Seress[4]

Eindhoven University of Technology[1], AT&T
Bell Laboratories[2], University of Giessen[3] and
Ohio State University[4].

1. Introduction

Let $G(n, m)$ be a partial linear space with $n^2 + m$ points and $n^2 + n + m$ lines of size n. This study of the structure of partial linear spaces $G(n, m) = (X, L)$ is motivated by a basic question in algebraic coding theory. Let $A(N, d, w)$ denote the size of the largest binary code with block length N, constant weight w and distance d. We identify a codeword with the set of coordinates that index the non-zero entries, and counting codewords through a given point we obtain

$$w \cdot A(n, d, w) \leq n \cdot A(n - 1, d, w - 1).$$

For $0 \leq m \leq n - 2$, it follows that

$$A(n^2 + m, 2(n - 1), n) \leq \lfloor \frac{n^2 + m}{n} A(n^2 + m - 1, 2(n - 1), n - 1) \rfloor = n^2 + n + m,$$

and equality holds if and only if $G(n, m)$ exists.

Let $B(n, m)$ be a partial linear space with $n^2 + n + m + 1$ points and lines, where there are $n + 1$ points on every line. Elementary counting arguments show that every point is on $n + 1$ lines, that there are m lines parallel to any given line, and that (dually) every point is not connected to m other points. If we delete a line and all points on that line, then $n^2 + n$ lines will contain n points, and there will be m exceptional lines containing $n + 1$ points. After deleting one point from every exceptional line we obtain a partial linear space $G(n, m)$. The purpose of this note is to prove the following converse

Theorem 1. *If $2 \leq m \leq n - 2$ and $n \geq 12m^3 + 6m^2 - m$, then it is possible to embed $G(n, m)$ into $B(n, m)$.*

Remarks (a). The case $m = 0$ is the well known fact that it is possible to embed an affine plane in a projective plane. For $m = 1$, we can prove that if $n \geq 5$, then it is possible to embed $G(n, 1)$ in $B(n, 1)$. We sketch the argument because it motivates the sequence of lemmas that serve to prove the theorem.

In the linear space $G(n, 1)$, every point is on $n + 1$ lines, except for a single point ω that is on n lines. Elementary counting reveals that a line L has $n + 1$ parallels if $\omega \in L$, and n parallels if $\omega \notin L$. Given a point $a \in X$, we let $a^\perp$ be the set of points that are not joined to a. Then $|\omega^\perp| = n$ and $|a^\perp| = 1$ if $a \neq \omega$. We need to prove that $\omega^\perp$ is a line, and we need to divide the remaining $n^2 + n$ lines into $n + 1$ parallel classes. More elementary counting proves that if L is a line and $a \notin L$, then the number of lines parallel to L through a is $|a^\perp \cap L|$ + 1 if $a \neq \omega$ and $|a^\perp \cap L|$ if $a = \omega$. We distinguish the collection $L^\perp$ of points $a \in X \backslash L$ that lie on two lines parallel to L.

Consider lines that do not contain ω. The key observation is that the line joining points $a, b \in L^\perp$ is parallel to L (this is Lemma 5 in its most basic form). The n lines parallel to L now cover every element of $L^\perp$ exactly twice, and if $n \geq 5$ this implies that there is a single line containing every point of $L^\perp$. The remaining $n - 1$ lines together with L constitute a parallel class. The argument for lines L that contain ω is slightly more complicated. Here the key observation is that there do not exist regular points a_1, a_2, a_3 in $L^\perp$ with the property that no line (a_i, a_j) is parallel to L. In this case $|L^\perp| = 2n - 1$ and it follows that these points are contained in the union of two lines one of which is $\omega^\perp$. Again the remaining $n - 1$ lines together with L can be shown to constitute a parallel class.

(b) We obtain the corollary that $A(26, 8, 5) = 30$, since Schellenberg (1975) has used a computer to prove that $B(5, 1)$ does not exist. We sketch a coding theoretic proof of this result. The partial linear space $B(5, 1)$ is a symmetric group divisible design. The incidence matrix A satisfies

$$A^T A = A A^T = \begin{bmatrix} 6 & 0 & & & & & & \\ 0 & 6 & & & & & & \\ & & 6 & 0 & & 1 & & \\ & & 0 & 6 & & & & \\ & & & & \ddots & & & \\ & & 1 & & & 6 & 0 \\ & & & & & 0 & 6 \end{bmatrix}$$

For every point $a \in X$ there is a unique point $a^\perp$ not joined to a. Dually, for every line L there is a line $L^\perp$ disjoint from L. It is easy to show that every 2×2 subblock of A looks like

$$
\begin{array}{cc} & a \;\; a^\perp \\ \begin{array}{c} L \\ L^\perp \end{array} & \begin{bmatrix} 0 & 0 \\ 0 & 0 \end{bmatrix} \end{array}, \qquad
\begin{array}{cc} & a \;\; a^\perp \\ \begin{array}{c} L \\ L^\perp \end{array} & \begin{bmatrix} 0 & 1 \\ 1 & 0 \end{bmatrix} \end{array}, \text{ or }
\begin{array}{cc} & a \;\; a^\perp \\ \begin{array}{c} L \\ L^\perp \end{array} & \begin{bmatrix} 1 & 0 \\ 0 & 1 \end{bmatrix} \end{array}
$$

and if we replace these blocks by 0, -1 and 1 respectively we obtain a 16×16 matrix S with 6 non-zero entries in each row and column, that satisfies $SS^T = S^TS = 6I$. (cf. Lamken, Mullin, and Vanstone, 1985). If we ignored signs, S would be the incidence matrix of a 2-(16, 6, 2) design.

We consider the ternary code C spanned by the signed blocks. Clearly C is self-orthogonal, and by considering the invariant factors of S we can show that C is self-dual. Standard arguments prove that the minimum weight in C is 6 and that the only codewords of minimum weight 6 are the signed blocks and their negatives. The possible weights in C are 6, 9, 12 and 15 so the Assmus-Mattson Theorem (see MacWilliams and Sloane 1977, Chapter 6) implies that the codewords of minimum weight 6 form a 3-design. But the signed blocks do not form a 3-design since $16\binom{6}{3}/\binom{16}{3}$ is not integral.

2. The Proof of the Main Theorem

We shall embed $G(n, m)$ in $B(n, m)$ by adjoining $n + 1$ new points $x_1, ..., x_{n+1}$ and a new line $L_\infty = \{x_1, ..., x_{n+1}\}$. In the partial linear space $G(n, m)$, every point $a \in X$ is on at most $n + 1$ lines, and we shall distinguish regular points that lie on $n + 1$ lines, from exceptional points that lie on fewer than $n + 1$ lines. Thus $X = R \cup E$ where R is the set of regular points and E is the set of exceptional points. Let $\deg(a)$ denote the number of lines through a, and let

$$\varphi(a) = n + 1 - \deg(a). \tag{1}$$

When embedding $G(n, m)$ in $B(n, m)$ an exceptional point $\omega \in E$ needs to be added to $\varphi(\omega)$ parallel lines. Counting point-line incidences gives

$$\sum_{\omega \in E} \varphi(\omega) = m, \tag{2}$$

so we need to distinguish m lines in L that will be added to the exceptional points. We also need to partition the remaining $n^2 + n$ lines into $n + 1$ parallel classes C_i so that we can adjoin the new point x_i to every line in C_i. This is exactly what it means to embed $G(n, m)$ in $\mathcal{B}(n, m)$.

Given a point $a \in X$ let

$$a^\perp = \{b \in X \mid a,b \text{ are not connected} \}.$$

Simple counting gives

$$| a^\perp | = m + \varphi(a)(n - 1). \tag{3}$$

Given a point $a \in X$ and a line $L \in \mathcal{L}$, define the L-weight of a to be $\zeta_L(a) = |a^\perp \cap L|$. The L-weight of a collection of points is then the sum of the L-weights of the constituent points. We have

$$\sum_{a \in X} \zeta_L(a) = \sum_{b \in L} |b^\perp| = nm + (\Sigma_{\omega \in L \cap E} \varphi(\omega))(n - 1). \tag{4}$$

We also extend the function φ to lines in L by setting $\varphi(L) = \Sigma_{b \in L} \varphi(b)$. Note that by (2) for each line L we have $\varphi(L) \le m$. Finally we let $L^\perp$ denote the set of points that are contained in at least two lines parallel to L.

Lemma 2. *For the partial linear space $G(n, m)$ we have*
(i) *every line $L \in \mathcal{L}$ has $n + m - 1 + \varphi(L)$ parallels,*
(ii) *if a is a regular point outside the line L, then there are $\zeta_L(a) + 1$ lines parallel to L containing a,*
(iii) *if ω is an exceptional point outside the line L then there are $\zeta_L(\omega) + 1 - \varphi(\omega)$ lines parallel to L containing ω.*

Proof. Counting point-line incidences.

Lemma 3. *If $\omega \in E$ and if the line $L \not\subset \omega^\perp$, then $\zeta_L(\omega) \le 2m$.*

Proof. If $\omega \in L$ then $\zeta_L(\omega) = 0$. If the exceptional point ω is outside L, then we obtain the upper bound by counting lines parallel to L. Let M be a line through ω that meets L. There are at least $(n - 2) - (m - \varphi(L) - $

$\varphi(\omega)$) regular points on M outside L, and Lemma 2 (ii) implies that there is a line parallel to L through each of these points. Lemma 2 (iii) implies that there are $\zeta_L(\omega) + 1 - \varphi(\omega)$ lines parallel to L through ω. Counting parallel lines, it follows from Lemma 2 (i) that

$$(n - 2) - (m - \varphi(L) - \varphi(\omega)) + \zeta_L(\omega) + 1 - \varphi(\omega) \leq n + m - 1 + \varphi(L)$$

which gives the required result.

Lemma 4. *Suppose that the line $L \not\subseteq \omega^\perp$, for all $\omega \in E$, and that a is a regular point outside L. If $m \geq 1$, then*

$$\sum_{b \in X_{a, L}} \zeta_L(b) \geq n\zeta_L(a) - (3m^2 - 2m\varphi(L)),$$

where $X_{a, L} = R \cap (\bigcup \{M \in L \mid a \in M \text{ and } M \| L\})$.

Proof. Let M be a line through a that meets L. By Lemma 2(i) and (ii)

$$\sum_{\substack{b \in (R \setminus L) \cap M \\ b \neq a}} \zeta_L(b) + 1 \leq n + m - 1 + \varphi(L) - \zeta_L(a) - 1$$

so that

$$\sum_{\substack{b \in (R \setminus L) \cap M \\ b \neq a}} \zeta_L(b) \leq m + \varphi(L) - \zeta_L(a) + |(E \setminus L) \cap M|.$$

Hence

$$\sum_{\substack{M \ni a \\ M \cap L \neq \emptyset}} \sum_{\substack{b \in (R \setminus L) \cap M \\ b \neq 0}} \zeta_L(b) \leq (n - \zeta_L(a))(m + \varphi(L) - \zeta_L(a)) + m - \varphi(L). \tag{5}$$

Now

$$\sum_{b \in a^\perp \cap R} \zeta_L(b) \leq |a^\perp \setminus L| (m - 1) = (m - \zeta_L(a))(m - 1) \tag{6}$$

and by Lemma 3

$$\sum_{\omega \in E} \zeta_L(\omega) \leq 2m(m - \varphi(L)). \tag{7}$$

Together (4), (5), (6), and (7) give

$$\sum_{b \in X_{a, L}} \zeta_L(b) \geq nm + \varphi(L)(n - 1) - (n - \zeta_L(a))(m + \varphi(L) - \zeta_L(a))$$

$$- (m - \varphi(L)) - (m - \zeta_L(a))(m - 1) - 2m(m - \varphi(L))$$

$$= (n + \varphi(L) + m - 1 + m - \zeta_L(a))\zeta_L(a) - (3m^2 - 2m\varphi(L)) \qquad (8)$$

which gives the required result.

Lemma 5. *Suppose $n > 10m^3 + 7m^2$, and that the line $L \not\subset \omega^\perp$, for all $\omega \in E$. Let $a_1, ..., a_s$ be regular points in $L^\perp$ with the property that no line (a_i, a_j) is parallel to L, and let $S = \sum_{i=1}^{s} \zeta_L(a_i)$. Then $S \le m + \varphi(L)$.*

Proof. Let $A_L = \{a_1, ..., a_s\}$ be a counterexample to the lemma that is minimal with respect to inclusion. By Lemma 4

$$\sum_{i=1}^{s} \sum_{b \in X_{a_i,L}} \zeta_L(b) \ge Sn - s(3m^2 - 2m\varphi(L)). \qquad (9)$$

Given a point b, let $r(b)$ be the number of times $\zeta_L(b)$ occurs as a summand on the left hand side of (9). If $\Omega = \bigcup_{i=1}^{s} X_{a_i,L}$, then

$$\sum_{\substack{b \in \Omega \\ r(b) \ge 1}} (r(b) - 1)\zeta_L(b) + \sum_{b \in \Omega} \zeta_L(b) \ge Sn - s(3m^2 - 2m\varphi(L)).$$

For any $r > 0$, we have $(r - 1)m \le \binom{r}{2}m$. Since there are no exceptional points in Ω we obtain

$$\sum_{b \in \Omega} \binom{r(b)}{2}m + \sum_{b \in X} \zeta_L(b) \ge Sn - s(3m^2 - 2m\varphi(L)). \qquad (10)$$

There are $S + s$ parallels through the points a_i, and any pair of these parallels intersect in at most one point. Thus

$$\sum_{b \in \Omega} \binom{r(b)}{2}m \le \binom{S + s}{2}m,$$

and we may rewrite (10), using (4) as

$$s(3m^2 - 2m\varphi(L)) + m\binom{S + s}{2} - \varphi(L) \ge (S - m - \varphi(L))n. \qquad (11)$$

Minimality of the set A_L implies $s \le m + \varphi(L) + 1$ and that

$$S \le \zeta_L(a_i) + \sum_{j \ne i} \zeta_L(a_j) \le m + (m + \varphi(L)) = 2m + \varphi(L).$$

If $S = m + \varphi(L) + 1$ then we estimate the left hand side of (11) by taking $s = m + \varphi(L) + 1$, and obtain $n \leq 10m^3 + 7m^2$ which contradicts the initial hypothesis.   If $m + \varphi(L) + 2 \leq S \leq 2m + \varphi(L)$, then we estimate the left hand side of (11) from above by taking $s = m + \varphi(L) + 1$ and $S = 2m + \varphi(L)$, and we estimate the right hand side of (11) from below by taking $S = m + \varphi(L) + 2$.   We obtain $n < 8m^3 + 2m^2$ which contradicts the initial hypothesis.

Lemma 6.   *Suppose $n > 10m^3 + 7m^2$, and that the line $L \not\subseteq \omega^\perp$ for any $\omega \in E$.   Then for each regular point $a \in L^\perp$, there exists $M \in L$ such that $a \in M, M \| L$ and*

$$\sum_{b \in M \cap R} \zeta_L(b) \geq n - 5m^2.$$

Proof.   Consider a set $A_L = \{a_1, ..., a_s\}$ of regular points in $L^\perp$ that is maximal with respect to the property that no line (a_i, a_j) is parallel to L. Let $M_1, ..., M_{s'}$ be the lines through points a_i that are parallel to L. Maximality of the set A_L implies that the union of the lines M_i contains every regular point in $L^\perp$.
By Lemma 5 we have

$$\sum_{i=1}^{s} \zeta_L(a_i) \leq m + \varphi(L),$$

and since the points a_i are in $L^\perp$, so that $\zeta_L(a_i)$ is at least one, we have

$$s' = \sum_{i=1}^{s} (\zeta_L(a_i) + 1) \leq s + m + \varphi(L) \leq 2(m + \varphi(L)). \tag{12}$$

Given a regular point $a \in L^\perp$, we prove that there is a line M_i through a, and a regular point $c \in L^\perp \cap M_i$ such that $M_i = (a, c)$ is the only line M_j that passes through c.   A line M_j that misses a intersects the parallels through a in at most $\zeta_L(a) + 1$ (regular) points, so we obtain

$$\sum_{M_j \not\ni a} \sum_{b \in X_{a, L} \cap M_j} \zeta_L(b) \leq (\zeta_L(a) + 1)(s' - 1)m.$$

Since the right hand side of this is certainly less than $\zeta_L(a)n - 3m^2 + 2m \varphi(L)$, the existence of the point c and the line M_i now follows from Lemma 4.

Finally Lemma 4 and the same reasoning as above in a slightly stronger form, because of the special nature of the point c yield

$$\sum_{b \in (a,c) \cap R} \zeta_L(b) \;\geq\; \sum_{b \in X_{c,L}} \zeta_L(b) - \sum_{j \neq i} \sum_{b \in X_{c,L \cap M}} \zeta_L(b)$$

$$\geq \zeta_L(c)n - 3m^2 + 2m\varphi(L) - \zeta_L(c)(s'-1)m \;\geq\; n - 5m^2$$

as required.

Given a line L, let $\mathcal{M}(L)$ be the set of lines M parallel to L for which $\sum_{b \in M \cap R} \zeta_L(b)$ (the regular L-weight) is at least $n - 5m^2$.

Lemma 7. *Suppose $n \geq 12m^3 + 6m^2 - m$, and that the line $L \not\subseteq \omega^\perp$ for any $\omega \in E$. Then $|\mathcal{M}(L)| = m + \varphi(L)$.*

Proof. If M is a line parallel to L, then the regular L-weight of M is at most equal to the number of lines parallel to L that intersect M in one point, and so

$$1 + \sum_{b \in M \cap R} \zeta_L(b) \leq n + m - 1 + \varphi(L).$$

Therefore the regular L-weight of any line parallel to L is bounded above by $n + 2m - 2$. Now we estimate the total L-weight of the regular points from above and below. Since every point a in $L^\perp$ is on some line in $\mathcal{M}(L)$ we have

$$\sum_{a \in R} \zeta_L(a) \leq |\mathcal{M}(L)|(n + 2m - 2). \tag{13}$$

On the other hand, it follows from (4) and Lemma 3 that

$$\sum_{a \in R} \zeta_L(a) \geq \sum_{a \in X} \zeta_L(a) - 2m(m - \varphi(L)) = nm + \varphi(L)(n - 1) - 2m(m - \varphi(L)).$$

$$\tag{14}$$

Together (13) and (14) imply $|\mathcal{M}(L)| \geq m + \varphi(L)$.

If $|\mathcal{M}(L)| \geq m + \varphi(L) + 1$, then we consider a subset A of $\mathcal{M}(L)$ containing $m + \varphi(L) + 1$ lines. The argument used to prove (11) in Lemma 5 gives

$$(n - 5m^2)(m + \varphi(L) + 1) \leq \sum_{M \in A} \sum_{b \in M \cap R} \zeta_L(b) \leq \sum_{b \in X} \zeta_L(b) + \binom{m + \varphi(L) + 1}{2} m,$$

and using (4) we obtain a contradiction to the initial hypothesis $n \geq 12m^3 + 6m^2 - m$.

Lemma 8. *Suppose $n \geq 12m^3 + 6m^2 - m$, and that the line $L \not\subseteq \omega^\perp$ for any $\omega \in E$. Then every regular point a in $L^\perp$ is contained in at least $\zeta_L(a)$ lines of $\mathcal{M}(L)$. Furthermore, the number of regular points in $L^\perp$ that are contained in $\zeta_L(a) + 1$ lines of $\mathcal{M}(L)$ is at most $(m - \varphi(L))2m + \varphi(L)$.*

Proof. Given a regular point a in $L^\perp$, let $\lambda(a)$ be the number of lines in $\mathcal{M}(L)$ that contain a. To prove that $\lambda(a) \geq \zeta_L(a)$ we estimate the L-weight of $X_{a, L}$ from above and below. By Lemma 4 we have

$$\sum_{b \in X_{a, L}} \zeta_L(b) \geq \zeta_L(a)n - (3m^2 - 2m\varphi(L)). \tag{15}$$

By Lemma 6, every regular point in $L^\perp$ is on some line in $\mathcal{M}(L)$. Furthermore, in the proof of Lemma 7 we argued that the L-weight of any line parallel to L is bounded above by $n + 2m - 2$. We obtain

$$\sum_{\substack{b \in X_{a, L}}} \zeta_L(b) \leq \sum_{\substack{a \in M \in \mathcal{M}(L) \\ b \in M \cap R}} \zeta_L(b) + \sum_{\substack{a \notin M \in \mathcal{M}(L) \\ b \in M \cap X_{a, L}}} \zeta_L(b)$$

$$\leq \lambda(a)(n + 2m - 2) + (m + \varphi(L) - \lambda(a))(\zeta_L(a) + 1 - \lambda(a))m.$$

Together (15) and (16) imply $\lambda(a) \geq \zeta_L(a)$.

Counting point-line incidences we obtain

$$(m + \varphi(L))n \geq \sum_{a \in L^\perp \cap R} \lambda(a) \tag{17}$$

using Lemma 7. Making allowance for exceptional points in $L^\perp$ we obtain

$$\sum_{a \in L^\perp \cap R} \zeta_L(a) \geq \sum_{a \in X} \zeta_L(a) - (m - \varphi(L))2m = nm + \varphi(L)(n - 1) - (m - \varphi(L))2m$$

$$\tag{18}$$

using (4). Together (17) and (18) give

$$\sum_{a \in L^\perp \cap R} (\lambda(a) - \zeta_L(a)) \leq (m - \varphi(L))2m + \varphi(L)$$

which proves the lemma.

Lemma 9.   *Suppose $n \geq 12m^3 + 6m^2 - m$, and that the line $L \not\subseteq \omega^\perp$ for any $\omega \in E$. If $M \in \mathcal{M}(L)$ then M contains at most $5m^2 + m + \varphi(L) - 1$ points not in $L^\perp$.*

Proof.   Since $M \in \mathcal{M}(L)$ we obtain

$$\sum_{a \in L^\perp \cap R \cap M} \zeta_L(a) \geq n - 5m^2. \tag{19}$$

Next we count pairs (b, B), where B is a line in $\mathcal{M}(L)$ other than M and $b \in M \cap B$. By Lemma 8 we have

$$\sum_{a \in L^\perp \cap R \cap M} (\zeta_L(a) - 1) \leq m + \varphi(L) - 1. \tag{20}$$

We subtract (20) from (19) to obtain $\left| L^\perp \cap R \cap M \right| \geq n - (5m^2 + m + \varphi(L) - 1)$ which proves the lemma.

Lemma 10.   *Suppose $n \geq 12m^3 + 6m^2 - m$, and that the line $L \not\subseteq \omega^\perp$ for any $\omega \in E$. If $\omega \in L^\perp \cap E$, then ω is contained in at least $\zeta_L(\omega) - \varphi(\omega)$ lines of $\mathcal{M}(L)$.*

Proof.   There are $n + m + \varphi(L) - 1$ lines parallel to L, and $m + \varphi(L)$ of these lines belong to $\mathcal{M}(L)$, so that $n - 1$ lines, say $L_1, \ldots, L_{n-1}$, remain. Since ω is contained in $\zeta_L(\omega) - \varphi(\omega) + 1$ lines parallel to L, we need to prove that ω is on at most one line L_i.

Now suppose that ω is on at least two lines L_i. We derive a contradiction by estimating from above and below, the number of pairs (a, M) for which $a \in X \backslash L^\perp, M \in \mathcal{M}(L)$, and $a \in M$. Lemma 9 provides the upper bound

$$(m + \varphi(L))(5m^2 + m + \varphi(L) - 1). \tag{21}$$

Let Z be a line through ω that meets L, that contains no exceptional points outside L other than ω, and no points $a \in L^\perp \cap R$ that are incident with $\zeta_L(a) + 1$ lines of $\mathcal{M}(L)$. Every point of $Z \backslash (L \cup \{\omega\})$ is incident with at least one line parallel to L. Since $\left| Z \backslash (L \cup \{\omega\}) \right| = n - 2$, there exists a point $a \in Z \backslash (L \cup \{\omega\})$ that is contained in $\zeta_L(a) + 1$ lines of $\mathcal{M}(L)$. Hence $a \notin L^\perp$. By counting the number of lines Z of the above type and using the fact that all of them contain at least one point a of the above type we obtain the lower bound

$$n - |\omega^\perp \cap L| - |\{a \in E \mid a \neq \omega, a \notin L\}| - |\{a \in L^\perp \mid a \text{ is on } \zeta_L(a) + 1 \text{ lines of } \mathcal{M}(L)\}|$$
$$\geq n - 2m - (m - \varphi(L) - 1) - (2m^2 - 2m\varphi(L) + \varphi(L)), \tag{22}$$

using Lemmas 3 and 8. Now (21) and (22) provide a contradiction to the hypothesis $n \geq 12m^3 + 6m^2 - m$.

Let $\mathcal{L}^*$ be the set of lines L for which $L \not\subseteq \omega^\perp$ for any $\omega \in E$. Given $L \in \mathcal{L}^*$ let $\pi(L) = \{M \in \mathcal{L} \mid M = L, \text{ or } M \| L \text{ and } M \notin \mathcal{M}(L)\}$.

Lemma 11.   *Suppose $n \geq 12m^3 + 6m^2 - m$.   Then the function π partitions $\mathcal{L}^*$ into $n + 1$ equivalence classes, each consisting of n pairwise disjoint lines.*

Proof.   Given $L \in \mathcal{L}^*$, Lemma 7 implies $|\mathcal{M}(L)| = m + \varphi(L)$, so that $|\pi(L)| = n$. Lines in $\pi(L)$ are pairwise disjoint, since Lemmas 8 and 10 imply that a point in $L^\perp$ is contained in at most one line of $\pi(L)$. We also need to check that every line $M \in \pi(L)$ is in $\mathcal{L}^*$. If ω is an exceptional point, and if $M \subseteq \omega^\perp$, then M is disjoint from the $n - 1$ parallels in $\pi(L)$, and the $n + 1 - \varphi(\omega)$ lines through ω parallel to L. There are now too many lines disjoint from M.

Finally we prove that π defines a partition of $\mathcal{L}^*$. Let $M \in \pi(L)$, $K \in \pi(M)$, and suppose that $K \notin \pi(L)$. The line K contains at least $n - m$ of the n^2 points covered by the lines in $\pi(L)$. It follows that there are at least $n - m$ lines in $\pi(L)$ that are disjoint from M and that do not appear in $\pi(K)$. Again there are too many lines disjoint from M.

Lemma 12.   *Suppose $n \geq 12m^3 + 6m^2 - m$.   If ω is an exceptional point, then there are $\varphi(\omega)$ lines contained in $\omega^\perp$.*

Proof.   Since $|\omega^\perp| = m + \varphi(\omega)(n - 1)$, there are at most $\varphi(\omega)$ lines contained in $\omega^\perp$. The total number of lines contained in $\omega^\perp$ for some exceptional point ω is then at most $\Sigma_{\omega \in E} \varphi(\omega) = m$. Since $|\mathcal{L}| = n^2 + n + m$, and since Lemma 11 implies $|\mathcal{L}^*|$ is a multiple of n, we see that the only possibility is $|L^*| = n^2 + n$. This means that there are exactly $\varphi(\omega)$ lines in $\omega^\perp$.

The proof of the main theorem is now complete. We adjoin $n + 1$ points $x_1, ..., x_{n+1}$ to X, and a new line $L_\infty = \{x_1, ..., x_{n+1}\}$ to L. The point x_i is added to every line in the i-th equivalence class of L^*, and each exceptional point ω is added to the lines in $\omega^\perp$.

Acknowledgement.   The first author gratefully acknowledges support from A.T. & T. Bell Laboratories, and the fourth author gratefully acknowledges support from the Dutch organisation for Scientific Research (NWO).

References

E.R. Lamken, R.C. Mullin and S.A. Vanstone (1985), Some non-existence results on twisted planes related to minimum covers, *Congr. Numer.* **48**, 265-275.
F.J. MacWilliams and N.J.A. Sloane (1977), *The Theory of Error Correcting Codes*, North-Holland, Amsterdam.
P.J. Schellenberg (1975), A computer construction for balanced orthogonal matrices, *Congr. Numer.* **14**, 513-522.

A non-degenerate generalized quadrangle with lines of size four is finite

A.E. Brouwer
Eindhoven University of Technology

For finite generalized quadrangles of order (s, t) (that is, with $s + 1$ points on each line and $t + 1$ lines on each point) one has the inequality $t \leq s^2$ when $s > 1$.   However, the proof uses either counting or the spectrum of matrices, and it is not clear how to generalize it to possibly infinite t.   Thus we have the open problem (posed by Tits many years ago): *Must a generalized quadrangle (or, more generally, a generalized $2m$-gon) of order (s, t) have finite t when s is finite and $s > 1$?*   For generalized $2m$-gons with $2m > 4$ nothing is known.   For generalized quadrangles the case $s = 2$ was settled by Cameron (1981) (cf. Brouwer et al. 1989, p.32) and the case $s = 3$ by Kantor (personal communication, Como 1984;   see also Cameron 1984).   Kantor described his proof as complicated and group theoretical and asked for a simpler combinatorial argument.   Shortly afterwards I found the argument presented below.   For $s > 3$ nothing is known.

Theorem. *Let $Q = (X, L)$ be a non-degenerate generalized quadrangle with lines of size four and at least one line.   Then Q is finite (and hence known).*

Proof.   Each point is incident with the same number of lines, t say, where our aim is to show that t is finite.   The hypothesis implies that there are two disjoint lines, L and M say.   Consider the 16-set Y that is the union of four lines meeting both L and M.

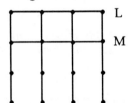

(i) *Either Y induces a 4×4-grid (generalized quadrangle of order $(3, 1)$), or $Y\backslash(L \cup M)$ induces a $K_{4,4}$ in the collinearity graph of Q.*

Indeed, no line can meet $Y\backslash(L \cup M)$ in precisely three points.   For, if L, M, N are three lines with three transversals, and p, q, r are the points on these lines not on the transversals, then p, q, r are pairwise collinear, and hence on a fourth transversal.   Therefore, if Y does not induce a 4×4-grid, then $Y\backslash(L \cup M)$ induces a regular graph on 8 points, of valency 4 and without triangles; this is necessarily $K_{4,4}$.

(ii) *If Y does not induce a grid, then* $t + 1 \leq 8$.

To show this, let us label the 16 points of Y with $a_1, ..., d_4$ as shown below.

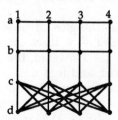

Let $N = \{a_1, p_1, p_2, p_3\}$ be a line meeting Y in a_1 only and not meeting one of the six lines $c_i d_j$ ($i \neq j$, $i,j = 2, 3, 4$). The set $Z = \{c_2, c_3, c_4, d_2, d_3, d_4\}$ is partitioned into three pairs $p_i^\perp \cap Z$, and at least one of these pairs must contain two adjacent points c_j, d_k, a contradiction. Thus there is no such line N, and $t + 1 \leq 8$.

From now on assume that Y induces a grid.

Label the points $p \in X \backslash Y$ with the 4-cocliques $L_p = p^\perp \cap Y$ they determine. We may identify these labels with permutations of 4 symbols. Denote collinearity by $\sim$.

(iii) *Let p, q be two collinear points in* $X \backslash Y$ *such that* $L_p \cap L_q \neq \varnothing$. *Then the line pq meets Y in the single point* $L_p \cap L_q$ *and in particular* L_p *and* L_q, *viewed as permutations, have the same parity.*

This is obvious.

(iv) *Let A, B be two disjoint 4-cocliques in Y with the same parity. If* $L_p = A$ *for a point* $p \in X \backslash Y$, *then* $p \sim x$ *for all but at most two points* $x \in X \backslash Y$ *with* $L_x = B$.

Choose $q \in B$. For any such x, the line qx contains two further points y, z and $|A \cap L_y| = |A \cap L_z| = 1$ (and L_y, L_z are determined by q, B, independent of x) so that $p \sim y$ for at most one choice of X, $p \sim z$ for at most one choice of X, and never $p \sim q$. Consequently $p \sim x$ for all but at most two points x with $L_x = B$.

(v) *Let A, B, C be three pairwise disjoint 4-cocliques in Y with the same parity. Then at least one of these is the label of at most five points.*

For, let $L_x = A, L_y = B, x \sim y$. Then x and y are both adjacent to all but at most four points with label C, but the line xy contains at most one point with label C.

Consider the partial linear space on $X \setminus Y$ with lines of size three obtained by taking all lines meeting Y in one point. From the above remarks it follows that each connected component of $X \setminus Y$ in this space is a cover of $T[3; 4]$, the transversal design on 12 points obtained by taking the even permutations on 4 symbols as points, and the sets of such permutations containing a given symbol at a given position as blocks. It follows that if A, B are permutations with the same parity, then there are as many points with label A as there are with label B because fibres in a cover of a connected graph have the same cardinality.

Conclusion. The total number of points of the generalized quadrangle is $16 + 12m$ with $0 \le m \le 10$, and in particular it is finite. $\square$

The finite non-degenerate generalized quadrangles with lines of size four were classified in Dixmier and Zara (1976). There are unique examples with $t = 1, 5, 9$ and two (dual) examples with $t = 3$. See also Payne and Thas (1984).

References

A.E. Brouwer, A.M. Cohen and A. Neumaier (1989), *Distance Regular Graphs*, Springer-Verlag, Berlin.

P.J. Cameron (1981), Orbits of permutation groups on unordered sets, II, *J. London Math. Soc.* **23**, 249-264.

P.J. Cameron (1984), Infinite versions of some topics in finite geometry, *Geometrical Combinatorics*, Research Notes in Math. **114**, 13-20, (ed. F.C. Holroyd and R.J. Wilson), Pitman, London.

S. Dixmier and F. Zara (1976), Etude d'un quadrangle généralisé autour de deux de ses points non liés, Preprint.

S. Payne and J. A. Thas (1984), *Finite Generalized Quadrangles*, Pitman, London.

On abelian collineation groups of finite projective planes

J.M.N. Brown
York University

Our main results are Theorems 1 and 2 below.

Theorem 1. *Let G be an abelian collineation group of a finite projective plane. Let $\mathcal{P}$ and L denote point and line orbits of G.*
   Let z denote the number of points of $\mathcal{P}$ on a line of L.
   Let z^* denote the number of lines of L on a point of $\mathcal{P}$.
Then one of the following holds:
(a) $|\mathcal{P}| = |L|$;

(b) $|L|$ *divides* $|\mathcal{P}|$ *and* $z = \dfrac{|\mathcal{P}|}{|L|}$ *and* $z^* = 1$;

(c) $|\mathcal{P}|$ *divides* $|L|$ *and* $z = 1$ *and* $z^* = \dfrac{|L|}{|\mathcal{P}|}$;

(d) $z = z^* = 0$.

Remark.   Theorem 1 is not true for some non-abelian collineation groups.   For example, the non-abelian group $G = PGL(3, q)$ acting naturally on $PG(2, q^3)$ (or on the Figueroa plane of order q^3) has a point orbit $\mathcal{P}$ and a line orbit L satisfying:

$$\begin{aligned}
|\mathcal{P}| &= (q^2 + q + 1)(q + 1)q(q - 1), \\
|L| &= (q + 1)q^3(q - 1)^2, \\
z &= q^2 + q + 1, \\
z^* &= q^2(q - 1).
\end{aligned}$$

The point orbit $\mathcal{P}$ is the set of all points which are incident with exactly one line of a G-invariant order q subplane and the line orbit L is the set of lines incident with no point of this sub-plane.   See, for example, Brown (1983, §6).   The conclusion of Theorem 1 fails for this non-abelian example.

Theorem 2. *Let G be an abelian collineation group of a projective plane of finite order n with the property that every point orbit is incident with every line orbit.   Then either:*
(a)   *all orbits (point or line) have the same size; or*
(b)   (i)   *there are precisely two orbit sizes;*

(ii)   *the smaller orbit size is $n + \sqrt{n} + 1$, the larger orbit size is $k(n + \sqrt{n} + 1)$ for some $k \mid n - \sqrt{n}$, the larger orbit size equals the group order;*

(iii)   *there is precisely one point (line) orbit of size $n + \sqrt{n} + 1$,*

(iv)   *there is a Baer sub-plane such that the point (line) orbit of size $n + \sqrt{n} + 1$ consists of the set of all points (lines) of this Baer sub-plane;*

(v)   *the Baer sub-plane in (iv) is the Fix of the stabilizer of any one of its points or lines.*

Remark. Both cases (a) and (b) can occur for Pappian planes $PG(2, q^2)$. Case (a) occurs if G is a full Singer group or if G is the group generated by $\tau = \sigma^{q^2 - q + 1}$ where σ is a Singer cycle of $PG(2, q^2)$. See Hirschfeld (1976, Theorem 4.5, 1979, Theorem 4.3.6). Case (b) with $|G| = 2(n + \sqrt{n} + 1)$ occurs if the previous group is extended by a Baer involution whose set of fixed points is one of the point orbits of $\langle \tau \rangle$.

Proposition 3. *Let G be an abelian collineation group of a finite projective plane. For any point P and $g \in G$, there is an equality $G_P = G_{Pg}$ between point stabilizers and the point orbit PG is a subset of the set of points of $Fix(G_P)$.*

Proof. Let $x \in G$. Then $x \in G_P \Leftrightarrow P^x = P \Leftrightarrow P^{xg} = Pg \Leftrightarrow P^{gx} = Pg \Leftrightarrow x \in G_{Pg}$. This proves the equality. The equality implies that every Pg is fixed by every element of G_P.

Proposition 4. *Let G be an abelian collineation group of a finite projective plane. For any line l and $g \in G$, there is an equality $G_l = G_{lg}$ between line stabilizers and the line orbit lG is a subset of the set of lines of $Fix(G_l)$*

Proof. This is the dual of Proposition 3.

Proposition 5. *Let G be an abelian collineation group of a finite projective plane. Let $\mathcal{P}$ and $\mathcal{L}$ denote point and line orbits of G. Let $P \in \mathcal{P}$ and $l \in \mathcal{L}$. Let z denote the number of points of $\mathcal{P}$ on a line of $\mathcal{L}$. Let z^* denote the number of lines of $\mathcal{L}$ on a point of $\mathcal{P}$. Then*

(a) $z |\mathcal{L}| = z^* |\mathcal{P}|$;

(b) $z \geq 2 \Rightarrow G_P \subseteq G_l$ *and* $|G_P| \mid |G_l|$ *and* $|\mathcal{L}| \mid |\mathcal{P}|$;

(c) $z^* \geq 2 \Rightarrow G_l \subseteq G_P$ and $|G_l| \mid |G_P|$ and $|P| \mid |L|$.

Proof. The equality in (a) follows from counting incident point line pairs (P, l) for which the point $P \in P$ and the line $l \in L$.

Suppose $z \geq 2$. Then there exist $P \in P, l \in L, g \in G$ such that $P, Pg \in l$ and $P \neq Pg$. Thus $l = PgP$. Thus g' fixes $P \Rightarrow$ (by Proposition 3) g' fixes both P and $Pg \Rightarrow g'$ fixes l. Thus $G_P \subseteq G_l$ and thus $|G_P| \mid |G_l|$. The second divisibility claim of (b) follows from the first divisibility claim of (b) and from $|P| = |G| / |G_P|$ and $|L| = |G| / |G_l|$. This proves (b). Claim (c) follows from (b) by duality.

Proof of Theorem 1. If $z, z^* \geq 2$, then (a) holds by Proposition 5(b, c). If $z \geq 2$ and $z^* = 1$, then (b) holds by Proposition 5(a, b). If $z^* \geq 2$ and $z = 1$, then (c) holds by Proposition 5(a, c). If $z = 1$ and $z^* = 1$, then (a) holds by Proposition 5(a). Finally, by Proposition 5(a), $z > 0 \Leftrightarrow z^* > 0$. So the only other case is $z = z^* = 0$.

Corollary 6. Let G be an abelian collineation group of a finite projective plane of order n. If there is a line orbit whose size is greater than the size of every point orbit, then lines of that line orbit meet exactly $n + 1$ point orbits.

Proof. Let $L = lG$ be a line orbit whose size exceeds the size of every point orbit. Suppose the number of point orbits is m. Let $P_1, P_2, \ldots, P_m$ be distinct point orbits. Let $z_i = $ the number of points of P_i incident with l. Then

$$n + 1 = \sum_{i=1}^{m} z_i .$$

But $|L| > |P_i|$ for all i implies that only case (c) or (d) of Theorem 1 can hold. Thus each z_i is one or zero. This fact together with the equality implies that exactly $n + 1$ of the z_i are non-zero. Thus l is incident with points from exactly $n + 1$ orbits.

Corollary 7. Let G be an abelian collineation group of a finite projective plane of order n. If the number of point orbits is at most n or the number of line orbits is at most n, then the size of the largest point orbit equals the size of the largest line orbit.

Proof.   By Brauer (1941) or Dembowski (1958) or Hughes (1957) or Parker (1957), the number of point orbits is at most n and the number of line orbits is at most n.   If the conclusion of Corollary 7 fails, then the hypothesis of either Corollary 6 or its dual holds.   Then the number of point or, respectively, line orbits is at least $n + 1$. Contradiction.

Proposition 8. *Let G be an abelian collineation group of a finite projective plane.   Let $\mathcal{P}$ be a point orbit which is incident with every line orbit and let L be a line orbit for which $|\mathcal{P}| < |L|$.   Then either there is a line l such that $\mathcal{P}$ contains every point incident with l or*

> $\mathcal{P}$ = *the set of points of a Baer sub-plane π_0*
> = *the set of <u>all</u> points fixed by the stabilizer of any point of π_0*
> = *the set of <u>all</u> points fixed by the stabilizer of any line of π_0*

and the set of lines of π_0 is a line orbit.

Proof.   Suppose the first conclusion is false, then $\mathcal{P}$ is a blocking set.   Let $P \in \mathcal{P}$ and $l \in L$.   Only case (c) of Theorem 1 can hold and it holds with $z^* > 1$.   Thus Proposition 5(c) implies

$$G_l \subseteq G_P$$

and equality cannot hold because $|G_P| = |G| / |\mathcal{P}| > |G| / |L| = |G_l|$.   Thus by Proposition 3:

$$\text{Fix}(G_l) \supseteq \text{Fix}(G_P) \supseteq \mathcal{P}.$$

Thus $\text{Fix}(G_P)$, which is a closed configuration, includes a blocking set which must include a subset of four points no three collinear.   Thus $\text{Fix}(G_P)$ is a subplane π_0.   It is a proper sub-plane because otherwise G_P fixes all points and lines which implies $G_P \subseteq G_l$ contradicting $|G_P| > |G_l|$.   So $n + \sqrt{n} + 1 \geq$ (by Bruck (1955, Lemma 3.1)) | the set of all points of $\text{Fix}(G_P)| \geq |\mathcal{P}| \geq$ (by Bruen (1971)) $n + \sqrt{n} + 1$.   Thus equalities $n + \sqrt{n} + 1 = $ | the set of points fo $\text{Fix}(G_P)| = |\mathcal{P}|$ hold.   Thus $\mathcal{P} = $ the set of all points of the Baer sub-plane $\pi_0 = $ the full set of fixed points of G_P for any $P \in \mathcal{P}$.   Now, (by Brauer (1941) or Dembowski (1958) or Hughes (1957) or Parker (1957)) G restricted to the Baer sub-plane π_0 has the same number of orbits on points as it has orbits on lines.   For points this number is one.   So the lines of this Baer subplane π_o form an orbit. By Proposition 5(b, c), $G_P = G_l$ for any point P and line l of the Baer sub-plane π_o.

Proof of Theorem 2. If some point orbit $\mathcal{P}$ contains every point incident with some line, then $\mathcal{P}$ must be the set of all points of the plane. Then the number of line orbits equals the number of point orbits equals one. So there is only one line orbit and this orbit is the set of all lines of the plane. Thus condition (a) holds. So we may assume that for every point orbit $\mathcal{P}$ and every line l there is a point incident with l and not in $\mathcal{P}$. By duality, we may also assume that for every line orbit L and every point P there is a line incident with P and not in L.

Without loss of generality, we may further assume that there is a line orbit L such that all point or line orbits have size at most $|L|$. Then by Proposition 8 every point orbit has size either $|L|$ or $n + \sqrt{n} + 1$.

If every point orbit size is $|L|$ then $|L|$ is the average point orbit size. But then $|L|$ is the average line orbit size. But $|L|$ is the maximum line orbit size. Thus all line orbits have size $|L|$, So conclusion (a) is true.

If every point orbit size is $n + \sqrt{n} + 1$ then the average point orbit size is $n + \sqrt{n} + 1$. But then $n + \sqrt{n} + 1$ is the average line orbit size and also is Bruen's lower bound for each line orbit size. Thus all line orbits have size $n + \sqrt{n} + 1$. This is a contradiction if $|L| > n + \sqrt{n} + 1$. Otherwise, conclusion (a) is true.

If both possible point orbit sizes $|L|$ and $n + \sqrt{n} + 1$ occur, then there is a point orbit $\mathcal{P}$ such that all line and point orbit sizes have size at most $|\mathcal{P}|$. Then by the dual of the initial part of this proof, either conclusion (a) is true or there are two line orbit sizes $|\mathcal{P}|$ $(=|L|)$ and $n + \sqrt{n} + 1$ and both sizes occur. In the two orbit size case suppose there are distinct point orbits of size $n + \sqrt{n} + 1$. Let π' and π'' denote Baer subplanes whose point sets $\mathcal{P}'$ and $\mathcal{P}''$ are distinct point orbits of size $n + \sqrt{n} + 1$. Let L' and L'' be the line sets of π' and π''. By Proposition 8, L' and L'' are line orbits. They are distinct line orbits because $\mathcal{P}'$ and $\mathcal{P}''$ (and thus π' and π'') are distinct. Thus L' and L'' are disjoint (because distinct orbits are disjoint). Let $P'' \in \mathcal{P}''$ and $P' \in \mathcal{P}'$. Let $l' \in L'$ and $l'' \in L''$. Because G is abelian, it follows that $G_{P''}$ permutes the orbits of $G_{P'}$. Thus $G_{P''}$ permutes (fixed point freely by Propositon 8) the set $\mathcal{P}'$ of fixed points of $\mathcal{G}_{P'}$. Similarly $\mathcal{G}_{P'}$ ($= \mathcal{G}_{l'}$) permutes (fixed line freely) the set L'' of lines fixed by $\mathcal{G}_{P''}$ ($= \mathcal{G}_{l''}$). Then $\mathcal{P}'$ has at most one point on each line of L''. (Otherwise a line of L'' would be fixed by $\mathcal{G}_{P'}$ and thus would be a line of $L' \cap L'' = \varnothing$). Since $\mathcal{G}_{P''}$ fixes the lines of L'', $\mathcal{G}_{P''}$ cannot move any point of $\mathcal{P}'$. So $\mathcal{G}_{P''}$ both acts fixed point freely on $\mathcal{P}'$ and fixes every point of $\mathcal{P}'$. Thus $\mathcal{G}_{P''}$ contains only the identity collineation. Thus $|P''^{\mathcal{G}}| = |\mathcal{P}''|$ is the maximum orbit size. Contradiction. So there is a unique point (line) orbit of smallest size.

For any point P in an orbit of maximum size, $|$ set of points of Fix$(\mathcal{G}_P)| \geq$ (by Proposition 3) $|P^{\mathcal{G}}| > n + \sqrt{n} + 1$ and Fix$(\mathcal{G}_P)$ is a subplane. Thus, by Bruck (1955), Fix$(\mathcal{G}_P)$ is the entire plane. Thus $\mathcal{G}_P$ is the identity group. Thus $|\mathcal{G}| = |\mathcal{P}^{\mathcal{G}}| = |\mathcal{G}_{P'}| |\mathcal{G}| / |\mathcal{G}_{P'}| = k(n + \sqrt{n} + 1)$ for $k = |\mathcal{G}_{P'}|$. Finally $|\mathcal{P}^{\mathcal{G}}| = k(n + \sqrt{n} + 1) | (n^2 + n + 1) - (n + \sqrt{n} + 1) = (n - \sqrt{n})(n + \sqrt{n} + 1)$ so that k divides $n - \sqrt{n}$.

This proves conclusions (b)(i, ii, iii). Conclusions (b)(iv, v) follow from Proposition 8 and its dual.

Acknowledgement. This research was supported by Grant Number A8027, Natural Sciences and Engineering Research Council of Canada.

References

R. Brauer (1941), On the connections between the ordinary and the modular characters of groups of finite order, *Ann. of Math.* **42**, 926-935.

J.M.N. Brown (1983), On constructing finite projective planes from groups, *Ars Combin.* **16-A**, 61-85.

R.H. Bruck (1955), Difference sets in a finite group, *Trans. Amer. Math. Soc.* **78**, 464-481.

A.A. Bruen (1971), Blocking sets in finite projective planes, *SIAM J. Appl. Math.* **21**, 342-344.

P. Dembowski (1958), Verallgemeinerungen von Transitivitätsklassen endlichen projektiver Ebenen, *Math. Z.* **69**, 59-89.

J.W.P. Hirschfeld (1976), Cyclic projectivities in $PG(n, q)$, *Teorie Combinatorie, Volume* 1, 201-211, Accademia Nazionale dei Lincei, Rome.

J.W.P. Hirschfeld (1979), *Projective Geometries over Finite fields*, Oxford University Press, Oxford.

D.R. Hughes (1957), Collineations and generalised incidence matrices, *Trans. Amer. Math. Soc.* **86**, 284-296.

E.T. Parker (1957), On collineations of symmetric designs, *Proc. Amer. Math. Soc.* **8**, 350-351.

More geometry for Hering's 3^6 : $SL(2, 13)$

F. Buekenhout
Free University of Brussels

1. Introduction

We start from the following facts. First, the existence of a subgroup of $GL(6, 3)$, isomorphic to $SL(2, 13)$ and acting transitively on the 1-dimensional subspaces. This is due to Hering (1969). Next, there is a sharper inclusion namely $SL(2, 13) \leq Sp(6, 3) \leq GL(6, 3)$. We do not know who obtained this first, but its projective counterpart is mentioned in Conway et al. (1985) and Kleidman (to appear).

From there on, we can prove rather smoothly all the other facts mentioned by Hering (1969, 1985) and in particular, his three remarkable spreads. Whilst doing so we get a bit more. For instance, all components of the above spreads are singular subspaces of the symplectic polar space. Next we are adding another contribution to the many ways in which a piece of projective geometry can be used to produce affine geometry. Here, our way to do so, is to use two partial spreads with no common component, such that the unions of the components of each one be equal. We apply this to Hering's spreads and we find new geometries to add to those he produced already.

2. Hering's geometry for $PSL(2, 13)$ in $PG(5, 3)$

Let P be a projective space $PG(5, 3)$ equipped with a symplectic polarity $\perp$ and let $PSp(6, 3)$ be the group leaving this structure invariant. By Conway et al. (1985) and Kleidman (to appear), the subgroups of $PSp(6, 3)$, isomorphic to $PSL(2, 13)$ are conjugate. Hence, each $PSL(2, 13)$ is contained in a unique $PSp(6, 3)$. This means that it is natural to fix both P and $\perp$ before we introduce $PSL(2, 13)$. Next let G be a subgroup of $PSp(6, 3)$ isomorphic to $PSL(2, 13)$. Let us make use of the fact that G acts transitively on the 364 points of P (and with the help of $\perp$) on its 364 hyperplanes.

The main facts gathered on this action are summarised as follows. They are essentially due to Hering (1969, 1985).

Theorem 1. *The permutation group G of degree 364 has a point stabiliser of order 3 and systems of blocks of imprimitivity S_1, S_2, S_3 such that:*
(a) *S_1 and S_2 are 1-spreads, and S_3 is a 2-spread;*

(b) *components of S_1 and S_2 are singular lines with respective stabilisers D_{12}, Alt(4);*

(c) *components of S_3 are singular planes with stabiliser 13:3;*

(d) *there are four other non-trivial systems of blocks of imprimitivity, three of which have blocks of two points contained in a component of S_1 and the fourth has a block of 26 points consisting of two components of S_3 with block-stabiliser 13:6;*

(e) *for each line $l \in S_i$, $i = 1, 2$, the 3-space $l^\perp$ contains 4 lines of S_j, $j = 1, 2$, $j \neq i$, namely the 4 lines of S_j intersecting l;*

(f) *there are no quadrangles whose sides are lines of $S_1 \cup S_2$.*

Proof. (1) The maximal subgroups of $PSL(2, 13)$ are well known since Dickson at least.   We refer to Conway et al. (1985) for simplicity. Using these data, it is not difficult to obtain the full subgroup structure of G which is displayed in Figure 1.

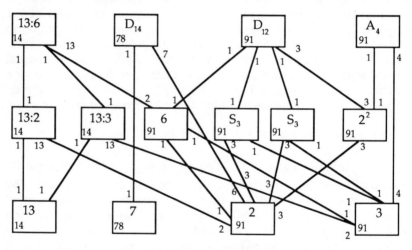

Figure 1.   The subgroup structure of PSL(2, 13).

This information can also be obtained on CAYLEY.   We use this in order to find the systems of blocks of imprimitivity.   Indeed, if G_p is the stabiliser of a point $p \in P$ and if $G_p < H < G$ where H is a subgroup, then the orbit of p under H is a block of imprimitivity for G.

(2) Since $|G| = 14.13.6$ it is obvious that a point stabiliser G_p has order 3.

(3)   The normaliser $N_G(G_p)$ is D_{12} (see Figure 1).   Hence G_p fixes 4 points.   Also these points constitute a projective subspace of P and hence they constitute a line.   All such lines give us the 1-spread S_1.

(4) Let $l \in S_1$ be fixed by G_p and let $p \neq q$ be the points on l. If $p \in q^\perp$, then G_p fixes the 3-space $p^\perp \cap q^\perp$ which is disjoint from l. But this space has 40 points so G_p would fix one more point in it, a contradiction. Finally the lines in S_1 are singular.

(5) If l is the set of fixed points of G_p, then G_p fixes $x^\perp$ for each $x \in l$ and $l \leq x^\perp$, by (4), so G_p fixes $l^\perp$. We have $l \leq l^\perp$, again by (4). In $l^\perp$, l is on four planes π_i, i = 1, 2, 3, 4. So G_p leaves at least one of these planes, say π_1 invariant. In π_1, G_p induces a group of perspectivities with axis l, hence it has centre $p_1 \in l$. Now $N_G(G_p) = D_{12}$ leaves l and $l^\perp$ invariant, and acts transitively on the four points of l. So each of them should be a centre for the action of G_p on some π_i. Therefore G_p leaves each π_i invariant and D_{12} permutes them transitively. The π_i are non-singular; otherwise $l^\perp$ would be singular while the rank of the polar space is only 3.

(6) G_p fixes 4 lines through p. Indeed, the union of the invariant lines through p is a projective subspace in which G_p induces a group of perspectivities of centre p. Hence G_p fixes all points of a hyperplane in it and we know already, by (3), that this is a line. Hence we get the statement.

(7) If l is the set of fixed points of G_p, then the four lines through p, invariant under G_p, are contained in $l^\perp$. Indeed, by (5), the plane containing these four lines is necessarily one of the planes of π_i.

(8) The four lines through p invariant under G_p, are singular. This is immediate, after (7).

(9) If $l \in S_1$, then $l^\perp$ contains a unique line of S_1, namely l.

If $l' \neq l, l' \in S_1, l' \leq l^\perp$ and x' is a point on l' then $x'^\perp$ contains l; so the plane $\langle l, x' \rangle$ is singular and this is one of the planes π_i of (5) where it was seen that these planes cannot be singular.

(10) It can easily be shown that each hyperplane of P contains 10 lines of S_1.

(11) Let $p \in P$ and let A be a subgroup $Alt(4)$ of G containing G_p while $A(p)$ is the orbit of p under A. Then $A(p)$ is a line and these lines constitute a 1-spread S_2. Assume the contrary. Then $A(p)$, which is a set of 4 points, generates a 3-space W because the involutions of A are not allowed to fix points in P and so $\langle A(p) \rangle$ cannot be a plane of 13 points. Let w be the number of transforms of W under G and let W contain α transforms of $A(p)$. Then $w\alpha = 91 = 13 \cdot 7$. Since $PSL(2, 13)$ has no transitive action on 7 or 13 objects (see Figure 1), there follows $w = 91, \alpha = 1$. In W there are 8 points linearly independent of any 3 of $A(p)$ and A acts upon these. The stabiliser G_p cannot act semiregularly on those 8 points; hence it fixes one of them and then $\alpha \geq 2$, a contradiction.

Since $A(p)$ is a block of imprimitivity by (1), it is obvious that these lines constitute a 1-spread.

(12) The lines of S_2 are singular. Let $m \in S_2$ and $p \in m$. Then G_p leaves m invariant and we can apply (8).

(13) Let $l \in S_1$. Then the lines of S_2 in $l^\perp$ are the four lines of S_2 intersecting l. Indeed, for $p \in l$, G_p leaves these four lines invariant and such invariant lines are in $l^\perp$ by (7). If there was a line $m \in S_2$ in $l^\perp$ with $m \cap l = \phi$ then we would see, as in (9), that each plane on l in $l^\perp$ is singular, contradicting (5).

(14) Let $m \in S_2$. Then the lines of S_1 in $m^\perp$ are the four lines of S_1 intersecting m.

If $l \in S_1$ and $l \subseteq m^\perp$ then, applying duality, $m \subseteq l^\perp$; so by (13), m and l have a common point. This condition is also sufficient. Hence, the statement follows.

(15) If there is a quadrangle whose sides are in $S_1 \cup S_2$, then it generates a 3-space U. Now U is invariant by subgroups D_{12} and $Alt(4)$, both maximal in G. Hence G is leaving U invariant, a contradiction.

(16) Consider a subgroup $H = 13 : 3$ in G. It has orbits on planes whose size is 39, 13, 3, or 1. The number of singular planes is 1120, which is equal to 2 (mod 13). If H has no invariant plane then its normaliser 13:6 permutes its orbits of length 3 in pairs; so there are an even number of these. Therefore we see that there are at least 66 planes in orbits of 3 for H. But any 2 such planes are disjoint; hence $66 \times 13 = 858 < 364$, a contradiction. Consequently, H has an invariant plane α and by the action of 13:6 which cannot fix a plane, it has at least two invariant planes α, α'.

Now α is a block of imprimitivity as well as $\alpha \cup \alpha'$. A subgroup T of order 3 in H fixes 4 points, and hence 4 planes invariant under H or a conjugate of H. But T is contained in two conjugates if H (see Figure 1), hence it leaves invariant two invariant planes of H; that is, H has exactly two invariant planes.

Thus α and its transforms under G constitute a 2-spread S_3 of singular planes.

(17) Now (1), (2), (4) and (5) are established, and (3) is easily achieved by (1). There are subgroups K of G with $G_p < K < G$, K of order 6 and then $K < D_{12}$ for some D_{12}; so K produces an orbit of size 2 on p which is a block of imprimitivity contained in a line of S_1. If $G_p < K < G$

with K = 13:3 or 13:6, we saw in (16) that the resulting block of imprimitivity must be a plane or the union of two planes.

Remark.   There are some more known facts that one would like to prove in the preceding style:
(a)   the subgroup of $PSL(6, 3)$ leaving one of S_1, S_2, S_3 invariant, is $PSL(2, 13)$;
(b)   no collineation of P is mapping S_1 onto S_2;
(c)   the symplectic polarity $\perp$ of P can be reconstructed from one of S_1, S_2, S_3.
    We now produce an additional piece of information, useful for geometries.

Lemma 2.   *If $\alpha \in S_2$ and if x is a line of α, there is a unique $l \in S_1$ and a unique $m \in S_2$ such that $l^\perp \cap \alpha = x = m^\perp \cap \alpha$.*

Proof.   We saw earlier (part 9) in the proof of Theorem 1 that no $l^\perp$ can contain α and the argument goes over to an $m^\perp$: its stabiliser $Alt(4)$ must act transitively on the 4 planes in $m^\perp$, containing m otherwise there is an involution fixing one.   If $l^\perp \cap \alpha$ is a line V, then, applying duality, $V^\perp$ is a 3-space containing $\alpha^\perp = \alpha$ and l, so α and l have a common point.   Conversely, if $\alpha \cap l$ is a point, then α and l generate a 3-space U; so the line $U^\perp$ is in $\alpha \cap l^\perp$.   On each point of α there is a unique line of S_1; so, dually, each line of α is on a unique $l^\perp$. The argument is valid for S_2 as well.

3. Hering's geometries

From the preceding situation, Hering (1969, 1985) gets 5 remarkable geometries.   Here we shall produce more such geometries.   Each of these admits Hering's 2-transitive group $3^6 : SL(2, 13)$ as group of flag-transitive automorphisms.   Starting with $P = PG(5, 3)$, $\perp$ and $G = PSL(2, 13)$ as in section 2, we consider an affine space $A = AG(6, 3)$ such that P is the hyperplane at infinity of A.
    In P we dispose of 5 interesting families of non-trivial subspaces, namely $S_1, S_2, S_3, S_1^\perp = S_4, S_2^\perp = S_5$, where the two latter consist of 3-spaces.
    In A we dispose likewise of 6 families of affine subspaces namely A_0, the set of points, and A_i the set of subspaces intersecting P in a member of S_i.

There is a natural way to define incidence either on US_i or UA_i by the principle of maximal intersection due to J. Tits: $x \in A_i$ and $y \in A_j$ are incident if $x \cap y$ is maximal among all intersections $a_i \cap a_j$, $a_i \in A_i$, $a_j \in A_j$ (and similarly for the S_i).

Of course, it would be rather naïve to consider all six families A_i, $i = 0, \ldots, 5$ together, to build a rank 6 geometry, but the above construction gives us a unique pregeometry $H = \Gamma(A_0, A_1, \ldots, A_5)$ and a unique $H^\infty = \Gamma(S_1, \ldots, S_5)$ which are not quite geometries because they are not firm and not residually connected. If we look for a genuine geometry Γ built on the above principles (sets A_i, incidence as described), then Γ is a truncation of H; that means, we remove some A_i's from H.

Lemma 3. *If $\Gamma = \Gamma(A_i, i \in I, I \subseteq \{0, \ldots, 5\})$ is a geometry then each of the following subsets of $\{0, \ldots, 5\}$ is not contained in I:* $\{1, 4\}$, $\{2, 5\}$, $\{1, 2, 3\}$, $\{1, 3, 5\}$, $\{2, 3, 4\}$, $\{3, 4, 5\}$.

Proof. We proceed by contradiction. In H^∞, each element in S_1 is incident to a unique element in S_4, by (9) of the proof of Theorem 1 and vice versa. Therefore, a residue of type $\{1, 4\}$ in Γ cannot be residually connected. So $\{1, 4\}$ is not in I. The same argument works for $\{2, 5\}$. That argument applies again for $\{1, 2, 3\}$: an incident pair in $S_1 \cup S_2$ is incident to a unique element of S_3. Consider $\{1, 3, 5\}$. If $s \in S_3$, $v \in S_5$ are incident, they intersect in a line l. Now $v^\perp \in S_2$ and $v^\perp \neq l$ because the stabiliser of s in G acts transitively on the lines of s and they cannot all be in S_2. But s and l, by duality, have a common point p, which is incident to s and v. Now it is easy to see that a residue of type $\{1, 3\}$ has 4 connected components of two elements each. Therefore a residue of type $\{1, 3\}$ in Γ cannot be connected. The case $\{2, 3, 4\}$ works similarly and $\{3, 4, 5\}$ is brought back to $\{3, 1, 2\}$ by duality.

The maximal subsets of $\{0, \ldots, 5\}$ satisfying the conditions of Lemma 3 indeed provide geometries, optimal in some sense, which we now explain.

Theorem 4. *The maximal truncations of Hering's pregeometry H providing firm, residually connected geometries are described below. The Hering group $3^6: SL(2, 13)$ acts flag-transitively on each of them. Each geometry has the diagram and parameters described here.*

(1) $\Gamma(A_0, A_1, A_2)$

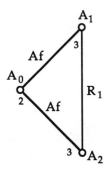

$$
\begin{array}{ccc}
S_1 & R_1 & S_2 \\
\circ & \!\!\!\!\rule[0.3em]{6em}{0.4pt}\!\!\!\! & \circ \\
3 & & 3 \\
D_{12} & & A_4 \\
91 & & 91
\end{array}
$$

PSL(2, 13)

(2) $\Gamma(A_0, A_1, A_3)$

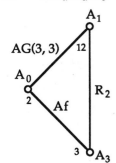

$$
\begin{array}{ccc}
S_1 & R_2 & S_3 \\
\circ & \!\!\!\!\rule[0.3em]{6em}{0.4pt}\!\!\!\! & \circ \\
12 & & 3 \\
D_{12} & & 13{:}3 \\
91 & & 91
\end{array}
$$

PSL(2, 13)

(3) $\Gamma(A_0, A_2, A_3)$

$$
\begin{array}{ccc}
S_2 & R_3 & S_3 \\
\circ & \!\!\!\!\rule[0.3em]{6em}{0.4pt}\!\!\!\! & \circ \\
12 & & 3 \\
A_4 & & 13{:}3 \\
91 & & 28
\end{array}
$$

PSL(2, 13)

(4) $\Gamma(A_0, A_2, A_4)$

$$
\begin{array}{ccccc}
A_0 & N_1 & A_2 & R_1^* & A_4 \\
\circ & \rule[0.3em]{2em}{0.4pt} & \circ & \rule[0.3em]{2em}{0.4pt} & \circ \\
8 & & 3 & & 3
\end{array}
$$

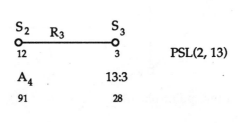

$3^4 :_2 D_{12}$

A 4-net

(5) $\Gamma(A_0, A_1, A_5)$

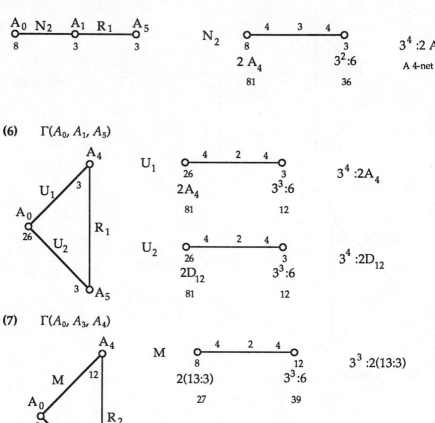

(6) $\Gamma(A_0, A_1, A_5)$

(7) $\Gamma(A_0, A_3, A_4)$

M is the (point plane)-truncation of $AG(3, 3)$

(8) $\Gamma(A_0, A_3, A_5)$

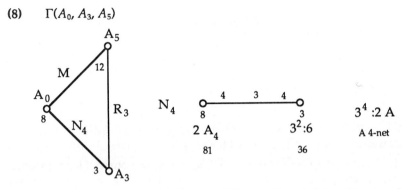

Proof.   (1) The maximal subsets of $\{0, ..., 5\}$ satisfying Lemma 3 are indeed $\{0, 1, 2\}, \{0, 1, 3\}, ..., \{0, 3, 5\}$ and so we are dealing with the right candidates.

(2)   The rest is a matter of straightforward checking as far as the diagrams, orders, residues and stabilisers are concerned.   Firmness is an immediate consequence.   The connectedness of the rank 2 residues is almost always obvious.   For R_1, R_2, R_3 it follows from the fact that any two of the subgroups $D_{12}, A_4,$ 13:3 generate $PSL(2, 13)$.   The connectedness of the full incidence graph is again straightforward. Also, it is straightforward to check the flag-transitivity of the group.

Remarks.   (1) The geometries (4) and (5) are due to Hering (1985).   All others are new to the best of our knowledge.

(2)   Are the 4-nets $N_1, ..., N_4$ isomorphic?   This is unlikely for N_1, N_2 but how about N_1 and N_3, N_2 and N_4?

(3) R_1 and R_2 are not isomorphic.   This can be shown either in a (too) complicated way using the spreads and the classification of the 2-transitive groups or with the help of CAYLEY.

(4)   The parameters of R_2 are as follows

$$\overset{\displaystyle 4 \quad\quad 2 \quad\quad 3}{\underset{\displaystyle \underset{91}{12} \qquad\qquad\qquad \underset{28}{3}}{\circ\!\!-\!\!\!-\!\!\!-\!\!\!-\!\!\!-\!\!\!-\!\!\!-\!\!\!-\!\!\!-\!\!\circ}}$$

Indeed $PSL(2, 13) = G$ and the objects of R_2 an identified with $S_1 \cup S_3$. If $s \in S_3$, there is a "companion" plane s' transformed by s by the normaliser of the stabiliser of s in G.   Now an element of order 3 leaving s invariant, fixes s' also so we see that for each $p \in s$, the line of S_1 on p, intersects s'.   This shows that the gonality of R_2 is 2.   From this situation we see easily that there are 27 planes of S_3 at distance 2 from s and so that the line-diameter of R_2 is 3.   This already makes the point-diameter d_p satisfy $3 \le d_p \le 4$ and the smallest bound is clearly too small.

(5) Similarly, we have R_3

$$\begin{array}{ccccc} & 4 & 2 & 3 & \\ \circ\!\!\!\!&\!\!\!\!-\!\!\!\!&\!\!\!\!-\!\!\!\!&\!\!\!\!-\!\!\!\!&\!\!\!\!\circ \\ 12 & & & & 3 \\ 91 & & & & 28 \end{array}$$

(6) A computation on CAYLEY shows that R_1 has shape

$$\begin{array}{ccccc} & 6 & 4 & 6 & \\ \circ\!\!\!\!&\!\!\!\!-\!\!\!\!&\!\!\!\!-\!\!\!\!&\!\!\!\!-\!\!\!\!&\!\!\!\!\circ \\ 12 & & & & 3 \\ 91 & & & & 91 \end{array} \ .$$

(7) The geometries (1), ... (8) have, of course, some interesting truncations. $\Gamma(a_0, a_1)$ is one of Hering's linear spaces, with lines of 9 points, $\Gamma(a_0, a_2)$ is the other one and $\Gamma(a_0, a_3)$ is, of course, Hering's non-desarguesian plane of order 27. We shall not go into further discussion of such truncations.

4. Generalising

We shall now briefly consider the basic ideas leading to the geometries of the preceding section. We use no polarity, but this could be done with the same obvious richness of consequences. So another field opens. Let P be a projective space $PG(d, q)$ -actually it need not be finite- and a *partial t_1-spread* S_1 together with a *partial t_2-spread* S_2 in $P(1 \le t_1, t_2)$ with the property that:

(i) the union of the components of S_1 is equal to the union of the components of S_2;

(ii) a component of S_1 and a component of S_2 intersect in at most one point.

Examples. (1) The two reguli of a quadric in $PG(3, q)$.

(2) Two 1-spreads with no common line in $PG(3, q)$. These exist because packings of lines exist (see Hirschfeld 1985, Ch. 17).

From S_1, S_2 we build a rank 3 geometry. Consider an affine space A admitting P as hyperplane at infinity. Let A_0 be the set of points of A and A_i $(i = 1, 2)$ the set of affine spaces intersecting P in a component of S_i. Let incidence be defined as in section 3, by the principle of maximal intersection. Thus we get a pregeometry $\Gamma = \Gamma(A_0, A_1, A_2)$. Here Γ is obviously firm. It is residually connected unless, S_1 and S_2 have proper subsets with the same properties (i) and (ii). The diagram of Γ is of shape

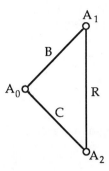

where:

    B is the (point-line)-truncation of $AG(t_2 + 1, q)$

    C is the (point-line)-truncation of $AG(t_1 + 1, q)$

R is $\Gamma(S_1, S_2)$ where incidence is again defined by maximal intersection.

Example. If we start with the above example for S_1, S_2 we get

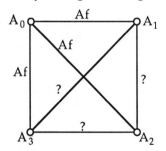

Can we get such a rank 4 geometry? Yes, if we start with convenient spreads S_1, S_2, S_3 such that two components of distinct S_i meeting in a point, have a common line and three components of distinct S_i have at most one common point.

    The resulting geometry belongs to a diagram

We know of no example realising the required conditions.

References

J.H. Conway, R.T. Curtis, S.P. Norton, R.A. Parker and R.A. Wilson (1985), *An ATLAS of Finite Groups*, Oxford University Press, Oxford.

C. Hering (1969), Eine nicht-desarguessche zweifach transitive affine Ebene der Ordnung 27, *Abh. Math. Sem Univ. Hamburg* **34**, 203-208.

C. Hering (1985), Two new sporadic doubly transitive linear spaces, *Finite Geometries*, (ed. C.A. Baker and L.M. Batten), Lecture Notes in Pure and Applied Math. **103**, 127-129, Dekker, New York.

J.W.P. Hirschfeld (1985), *Finite Projective Spaces of Three Dimensions*, Oxford University Press, Oxford.

P.B. Kleidman (to appear), *The Low-dimensional Finite Classical Groups and their Subgroups*, Longman Research Notes.

Presentations for certain finite quaternionic reflection groups

A. M. Cohen
C.W.I, Amsterdam and University of Utrecht

1. Introduction

One of Coxeter's highly remarkable discoveries is that the diagrams bearing his name can be interpreted as presentations (by means of generators and relations) for real linear groups generated by reflections having roots (that is, -1 eigenvalues) at angles indicated by the diagram.

Given such a finite linear reflection group, there is a natural and canonical way to isolate a fundamental domain in the reflection space bounded by reflecting hyperplanes. The associated diagram is then obtained by taking for nodes these hyperplanes, joining two nodes if the reflections corresponding to these hyperplanes do not commute and labelling the resulting edge by the order of the product of these two reflections. (This number is of course directly related to the angle between the roots and also the angle between the two hyperplanes.) Conversely, given a diagram, a realization by means of hyperplanes having the right angles leads to a reflection group, namely the group generated by all reflections in these hyperplanes.

In 1953, Shephard (1953) provided similar diagrams for finite unitary reflection groups. The fundamental domain no longer played a role, but the nodes still represented roots of a generating set of reflections, edges still corresponded to non-orthogonal roots (and therefore non-commuting reflections), and the edge labelling kept its meaning. The Coxeter diagrams of finite real reflection groups are free of circuits. Shephard's connected diagrams corresponding to unitary reflection groups contain a single circuit, which is a triangle.

Fourteen years later, Coxeter (1967) provided presentations for these unitary reflection groups using Shephard's diagrams. The presentation for the Coxeter group related to these diagrams had to be extended by a single relation coming from the triangle (here we assume, without loss of generality, that the diagram is connected - otherwise study of the group and its reflection representation can be reduced to the factors of a direct product decomposition). Upon adding a number to the diagram that is to be interpreted as a label of the triangle, the presentation of the corresponding reflection group could still be read off from the diagram.

Now that Hoggar (1982) has provided diagrams for the most interesting finite quaternionic reflection groups, the question arises of producing a presentation that can, again, be read off directly from the diagram. In the present note, we report on some attempts to this end.

For each of the groups under study, we exhibit a presentation, largely based on some rules of thumb for reading off relations from the diagram. One of these is a relation obtained from identifying subgroup centres which McMullen once suggested to Coxeter for unitary groups; I am grateful to Leonard Soicher for pointing out such a rule to me for the quaternionic case. Two presentations remain quite peirastic: we have found no proper mnemonic device to distill a presentation from the diagram.

The presentations are useful in the study of subgroups generated by reflections corresponding to subsets of nodes of the diagram: presentations for these subgroups can be obtained by removal of all relations involving a reflection corresponding to an excluded node. Thus, we provide a good starting point for the geometric study of the permutation representations on these and (other) reflection subgroups. Extrapolating from the complex case, one may also venture to predict a use of these presentations for a possible classification of discrete quaternionic groups.

Two out of the seven groups (viz. $W(S_1)$ and $W(U)$) are 3-transposition groups: the order of the product of any two reflections is either 2 or 3. In this context, the presentations for these two groups described below are well known, see for instance Hall (1990) and Zara (1985).

2. Preliminaries

A *quaternionic reflection group* is a quaternionic linear group, that is, a subgroup of the general quaternionic linear group $GL(n, \mathbf{D})$ for some integer n generated by *reflections* (that is, elements of $GL(n, \mathbf{D})$ having a fixed space of dimension $n - 1$); the number n will be referred to as the *dimension* of the reflection group. We recall that a subgroup G of $GL(n, \mathbf{D})$ is called *primitive* if the only set $\{V_1, ..., V_t\}$ of subspaces $\mathbf{D}^n$ decomposing $\mathbf{D}^n$ and stable under G occurs for $t = 1$.

The group G is called *complex* or *real* if it is conjugate in $GL(n, \mathbf{D})$ to a subgroup of $GL(n, \mathbf{C})$ or $GL(n, \mathbf{R})$, respectively. Finite real groups are dealt with extensively in Bourbaki (1968), complex reflection groups have been classified in Shephard and Todd (1954) (see also Cohen (1976)). As finite complex linear groups leave invariant a unitary form, they are also called unitary instead of complex.

The finite quaternionic groups have been classified in Cohen (1980). In studying presentations, we shall content ourselves with considering the most interesting examples, namely those finite quaternionic reflection groups (of dimension ≥ 3) whose complexifications are primitive complex linear groups. There are seven of them. Hoggar (1982) has worked out certain minimal sets of generating reflections for these groups and has drawn the diagrams based on the corresponding roots. Here, we shall provide presentations of these groups on the basis of these (labelled) graphs.

3. The groups involved

The primitive quaternionic reflection groups of dimension ≥ 3 whose complexifications are primitive are the following (cf. Cohen 1980):

dim n	root system	reflection group	its order	selected subgroup	its index
3	Q	$2 \times PSU(3, 3)$	$2^6 \cdot 3^6 \cdot 7$	$W(J_3(4))$	36
3	R	$2 \cdot HJ$	$2^8 \cdot 3^3 \cdot 5^2 \cdot 7$	$W(J_3(5))$	560
4	S_1	$(\mathbf{D_2} \circ \mathbf{D_8} \circ \mathbf{D_8}) \cdot G(3, 3, 3)$	$2^8 \cdot 3^3$	$G(3, 3, 3)$	128
4	S_2	$(\mathbf{D_2} \circ \mathbf{D_8} \circ \mathbf{D_8}) \cdot G(3, 3, 4)$	$2^{10} \cdot 3^4$	$G(4, 4, 3)$	864
4	S_3	$(\mathbf{D_2} \circ \mathbf{D_8} \circ \mathbf{D_8}) \cdot \Omega^-(6, 2)$	$2^{13} \cdot 3^4 \cdot 5$	$W(S_2)$	40
4	T	$(\circ^3 SL(2, 5)) Sym_3$	$2^8 \cdot 3^4 \cdot 5^3$	$G(5, 5, 3)$	17280
5	U	$2 \times PSU(5, 2)$	$2^{11} \cdot 3^5 \cdot 5 \cdot 11$	$W(S_1)$	3960

Here, each line corresponds to a quaternionic reflection group. Its dimension is listed in the first column, the name given in Cohen (1980) for its 'root system', that is, a suitably chosen set of root reflections (at least one for each reflection) in the group, is listed next. Analogously to the real case, the reflection group corresponding to the root system X is denoted by $W(X)$ (see the fifth column for examples). The third column contains a description of the isomorphism types; $\circ^3 SL(2, 5)$ stands for the central product of three copies of the special linear group $SL(2, 5)$ on the 2-dimensional vector space over the field with 5 elements. The fourth column provides a reflection subgroup with respect to which a presentation will be given below.

The diagrams found by Hoggar (1982) are depicted in Figure 1. For the diagrams associated with S_2 and S_3 we have made a slight adaption; Hoggar's diagrams for those two cases can be obtained from ours by replacing the reflection c with aca. The absence of a label at an edge or triangle is to be interpreted as the label 3.

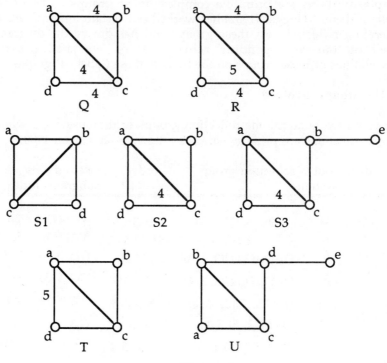

Figure 1.

4. The complex reflection groups

Let us briefly recall Coxeter's 'complex' results (cf. Coxeter (1967)). The usual Coxeter group presentation related to these diagrams had to be extended by a single relation coming from the triangle (here we assume, without loss of much generality, that the diagram is connected - otherwise study of the group and its reflection representation can be reduced to the factors of a direct product decomposition). To this end an additional number was introduced that could be interpreted as the label of the triangle. Now, if the reflections corresponding to the nodes of the diagram are denoted by a, b, c and the triangle label is m, the additional relation reads

$$(abcb)^m = 1.$$

Since at least two edges of the triangle had label 3, the additional relation is independent of the order in which the nodes are read off from the diagram. We provide the two most important series as examples.

4.1 $J_3(m)$

Here, only $m = 4, 5$ are relevant. The above rule leads to the following presentation of $W(J_3(m))$:

generators: a, b, c;

relations: $a^2 = b^2 = c^2 = (ab)^4 = (ac)^3 = (bc)^3 = 1,$

$$(abcb)^m = 1.$$

Such a presentation is checked by use of a Todd-Coxeter coset enumeration[1]. Here the enumeration has taken place with respect to the subgroup $\langle a, b \rangle$.

If $m = 4$, coset enumeration outputs 42 cosets. The element $(abc)^7$ has order 2 and lies in the centre.

If $m = 5$, the output yields 270 cosets. The group is then isomorphic to the central extension $6 \cdot Alt_6$ of the alternating group on 6 letters. Its centre is cyclic of order 6. The element abc has order 30. The element $(abc)^5$ generates the centre.

4.2 $G(m, m, 3)$

The imprimitive complex reflection groups we shall need are $G(m, m, 3)$ for $m = 3, 4, 5$ in the notation of Shephard and Todd (1954). They have the following presentation.

generators: a, b, c;

relations: $a^2 = b^2 = c^2 = (ab)^3 = (ac)^3 = (bc)^m = 1,$

$$(abcb)^3 = 1.$$

Thus, the underlying diagram is a triangle one of whose sides has label m, while the label of the triangle equals 3. If $m = 3$, the centre has order 3 and is generated by $(abc)^2$.

4.3 Five generators in four dimensions

Among the primitive complex reflection groups, there is a single 4-dimensional one that cannot be generated by 4 reflections. A diagram corresponding to five generating reflections is given in Figure 2.

---

[1]The computations have been performed in CAYLEY and MAPLE on the computers of t Dutch Computer Algebra Centre of Expertise CAN.

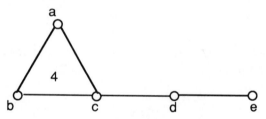

Figure 2.

Coxeter (1967) presented - among others - the following relation due to McMullen

$$[(abcd)^5, e] = 1.$$

Since the four reflections a, b, c, d generate an irreducible linear 4-dimensional group, the centre of the group they generate will be the centre of the whole group, whence also commute with the fifth generator e. The above relation emerges from the rule that in an n-dimensional linear group generated by a set Y of $n + 1$ reflections, the centres of the groups generated by a subset of Y of size n should commute with each member of Y. In the next section, we shall abide with this rule and the variation that central involutions obtained in this way should be identified.

5. Presentations of the quaternionic groups

We now come to the 7 presentations corresponding to the diagrams of Figure 1. They maintain the property that for each subset X of the nodes, a presentation of the corresponding reflection groups (generated by the reflections in X) can be obtained by disregarding all generators and relations containing a generator outside of X.

5.1 Q

The group $W(Q)$ is 3-dimensional, but cannot be generated by 3 reflections. The coset enumeration for the relations obtained from the 3-generator subgroups did not complete. The 3-generator subgroups $\langle a, b, d \rangle$ and $\langle b, c, d \rangle$ are both isomorphic to the well-known Coxeter group $W(B_3)$, and so have central involutions $(b a d)^3$ and $(b c d)^3$, respectively. But also $\langle a, c d \rangle \cong W(J_3(4))$ has a central involution: $(a c d)^7$, see §4.1. Now the centre identification rule comes into effect: we identify these central involutions, found in 3-generator subgroups. The following presentation results:

generators: $\qquad a, b, c, d;$

relations: $\qquad a^2 = b^2 = c^2 = d^2 = 1,$

$$(ab)^4 = (ac)^3 = (ad)^3 = (bc)^3 = (bd)^2 = (cd)^4 = 1,$$

$$(abcb)^3 = (acdc)^4 = 1,$$

$$(bad)^3 = (bcd)^3 = (acd)^7.$$

Coset enumeration with respect to the subgroup $\langle a, c, d \rangle \cong W(J_3(4))$ gives that the latter subgroup has index 36 in the presented group. Thus the presented group has the same order as its homomorphic image $W(Q)$ and so is isomorphic to it. We conclude that the above is a presentation for $W(Q)$. The centre of order 2 in the presented group is generated by $ababdabad$.

5.2 R

For R, observations similar to those for Q seem to apply. The subgroup generated by a, c, d is the complex reflection subgroup $W(J_3(5))$, so the involution $(a\,d\,c)^{15}$ is central and can again be identified with central involutions $(b\,a\,d)^3$ and $(b\,c\,d)^3$. Thus we are lead to consider the following presentation:

generators: $\qquad a, b, c, d;$

relations: $\qquad a^2 = b^2 = c^2 = d^2 = 1,$

$$(ab)^4 = (ac)^3 = (ad)^3 = (bc)^3 = (bd)^2 = (cd)^4 = 1,$$

$$(abcb)^3 = (acdc)^5 = 1,$$

$$(bad)^3 = (bcd)^3 = (adc)^{15}.$$

But a new phenomenon presents itself: The reflection group $W(R)$ is perfect, whereas the above presentation has relations of even length only, so that the subgroup generated by all products in a, b, c, d of even length is a subgroup of index 2. (Over a commutative field, of course, the reflections would have had determinant -1 and so never generate a perfect group.) The coset enumeration however does complete giving index 1120 over $\langle a, c, d \rangle$, twice the index the reflection subgroup $W(J_3(5))$ generated by the corresponding reflections has in $W(R) \cong 2 \cdot HJ$. Thus a single relation of odd length would suffice, or for that matter, a way to express a reflection as a product of commutators. The last line of the following presentation does that job (admittedly not in a very pretty way) providing a presentation for $W(R)$:

generators: $\qquad\qquad\qquad\qquad a, b, c, d;$

relations: $\qquad\qquad\qquad\qquad a^2 = b^2 = c^2 = d^2 = 1,$

$$(ab)^4 = (ac)^3 = (ad)^3 = (bc)^3 = (bd)^2 = (cd)^4 = 1,$$

$$(abcb)^3 = (acdc)^5 = 1,$$

$$acdcadcdacdacdacdcadcdacababcdacdabcdacdabcab = 1.$$

5.3 S_1

The group $W(S_1)$ is 4-dimensional. Thus, no additional rules can be derived from identification of central elements of subgroups.

Since the relations for the complex reflection subgroups did not lead to completion of the coset enumeration, an additional rule is called for. Observing that the diagram has a single 4-circuit, we have extended the labelling of triangles to one for all circuits. Here we label the full circuit by a 3 (not written in the diagram of Figure 1) to remind us of the additional relation $((aba)(dcd))^3 = 1$.

Thus we arrive at the following presentation:

generators: $\qquad\qquad\qquad\qquad a, b, c, d;$

relations: $\qquad\qquad\qquad\qquad a^2 = b^2 = c^2 = d^2 = 1,$

$$(ab)^3 = (ac)^3 = (ad)^2 = (bc)^3 = (bd)^3 = (cd)^3 = 1,$$

$$(abcb)^3 = (acdc)^3 = 1,$$

$$(badcda)^3 = 1.$$

Enumeration of cosets of $\langle b, c, d \rangle$ yields index 128. Since $\langle b, c, d \rangle \cong G(3, 3, 3) \cong 3^2 \cdot Sym_3$, this yields the right order for $\langle a, b, c, d \rangle$ to coincide with $W(S_1)$. Consequently, the above is indeed a presentation of $W(S_1)$.

5.4 S_2

Proceeding analogously to the former case, we find that the following is a presentation for $W(S_2)$:

generators: $\qquad\qquad\qquad\qquad a, b, c, d;$

relations: $\qquad\qquad\qquad\qquad a^2 = b^2 = c^2 = d^2 = 1,$

$$(ab)^3 = (ac)^3 = (ad)^3 = (bc)^3 = (bd)^2 = (cd)^3 = 1,$$

$$(abcb)^3 = (acdc)^4 = 1,$$

$$(adbcbd)^3 = 1.$$

We observe that the element $abcd$ has order 24 and that its 12-th power generates the centre (a group of order 2). This fact plays a role in the presentation for $W(S_3)$ below.

5.5 S_3

The diagram of S_3 contains 5 nodes, while the group $W(S_3)$ is 4-dimensional. Thus the central element $(abcd)^{12}$ of $W(S_3)$ is central in the whole group. Adding this relation to the usual rules for the presentation, we obtain:

generators: $\qquad\qquad a, b, c, d, e$

relations: $\qquad\qquad a^2 = b^2 = c^2 = d^2 = e^2 = 1,$

$$(ab)^3 = (ac)^3 = (ad)^3 = (ae)^2 = (bc)^3 = (bd)^2 = (be)^3$$

$$= (cd)^3 = (de)^2 = (ce)^2 = 1,$$

$$(abcb)^3 = (acdc)^4 = 1,$$

$$(adbcbd)^3 = 1,$$

$$[e, (abcd)^{12}] = 1.$$

Enumeration of cosets with respect to the subgroup $\langle a, b, c, e, (adcd)^2 \rangle$ showed that the group with this presentation has isomorphism type $2 \cdot 2 \cdot ((2^8 \cdot 2^6) \times 2) \cdot P S\Omega^-(6, 2)$, while $W(S_3) \cong 2 \cdot 2^6 \cdot P S\Omega^-(6, 2)$. Since $W(S_3)$ is perfect, we should add a relation of odd length; we take the relation $(abcde)^5 = 1$. But then the resulting group still has a normal 2-subgroup of order 2^8 which does not occur in $W(S_3)$. This shows that the rules of thumb introduced so far do not suffice. To kill the latter normal 2-group, we had to add one more relation. The resulting presentation for $W(S_3)$ is:

generators: $\qquad\qquad a, b, c, d, e$

relations: $\qquad\qquad a^2 = b^2 = c^2 = d^2 = e^2 = 1,$

$$(ab)^3 = (ac)^3 = (ad)^3 = (ae)^2 = (bc)^3 = (bd)^2 = (be)^3 \quad = (cd)^3 = (de)^2 = (ce)^2 = 1,$$

$$(abcb)^3 = (acdc)^4 = 1,$$

$$(adbcbd)^3 = 1,$$

$$[e, (abcd)^{12}] = 1,$$

$$dc(acd)^4cd = (be(abc)^2ebadc)^6,$$

$$(abcde)^5 = 1.$$

5.6 T

Exploiting the 4-circuit rule for the 4-dimensional group $W(T)$, the following presentation has been found:

generators: a, b, c, d

relations: $a^2 = b^2 = c^2 = d^2 = 1,$

$$(ab)^3 = (ac)^3 = (ad)^5 = (bc)^3 = (bd)^2 = (cd)^3 = 1,$$

$$(abcb)^3 = (acdc)^3 = 1,$$

$$(adbcbd)^3 = 1.$$

Coset enumeration with respect to $\langle a, c, d \rangle \cong G(5, 5, 3)$ gives index 17280 $= 2^7 3^3 5$. Since $G(5, 5, 3)$ has order $5^2 \cdot 3!$, the order of the presented group is $2^7 3^4 5^3$, which coincides with the order of $W(T)$. Therefore, the above is indeed a presentation of $W(T)$.

5.7 U

Finally we treat the 5-dimensional group $W(U)$. The 4-circuit rule suffices to find a satisfactory presentation for $W(U)$:

generators: a, b, c, d, e

relations: $a^2 = b^2 = c^2 = d^2 = e^2 = 1,$

$$(ab)^3 = (ac)^3 = (ad)^2 = (ae)^2 = (bc)^3 = 1,$$

$$(bd)^3 = (be)^2 = (cd)^3 = (de)^3 = (ce)^2 = 1,$$

$$(abcb)^3 = (bcdc)^3 = 1,$$

$$(cdabad)^3 = 1.$$

The fact that the presented group is isomorphic to $W(U)$ follows from the fact that the index 3960 has been found in enumerating the cosets of the subgroup $\langle a, b, c, d \rangle$ (isomorphic to $W(S_1)$) of the presented group.

References

N. Bourbaki (1968), *Groupes et Algèbres de Lie, Chap.* IV, V *et* VI, Hermann, Paris.

A.M. Cohen (1976), Finite complex reflection groups, *Ann. Sci. École Norm. Sup.* **4**, 379-436.

A.M. Cohen (1980), Finite quaternionic reflection groups, *J. Algebra* **64**, 293-324.

H.S.M. Coxeter (1967), Finite groups generated by unitary reflections, *Abh. Math. Sem. Univ. Hamburg* **31**, 125-135.

J.I. Hall (1990), 3-transposition groups with non-central normal 2-subgroups, *J. Algebra*, to appear.

S.G. Hoggar (1982), *t*-designs in projective spaces, *European J. Combin.* **3**, 233-254.

G.C. Shephard (1953), Unitary groups generated by reflections, *Canad. J. Math.* **5**, 364-383.

G.C. Shephard and J.A. Todd (1954), Finite unitary reflection groups, *Canad. J. Math.* **6**, 274-304.

F. Zara (1985), Classification des couples fischerien, Ph.D. Thesis, University of Picardie, III, CNRS no. 944.

Geometric hyperplanes of embeddable Lie incidence geometries

B.N. Cooperstein and E.E. Shult
University of California at Santa Cruz and
Kansas State University

1. Introduction

Very large subspaces play a key role in determining the structure of a space and in providing interesting examples of spaces not easily defined in other ways. A celebrated criterion, due to Teirlinck (1980), that a linear space be a projective space involves only the way certain geometric hyperplanes are distributed. In a classic construction, Veldkamp used the geometric hyperplanes of a polar space to provide an embedding of the latter in projective space (1960). Indeed, this idea has been exploited in a recent elementary characterization of all non-degenerate polar spaces of (possibily infinite) rank at least 4 due to to Cuypers et. al. (to appear). Among the locally co-triangular graphs, (classified by Hall and Shult (1985)), one encounters a large family formed by removing a geometric hyperplane from a non-degenerate polar space over $GF(2)$. An adaption of Hall's proof characterizing that family to the case of thick lines of arbitrary cardinality was utilized by Cohen and Shult (1990) to show that affine polar spaces are also non-degenerate polar spaces with a geometric hyperplane removed. (A very beautiful, recent generalization of this theorem due to Cuypers and Pasini (personal communication) shows that if the gamma space hypothesis is dropped, the geometries that result are still homomorphic images of affine polar spaces.) Indeed, it seems likely that any future characterizations of Lie incidence geometries built around affine planes (rather than the usual projective planes) must involve the conclusion that such a geometry is a homomorphic image of a classical Lie incidence geometry with a geometric hyperplane removed. This eventuality itself motivates the study of geometric hyperplanes of the Lie incidence geometries.

2. Geometric hyperplanes and projective embeddings

For the purpose of this paper, a point-line geometry $(\mathcal{P}, \mathcal{L})$ is a rank two incidence system of points and lines with each line incident with at least two points and no repeated line — that is, lines may be regarded as sets of points. A *subspace* is a subset of the set $\mathcal{P}$ of points which contains any line which meets it in at least two points. A *geometric hyperplane* is a proper subspace which intersects non-trivially every line. A geometric hyperplane of a projective space (the truncation of a projective geometry to projective points and lines) is an ordinary *projective hyperplane.*

A *projective embedding* $e: \Gamma \to P$, of a point-line geometry $\Gamma = (\mathcal{P}, L)$ into a projective space $P = (\mathcal{P}', L')$ is an injection $e: \mathcal{P} \to \mathcal{P}'$ whose restriction to any line of L is a (full) line of L', and whose image points $e(\mathcal{P})$ span P.   (In some settings this is called "full embedding", but since the term "weak embedding" is already employed in the literature to indicate a projective embedding which may not be full, and since all embeddings considered here are into projective spaces, we shall simply drop the term "full" and "projective" when referring to embeddings in this paper.)   If H' (viewed as a subset of $\mathcal{P}'$) is an ordinary projective hyperplane of P, then $H = e^{-1}(H' \cap e(\mathcal{P}))$ is a geometric hyperplane of Γ.   In this case we say that the *geometric hyperplane H arises from an embedding.*

3. Ronan's Criterion and its application to parapolar spaces

When does a given geometric hyperplane H of a point-line geometry $\Gamma = (\mathcal{P}, L)$ arise from an embedding?   In a fundamental paper, Ronan (1987) gave a complete answer to this question.   To understand this completely, we must first introduce the notion of a *universal embedding.*   Suppose $e': \Gamma \to P'$ and $e: \Gamma \to P$ are embeddings of Γ.   A *morphism of embeddings* is a mapping t of the point set of P', which is induced by a linear surjection of their underlying vector spaces, and such that e on $\mathcal{P}$ is a composition of of e' and t.   With a slight abuse of notation, we denote this morphism by $t: e' \to e$.   Given an embedding $e: \Gamma \to P$, an embedding $e^*: \Gamma \to P^*$ is said to be *universal for e* if there is a morphism of embeddings $u: e^* \to e$ such that, given any morphism of embeddings $t: e' \to e$, there exists a morphism $t^*: e^* \to e'$, so that u is the composition of t and t^*.   We then have

Proposition 1 (Ronan 1987).   *Let $e: \Gamma \to P$ be an embedding of $\Gamma = (\mathcal{P}, L)$ into the projective space P.   Then there is a natural embedding $e^*: \Gamma \to P^*$, where P^* is the projective space of the zero homology space $H_0(\mathcal{F}_e)$ of the constant sheaf $\mathcal{F}_e$ associated with the embedding e.*

(The reader should consult Ronan's paper for definitions and details of the construction.)   An immediate consequence, of course, is that given any embedding e, a universal embedding e^* for e always exists.

Now let H be a geometric hyperplane of $\Gamma = (\mathcal{P}, L)$ and let $e: \Gamma \to P$ be an embedding and let V be the ambient vector space for P. An H-chain is a mapping $\phi: \mathcal{P} - H \to V$ such that

(a) $\langle \phi(x) \rangle = e(x)$ for each point x in $\mathcal{P} - H$, and

(b)   for each line L of L with $L \cap H = \{p\}$ and points x and y of $L - \{p\}$, the vector $\phi(x) - \phi(y)$ lies in the 1-space $e(p)$ of V.

Proposition 2. (Ronan's Criterion (1987)). *Let Γ, H and $e: \Gamma \to P$ be as in the previous paragraph. Then the geometric hyperplane arises from the embedding e^* universal for e if and only if an H-chain exists for Γ, H and e.*

Let $\mathfrak{R}$ be the full collection of point-line geometries $\Gamma = (\mathcal{P}, L)$ such that for every geometric hyperplane H and every embedding $e: \Gamma \to P$, H arises from the embedding e^* universal for e. Obviously projective spaces of finite rank belong to $\mathfrak{R}$. Not quite so obvious is that all embeddable non-degenerate polar spaces of finite rank at least 3 belong to $\mathfrak{R}$. It follows from the theorems of Dienst (1980), Buekenhout and Lefèvre (1974) and Lefèvre-Percsy (1981) that any embedding of a non-degenerate polar space of finite rank at least 2 is its natural embedding, with the usual proviso that those arising from symplectic geometries over fields in characteristic 2 have, as a second universal embedding, their realisation as an orthogonal polar space over the same field, in one dimension higher. Of course, the theorems actually show that the embeddings are dominated embeddings — that is, P is endowed with a polarity with respect to which all embedded points $e(\mathcal{P})$ are absolute — and then recourse to Tits' (1974) fundamental Lemma 8.6 completes the result. (Very recently, Johnson (1990) has shown that embeddings of polar spaces of arbitrary rank are also dominated. This advance figures in Cuypers et. al (to appear) polar space classification.)

Now if $e: \Gamma \to P$ is an embedding of a non-degenerate polar space of finite rank at least 3 (assumed to be the universal one of type $\Omega(2n + 1, K)$ if Γ is type $Sp(2n, K)$, char $K = 2$) and H is a geometric hyperplane of Γ, then H itself is a polar space and there are two cases.

(1) If H is degenerate, then the fact that it is a geometric hyperplane implies $H = p^\perp$ for some point p, so H arises from the projective hyperplane $V \cap e(p)^\perp$ in this case.

(2) H is non-degenerate with rank at least 2, so by Buekenhout-Lefèvre-Dienst theory, the restriction of e to H yields a natural embedding of H as a non-degenerate polar space. This is a known list

and in all cases $\langle e(H) \rangle$ is a hyperplane of P.   (This is argued in Cohen and Shult (1990 Proposition 5.2).   Thus we see

Proposition 3. *All embeddable non-degenerate polar spaces of finite rank at least 3 belong to $\mathfrak{R}$.*

What is the significance of $\mathfrak{R}$?   Suppose H is a geometric hyperplane of $\Gamma = (P, L)$, and e is a universal embedding of Γ.   Select any point p in $P - H$.   We assign a non-zero vector $\phi(p)$ in $e(p)$, and attempt to extend ϕ to an H-chain $\phi: P - H \rightarrow V$, the ambient vector space of $P = \langle e(P) \rangle$.   Clearly property (ii) of the definition of H-chain shows that for every neighbour q of p in the collinearity graph on $H' = P - H$, $\phi(q)$ is uniquely determined by the assignment of vector $\phi(p)$.   In turn ϕ is determined on all neighbours of q and so on.   This process consistently defines an H-chain on the connected component of p in the collinearity graph H', provided it can be consistently carried out on all circuits in this component.   But if a circuit C in H' is the simple sum mod 2 of two other circuits C_1 and C_2 ("simple" here means that the shared edges of C_1 and C_2 form a simple path: we denote this by $C = C_1 + C_2$) and ϕ can be consistently defined on C_1 and C_2 so that (i) and (ii) are satisfied, then ϕ is consistently defined on C. But if X is a subspace of Γ which belongs to $\mathfrak{R}$, then by proposition 2, an H-chain exists for $X - H$, and so ϕ can be consistently defined on any circuit C' of H' contained within $X - H$.   We thus observe

Proposition 4. *Let H be a geometric hyperplane of the point-line geometry $\Gamma = (P, L)$ and suppose $e: \Gamma \rightarrow P$ is an embedding.   We assume that*
(N) there exists a family $\mathcal{F} = \{X_i\}$ of subspaces belonging to the class $\mathfrak{R}$ such that every circuit in $P - H$ is built up as simple sums of circuits C_i which individually lie in sets $X_i - H$, for some X_i in $\mathcal{F}$.
Then H arises from the embedding e^* universal for e.

A point-line geometry Γ is called a *parapolar space* if
(PP1) $p^\perp$ is a subspace for each point p, (that is, Γ is a *gamma space*);
(PP2) for no line L is $L^\perp$ a singular subspace;
(PP3) for any two points p and q at distance 2, $p^\perp \cap q^\perp$ is a single point or a non-degenerate polar space of rank at least 2.

It is well known that in such a space the convex closure of a distance two pair $\{p, q\}$ is a *symplecton*, that is, a convex subspace which is a non-degenerate polar space of finite rank at least 3. (This was first proved in Cooperstein (1977) for strong parapolar spaces and generalised to parapolar spaces by Buekenhout (1983) and Cohen (1983), and to weak parapolar spaces by Shult (1986)). Thus, in the collinearity graph on the points of a parapolar space, (1) every triangle lies in a singular subspace and (2) every 4-circuit lies in a symplecton. Then, in view of Propositions 3 and 4, we have

Proposition 5. *Let Γ be an embedded parapolar space and let H be a geometric hyperplane of Γ such that every circuit of the collinearity graph on $\mathcal{P} - H$ is a simple sum of triangles and 4-cicuits. Then the hyperplane H arises from an embedding.*

4. Applications to certain Lie geometries

Let G be the geometry whose objects are the residues of co-dimension 1 in a spherical building Δ with diagram D. Following Cooperstein (1977), a *Lie incidence geometry*, $\Gamma = (\mathcal{P}, L)$, is the geometry whose points are all objects of G of a fixed type (and therefore associated with a fixed node n of D), and whose lines are all flags of G whose type T_n is given by the set of nodes neighbouring n in diagram D. Such geometries are not always embeddable (for example Grassmann spaces $A_{n, d}(K)$, where K is a non-commutative division ring.) But normally, they do possess embeddings — at least when defined over a field. For example, in the case that $K = C$, the complex numbers, an embedding $e: \Gamma \to P$ is obtained where P is the projective space obtained from the Lie-algebra module V whose highest weight w_n satisfies $(w_n, \alpha) = 1$, or 0, according as the fundamental root α is, or is not that associated with the node n. If v is an eigen-space affording the highest weight representation of a Cartan subalgebra and G is the associated Chevally group, then all points $\mathcal{P}$ are embedded onto the G-orbit v^G. If P is the parabolic subgroup represented by the stabiliser in G of a flag F of type T_n incident with $p = e^{-1}(v)$, then it must be shown that the subspace $W = \langle v^P \rangle$ is indeed 2-dimensional — so that a line of L is represented by a projective line of P. But this follows from the definition of type T_n, particularly that the nodes neighbouring n separate n from the remaining nodes, the action of G on weight spaces and Dynkin's algorithm (which implies $w_n - 2\alpha_n$ is not a weight of V). It would seem that a similar argument should be available for other fields, perhaps utilizing Curtis' notion of weights (as in Smith (to appear)).

Whatever the status of a general proof concerning the existence of embeddings of Lie incidence geometries, there are many cases in which an algebraic construction of an appropriate G-module always provides such an embedding of the Lie incidence geometry.   For example, Grassmann spaces $A_{n,d}(k)$, where k is a field, are embeddable in the projective space $P(W)$, where W is the d-fold wedge product of $V = k^{(n)}$ with itself.   Similarly, the classical dual polar spaces of type $C_{n,n}(k)$, and the half-spin geometries of type $D_{n,n}(k)$ have embeddings in $P(V)$ where V is a certain left ideal in the Clifford algebra and $E_{6,1}(k)$ embeds in $P(V)$, where V is the exceptional 27-dimensional Jordan algebra on which the group $G = E_6(k)$ acts as a group of automorphisms.   (We shall see in the next section that all these and several other embeddings of Lie incidence geometries are "universal" — that is, universal for themselves in the terminology above.)

Now, $A_{n,d}(k)$, $D_{n,n}(k)$ and $E_{6,1}(k)$ are all embeddable parapolar spaces (actually "strong" parapolar spaces in the sense that the alternative $|p^\perp \cap q^\perp| = 1$ never arises for a distance-2-pair $\{p, q\}$ of points in these geometries).   Thus, by Proposition 4, one need only study the circuitry in the collinearity graph $H' = \mathcal{P} - H$ to determine whether a hyperplane H arises from an embedding.   The problem, of course, is that we are dealing with an *unknown* hyperplane H and still wish to show, nevertheless, that every circuit of C is built up from simple sums of triangles and 4-circuits.   By way of contradiction, one is led therefore, to consider a circuit C, which simply cannot be expressed as $C = C_1 + C_2$ where $|C_i| < |C|$, $i = 1, 2$.   Clearly $|C| <$ 2diam$(\mathcal{P} - H) + 1$, and so diam$(\mathcal{P} - H)$ must be bounded in order to determine the various cases for $|C|$ which must be considered.   A few results are listed in the following table.

Table 1.

GEOMETRY	DIAM($\mathcal{P}$-H)
Classical Polar Spaces $C_{n,n}(k)$	≤ 3
Grassmann Spaces $A_{n,d}(k)$	$\leq 2d - 2$
(K a division ring, $3 \leq d \leq (n + 1)/2$)	
Half-spin Geometry $D_{5,5}(k)$	≤ 4
$E_{6,1}(k)$	≤ 4

The absence of a minimal indecomposable circuit in the appropriate range of cardinalities could be shown in the cases encompassed by

Theorem 6. *Let H be a geometric hyperplane of a Lie incidence geometry Γ of type $A_{n,2}(k)$, $A_{n,3}(k)$, $D_{5,5}(k)$ or $E_{6,1}(k)$. Then H arises from an embedding.*

5. The universality of certain embeddings

The universality of the embedding $e: \Gamma \to P(W)$ where Γ is the Grassmann space of type $A_{n,d}(k)$ associated with the vector space $V = k^{(n)}$ and W is the d-fold wedge product of V with itself, was proved by Wells (1983).

In the case that $\Gamma = (\mathcal{P}, L)$ is the *truncation* of the geometry of a spherical building whose objects are the maximal parabolic subgroups (this happens when points are the objects associated with an *end node* of the Dynkin diagram, and lines with its unique neighbour in the diagram), then for the module W, whose highest weight is w_n, there is a well-defined epimorphism between $H_0(\mathcal{F}_e) \to H_0(\mathcal{F})$ where, as in the remarks at the beginning of section 3, e is the embedding $\Gamma \to P(W)$ afforded by module W, $\mathcal{F}_e$ its corresponding sheaf and $\mathcal{F}$ the fixed point sheaf in the sense of Ronan and Smith (1985). In many cases it has been shown that $H_0(\mathcal{F}) \simeq W$. Thus, if it can be shown in any of these cases that the epimorphism between the zero homologies of $\mathcal{F}_e$ and $\mathcal{F}$ is actually an isomorphism, then one has $H_0(\mathcal{F}_e) \simeq W$ from which it follows that e is a universal embedding.

An argument showing when this epimorphism of zero homologies of a sheaf and its truncated sheaf is an isomorphism was sketched for us by Stephen Smith at Oberwolfach. In order to state his result, it would be helpful to review some terminology from Ronan and Smith (1985).

Let Δ be a flag complex of a geometry over I. As usual, a *sheaf* $\mathcal{F}$ over Δ is the assignment of an abelian group $\mathcal{F}_s$ at every simplex s in Δ, with connecting morphisms $\phi_{xy}: \mathcal{F}_x \to \mathcal{F}_y$, whenever y is a face of x, such that ϕ_{xz} is the composition of ϕ_{yz} and ϕ_{xy} whenever z is in turn a face of y. A sheaf is a *subsheaf of the constant sheaf* $\mathcal{K}_W$ if its values (terms) are subspaces of W and all connecting morphisms are containment maps. A subsheaf $\mathcal{F}$ over Δ' of the constant sheaf $\mathcal{K}_W$ is said to be *universal* if $H_0(\Delta', \mathcal{F}) \simeq W$. If v is a vertex, the complex Δ^v of all simplices equal to or above v is called the *residual complex* of v. It admits the residual sheaf $\mathcal{F}^v$, whose value at $x \in \Delta^v$ is just the value of $\mathcal{F}$ at x. We can now state

Theorem 7 (Smith's argument).   *Let Δ be a flag complex of geometry Γ' over the type set $I = \{1, 2, ..., n\}$, let W be a vector space, and let $\mathcal{F}$ be a chamber-generated sheaf over Δ which is a subset of the constant sheaf $\mathcal{K}_W$. We fix two types, $J = \{1, 2\}$, and let $\Gamma = (\mathcal{P}, L) = (_1\Gamma', {}_2\Gamma')$ be the truncation of Γ' to J.   We make these three assumptions:*

(T1)   The mappings of points and lines into subspaces of W given by $p \to \mathcal{F}_p$, $L \to \mathcal{F}_L$, $(p, L) \in \mathcal{P}{\times}L$, defines an embedding $e\colon \Gamma \to P(W)$.

(T2)   For each vertex v in Δ, the residual sheaf $\mathcal{F}^v$ is universal.

(T3)   The flag complex Δ is uniquely determined by its truncation (of type J) to the flag complex of Γ in the sense that for each index $i \in I$, there is a canonically defined system $\mathcal{X}_i$ of subspaces of Γ and a definition of incidence in terms of the intersection of these subspaces, so that the resulting geometry over I has a flag complex isomorphic to Δ.

Then the zero homology of the truncated sheaf $_J\mathcal{F} = \mathcal{F}_e$ over the flag complex of Γ is isomorphic to $H_0(\Delta, \mathcal{F})$, the zero homology of the sheaf $\mathcal{F}$ over Δ, the flag complex of Γ'.

Now the isomorphism $H_0(\Delta, \mathcal{F}) \simeq W$ has been shown for the modules W given in the following table.

Table 2.

Lie Geometry	Module	Dimension	Reference
$F_{4,1}(k)$	M_1	26	Cohen and Smith (Personal communication)
$F_{4,4}(k)$	M_2	52	H. Volklein (Personal communication)
$E_{6,1}(k)$	M_3	27	Ronan and Smith (1985)
$E_{7,1}(k)$	M_4	56	Ronan and Smith (1985)
$E_{8,1}(k)$	M_5	248	H. Volklein (Personal communication)

Theorem 8. *The standard embeddings of the Lie incidence geometries $F_{4,1}(k)$, $F_{4,4}(k)$ (char k odd), $E_{6,1}(k)$, $E_{7,1}(k)$, $E_{8,1}(k)$, k a field, associated with the modules $M_1, ..., M_5$ of table 2, are all universal embeddings.*

In all cases (T1)-(T3) hold so Theorem 7 applies.

6. Description of some of the hyperplanes

Knowing that a geometric hyperplane arises from an embedding occasionally allows one an internal description of the hyperplane as a geometry. In other cases it is sometimes possible to describe some geometric hyperplanes even when it is not known that they arise from an embedding.

The case $A_{n,2}(k)$. Here 2-subspaces of $V = k^{(n)}$ are the points $\mathcal{P}$. Each geometric hyperplane H is associated with a class k^*B of scalar multiples of a symplectic form B. H consists of all B-isotropic 2-spaces. In the case that B is non-degenerate, H has the structure of a *polar Grassmann space*: points are all isotropic 2-spaces, lines are singular 1-space-3-space flags, so H is a Lie incidence geometry of type $C_{n,2}(k)$.

The case $E_{6,1}(k)$. The group of automorphisms acts on the standard 27-dimensional module in 3 orbits on 1-spaces and similarly in 3 orbits on the dual module so that there are just 3 classes of geometric hyperplanes. One of these is stabilised by the subgroup $F_4(k)$, but H is not itself that standard meta-symplectic space. Instead, H is obtained from the meta-symplectic space embeddable in the 26-dimensional module (so that symplecta are type $Sp(6, k)$), by enlarging the set of lines to include the hyperbolic lines of each symplecton. The resulting geometric hyperplane then has diameter only 2, rather than 3 as expected for a meta-symplectic space.

The case $D_{n,n}(k)$. For each $n \geq 4$, there exists two classes of geometric hyperplanes, $H_n{}^0$ and $H_n{}^1$, which for the cases $n = 4$ and 5, comprise all geometric hyperplanes. Each half-spin geometry Γ of type $D_{n,n}(k)$ possesses a unique class $\mathcal{D}$ of convex subspaces of type $D_{n-1,n-1}(k)$. For each subspace X in $\mathcal{D}$, there is an automorphism σ of it's associated polar space which exchanges the two classes of maximal singular subspaces of the polar space. Then in the half-spin geometry X, σ take the points of X to a class $\mathcal{M}$ of maximal singular subspaces of X. Then σ takes a geometric hyperplane H of X to a subset $\sigma(H)$ of $\mathcal{M}$ having the property that if L is any line of X, then all members of $\mathcal{M}$ which contain L, or exactly one such member of $\mathcal{M}$ over L belongs to $\sigma(H)$. Now each singular subspace M of X is contained as a hyperplane in a unique maximal singular subspace M' of Γ. We now define $H(H, X)$ to be the union of the singular subspaces M' of Γ as M ranges over the subset $\sigma(H)$ of $\mathcal{M}$. It is a theorem that $H(H, X)$ is a geometric hyperplane of Γ.

Now let $H_4{}^0$ and $H_4{}^1$ be the two classes of geometric hyperplanes of $D_{4,4}(k)$ which under the triality map correspond to the two geometric hyperplanes of the polar space $D_{4,1}(k)$ of type $p^\perp$ and $\Omega(7, k)$. Inductively for Γ of type $D_{n,n}(k)$ as in the previous paragraph, define

$$H_n{}^j = \{H(H, X) \mid X \in \mathcal{D}, H \text{ a geometric hyperplane of } H_{n-1}{}^j(X)\}, j = 0, 1.$$

Then

Theorem 9. *The subspaces of $\Gamma = D_{n,n}(k)$ in $H_n{}^0$ and $H_n{}^1$ comprise two isomorphism classes of geometric hyperplanes. If $n \leq 6$, any geometric hyperplane H of Γ which contains a subspace in $\mathcal{D}$ must belong to one of these two classes. If $n = 4$, or 5, the classes exhaust all geometric hyperplanes of Γ.*

Remark 10. Obviously there are many open questions concerning the existence of embeddings, their universiality and whether geometric hyperplanes arise from such embeddings, beyond those considered here. The authors hope at least that this, mostly expository note, may serve as an introduction to this interesting subject.

References
F. Buekenhout (1983), Cooperstein's theory, *Simon Stevin* **57**, 125-140.
F. Buekenhout and C. Lefèvre (1974), Generalised quadrangles in projective spaces, *Arch. Math.* **25**, 540-552.
A. Cohen and E. Shult (1990), Affine polar spaces, *Geom. Dedicata* **35**, 43-76.
B.N. Cooperstein (1977), A characterization of some Lie incidence geometries, *Geom Dedicata* **6**, 205-258.
H. Cuypers, P. Johnson and A. Pasini (submitted), Synthetic foundations of polar geometry, Preprint.
K.J. Dienst (1980), Verallgemeinene Vierecke in projectiven Räumen, *Arch. Math.* **45**, 177-186.
J.I. Hall and E. Shult (1985), Locally cotriangular graphs, *Geom. Dedicata*, **18**, 113-159.
P. Johnson (submitted), Fully embedded polar spaces, Preprint.
C. Lefèvre-Percsy (1981), Polar spaces embedded in a projective space, *Finite Geometries and Designs*, (ed. P. Cameron, J.W.P. Hirschfeld and D.R. Hughes), London Math. Soc. Lecture Note Series **49**, 216-220, Cambridge University Press, Cambridge.
M. Ronan (1987), Embeddings and hyperplanes of discrete geometries, *European J. Combin.* **8**, 179-185.
M. Ronan and S. Smith (1985), Sheaves on buildings and modular representations of Chevally groups, *J. Algebra* **96**, 319-346.

S. Smith (submitted), Weights and covers of chamber systems, Preprint.

E. Shult (1986), Hanssen's principle, Preprint.

L. Tierlinck (1980), On projective and affine hyperplanes, *J. Combin. Theory Ser. A* **28**, 290-306.

J. Tits (1974), *Buildings of spherical type and finite BN-Pairs*, Lecture Notes in Math. **386**, Springer-Verlag, Berlin.

F.D. Veldkamp (1960), Polar geometry I-V, *Indag. Math.* **21**, 512-522.

A.L. Wells (1983), Universal projective embeddings of the Grassmannian, half-spinor, and dual orthogonal geometries, *Quart. J. Math. Oxford* **34**, 375-386.

Pointsets in partial geometries

F. De Clerck[1], A. Del Fra[2] and D. Ghinelli[2]

State University of Ghent[1] and
University of Rome I[2]

1. Introduction

1.1 Definitions

A (finite) partial geometry of order (s, t) with incidence parameter α (briefly: a $pG_\alpha(s, t)$), is an incidence structure with $s + 1$ points on a line, $t + 1$ lines on a point, two points on at most one line, such that for all antiflags (P, l) the number of points on l collinear with P is a constant $\alpha \neq 0$. Obviously we must have $1 \le \alpha \le \min (s + 1, t + 1)$.

Introduced by Bose (1963), these geometries include many incidence structures, widely studied in the last 30 years: a $pG_1(s, t)$ is a generalized quadrangle $GQ(s, t)$ (Payne and Thas (1984)); a $pG_{s+1}(s, t)$ is a 2-$(v, s + 1, 1)$ design (or a Steiner system $S(2, s + 1, v)$) (Hughes and Piper (1985)); a $pG_t(s, t)$ is a net of order $s + 1$ and degree $t + 1$ (Bruck (1963)). The dual of a $pG_\alpha(s, t)$ is clearly a $pG_\alpha(t, s)$, hence for $\alpha = t + 1$, partial geometries are dual Steiner systems and for $\alpha = s$ they are dual nets.

1.2 Remarks

If $S = (\mathcal{P}, \mathcal{B}, I)$ is a partial geometry $pG_\alpha(s, t)$, then $|\mathcal{P}| = v = (s + 1)(st + \alpha)/\alpha$ and $|\mathcal{B}| = b = (t + 1)(st + \alpha)/\alpha$; moreover the pointgraph $\Gamma(S)$ is strongly regular with parameters $v, k = s(t + 1), \lambda = s - 1 + t(\alpha - 1), \mu = \alpha(t + 1)$ (Bose (1963)). From the theory of strongly regular graphs follows that the existence of a $pG_\alpha(s, t)$ implies the following conditions:

(1) $\alpha \mid (s + 1)st$ and $\alpha \mid (t + 1)st$.

(2)   The adjacency matrix A of $\Gamma(S)$ has three eigenvalues $\chi_1 = -t - 1$, $\chi_2 = s - \alpha, \chi_3 = s(t + 1)$. The multiplicities of these eigenvalues are integers, which yields the integrality condition

$$\alpha(s + t + 1 - \alpha) \mid st(s + 1)(t + 1).$$

(3)   The Krein inequalities for strongly regular graphs are satisfied, hence

$$(s + 1 - 2\alpha)t \le (s - 1)(s + 1 - \alpha)^2.$$

2. The known models of proper partial geometries

In this section we will give a short overview of the known models of the so-called proper partial geometries, which have the property that $1 < \alpha < \min(s, t)$.   For the non-proper ones we refer to the literature, for instance to Payne and Thas (1984) for the generalized quadrangles and Hughes and Piper (1985) for the designs.

(1)   The partial geometry S($\mathcal{K}$). This infinite family was constructed in Thas (1973, 1974), and independently in Wallis (1973).   Let $\mathcal{K}$ be a maximal arc of degree d (or a $\{qd - q + d; d\}$-arc) in a projective plane π of order q (this is a set of $qd - q + d$ points of π such that every line of π intersects it in 0 or d points, see for instance Hirschfeld (1979) for the definitions and constructions).   We define the incidence structure $S(\mathcal{K}) = (\mathcal{P}, \mathcal{B}, I)$. The points of $S(\mathcal{K})$ are the points of π which are not contained in $\mathcal{K}$. The lines of $S(\mathcal{K})$ are the lines of π which are incident with d points of $\mathcal{K}$. The incidence is the one of π.   Then, for $1 < d < q$, $S(\mathcal{K})$ is a partial   geometry with $t = q - q/d, s = q - d, \alpha = q - q/d - d + 1$.

Remark.   As there exist $\{2^{h+m} - 2^h + 2^m; 2^m\}$-arcs, whenever $0 < m < h$, in $PG(2, 2^h)$, there exists a class of partial geometries S($\mathcal{K}$) with parameters $s = 2^h - 2^m, t = 2^h - 2^{h-m}, \alpha = (2^m - 1)(2^{h-m} - 1)$.   Such a geometry is a generalized quadrangle if and only if $h = 2$, and then it is the unique quadrangle of order 2.

(2)   The partial geometry $T_2^*(\mathcal{K})$. This infinite family was constructed by Thas, details can be found in Thas (1973, 1974).   Let $\mathcal{K}$ be a maximal arc of degree $d, d > 1$, in the projective plane $PG(2, q)$ over $GF(q)$ ($q = p^h, p$ prime).   We define an incidence structure $T_2^*(\mathcal{K}) = (\mathcal{P}, \mathcal{B}, I)$ as follows.   Let $PG(2, q)$ be embedded as a plane π in $PG(3, q)$.   The points of $T_2^*(\mathcal{K})$ are the points of $PG(3, q) \backslash \pi$.   The lines of $T_2^*(\mathcal{K})$ are the lines of $PG(3, q)$ which are not contained in π and which meet $\mathcal{K}$ (necessarily in a unique point).   The incidence is the one of $PG(3, q)$.   Then $T_2^*(\mathcal{K})$ is a partial geometry with $t = (q + 1)(d - 1), s = q - 1, \alpha = d - 1$.

Remark.   The partial geometry $T_2^*(\mathcal{K})$ arising from a maximal arc of degree $2^m, 0 < m < h$, in $PG(2, 2^h)$ has parameters $s = 2^h - 1, t = (2^h + 1)(2^m - 1), \alpha = 2^m - 1$.   This is a generalized quadrangle if and only if $m = 1$, hence if and only if $\mathcal{K}$ is a complete oval.

(3)   The partial geometries $PQ^+(4n - 1, 2)$.   In De Clerck et al. (1980) an infinite class of partial geometries is constructed as follows.   Define a spread Σ of the non-singular hyperbolic quadric $Q^+ = Q^+(4n - 1, 2)$, $n \geq 2$, in $PG(4n - 1, 2)$ to be a (maximal) set of $2^{2n-1} + 1$ disjoint $(2n - 1)$-dimensional spaces on Q^+ (Dye (1977)).   Let Σ be a spread of $Q^+ = Q^+(4n - 1, 2)$ and let Ω be the set of all hyperplanes of the elements of Σ. Consider the incidence structure $PQ^+(4n - 1, 2) = (\mathcal{P}, \mathcal{B}, I)$ with $\mathcal{P}$ the set of points of $PG(4n - 1, 2)$ not on the quadric, $\mathcal{B} = \Omega$ and $x I L, x \in \mathcal{P}$ and $L \in \mathcal{B}$, if and only if x is contained in the polar space L^* of L with respect to Q^+.   One can prove that $PQ^+(4n - 1, 2)$ is a partial geometry with $s = 2^{2n-1} - 1, t = 2^{2n-1}, \alpha = 2^{2n-2}$.

If $n = 2$, then the parameters of $PQ^+(7, 2)$ are $s = 7, t = 8, \alpha = 4$. Cohen (1981) was the first to construct a partial geometry with these parameters using the root system E_8.   In Haemers and van Lint (1982) a partial geometry with parameters $s = 8, t = 7, \alpha = 4$ was constructed using coding theory.   Kantor (1982a) proved that $PQ^+(7, 2)$ and the dual of the geometry of Haemers-van Lint are isomorphic.   Later on Tonchev (1984) showed with the help of a computer that the model of Cohen and the dual of the geometry of Haemers-van Lint are isomorphic.   In De Clerck et al., (1988) this isomorphism is proved without the use of a computer.

Remark that non-isomorphic spreads of the quadric $Q^+(4n - 1, 2)$ will produce non-isomorphic partial geometries. If $2n - 1$ is composite then $Q^+(4n - 1, 2)$ has non-isomorphic spreads, and probably this is true for all $n > 2$ (Kantor (1982b)).

Remark.   For $q = 3$ a more or less analogous construction is given in Thas (1981) to obtain a partial geometry with parameters $s = 3^{2n-1} - 1, t = 3^{2n-1}, \alpha = 2 \cdot 3^{2n-2}$.

Up to now it is only known that $Q^+(7,3)$ has a spread. This yields a geometry $PQ^+(7, 3)$ with parameters $s = 26, t = 27, \alpha = 18, (v = 1080, b = 1120)$.

(4)   The sporadic partial geometries.   In van Lint and Schrijver (1981) a sporadic $pG_2(5, 5)$ is constructed.   Another construction of this geometry is given in Cameron and van Lint (1982).

In Haemers (1981) another sporadic proper partial geometry is constructed.   This partial geometry has parameters $s = 4, t = 17, \alpha = 2$. The pointgraph Γ however was known before (Hubaut (1975)).

3. Bounds for the size of pointsets in a partial geometry

3.1   A first bound

Let K be a nonempty set of points of a finite $pG_\alpha(s, t)$, say S.   We denote by L^+ the set of lines meeting K (*intersecting lines*) and by L^- the set of lines disjoint from K (*external lines*).

We consider the positive integer

$$n = \min_{l^+ \in L^+} |l^+ \cap K|. \tag{1}$$

Bounds for the size k of K can be expressed in terms of s, t and n.

K is called *a set of class* $(0, n)$ or briefly *a* $(0, n)$-*set* in S if every line intersecting K in at least 1 point, intersects it in exactly n points.   If there are no external lines, then K is called an n-*ovoid* (Thas (1989)).

Theorem 1. *Let K be a nonempty subset of the pointset of S and let n be as in (1).   Then the size k of K satisfies*

$$k \geq n[(n - 1)t + \alpha]/\alpha, \tag{2}$$

where if $n \neq 1$, equality holds if and only if the lines of S induce on K a subgeometry $pG_\alpha(n - 1, t)$.   If $L^- = \varnothing$ the bound in (2) can be improved. Namely:

$$k \geq n(st + \alpha)/\alpha, \tag{3}$$

and equality holds if and only if K is an n-ovoid.

Proof. The proof is similar to that of Del Fra et al. (1990b).

Let l be a line of S meeting K in precisely n points, say $P_1, \dots, P_n$. Clearly K contains at least these n points and has at least $n - 1$ further points Q on each of the t lines $x \neq l$ on any of the points P_i.   Since each of the points Q is on exactly α lines meeting l, each Q is counted at most α times hence the number of points Q is at least $n(n - 1)t/\alpha$ and we have (2).   We remark that if $n < \alpha$, then (2) can be improved by

$$k \geq n + t(n - 1), \tag{4}$$

a bound obtained by considering the points of K collinear with a fixed point on an intersecting line.

If $L^- = \emptyset$ this bound can be improved by counting flags (P, l), $P \in K$: $|k| \cdot (t + 1) \geq b \cdot n$ hence

$$k \geq n(st + \alpha)/\alpha, \tag{5}$$

where equality holds if and only if K is an n-ovoid. □

3.2 Another bound

The lower bounds given by (2) and (3) are not always the best possible ones. To give another lower bound which is better than the first for some parameter values, we need to use the following theorem. The proof uses the so-called Higman-Sims technique and is completely analogous to the case of generalized quadrangles (see result 1.10.1 of Payne and Thas (1984)).

Theorem 2. *Let $K \neq \emptyset$ be a subset of points in a $pG_\alpha(s, t)$, say $S = (\mathcal{P}, \mathcal{B}, I)$, $\alpha \neq s + 1$. Then the average number $\Sigma_{P \in K} |P^\perp \cap K| / k$ of points of K collinear with a point of K satisfies*

$$\Sigma_{P \in K} \frac{|P^\perp \cap K|}{k} \leq s - \alpha + 1 + \frac{k\alpha}{s + 1}, \tag{6}$$

and equality holds if and only if

$$|P^\perp \cap K| = \begin{cases} s - \alpha + 1 + \dfrac{k\alpha}{(s + 1)} & \text{for all } P \in K, \\[2mm] \dfrac{k\alpha}{(s + 1)} & \text{for all } P \notin K. \end{cases} \tag{7}$$

Proof. Let $B = J - A - I$ be an adjacency matrix for the complement $\Gamma^c(S)$ of the pointgraph $\Gamma(S)$; B is a real symmetric (v, v)-matrix with $v = (s + 1)(st + \alpha)/\alpha$. The matrix B has 3 eigenvalues

$$\lambda_1 = -s + \alpha - 1, \qquad \lambda_2 = t, \qquad \lambda_3 = st\left(\frac{s + 1}{\alpha} - 1\right).$$

Obviously λ_3 is the constant row (and column) sum of B and represents the number of points in S noncollinear with a fixed point of S.

Without loss of generality, we may assume that the first k rows and columns of B correspond to the points of K; clearly this gives a decomposition of B into four submatrices B_i

$$B = \begin{bmatrix} B_1 & B_2 \\ B_3 & B_4 \end{bmatrix}$$

where B_1 is of order k, B_4 is of order $v - k$, while $B_3 = B_2{}^T$ is a $(v - k, k)$-matrix. We observe that B_1 and B_4 are the adjacency matrices of the complements of the pointgraphs $\Gamma(K)$ and $\Gamma(\mathcal{P} - K)$ induced by Γ on K and $\mathcal{P} - K$, respectively.

We denote by $\sigma(M)$ the sum of the elements of a matrix M. In order to apply the Higman-Sims technique, we consider the matrix

$$B^\Delta = \begin{bmatrix} \dfrac{\sigma(B_1)}{k} & \dfrac{\sigma(B_2)}{k} \\[2mm] \dfrac{\sigma(B_3)}{v - k} & \dfrac{\sigma(B_4)}{v - k} \end{bmatrix}$$

of the average row-sums of the B_i's.

If $\mu_1 \le \mu_2$ are the eigenvalues of B^Δ, then it is known that $\lambda_1 \le \mu_1 \le \mu_2 \le \lambda_3$. Moreover, if $\lambda_1 = \mu_1$ and $y = (y_1, y_2)^T$ is an eigenvector of B^Δ associated to μ_1, then $x = (x_1, \ldots, x_v)^T$ with $x_1 = \ldots = x_k = y_1$ and $x_{k+1} = \ldots = x_v = y_2$ is an eigenvector of B associated to $\lambda_1 = \mu_1$. From this it is not difficult to prove the statement, since B^Δ has eigenvalues $\mu_2 = \lambda_3$ and $\mu_1 = Tr(B^\Delta) - \mu_2 = \rho - k(\mu_2 - \rho)/(v - k)$ with $\rho = k - \Sigma_{P \in K} |P^\perp \cap K|/k$. ☐

Theorem 3.   *Let $K \ne \varnothing$ be a subset of points in a $pG_\alpha(s, t)$ and let n be the minimum defined by* (1). *Then*

$$k \ge (s + 1)[n(t + 1) - (s + t + 1 - \alpha)]/\alpha, \tag{8}$$

and equality holds if and only if K is a $(0, n)$-set and moreover for every point (7) *holds, that is:*

$$|P^\perp \cap K| = \begin{cases} n(t + 1) - t & \text{for all } P \in K \\ n(t + 1) - (s+t + 1 - \alpha) & \text{for all } P \notin K. \end{cases} \tag{9}$$

Proof. First we assume $\alpha = s + 1$. Let l be a line in L^+ and let Q be a point of $l \cap K$. Then K contains at least the points of $l \cap K$ and at least $(n - 1)t$ points, different from Q, on the t lines through Q different from l. Thus $k \geq t(n - 1) + n = n(t + 1) - t$ which is (8) for $\alpha = s + 1$.

Let $\alpha \neq s + 1$ so that Theorem 2 applies, and the proof is now similar to that of Del Fra et al (1990b).

Remarks
(1)   Comparing (2) and (8) we see that the lower bound given by (2) is better than that of (8) if and only if

$$n < (s + t + 1 - \alpha)/t. \tag{10}$$

The two bounds are equal when $n = s + 1$ or $n = (s + t - \alpha + 1)/t$. If $n = s + 1$, then (2) implies $k = |S| = (s + 1)(st/\alpha + 1)$. Hence $n = s + 1$ if and only if $K = S$, which is an uninteresting case.

We note that for Steiner systems ($\alpha = s + 1$) (2) is never better than (8) and for dual nets ($\alpha = s$) (2) is better than (8) only in the trivial case $n = t = 1$.

(2)   A result similar to Theorem (3) in the particular case that K is a subgeometry $pG_{\alpha'}(s', t')$ is in De Clerck (1984).

(3)   If K is a $(0, n)$-set, then a similar proof will yield

$$k \geq (s + 1)[n(t + 1) - (s + t + 1 - \alpha)]/\alpha = b_2, \tag{11}$$

and equality holds if and only if for every P of S not in K, the number of lines on P missing K is a constant

$$e = (s + t + 1 - \alpha)/n, \tag{12}$$

hence n divides $s + t - \alpha + 1$.

(4)   From Theorem 1 we also have for the size k of a $(0, n)$-set the bound

$$k \geq n[(n - 1)t + \alpha]/\alpha = b_1,$$

and b_1 is better than the bound b_2 above if and only if

$$n < (s + t + 1 - \alpha)/t. \tag{13}$$

If $n = (s + t + 1 - \alpha)/t$, then $b_1 = b_2 = (1 + s)(s + t + 1 - \alpha)/t\alpha$. If k equals this bound, then K is a sub-$pG_{\alpha}((s + 1 - \alpha)/t, t)$.

(5)   From Theorem 3 and the above remarks we have immediately
the following corollary.   Let $K \neq \emptyset$ be any subset of points in a $pG_\alpha(s, t)$
and let L^+ be the set of secant lines. If $n = \min_{l^+ \in L^+} |l^+ \cap K|$ and $k = |K| = (s + 1)[n(t + 1) - (s + t + 1 - \alpha)]/\alpha$, then K is a $(0, n)$-set and each
point $P \notin K$ is on $e = (s + t + 1 - \alpha)/n$ lines missing K.
(6)   Theorems 1 and 3 have important applications in the study of
extended partial geometries (connected structures such that the
residue in each point is a partial geometry).   In Del Fra and Ghinelli
(1990a) inequalities (2), (3) and (8) are used to give bounds for the
diameter of an extended $pG_\alpha(s, t)$.   Other applications are in Del Fra et
al. (1990a).

3.3   Bounds with respect to two subsets

The next results are applied in Del Fra and Ghinelli (1990b) to
characterize extended partial geometries with maximum diameter.

Theorem 4.   *Let S be a $pG_\alpha(s, t)$ and let H and K be two nonempty
subsets of points in S such that no line of S can meet both H and K.
We denote by L_H^+ and L_K^+ the sets of lines of S intersecting H and K,
respectively.   For $I = H, K$ we set*

$$n_I = \min_{l \in L_I^+} |l \cap I|, \qquad n^*_I = \max_{l \in L_I^+} |l \cap I|. \qquad (14)$$

Then

$$(s + 1)(t + 1)(n_H n_K^* + n_H^* n_K) \leq (s + 1)(s + t - \alpha + 1)(n_H^* + n_K^*) +$$

$$(st + \alpha)n_H^* n_K^*. \qquad (15)$$

Furthermore, if v_I and v_I^* $(I = H, K)$ satisfy

$$v_I \leq n_I, \text{ and } n_I^* \leq v_I^*, \qquad (16)$$

then

$$(s + 1)(t + 1)(v_H n_K^* + n_H^* v_K) \leq (s + 1)(s + t + 1 - \alpha)(v_H^* + v_K^*) +$$

$$(st + \alpha)n_H^* n_K^*. \qquad (17)$$

Proof.   Counting flags (Q, l) with Q in $I = H, K$, we obtain

$$|L_I^+| \geq |I|(t + 1)/n_I^*.$$

By (8) we have

$$|I| \geq (s + 1)[n_I(t + 1) - s - t - 1 + \alpha]/\alpha;$$

Since $|L^+_H| + |L^+_K| \leq (t + 1)(1 + st/\alpha)$, we obtain

$$\frac{s + 1}{\alpha}[n_H(t + 1) - s - t - 1 + \alpha]\frac{t + 1}{n_H^*} + \frac{s + 1}{\alpha}[n_K(t + 1) - s - t - 1 + \alpha]\frac{t + 1}{n_K^*}$$
$$\leq (t + 1)(st + \alpha)/\alpha;$$

hence

$$(s + 1)(t + 1)\left(\frac{n_H}{n_H^*} + \frac{n_K}{n_K^*}\right) \leq (s + 1)(s + t + 1 - \alpha)\left(\frac{1}{n_H^*} + \frac{1}{n_K^*}\right) + st + \alpha,$$

which implies (15).   A similar proof, using (16), yields (17).   □

Corollary 5. *Let H and K be two nonempty $(0, n_H)$ and $(0, n_K)$ subsets of points in a $pG_\alpha(s, t)$, say S, such that no line of S can meet both H and K.   Then*

$$n_H n_K(st + 2s + 2t + 2 - \alpha) \leq (s + 1)(s + t + 1 - \alpha)(n_H + n_K). \qquad (18)$$

Proof.   This is the special case of (15) for $n_I = n_I^*$, $I = H, K$.   □

4. Arithmetic conditions for $(0, n)$-sets

From now on S will be a $pG_\alpha(s, t)$ and K a $(0, n)$-set in S with k points, and $1 \leq n \leq s$ (to avoid the trivial case that K is the set of all points of S, corresponding to $n = s + 1$).   As in the previous section, L^+ and L^- will be the sets of intersecting lines and external lines, respectively. Clearly:

$$|L^+| + |L^-| = (st/\alpha + 1)(t + 1). \qquad (19)$$

Counting flags (Q, l) with Q in K, we see that

$$k = n|L^+|/(t + 1). \tag{20}$$

From (19) and (20) we obtain

$$k = n(st + \alpha)/\alpha - n|L^-|/(t + 1). \tag{21}$$

We denote by θ

$$\theta = n|L^-|/(t + 1). \tag{22}$$

Obviously

$$\theta \geq 0 \quad \text{or equivalently} \quad k \leq n(st + \alpha)/\alpha, \tag{23}$$

and we have equality if and only if K is an n-ovoid. Since $k = (n/\alpha)(st + \alpha) - \theta$, it is natural to call θ the *deficiency* of K. Remark that θ will be an integer if and only if α divides nst.

Condition (23) and Theorem 1 are equivalent to the following theorem.

Theorem 6. *The deficiency* $\theta = n|L^-|/(t + 1) = (n/\alpha)(st + \alpha) - k$ *of a* $(0, n)$-*set* K *satisfies*

$$0 \leq \theta \leq nt(s - n + 1)/\alpha. \tag{24}$$

Furthermore, $\theta = 0$ *if and only if* K *is an n-ovoid, and* $\theta = nt(s - n + 1)/\alpha$ *if and only if the lines of S induce on K a subgeometry* $pG_\alpha(n - 1, t)$.

If P is a point of S not on K, we denote by $e(P)$ the number of external lines on P (so $t + 1 - e(P)$ is the number of intersecting lines on P).

Theorem 7. *For every intersecting line l^+ of a $(0, n)$-set K in a $pG_\alpha(s, t)$, the number of external lines meeting l^+ is a constant h which does not depend on l^+, namely*

$$h = \sum_{P \in l^+ - K} e(P) = st + \alpha - \frac{k\alpha}{n} = \frac{\theta\alpha}{n} = \frac{|L^-|\alpha}{(t + 1)}. \tag{25}$$

Furthermore h satisfies

$$0 \leq h \leq t(s - n + 1), \tag{26}$$

and $h = 0$ *if and only if* K *is an* n-*ovoid;* $h = t(s - n + 1)$ *if and only if the lines of* S *induce on* K *a subgeometry* $pG_\alpha(n - 1, t)$.

Proof. This is rather similar to the proof of Theorem 3.2 of Del Fra et al. (1990b), we therefore omit it. □

Corollary 8. *As an immediate consequence (see also (21)) we have that if there exists a* $(0, n)$-*set* K *in a* $pG_\alpha(s, t)$, *then*

$$n \mid k\alpha, \quad n \mid \theta \, \alpha. \tag{27}$$

Moreover

$$t + 1 \mid |L^-|\alpha \quad and \quad t + 1 \mid |L^+|\alpha. \tag{28}$$

Remarks

(1)   In a similar way, if $L^- \neq \varnothing$, we choose a line $l^- = \{P_1, \ldots, P_{s+1}\}$. Since S is a $pG_\alpha(s, t)$, counting the points of K on the $s + 1$ pencils of secant lines on the points P_i, we obtain

$$k = \sum_{i=1}^{s+1} \frac{n}{\alpha} (t + 1 - e(P_i)) = \frac{n}{\alpha} \left((t + 1)(s + 1) - \sum_{i=1}^{s+1} e(P_i) \right).$$

This implies that the number

$$h' = \Sigma_{P \in l^-} e(P) = (s + 1)(t + 1) - k\alpha/n, \tag{29}$$

does not depend on l^-. Since l^- is counted in the $\Sigma_{P \in l^-} e(P)$ exactly once for each $P_i \in l^-$, we see that $h' - (s + 1)$ is *the total number of external lines different from* l^- *meeting a chosen* l^-.   Obviously

$$k = n[(s + 1)(t + 1) - h']/\alpha. \tag{30}$$

Hence, if $L^- \neq \varnothing$, then

$$h' = h + s + t + 1 - \alpha, \tag{31}$$

and it easily follows that

$$h' > h > 0 \quad and \quad h' > s + t + 1 - \alpha. \tag{32}$$

(2)   As a consequence of (31), we have the following expressions for the deficiency $\theta = nh/\alpha$, for $|L^-| = (t + 1)h/\alpha$, and for $|L^+|$ as functions of h':

$$\theta \;=\; n(h' - s - t - 1 + \alpha)/\alpha, \tag{33}$$
$$|L^-| \;=\; (t + 1)(h' - s - t - 1 + \alpha)/\alpha, \tag{34}$$
$$|L^+| \;=\; (t + 1)[(t + 1)(s + 1) - h']/\alpha. \tag{35}$$

Furthermore, since now $h \geq 1$ we have from (26) and (31)

$$s + t + 2 - \alpha \;\leq h' \leq\; t(s - n + 2) + s + 1 - \alpha, \tag{36}$$

and $h' = t(s - n + 2) + s + 1 - \alpha$ if and only if the lines of S induce on K a subgeometry $pG_\alpha(n-1,t)$.

(3)   The inequality in Theorem 2 is clearly equivalent to

$$\theta \;\leq (s + 1 - n)(s + t + 1 - \alpha)/\alpha \tag{37}$$

(see also Remark 3 in section 3).   Equality holds if and only if $e(P)$ is a constant, namely $(s + t + 1 - \alpha)/n$ (see (12)).   This upper bound for θ is better than that derived from (6) when $n > (s + t + 1 - \alpha)/t$.   Similarly, we have for $h = \theta\alpha/n$ and $h' = h + s + t + 1 - \alpha$ the obvious upper bounds improving (26) and (36) when $n > (s + t + 1 - \alpha)/t$, namely

$$h \leq (s + 1 - n)(s + t + 1 - \alpha)/n, \quad h' \leq (s + 1)(s + t + 1 - \alpha)/n, \tag{38}$$

where equality holds if and only if $e(P)$ is the constant $(s + t + 1 - \alpha)/n$.

We conclude this section proving a result which has applications in the classification of extended partial geometries with maximum diameter (see Del Fra and Ghinelli (1990a, 1990b)).

Theorem 8.   *Let S be a dual 2-design (with $\lambda = 1$).   Assume that in S there are a $(0, n)$-set K and a $(0, s - n)$-set $H(0 < n < s)$ such that no line of S meets both K and H.   Then $t = 1$.*

Proof.   Let l and l' be two secant lines of K and H, respectively.   Since S is a dual 2-design, there is a unique point $P = l \cap l'$.   The lines different from l through the n points of $l \cap K$ are exactly the lines different from l' through the n points of $l' \cap (\mathcal{P} - H - \{ P\})$.   Therefore

$$h'_K = \sum_{P \in l'} e(P) = (s - n)(t + 1) + n + 1 = st + s - nt + 1 \qquad (39)$$

while

$$h'_H = \sum_{P \in l} e(P) = n(t + 1) + s - n + 1 = nt + s + 1. \qquad (40)$$

It follows from (30) and (39) that

$$|K| = \frac{n}{t + 1}[(s + 1)(t + 1) - st - s + nt - 1] = \frac{n}{t + 1} t(n + 1).$$

Similarly, from (30) and (40) we obtain

$$|H| = \frac{s - n}{t + 1}[(s + 1)(t + 1) - nt - s - 1] = \frac{s - n}{t + 1} t(s - n + 1).$$

If L_H^+ and L_K^+ are the sets of lines secant to H and K respectively, then

(see (20))

$$|L_K^+| = \frac{n}{t + 1} t(n + 1) \frac{t + 1}{n} = t(n + 1),$$

$$|L_H^+| = \frac{s - n}{t + 1} t(s - n + 1) \frac{t + 1}{s - n} = t(s - n + 1). \qquad (41)$$

Since $|L_K^+| + |L_H^+| \leq (t + 1)(st + \alpha)/\alpha$ and $\alpha = t + 1$ we obtain from (41)

that $t(n + 1) + t(s - n + 1) \leq st + t + 1$ or $t(s + 2) \leq st + t + 1$, which
implies $2t \leq t + 1$. This yields $t = 1$. $\square$

5. (0, s)-sets in a pG$_\alpha$(s, t)

In this section we classify the $(0, s)$-sets in a pG$_\alpha(s, t)$, say S. We will
suppose $s \geq 2$. For the case of the generalized quadrangles we refer to
Del Fra et al. (1990b).

The classification of the $(0, s)$-sets is a corollary of a theorem on
partial subgeometries of a partial geometry, as it appeared in De Clerck
(1978, 1984).

Theorem 9. *If S' is a sub-pG$_\alpha$(s', t') of a pG$_\alpha$(s, t) S then $s = s'$ or $s \geq s't'$
$+ \alpha - 1$, and dually $t = t'$ or $t \geq s't' + \alpha - 1$.*

Remarks

(1)   A geometric interpretation of the equalities in the above theorem follows from the proof.   For instance $s = s't' + \alpha - 1$, if and only if each point of S not in S' is on one line of S', hence dually $t = s't' + \alpha - 1$ if and only if every line of S not belonging to S' intersects S' in exactly one point.

(2)   For a more general theorem than the above one, namely a restriction on the parameters of a sub-$pG_{\alpha'}(s', t')$ of a $pG_\alpha(s, t)$ we refer to De Clerck (1978, 1984).

Theorem 10.   *Let $S = (\mathcal{P}, \mathcal{B}, I)$ be a $pG_\alpha(s, t)$, $\alpha \neq 1$.   If $K (|K| = k)$ is a $(0, s)$-set of S, then one of the following cases occurs:*

(1) $\mathcal{P}\backslash K$ *is an ovoid, here $k = s(st + \alpha)/\alpha$.*

(2) $\mathcal{P}\backslash K$ *is the set of $s + 1$ points on one line.   In this case $\alpha = t + 1$.*

(3) $\mathcal{P}\backslash K$ *and the lines not intersecting K form a sub-$pG_\alpha(s, t')$ with $t' = (t + 1 - \alpha)/s$.   In this case*

$$k = (s + 1)(t(s - 1) - 1 + \alpha)/\alpha.$$

Proof.   Let K be a $(0, s)$-set in S.   If $L^- = \emptyset$ (hence if K is an s-ovoid), then the complement $K^c = S - K$ of K is an ovoid (or a 1-ovoid) and conversely it is clear that the complement of an ovoid is an s-ovoid.

Let $L^- \neq \emptyset$.   Then every line intersecting K^c in at least 2 points intersects K^c in $s + 1$ points.   It is clear that if there is only one exterior line, that α should be equal to $t + 1$.   However, if there is more than one exterior line of K, it is easy to prove that the incidence structure (K^c, L^-, I) is a partial subgeometry $pG_\alpha(s, t')$.   See for instance De Clerck (1978) for a proof, which is in fact completely analogous to the proof for generalized quadrangles (see 2.3.1 of Payne and Thas (1984)).   Hence $t \geq st' + \alpha - 1$, but since every line intersecting K has (exactly) one point in common with this partial subgeometry, equality holds. □

6. Examples of $(0, n)$-sets

In the last part of this paper we shall give examples of $(0, n)$-sets in some of the known proper $pG_\alpha(s, t)$.   Most of those examples are special cases of the following general construction.

Theorem 11. *Let S be a* $pG_\alpha(s, t)$ *with* $\gamma\ (> 1)$ *distinct ovoids* $O_1, ..., O_\gamma,$ *pairwise intersecting in the same set X with $x = |X|$ points. Then, for every n, $1 < n \leq \gamma$, the set*

$$K = \bigcup_{i=1}^n (O_i - X), \tag{42}$$

is a $(0, n)$-set with

$$k = n(st/\alpha + 1 - x) \tag{43}$$

points.

6.1   Ovoids and $(0, n)$-sets in $S(\mathcal{K})$

An ovoid of the partial geometry $S(\mathcal{K})$ has $q + 1$ elements. The $q + 1$ points on a non-intersecting line of $\mathcal{K}$ define an ovoid of $S(\mathcal{K})$. Such ovoids are called linear. An ovoid of $S(\mathcal{K})$ can equivalently be defined as a set of $q + 1$ points of the plane not belonging to $\mathcal{K}$, such that any line of the plane connecting two such points is a non-intersecting line of $\mathcal{K}$. As soon as $\mathcal{K}$ is a maximal arc in the desarguesian plane $PG(2, q)$, any ovoid of $S(\mathcal{K})$ is linear. This is proved in Blokhuis and Wilbrink (1987) generalizing the theorem of Segre-Korchmáros (Segre and Korchmáros (1977)) where $\mathcal{K}$ was a conic union its nucleus; see also Bruen and Thas (1975). From this it follows that all the examples of type $S(\mathcal{K})$, with $\mathcal{K}$ a maximal arc in the desarguesian plane, have only linear ovoids. If $\mathcal{K}$ is a maximal arc of degree d in a plane of order q, $1 < d < q$, then every point not on $\mathcal{K}$ is incident with q/d linear ovoids. Hence, by the above lemma, n linear ovoids O_i, $1 < n \leq q/d$, through a point P of $S(\mathcal{K})$ will form a $(0, n)$-set $K = \bigcup_{i=1}^n (O_i - P)$ with nq points.

Suppose $q = 2^h\ (h > 2)$ and $d = 2^{h-1}$, then the pointgraph of $S(\mathcal{K})$ is $\overline{T(2^h + 2)}$, the complement of the triangular graph $T(2^h + 2)$. The ovoids of the partial geometry are the $2^h + 2$ (maximal) cliques of size $2^h + 1$ in the triangular graph $T(2^h + 2)$. Moreover through each vertex of $T(2^h + 2)$ pass two such cliques, which yield the $(2^{h-1} + 1)$ $(2^h + 1)$ $(0, 2)$ sets in $S(\mathcal{K})$.

Although the graphs $\overline{T(2^h + 2)}$ are uniquely defined by their parameters, this does not imply that a partial geometry with this graph is unique. For instance, in Mathon (1981) it is proved by computer that there exist exactly two partial geometries with parameters $t = 6$, $s = 4$, $\alpha = 3$ (and pointgraph $\overline{T(10)}$).   Both have the same number of ovoids and $(0, 2)$-sets, but the partial geometry which is not of type $S(\mathcal{K})$ is not embeddable in $PG(2, 8)$, hence the ovoids are not linear.

If $\mathcal{K}'$ is a maximal arc of degree d' in the plane π, such that $\mathcal{K}'$ is disjoint from the maximal arc $\mathcal{K}$ of degree d, then $\mathcal{K}$ is a $(0, d')$-set in $S(\mathcal{K})$.   Disjoint complete ovals (maximal arcs of degree 2) in a projective plane are known to exist, we have no knowledge of any construction if one of the maximal arcs is not a complete oval.

6.2   Ovoids and $(0, n)$-sets in $T_2^*(\mathcal{K})$

An ovoid in a partial geometry of type $T_2^*(\mathcal{K})$ has q^2 points.   Let π' be a plane different from the plane π (containing the maximal arc $\mathcal{K}$), and such that the line $l = \pi \cap \pi'$ is an exterior line of $\mathcal{K}$.   It is clear that the points of $\pi' \backslash l$ form an ovoid of $T_2^*(\mathcal{K})$.   On the other hand every set of n planes (different from the plane π) through l forms an n-ovoid $(n \leq q)$ of the partial geometry $T_2^*(\mathcal{K})$.   If the line $l = \pi \cap \pi'$ intersects $\mathcal{K}$ in the points $P_1, P_2, ..., P_d$, then the points and lines of $T_2^*(\mathcal{K})$ in π' define a net S of order $s' + 1 = s + 1 = q$ and degree $t' + 1 = d = \alpha + 1$. Each point Q of $T_2^*(\mathcal{K})$ which does not belong to S is collinear (in $T_2^*(\mathcal{K})$) with the $st' + \alpha = q(d - 1)$ points of a $(d - 1)$-ovoid O of S.   This set is the projection of $\mathcal{K} \backslash l$ from Q on the plane π'.

6.3   Ovoids and $(0, n)$-sets in $PQ^+(4n-1, 2)$

An ovoid in this model should have $st/\alpha + 1 = 2^{2n} - 1$ points. However, no such ovoids are known.   If one takes the $2^{2n} - 1$ hyperplanes in one element of the spread Σ, then these hyperplanes define a spread of the partial geometry $PQ^+(4n - 1, 2)$, hence they define an ovoid in the dual partial geometry.

Another non-isomorphic spread in $PQ^+(4n - 1, 2)$ can be constructed as follows.   Let M be a point on the quadric, and let ω be the element of the spread Σ containing M.   Let $\Omega(M)$ be the set of hyperplanes $l_i(M)$ $(1 \leq i \leq 2^{2n-1} - 1)$ of ω through M.   Note that $\Omega(M)$ is a set of $s = 2^{2n-1} - 1$ lines of the partial geometry $PQ^+(4n - 1, 2)$ which are pairwise non-concurrent.   The tangent hyperplane M^* of the quadric at M will intersect each of the 2^{2n-1} elements of the spread Σ different from ω in a hyperplane, which is a line of the partial geometry.   This set of 2^{2n-1} pairwise non-concurrent lines together with the set $\Omega(M)$ defines a spread of the partial geometry $PQ^+(4n - 1, 2)$, hence by dualizing we have constructed an ovoid in the dual partial geometry which is non-isomorphic to the ovoids coming from the hyperplanes in one element of the spread Σ.   For more information on these spreads and characterization theorems, we refer to De Clerck et al. (1988).

References

A. Blokhuis and H.A. Wilbrink (1987), A characterization of exterior lines of certain sets of points in $PG(2, q)$, *Geom. Dedicata* **23**, 253-254.

R.C. Bose (1963), Strongly regular graphs, partial geometries and partially balanced designs, *Pacific J. Math.* **13**, 389-419.

R.H. Bruck (1963), Finite nets II: uniqueness and embedding, *Pacific J. Math.* **13**, 421-457.

A. Bruen and J.A. Thas (1975), Flocks, chains and configurations in finite geometries, *Atti Accad. Naz. Lincei Rend.* **59**, 744-748.

P.J. Cameron and J.H. van Lint (1982), On the partial geometry $pg(6, 6, 2)$, *J. Combin. Theory Ser. A* **32**, 252-255.

A.M. Cohen (1981), A new partial geometry with parameters $(s, t, \alpha) = (7, 8, 4)$, *J. Geom.* **16**, 181-186.

F. De Clerck (1978), *Een kombinatorische studie van de eindige partiële meetkunden*, Ph.D. thesis, State University of Ghent.

F. De Clerck (1984), *Substructures of partial geometries*, Seminario di Geometrie Combinatorie, Università degli Studi de L'Aquila.

F. De Clerck, R.H. Dye, and J.A. Thas (1980), An infinite class of partial geometries associated with the hyperbolic quadric in $PG(4n - 1, 2)$ *European J. Combin.* **1**, 323-326.

F. De Clerck, H. Gevaert, and J.A. Thas (1988), Partial geometries and copolar spaces, in *Combinatorics '88*, to appear.

A. Del Fra and D. Ghinelli (1990a), Diameter bounds for an extended partial geometry, Preprint.

A. Del Fra and D. Ghinelli (1990b), Truncated D_n Coxeter complexes as EpGs with maximum diameter, Preprint.

A. Del Fra, D. Ghinelli, and D.R. Hughes (1990a), Extended partial geometries with minimal μ, Preprint.

A. Del Fra, D. Ghinelli, and S.E. Payne (1990b), $(0, n)$ sets in a generalized quadrangle, Preprint.

R.H. Dye (1977), Partitions and their stabilizers for line complexes and quadrics, *Ann. Mat. Pura Appl.* **114**, 173-194.

W. Haemers (1981), A new partial geometry constructed from the Hoffman-Singleton graph, *Finite Geometries and Designs* (ed. P.J.

Cameron et al.), London Math. Soc. Lecture Note Ser. **49**, 119-127, Cambridge University Press, Cambridge.

W. Haemers and J.H. van Lint (1982), A partial geometry $pg(9, 8, 4)$, *Ann. Discrete Math.* **15**, 205-212.

J.W.P. Hirschfeld (1979), *Projective Geometries over Finite Fields*, Clarendon Press, Oxford.

X.L. Hubaut (1975), Strongly regular graphs, *Discrete Math.* **13**, 357-381.

D.R. Hughes and F.C. Piper (1985), *Design Theory*, Cambridge University Press, Cambridge.

W.M. Kantor (1982a), Spreads, translation planes and Kerdock sets II, *SIAM J. Algebraic Discrete Methods* **3**, 308-318.

W.M. Kantor (1982b), Strongly regular graphs defined by spreads, *Israel J. Math.* **41**, 298-312.

R. Mathon (1981), The partial geometries $pg(5, 7, 3)$, *Congr. Numer.* **31**, 129-139.

S.E. Payne and J.A. Thas (1984), *Finite Generalized Quadrangles*, Pitman, London.

B. Segre and G. Korchmáros (1977), Una proprietà degli insiemi di punti di un piano di Galois caratterizzante quelli ormati dai punti delle singole rette esterne ad una conica, *Atti Acccad. Naz. Lincei Rend.* **62**, 1-7.

J.A. Thas (1973), Construction of partial geometries, *Simon Stevin* **46**, 95-98.

J.A. Thas (1974), Construction of maximal arcs and partial geometries, *Geom. Dedicata* **3**, 61-64.

J.A. Thas (1981), Some results on quadrics and a new class of partial geometries, *Simon Stevin* **55**, 129-139.

J.A. Thas (1989), Interesting pointsets in generalized quadrangles and partial geometries, *Lin. Algebra Appl.* **114/115**, 103-131.

V.D. Tonchev (1984), The isomorphism of the Cohen, Haemers-van Lint and De Clerck-Dye-Thas partial geometries, *Discrete Math.* **49**, 213-217.

J.H. van Lint and A. Schrijver (1981), Construction of strongly regular graphs, two-weight codes and partial geometries by finite fields, *Combinatorica* **1**, 63-73.

W.D. Wallis (1973), Configurations arising from maximal arcs, *J. Combin. Theory Ser. A* **15**, 115-119.

One diagram for many geometries

A. Del Fra[1], D. Ghinelli[1] and A. Pasini[2]
University of Rome I[1] and University of
Naples[2]

1. Introduction

We consider the following diagram of rank $n = m + 2$

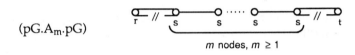

(pG.A_m.pG)

m nodes, $m \geq 1$

where r, s, t are finite parameters, $s > 1$, the stroke O——————O denotes the class of projective planes, as usual and ⟁⫽⟁ denotes the class of partial geometries, (we prefer to avoid the notation O═══════O of Hughes (to appear , this volume), as it resembles $\tilde{C}_3$ too closely. The class of partial geometries pG(s, t, α) of order s, t and index α will be denoted by the following symbol

$$\underset{s}{\vartriangleleft}\ \alpha\ \underset{t}{\vartriangleright}$$

as in Hughes (to appear , this volume).
    In particular

$$\underset{s}{\vartriangleleft}\ s+1\ \underset{t}{\vartriangleright}$$

is just the class of linear spaces of order s, t usually denoted as

$$\underset{s}{O}\overset{L}{———}\underset{t}{O}$$

whereas the following

$$\underset{s}{\vartriangleleft}\ t+1\ \underset{t}{\vartriangleright}$$

is the class of dual linear spaces of order s, t namely

$$\underset{s}{O}\overset{J}{———}\underset{t}{O}\ .$$

Clearly,

$$\underset{s}{\vartriangleleft}\ 1\ \underset{t}{\vartriangleright}$$

is a synonym of

$$\underset{s}{\vartriangleleft═══════}\underset{t}{\vartriangleright}\ .$$

Furthermore

$$\underset{s \quad\quad t}{\boxed{} \;\; t \;\; \boxed{}}$$

is the class of nets with lines of size $s + 1$ and $t + 1$ parallelism classes of lines, also denoted as follows

$$\underset{s \qquad\qquad t}{\circ \overset{N}{\longrightarrow} \circ}$$

whereas

$$\underset{s \quad\quad t}{\boxed{} \;\; s \;\; \boxed{}}$$

is the class of dual nets with $t + 1$ lines on each point and traversal blocks of size $s + 1$, often denoted as

$$\underset{s \qquad\qquad t}{\circ \overset{N^*}{\longrightarrow} \circ} \;\; \text{or} \;\; \underset{s \qquad\qquad t}{\circ \overset{\mathcal{N}}{\longrightarrow} \circ}.$$

Note that

$$\underset{t\text{-}1 \qquad\qquad t}{\circ \overset{N}{\longrightarrow} \circ} \;\; \text{and} \;\; \underset{s \qquad\qquad s\text{-}1}{\circ \overset{N^*}{\longrightarrow} \circ}$$

are the classes of affine planes of order t and dual affine planes of order s, respectively. We denote these classes by the following more standard notation:

$$\underset{t\text{-}1 \qquad\qquad t}{\circ \overset{Af}{\longrightarrow} \circ} \;\; \text{and} \;\; \underset{s \qquad\qquad s\text{-}1}{\circ \overset{Af^*}{\longrightarrow} \circ}$$

respectively. Therefore, we only use

$$\underset{s \qquad\qquad t}{\circ \overset{N}{\longrightarrow} \circ} \;\; \text{or} \;\; \underset{s \qquad\qquad t}{\circ \overset{N^*}{\longrightarrow} \circ}$$

to denote *proper nets* or *proper dual nets,* respectively; namely nets or dual nets other than affine planes or dual affine planes ($t < s + 1$ or $s < t + 1$, respectively).

A special notation is used in Ronan (1982, 1986) to denote the linear space of the points and lines of a $k + 2$-dimensional projective geometry, namely:

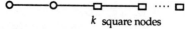

$$k \;\; \text{square nodes}$$

We prefer the following notation

$$\circ\!\!-\!\!-\!\!-\!\!\circ\!\!-\!\!-\!\!/\!\!/\,k \quad \text{or} \quad \circ\!\!-\!\!-\!\!-\!\!\circ\!\!-\!\!-\!\!/\!\!/ \;.$$

Clearly, k $/\!\!/\!\!-\!\!-\!\!\circ\!\!-\!\!-\!\!\circ$ or $/\!\!/\!\!-\!\!-\!\!\circ\!\!-\!\!-\!\!\circ$ will be the dual of the above. All partial geometries mentioned up to here are *improper*, a partial geometry pG(s, t, α) of order s, t and index α being *proper* (De Clerck to appear; Hughes to appear a, this volume) when $1 < \alpha < \min(s, t)$.

We mark types by underlined, positive integers as follows,

$$\underset{\underline{1}}{\overset{\alpha}{\circ}}\;/\!\!/\;\underset{\underline{2}}{\overset{}{\circ}}\!\!-\!\!-\!\!\underset{\underline{3}}{\circ}\;\cdots\;\underset{\underline{n\text{-}2}}{\circ}\!\!-\!\!-\!\!\underset{\underline{n\text{-}1}}{\overset{\alpha}{\circ}}\;/\!\!/\;\underset{\underline{n}}{\overset{}{\circ}}$$

We also use the words "point", "line", "plane", "solid" (or 3-space), ..., "dual plane" (n-3-space), "dual line" (n-2-space), "dual point" (n-1-space) when referring to elements of type $\underline{1}, \underline{2}, \underline{3}, \underline{4}, ..., \underline{n\text{-}2}, \underline{n\text{-}1}, \underline{n}$ respectively.

The Intersection Property (IP) of Buekenhout (1981) will often be assumed in this paper. Several equivalent formulations are known for that property, (see Biliotti and Pasini 1982). For instance, it is implicit in the proof of Theorem 6 of Buekenhout (1979) that the property Int is equivalent to IP in the case of a (residually connected) geometry Γ belonging to a string diagram

$$\underset{\text{points}}{\circ}\overset{\pi}{-\!\!-}\circ\overset{\pi}{-\!\!-}\circ\cdots\circ\overset{\pi}{-\!\!-}\circ$$

where $\circ\overset{\pi}{-\!\!-}\circ$ denotes the class of partial planes, as in Buekenhout (1979, 1981) (pG.A_m.pG is indeed a string diagram). If Γ is as above and x is an element of Γ, $\sigma(x)$ will denote the *shadow* of x, namely the set of points of Γ incident with x. We can state the property Int now:

(Int) Given any two elements x, y of Γ, if $\sigma(x) \cap \sigma(y) \neq 0$, then there is an element z of Γ incident with both x and y and such that $\sigma(z) = \sigma(x) \cap \sigma(y)$.

All geometries considered in this paper are assumed to be residually connected.

The main result of this paper will show that almost all geometries with Int belonging to pG.A_m.pG with $m \geq 3$ are known. We will give the precise statement of this theorem in §4. Of course, in order to do that we first must describe all known examples. This will be done in §2. However, we can state the following lemma immediately, from which the main theorem of this paper will follow. The proof of this lemma will be given in §3.

Lemma 1. *Let Γ be a geometry satisfying Int and belonging to* $pG.A_m.pG$ *with* $m \geq 3$. *Then no proper partial geometry can appear as a rank 2 residue in* Γ.

As a matter of fact, no example is known in $pG.A_2.pG$ either, having proper partial geometries as rank 2 residues; and it may be that we already know all the examples in this case too. However, even if a proof could be found, extending Lemma 1 to the case $m = 2$, a complete classification of this case would also need the solution of a difficult problem on linear spaces (see Doyen and Hubaut 1971; Kantor 1974).

On the contrary, a rich choice of examples is offered in the rank 3 case ($m = 1$; see Hobart and Hughes (1990); Hughes (1990), to appear a, this volume), some of them having proper partial geometries as rank 2 residues. This selection of examples is not completely wild of course; some features and constructions are met often. However, what we presently know is not sufficient to hazzard a guess of what a classification theorem would look like in this case.

2. The known examples

Most of the constructions and results presented in this section make no use of the finiteness assumption on the parameters r, s, t of $pG.A_m.pG$ made in §1. Nevertheless, we keep that assumption to prevent any possibile confusion in the next sections.

2.1   Projective geometries and their truncations

The following theorem is well known.

Theorem 2 (Buekenhout 1979; Tits 1981). *The geometries belonging to the following diagram are precisely the n-dimensional projective geometries*

(A_n)   o———o———o ····· o———o   (n nodes)

The *upper k-truncation* of a projective geometry Γ of dimension $n + k$ is the geometry Γ formed by taking all *i*-dimensional subspaces of Γ for all $i = 0, 1, ..., n - 1$ and dropping the rest. Clearly Γ belongs to the following diagram

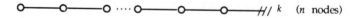

o———o———o ···· o———o———o———//^{k} (n nodes)

which is a specialization of the following one, of course:

(A_{n-1}.L) 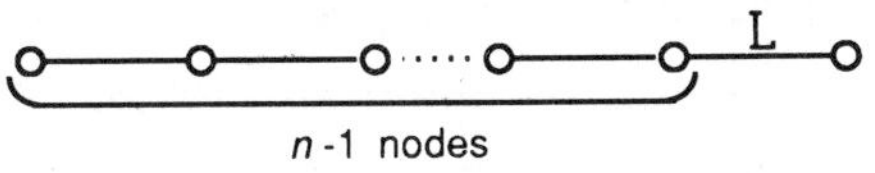

Theorem 3 (Buekenhout 1979). *Every geometry belonging to the diagram* A_{n-1}.L *above with* $n \geq 3$ *is an upper truncation of a projective geometry.*

Clearly projective geometries may be viewed as 0-truncations. so they are included in the statement of the previous theorem.

It is also clear how all the above can be dualized to lower k-truncations.

Theorem 3 has the following easy consequence.

Corollary 4. *The diagram* A_{n-1}.L *in fact coincides with the following, if* $n \geq 3$:

The *double h, k-truncation* of a projective geometry Γ of dimension $h + n + k$ is the geometry Γ formed by taking all i-dimensional subspaces of Γ for all $i = h, h + 1, \ldots, h + n - 1$ and dropping the rest. Clearly, property Int holds in Γ and, if $n \geq 3$, Γ belongs to the following diagram

(n nodes)

which is a specialization of the following one

(J .A_{n-2}.L) 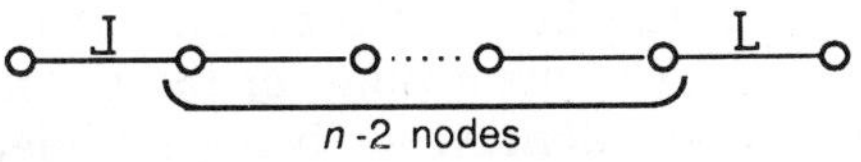

Theorem 5 (Sprague 1985). *A geometry belonging to the above diagram* J .A_{n-2}.L *is a double truncation of a projective geometry if and only if it satisfies property Int.*

Clearly, projective geometries are also included in Theorem 5, as they can be viewed as (0, 0)-truncations of themselves.

Remark 6. As we have assumed finite parameters in the preceding, all projective geometries considered above are finite, but it is clear that we can easily get rid of this assumption except that, in the infinite case,

the stroke $\circ\!\!-\!\!-\!\!-\!\!\circ\!\!-\!\!-\!\!-H/$ might also denote the point-line system of an infinite dimensional projective geometry and this infinite dimensional case would cause difficulties in Theorem 5. Hence that theorem should be stated more carefully in the infinite case, assuming that only diagrams of the following form are considered:

$$h \ /H\!\!-\!\!-\!\!-\!\!\circ\!\!-\!\!-\!\!-\!\!\circ\!\!-\!\!-\!\!-\!\!\circ\cdots\circ\!\!-\!\!-\!\!-\!\!\circ\!\!-\!\!-\!\!-\!\!\circ\!\!-\!\!-H/k$$

where h, k are non-negative integers.

2.2   Affine geometries and their truncations

The following theorem is due to Buekenhout (1969, 1979).

Theorem 7. *The geometries belonging to the following diagram are precisely the n-dimensional affine geometries:*

$$(\text{Af.A}_{n\text{-}1}) \qquad \circ\overset{\text{Af}}{-\!\!-\!\!-}\circ\!\!-\!\!-\!\!-\circ\cdots\circ\!\!-\!\!-\!\!-\circ\!\!-\!\!-\!\!-\circ \quad (n \text{ nodes})$$

Upper truncations can be defined for affine geometries as we have done for projective geometries. We obtain geometries belonging to the following diagram,

$$\circ\overset{\text{Af}}{-\!\!-\!\!-}\circ\!\!-\!\!-\!\!-\circ\cdots\circ\!\!-\!\!-\!\!-\circ\!\!-\!\!-\!\!-\circ\!\!-\!\!-H/$$

which is a specialization of the following one:

$$(\text{Af.A}_{n\text{-}2}.\text{L}) \qquad \circ\overset{\text{Af}}{-\!\!-\!\!-}\circ\!\!-\!\!-\!\!-\circ\cdots\circ\!\!-\!\!-\!\!-\circ\overset{\text{L}}{-\!\!-\!\!-}\circ$$

The next theorem can be proved by the same standard technique used in Theorem 7. The main point of the proof is in showing that the parallelism of lines, defined as 'being parallel inside a plane', is indeed an equivalence relation. This is easy if we are given the right 3-space as elements of the geometry we consider

$$\circ\overset{\text{Af}}{-\!\!-\!\!-}\circ\!\!-\!\!-\!\!-\circ\!\!-\!\!-\!\!-\circ\cdots\circ\!\!-\!\!-\!\!-\circ\overset{\text{L}}{-\!\!-\!\!-}\circ$$
$$\text{points} \quad \text{lines} \quad \text{planes} \quad \text{3-spaces}$$

Thus, we have the following

Theorem 8. *Every geometry belonging to the diagram $Af.A_{n-2}.L$ above, with $n \geq 4$, is an upper truncation of an affine geometry.*

The rank 3 case

(Af.L)

or even the following specialization of it

is harder and the conclusion of Theorem 8 fails to hold if referred to this case: some counterexamples are mentioned in Buekenhout (1979), arising from Steiner systems. However, all of them have lines with either 2, or 3 points. Indeed, we have the following:

Theorem 9 (Buekenhout 1969, 1979). *Let Γ be a geometry belonging to the diagram Af.L above and assume that the lines of Γ contain more than 3 points. Then Γ is an upper truncation of an affine geometry.*

It is clear that all the above can be dualized.
Furthermore, finiteness assumptions have no role in the above.

2.3   Polar spaces and their truncations

The following theorem is due to Tits (1981).

Theorem 10. *A geometry belonging to the following diagram:*

(C_n)

(n nodes)

is a polar space of (rank n) if and only if it satisfies Int.

The intersection property Int is essential in this context. Indeed, some C_n geometries exist which are not polar spaces (namely Int fails to hold in them). We do not go into details here; the reader may see Pasini (1988), or its up-to-date revision Lunardon and Pasini (to appear) for a survey of C_n geometries.

Let Γ be a polar space of rank $n + k$. The *lower k-truncation* of Γ is the geometry Γ obtained keeping all singular subspaces of Γ of rank greater than k and dropping the rest (we recall that points, lines, planes, ... have rank 1, 2, 3, ... respectively; also note that $\Gamma = \Gamma$ if and only if $k = 0$). The geometry Γ defined above satisfies Int and belongs to the following diagram

$$k \quad /\!H\!-\!\!-\!\!-\!\!O\!-\!\!-\!\!-\!\!O\!-\!\!-\!\!-\!\!O\cdots O\!-\!\!-\!\!-\!\!\square\!=\!\!=\!\!=\!\!\square \qquad (n \text{ nodes})$$

which is a specialization of the following one

(J.C_{n-1})

Theorem 11 (Ronan 1986). *Let Γ be a geometry belonging to the diagram* J.C_{n-1} *above, with $n \geq 4$. The geometry Γ is the lower truncation of a polar space if and only if it satisfies Int.*

    The statement of this theorem is not valid any more if $n = 3$ is allowed (J.C_2), even if we are assumed to be dealing with the following diagram

$$k \quad /\!H\!-\!\!-\!\!-\!\!O\!-\!\!-\!\!-\!\!\square\!=\!\!=\!\!=\!\!\square.$$

A couple of counterexamples are given in Ronan (1986). We notice that a property similar, but not equivalent, to Int is checked for those two examples (that property is also called Int by Ronan (1986)). However, it can be seen that our property Int also holds for them.

    Other examples for J.C_2 are considered in Hughes (to appear a, this volume).

Remark 12. Our finiteness assumption is quite irrelevant for all of the above, except that remarks similar to those of Remark 6 should now be made for Theorem 11.

2.4   Affine polar spaces and their standard quotients

Given a non-degenerate polar space Γ, a *hyperplane* of Γ is a proper subset of the set of points of Γ such that every line of Γ not contained in H has precisely one point in H (see Cohen and Shult (1990)). The *affine polar space* $\Gamma = \Gamma - H$ defined by H in Γ is the geometry obtained by dropping all singular subspaces of Γ contained in H (in particular all points and lines in H) and keeping the rest (Cohen and Shult (1990)).

    Let K_H be the pointwise stabilizer of H in Aut(Γ) and let A be a subgroup of K_H. Clearly, A can also be viewed as a subgroup of Aut(Γ) and defines a quotient Γ/A of Γ, in the meaning of Tits (1981). These quotients are called *standard quotients* of Γ. Of course, $A = 1$ is allowed (we have $\Gamma / A = \Gamma$ in this case). Other quotients of Γ may be considered, using subgroups of the stabilizer G_H of H in Aut(Γ) not contained in K_H. However, the intersection property of Int holds in Γ / A if and only if $A \leq K_H$ (see Cuypers and Pasini, submitted).

We are not going to discuss quotients of affine polar spaces in detail here. The reader may find more information on this topic in Cuypers and Pasini (submitted), together with a characterization of the minimal standard quotient Γ / K_H as the 'tangent geometry' of a polar space Γ^∞ which, afterwards, turns out to coincide with H itself.

If the polar space Γ, from which we start, has rank ≥ 3, then all affine polar spaces obtained from Γ and their standard quotients belong to the following diagram:

$(Af.C_{n-1})$ o—$\overset{Af}{\quad}$—o———o ···· ———o——o⇒o $n \geq 3$

Theorem 13 (Cuypers and Pasini submitted; Cuypers submitted). *Let Γ be a geometry belonging to the diagram $Af.C_{n-1}$ above. Then the property Int holds in Γ if an only if Γ is either an affine polar space or a standard quotient of an affine polar space.*

Remark 14. We recall that the assumption that all lines have at least three points is usually included in the definition of a polar space. However, polar spaces where some or all lines have two points only (that is, are *thin*) might be allowed in §2.3. On the contrary, we cannot allow them in this section. Similarly, degenerate projective geometries, where some or all lines are thin, may be allowed in §2.1, but only non-degenerate projective geometries may be considered if affine geometries have to be obtained from them.

Remark 15. No finiteness assumption is needed in Theorem 13 as far as the case $n \geq 4$ is concerned (Cuypers and Pasini submitted). On the contrary, the assumption that the geometry admits finite parameters (§1) is essential to the proof given in Cuypers (submitted) for the case $n = 3$ of Theorem 13. However, no counterexample is presently known for the statement of Theorem 13 in the infinite case with $n = 3$.

Remark 16. Property Int is essential to Theorem 13. Indeed, apart from non-standard quotients, which may occur and do not satisfy Int, one example is known which belongs to $Af.C_3$, does not satisfy Int and has nothing to do with affine polar spaces or their quotients, namely the so-called Neumaier geometry (see Buekenhout 1985).

2.5   Interlude: A common construction for different families of geometries

Let X, Y be subspaces of $PG(m, q)$ of dimensions x and y respectively, with $x - y + 1 \geq 3$ ($x = m$ and $y = -1$ are allowed). Let $\Gamma_m \big|_y^x$ be the geometry obtained taking the proper non-empty subspaces of $PG(m, q)$ that do not meet Y and that together with X span all of $PG(m, q)$.

The dimensions of the subspaces forming $\Gamma_m \big|_y^x$ and the rank n of $\Gamma_m \big|_y^x$ are as follows:

(1)   if $x < m$ and $y \geq 0$, then the dimensions are $m - x - 1$, $m - x$, ..., $m - y - 2$, $m - y - 1$ and the rank is $n = x - y + 1$;

(2)   if $x < m$ but $y = -1$, then the dimensions are $m - x - 1$, $m - x$, ..., $m - 1$ and the rank is $n = x - y$;

(3)   if $x = m$ but $y \geq 0$, then the dimensions are $0, 1, ..., m - y - 1$ and the rank is $n = x - y$;

(4)   if $x = m$ and $y = -1$, then the dimensions are $0, 1, ..., m - 1$ and the rank is $n = x - y - 1$ ($= m$).

Property Int holds in $\Gamma_m \big|_y^x$.

We have the following descriptions for $\Gamma_m \big|_y^x$, according to the values of x and y.

(A) $x = m$ and $y = -1$. We have $\Gamma_m \big|_y^x = PG(m, q)$.

(B) $x = m - 1$ and $y = -1$. We have $\Gamma_m \big|_y^x = AG(m - 1, q)$; note that $n = \Gamma_m \big|_y^x$.

(C) $x = m$ and $y = 0$. This is the dual of the above. Y is a point now, the 'point at infinity' of the dual affine geometry $\Gamma_m \big|_y^x$.

(D) $x < m - 1$ and $y = -1$. We now obtain *attenuated spaces* (Sprague 1981), also called *n-nets*. The diagram and the parameters are as follows:

$(N.A_{n-1})$

(E) $x = m$ and $y \geq 1$.   This is just the dual of the above; these geometries are sometimes calles *vector transversal designs*. However, we call them *dual attenuated spaces*, or also attenuated spaces for short, when no confusion will arise from this misuse.

The diagram and parameters are as follows:

$(A_{n-1}.N^*)$ $\underset{q}{\circ} \underset{q}{\!-\!\!-\!\!-\!\circ} \underset{q}{\!-\!\!-\!\!-\!\circ} \cdots \underset{q}{\circ} \underset{q}{\!-\!\!-\!\!-\!\circ} \overset{N^*}{\underset{q^{y+1}-1}{\!-\!\!-\!\!-\!\circ}}$

(F) $x = m - 1$ and $y = 0$.   These are the so-called *affine-dual-affine-spaces*, or *biaffine spaces*. They have the diagram as below:

$(Af.A_{n-2}.Af^*)$ $\underset{q-1}{\circ} \overset{Af}{\!-\!\!-\!\!-\!\circ} \underset{q}{\!-\!\!-\!\!-\!\circ} \cdots \underset{q}{\circ} \underset{q}{\!-\!\!-\!\!-\!\circ} \overset{Af^*}{\underset{q-1}{\!-\!\!-\!\!-\!\circ}}$

(G) $x = m - 1$ and $y \geq 1$.   We call these geometries *affine attenuated spaces*. They have diagram and parameters as follows:

$(Af.A_{n-2}.N^*)$ $\underset{q-1}{\circ} \overset{Af}{\!-\!\!-\!\!-\!\circ} \underset{q}{\!-\!\!-\!\!-\!\circ} \cdots \underset{q}{\circ} \underset{q}{\!-\!\!-\!\!-\!\circ} \overset{N^*}{\underset{q^{y+1}-1}{\!-\!\!-\!\!-\!\circ}}$

(H) $x < m - 1$ and $y = 0$.   This is just the dual of the above. The parameters are as follows:

$(N.A_{n-2}.Af^*)$ $\underset{q^{m-x}-1}{\circ} \overset{N}{\!-\!\!-\!\!-\!\circ} \underset{q}{\!-\!\!-\!\!-\!\circ} \cdots \underset{q}{\circ} \underset{q}{\!-\!\!-\!\!-\!\circ} \overset{Af^*}{\underset{q-1}{\!-\!\!-\!\!-\!\circ}}$

(I) $x < m - 1$ and $y \geq 1$.   We call these geometries *doubly attenuated spaces*. They have diagram and parameters as follows:

$(N.A_{n-2}.N^*)$ $\underset{q^{m-x}-1}{\circ} \overset{N}{\!-\!\!-\!\!-\!\circ} \underset{q}{\!-\!\!-\!\!-\!\circ} \cdots \underset{q}{\circ} \underset{q}{\!-\!\!-\!\!-\!\circ} \overset{N^*}{\underset{q^{y+1}-1}{\!-\!\!-\!\!-\!\circ}}$

Remark 17.   The previous construction works as well if we start from a projective geometry $PG(m, K)$ over any division ring K. We need only generalize the definition of net so that to cover the infinite case as well, in order to be able to give diagram descriptions. Such a general definition is indeed the one that Sprague (1981) refers to.

Remark 18.   Quotients of $\Gamma_m\big|_y^x$ can also be defined in some of the previous cases using suitable subgroups of the stabilizer of X and Y in $P\Gamma L(m, q)$.   Geometries still belonging to some of the previous diagrams can be obtained in this way. However, property Int does not hold in these quotient geometries.

Truncations. Upper truncations of attenuated spaces (dually, lower truncations of dual attenuated spaces) can be defined as for projective geometries (§2.1-2.3). These truncations belong to diagrams of the form:

$$\circ \overset{N}{\rule{1.5cm}{0.4pt}} \circ \rule{1cm}{0.4pt} \circ \cdots \circ \rule{1cm}{0.4pt} \circ \rule{1cm}{0.4pt} \circ \rule{1cm}{0.4pt} \circ \!\!/\!/ \; k \quad \text{(or dually)}$$

which is a specialization of the following

(N.A$_m$.L) $\quad \circ \overset{N}{\rule{1.5cm}{0.4pt}} \circ \rule{1cm}{0.4pt} \circ \cdots \circ \rule{1cm}{0.4pt} \circ \rule{1cm}{0.4pt} \circ \overset{L}{\rule{1.5cm}{0.4pt}} \circ$

(dually, we have J.A$_m$.N*). In particular, when $m = 1$ we have

(N.L) $\qquad\qquad \circ \overset{N}{\rule{1.5cm}{0.4pt}} \circ \overset{L}{\rule{1.5cm}{0.4pt}} \circ$

(or the dual J.N* of this).

Remark 19. We close this section with some comments on the two geometries H_q^m and H_q^{m*} considered in De Clerck and Thas (1978). The partial geometry H_q^m is nothing but $\Gamma_m \big|_{m-2}^m$, whereas the semipartial geometry H_q^{m*} is obtained from $\Gamma_m \big|_{-1}^{m-2}$ choosing points and lines in the most natural way, as follows

$$\underset{\text{points}}{\circ} \overset{N}{\rule{1.5cm}{0.4pt}} \underset{\text{lines}}{\circ} \rule{1cm}{0.4pt} \circ \cdots \circ \rule{1cm}{0.4pt} \circ$$

A nice characterizations of H_q^m has been obtained by De Clerck and Thas (1978).

2.6 Characterization theorems for some of the families constructed in the previous section

We will consider the following diagrams in this section:

(N.A$_{n-1}$) $\qquad \circ \overset{N}{\rule{1.5cm}{0.4pt}} \circ \rule{1cm}{0.4pt} \circ \cdots \circ \rule{1cm}{0.4pt} \circ \rule{1cm}{0.4pt} \circ$

(N.A$_{n-2}$.L) $\quad \circ \overset{N}{\rule{1.5cm}{0.4pt}} \circ \rule{1cm}{0.4pt} \circ \cdots \circ \rule{1cm}{0.4pt} \circ \overset{L}{\rule{1.5cm}{0.4pt}} \circ$

(Af.A$_{n-2}$.Af*) $\quad \circ \overset{Af}{\rule{1.5cm}{0.4pt}} \circ \rule{1cm}{0.4pt} \circ \cdots \circ \rule{1cm}{0.4pt} \circ \overset{Af*}{\rule{1.5cm}{0.4pt}} \circ$

$(Af.A_{n-2}.N^*)$ o——Af——o————o ····· o————o——N^*——o

In particular, when $n = 3$ we have the following

$(N.A_2)$ o——N——o————o

$(N.L)$ o——N——o——L——o

$(Af.Af^*)$ o——Af——o——Af^*——o

$(Af.N^*)$ o——Af——o——N^*——o.

It is clear that the following results can be stated in the dual form.

Theorem 20 (Sprague 1981). *Let Γ be a geometry belonging to the diagram* $N.A_{n-1}$, *with $n \geq 3$. Then Γ is an attenuated space if and only if property Int holds in it.*

Theorem 21 (Sprague 1981). *Let Γ be a geometry belonging to the diagram* $N.A_{n-2}.L$, *with $n \geq 4$. Then Γ is a truncation of an attenuated space if and only if property Int holds in it.*

The rank 3 case N.L looks much wilder, as usual. Some counterexamples to the statement of Theorem 21 exist in this case, and are mentioned in Sprague (1981)

Remark 22. No finiteness assumption is needed for the above results, except that some caution is necessary in Theorem 21 in the infinite case, as for $L.A_{n-2}.L$ and $C_{n-1}.L$ (see Remarks 6 and 12).

On the contrary, the assumption $s > 1$ made in §1 and implicitly confirmed at the very beginning of §2 is really essential for the above.

It amounts to saying that grids are not allowed as instances of o——N——o in this context. Actually, grids are explicitly excluded in the general definition of nets given in Sprague (op. cit).

After all, grids belong to C_2 and it is sensible in any case to avoid any overlapping between C_2 and o——N——o, as we have done for o——Af——o and o——N——o in §1.

Theorem 23 (Lefèvre-Percsy and Van Nypelseer (to appear); Van Nypelseer (to appear)). *Let Γ be a geometry belonging to the diagram*

Af.A_{n-2}.Af*, *with* $n \geq 3$. *Then* Γ *is a biaffine space if and only if property Int holds in it.*

Once again the above holds in the infinite case as well.

Theorem 24 (Del Fra et al. submitted). *Let* Γ *be a geometry belonging to the diagram* Af.A_{n-2}.N*, *with* $n \geq 3$ *and admitting (finite) parameters* $s - 1, s, t$ *as below with* $s \geq 4$:

$$\underset{s\text{-}1}{\circ}\overset{\text{Af}}{\underline{\hspace{1cm}}}\underset{s}{\circ}\underline{\hspace{1cm}}\underset{s}{\circ}\cdots\circ\underline{\hspace{1cm}}\underset{s}{\circ}\overset{N^*}{\underline{\hspace{1cm}}}\underset{t}{\circ}$$

Then Γ *is an affine attenuated space if and only if property Int holds in it.*

The finiteness assumption on s is essential this time.   At least, it is essentially exploited in the proof given in Del Fra et al. (submitted).

2.7   Attenuated polar spaces

Let Γ be a (non-degenerate, finite) polar space of rank $r \geq 3$, naturally embedded in a projective geometry $P = PG(m, q)$ (see Tits 1974).   Let X be a subspace of P of co-dimension x, with $1 \leq x \leq r - 2$.   We form a new geometry Γ/X taking as elements the singular subspaces of Γ that together with X span all of P.   The subspaces chosen in this way have dimensions $x - 1, x, ..., r - 1$.   The geometry Γ/X has rank $n = r - x + 1 \geq 3$, satisfies Int and has diagram and parameters as follows, according to whether $x = 1$ or $x > 1$:

(1) $x = 1$.   In this case Γ/X is just an affine polar space (see §2.4)

$$\underset{q\text{-}1}{\circ}\overset{\text{Af}}{\underline{\hspace{1cm}}}\underset{q}{\circ}\underline{\hspace{1cm}}\underset{q}{\circ}\cdots\circ\underline{\hspace{1cm}}\underset{q}{\circ}\underline{\hspace{1cm}}\overset{}{\underset{t}{\Longrightarrow}}\circ \qquad (t \text{ is the last}$$

parameter of Γ)

(2) $x > 1$.   We now obtain the following:

(N.C_{n-1}) $\underset{q^x\text{-}1}{\circ}\overset{\text{N}}{\underline{\hspace{1cm}}}\underset{q}{\circ}\underline{\hspace{1cm}}\underset{q}{\circ}\cdots\circ\underline{\hspace{1cm}}\underset{q}{\circ}\underline{\hspace{1cm}}\overset{}{\underset{t}{\Longrightarrow}}\circ \qquad (t \text{ is the last}$

parameter of Γ)

In this case we say that Γ/X is an *attenuated polar space*.

Clearly, the C_{n-1} residues are in any case the same as in the polar space Γ which we started from.

As in §2.5, the finiteness of P is quite inessential;   the same construction can be repeated in any classical polar space embedded in some $PG(m, K)$, where K is some division ring.

As in the case of affine polar spaces, quotients can be considered too; in particular, standard quotients (Int will hold in them).

We finish this section with a few examples of geometries belonging to the diagram $N.C_{n-1}$. We do not yet know if they are related to attenuated polar spaces.

Examples from co-polar spaces (Cuypers private communication). We recall that a *generalized Fischer space* is a partial linear space in which every two lines that intersect non-trivially generate a subspace that is isomorphic to an affine plane or the dual of an affine plane (Cuypers to appear, 1989a, 1989b). Generalized Fischer spaces that only contain dual affine planes are also called *co-polar spaces*. Let $\overline{S}$ be a copolar space of order q (that is, all lines of $\overline{S}$ have $q + 1$ points) and form a new space S taking the lines and planes of $\overline{S}$ as points and lines of S respectively. Consider the subspaces spanned by pairs of intersecting lines. We have two families of such subspaces, say $\mathcal{F}_1$ and $\mathcal{F}_2$. The subspaces in one of these families, say $\mathcal{F}_1$, are isomorphic with the partial geometry $PG(q^2 - 1, q, q)$. Take this family $\mathcal{F}_1$ as a set of planes. Then the points and the lines of S and these planes are the points, lines and planes respectively of a geometry belonging to $N.C_m$ with parameters as below

$$\underset{q^2-1}{\circ} \overset{N}{\rule{1cm}{0.4pt}} \underset{q}{\circ} \rule{1cm}{0.4pt} \underset{q}{\circ} \cdots \underset{q}{\circ} \rule{1cm}{0.4pt} \underset{q}{\Box} \rule{1cm}{0.4pt} \underset{t}{\circ}$$

where m and t depend on the copolar space $\overline{S}$ which we started from, of course. The parameters are the same as in Γ/X with X of co-dimension 2.

We call these geometries *quasi-attenuated spaces*.

One more example. Let us recall the following construction of $Q(4, 2)$. Start from a hyperoval I of $P_1 = PG(2, 4)$. Take the points of P_1 not in I as points and the lines of P_1 meeting I as lines. We obtain $Q(4, 2)$.

Now embed P_1 in $P_2 = PG(3, 4)$ as a plane. Take the points of the affine geometry $P_2 - P_1$ as new points, the lines of $P_2 - P_1$ meeting P_1 in a point of I as new lines and the planes of P_2 other than P_1 and meeting I as new planes. The geometry obtained in this way has diagram and parameters as follows

$$3 = \underset{\substack{2^2-1 \\ \text{points}}}{\circ} \overset{N}{\rule{1cm}{0.4pt}} \underset{\substack{2 \\ \text{lines}}}{\Box} \rule{1cm}{0.4pt} \underset{\substack{2 \\ \text{planes}}}{\circ}$$

as in Γ/X with $\Gamma = Q(8, 2)$ and X of co-dimension 2. However, property Int fails to hold in the example constructed above. Whence it cannot be an attenuated polar space.

This construction can be generalized as follows (Hughes private communication). Let Π be a partial geometry realized inside $P_1 = PG(2, q)$ (we had $q = 4$ and $\Pi = Q(4, 2)$ in the above). Note that this realization need not be an embedding in the usual sense (the partial geometries embeddable in $PG(m, q)$ in the usual sense are not many; see De Clerck 1978). Embed P_1 as a plane in $P_2 = PG(3, q)$ and repeat the above construction. We obtain a geometry with the following diagram

$$\underset{q\text{-}1}{\circ} \overset{N}{\rule{1cm}{0.4pt}} \underset{s}{\circ} \overset{\alpha}{\rule{1cm}{0.4pt}} \underset{t}{\circ}$$

where s, t, α are the parameters and the index of Π. For instance, if Π is a projective subplane of P_1 (e.g. a Baer subplane or a Fano subplane), then what we obtain belongs to $N.A_2$ and satisfies Int. Whence, it is nothing but an attenuated space, by Theorem 20. If $\Pi = AG(2, q)$, then we obtain

$$\underset{q\text{-}1}{\circ} \overset{N}{\rule{1cm}{0.4pt}} \underset{q\text{-}1}{\circ} \overset{Af}{\rule{1cm}{0.4pt}} \underset{q}{\circ}$$

(and Int fails to hold in this case). Here are other natural candidates for Π. Start from a maximal arc K of degree d in $p_1 = PG(2, q)$ and define $\Pi = S(K)$ (Thas 1973, 1974) taking the points of P_1 not in K as points and the lines of P_1 meeting K (in d points) as lines. Or $\Pi = \bar{S}(K)$, defined with the same points as above, but taking the lines of P_1 external to K as lines (a dual linear space is obtained now). Finally, $\Pi = c(K)$, defined as taking K as a set of points and the lines of $S(K)$ as lines (a linear space is obtained now). We let the reader find the diagrams of the rank 3 geometries with these choices of Π. We only warn that we should assume $d \geq 3$ when $\Pi = c(K)$ if we want to remain coherent with the assumption $s > 1$ made in §1.

We also describe a construction similar to the above, related to the construction of the partial geometry $T_2^*(K)$ (Thas 1973, 1974).

Let P_1 and K be as above and now embed P_1 as a plane in $P_3 = PG(4, q)$. Take as points, lines and planes the points of $P_3 - P_1$, the lines of P_3 not meeting P_1 and planes X of P_3 meeting K in only one point (whence they meet P_1 in one point only). We obtain a rank 3 geometry belonging to the following diagram

$$\underset{q}{\circ}\xrightarrow{\text{Af}^*}\underset{q\text{-}1}{\square}\xrightarrow{\quad\alpha\quad}\underset{t}{\square}$$

where $q - 1$, t, α are the parameters and the index of $T_2^*(K)$, namely $t = |K| - 1 = (q + 1)(d - 1)$ and $\alpha = d - 1$. If we also add the 3-spaces of P_3 meeting P_1 in lines that are secant for K, then a rank 3 geometry is obtained, of which the previous one is a truncation and with diagram and parameters as follows:

$$\underset{q}{\circ}\xrightarrow{\text{Af}^*}\underset{q\text{-}1}{\circ}\xrightarrow{\quad N\quad}\underset{d\text{-}1}{\circ}\xrightarrow{\quad L\quad}\underset{q}{\circ}$$

The residues of the points of this geometry are all the rank 3 geometries previously obtained for $\Pi = c(K)$.

2.8 Affine buildings of type $\tilde{C}_m$ and their quotients

The reader is referred to Tits (1986) for the definition of affine buildings. The following is a consequence of Theorem 1 of Tits (1981).

Theorem 25 (Tits 1981). *Let Γ be a geometry satisfying Int and belonging to the following diagram:*

$(\tilde{C}_m)$ $\quad\quad\square\!\!=\!\!\square\!\!-\!\!-\!\!\circ\cdots\circ\!\!-\!\!-\!\!\square\!\!=\!\!\square$ $\quad(n = m + 1 \text{ nodes})$

Then Γ is either a building of affine type $\tilde{C}_m$ or a quotient of such a building.

Once again, finiteness assumptions have nothing to do here. (All buildings of type $\tilde{C}_m$ are infinite (Tits 1974).

2.9 Resumé

We now list the most important examples considered in §2.
(1) projective geometries (§2.1);
(2) upper truncations of projective geometries (§2.1) and their duals;
(3) double truncations of projective geometries (§2.1) and their duals;
(4) affine geometries (§2.2) and their duals;
(5) upper truncations of affine geometries (§2.2) and their duals;
(6) polar spaces (§2.3) and their duals;
(7) lower truncations of polar spaces (§2.3) and their duals;
(8) affine polar spaces and their standard quotients (§2.4) and their duals;

(9) attenuated spaces (§2.5) and duals;

(10) upper truncations of attenuates spaces (§2.5) and their duals;

(11) biaffine spaces (§2.5);

(12) affine attenuated spaces (§2.5) and their duals;

(13) doubly attenuated spaces (§2.5)

(14) attenuated polar spaces and their standard quotients (§2.7) and duals;

(15) quasi-attenuated polar spaces (§2.7) and duals;

(16) buildings of affine type $\tilde{C}_m$ and their quotients (§2.8).

3. Preliminary results

In this section Γ will always be a geometry satisfying Int and belonging to the following specialization of the diagram pG.A_m.pG of §1

$$(A_k.pG) \qquad \underset{s}{\circ}\!\!-\!\!\!\!-\!\!\underset{s}{\circ}\cdots\underset{s}{\circ}\!\!-\!\!\!\!-\!\!\underset{s}{\square}\!\!/\!\!/\underset{t}{\square} \qquad \text{(k+1 nodes, k≥2)}$$

where s, t are finite parameters with $s > 1$, as in §1, and $n = k + 1$ is the rank of Γ.

   The words "point", "line", "plane", ..., "dual plane", "dual line", "dual point" are used according with the conventions stated in §1 and we keep the conventions of §1 for types.   Given an element x of type $\underline{i}$ < $\underline{n}$, the *upper residue* Γ_x^+ of x consists of the elements of Γ incident with x of type greater than $\underline{i}$.   The *lower residue* Γ_x^- of an element of type $\underline{i} > \underline{1}$ consists of the elements of Γ incident with x and of type less than $\underline{i}$.

Lemma 26   (Del Fra, Ghinelli and Pasini to appear).   *There is a positive number* $\alpha(\Gamma)$ *such that, for every dual plane* x *of* Γ, *the upper residue* Γ_x^+ *is a partial geometry of index* $\alpha(\Gamma)$.

This number $\alpha(\Gamma)$ will be called the *index* of Γ, and denoted as α, for short.

Remark 27.   The assumption $s > 1$ is essential for Lemma 26.   Indeed if $s = 1$, counterexamples exist (Del Fra et al. to appear).

   As Int holds in Γ by assumption, the elements of Γ of type $\underline{i} > \underline{1}$ may be identified with their shadows (see §1).   Therefore, we will use notation consistent with this identification;   for instance, given a point a and an element u of type $\underline{i} > \underline{1}$, we write $a \in u$ to mean that a is incident with u.

Lemma 28. *Let a, u be a point and a dual point of Γ, respectively. If $a \notin u$, then there is a dual point v of Γ such that $a \in v$ and $u \cap v$ is a dual line of Γ.*

Proof. By induction on n, allowing $k = 1$ in $A_k.pG$. If $n = 2$ ($k = 1$), there is nothing to prove. Let $n > 2$. Let d be the distance of a from u in the collinearity graph of Γ, and let $a = a_0, a_1, ..., a_d \in u$ be a minimal path from a to u in the collinearity graph of Γ. Let x_i be the line of Γ joining a_{i-1} to a_i ($i = 1, 2, ..., d$). Assume $d = 1$. Then in the residue $\Gamma_{a_1}^+$ of a_1 we find a dual point v containing x_1 and meeting u in a dual line, by the inductive hypothesis. Clearly, $a \in v$ and we are done. Assume $d > 1$ now. By the inductive hypothesis we find a dual point v in $\Gamma_{a_d}^+$ incident with with x_d and meeting u in a dual line $u \cap v$. Similarly, we find a dual point v' in $\Gamma_{a_{d-1}}^+$ incident with x_{d-1} and such that $v' \cap v$ is a dual line. As Γ_v^- is a projective geometry of rank $n - 1 \geq 2$, $u \cap v \cap v'$ is a dual plane. In the projective geometry $\Gamma_{v'}^-$ we see that a_{d-2} is collinear with some point of u, contradicting the minimality of d. Therefore $d = 1$ and we are done. $\square$

Given a point and a dual point not on d, let $\alpha_n(a, u)$ be the number of dual points containing a and meeting u in a dual line; given a line x not coplanar with a, let $\beta_n(a, x)$ be the number of points of x collinear with a (the index n reminds us of the rank n of Γ).

Lemma 29. *Assume there is a non-coplanar point-line pair. Then we have:*

(1) $\alpha_n(a, u) = (\alpha - 1)(1 + s + ... + s^{n-2}) + 1$

(2) $\beta_n(a, x) = \dfrac{(\alpha - 1)(1 + s + ... + s^{n-2}) + 1}{(\alpha - 1)(1 + s + ... + s^{n-3}) + 1}$

where a, u, x are chosen as above and α is the index of Γ.

Proof. By induction on n, allowing $k = 1$ ($n = 2$) in $A_k.pG$ to start the induction and substituting (2) with $\beta_2(a, x) = \alpha$ when $n = 2$. Clearly, we have $\alpha_2(a, u) = \beta_2(a, x) = \alpha$ when $n = 2$. In this case (1) holds trivially and (2) holds by the previous convention.

Assume $n > 2$. Let (a, x) be a non-coplanar point-line pair. Take a dual point u on x. Clearly, $a \notin u$, as a is not coplanar with x. By Lemma 28 , we have a dual point v on a such that $u \cap v$ is a dual line. As a and x are not coplanar, we have $x \not\subset v$. Hence, in the projective geometry Γ_u^- we see that $x \cap u \cap v$ is a point, say b_v. Clearly, b_v is collinear with a. Conversely, given a point b on x collinear with a, by the inductive hypothesis there are

$$\alpha_{n-1} = (\alpha - 1)(1 + s + \dots + s^{n-3}) + 1$$

dual points containing the line m on a and b meeting u in a dual line (note that, taking any plane y on x, no 3-space exists containing both y and m, as a and x are not coplanar; hence the inductive hypothesis can be applied).

Therefore we have $\alpha_n(a, u) = \beta_n(a, x)\alpha_{n-1}$.

Let us consider the following substructure $L_{u, a}$ of Γ_u^-. Take as 'points' and 'lines' respectively the dual lines and the dual planes contained in u and in some dual point containing a. The structure $L_{u, a}$ defined in this way is a linear space. Indeed, let w, w' be distinct 'points' of $L_{u, a}$. It is clear that $w \cap w'$ is the unique 'line' of $L_{u, a}$ through w and w'.

All dual planes contained in a 'point' of $L_{u, a}$ are evidently 'lines' of $L_{u, a}$. Therefore there are precisely $1 + s + \dots + s^{n-2}$ 'lines' on every 'point' in $L_{u, a}$.

Now let y be a 'line' of $L_{u, a}$. By definition, there is some dual point v containing both a and y. In the projective geometry Γ_v^- we find a dual line $\overline{w}$ containing both a and y. Furthermore, $\overline{w}$ is the unique dual line containing both a and y, by Int and because $a \notin y$ (indeed $a \notin u$). In the partial geometry Γ_y^+ there are precisely α pairs (v, w) with $\overline{w} \subseteq v, w = u \cap v, v$ a dual point and w a dual line. All dual lines w as above are clearly 'points' of $L_{u, a}$ on the 'line' y. Conversely, let w' be a 'point' of $L_{u, a}$ on the 'line' y and let v' be the dual point containing a and w'. The dual point v' is uniquely determined by Int, as $a \notin w'$ (indeed $a \notin u$). As $v' \supseteq y \subseteq \overline{w}$ and $a \in v' \cap \overline{w}$, we have $\overline{w} \subseteq v'$ by Int, as $a \notin y$. Hence (v', w') is one of the α

pairs we have found in Γ_y^+ before. This shows that the 'lines' of $L_{u,\,a}$ have precisely α 'points'. Hence $L_{u,\,a}$ has $(\alpha - 1)(1 + s + \ldots\ s^{n-2}) + 1$.

On the other hand, exploiting Int and the fact that $a \notin u$ as we have done, it is easily seen that the 'points' of $L_{u,\,a}$ are as many as the dual points on a meeting u in a dual line. Namely $\alpha_n(\,a,\,u)$ is the number of 'points' of $L_{u,\,a}$. This proves (1).

Relation (2) follows from (1) and from the relation $\alpha_n(a,\,u) = \beta_n(a,\,x)\alpha_{n-1}$, proved before.

One non-trivial detail still remains to be settled in order to finish the proof.

In the previous computation of $\alpha_n(a,\,u)$, the dual point u has been chosen on some line x not coplanar with a. We must show that, given any dual point u and any point a not in u, we can always find a line x in u not coplanar with a, provided that a non-coplanar point-line pair exists (as we have assumed).

We also prove this by induction. If $n = 2$, there is nothing to prove. Let $n > 2$ and let a, u be a point and a dual point, respectively, such that $a \notin u$ but a is coplanar with all the lines of u. Then a is collinear with all the points of u. Let b be a point of u and let x be the line on a and b. As all lines of u are coplanar with a, all the lines of u on b are coplanar with x, by Int. By the inductive hypothesis, one of the following holds in Γ_b^+:

(i) there is some plane w in u on b which is not contained in a common 3-space with x (or which does not contain x, if $x = 3$);

(ii) every plane and every line on b are contained in a common 3-space ($n \geq 4$ in this case).

Assume (i). We have

$$\beta_{n-1}(x,\,w) = \frac{(\alpha - 1)(1 + s + \ldots + s^{n-3}) + 1}{(\alpha - 1)(1 + s + \ldots + s^{n-4}) + 1} \quad \text{(if } n \geq 4\text{)};$$

$$\beta_{n-1}(x,\,w) = \alpha \quad \text{(if } n = 3\text{)}.$$

However, as x is coplanar with all the lines of u on b, we have $\beta_{n-1}(x,\,w) = s + 1$. This immediately contradicts the above relation for $\beta_{n-1}(x,\,w)$ in the case $n \geq 4$. Therefore $n = 3$ and $\alpha = s + 1$. This means that Γ belongs to the diagram $A_2.L$ (§2.1). Therefore Γ is a truncated projective geometry (Theorem 3; here we keep the convention stated in §2.1, considering projective geometries as 0-truncations of themselves). Hence no non-coplanar point-line pairs exist in Γ, contrary to our hypothesis on it. Thus (ii) occurs.

We now need to prove the following:

Subsidiary statement. Let Γ be a geometry in $A_k.pG$, satisfying Int, and assume that no non-coplanar point-line pair exists in Γ. Then Γ is a truncated projective geometry, of rank $n \geq 3$.

We also prove this by induction on rank $n = \text{rank}(\Gamma)$. If n - 2, there is nothing to prove. Otherwise, fix a point p of Γ. As above, one of (i), or (ii) holds in Γ_p^+. If (ii) holds, then Γ_p^+ is a truncated projective geometry by the inductive hypothesis and Γ is a truncated projective geometry (of rank $n \geq 4$) by Theorem 3. If (i) occurs, then we obtain that Γ is a truncated projective geometry of rank 3, as we have done for Γ above. The statement is proved.

By this statement, we have that Γ_b^+ is a truncated projective geometry if (ii) occurs for it. Hence Γ is a truncated projective geometry by Theorem 3 and we have the contradiction again.
The proof is finished now. $\square$

Also, we have obtained the following as a by-product.

Lemma 30. *If Γ is not a (possibly truncated) projective geometry, then for every dual point u and for every point $a \notin u$, there are lines of u that are not coplanar with a.*

Remark 31. The possibility that some dual point is incident with all points is clearly excluded by Int.

Proposition 32. *Assume that Γ is not a (possibly truncated) projective geometry. Then α divides s. Also, if $n \geq 4$, then $\alpha = 1$ or s.*

Proof. By Lemma 30, we are in the hypotheses of Lemma 29, so the divisibility conditions implicit in (2) of that lemma can be exploited. If $n = 3$, (2) of Lemma 29 yields $\beta_3(a, x) = s + 1 - s/\alpha$ Whence α divides s. Define $h = s/\alpha$. If $n = 4$, by (2) of Lemma 29 we have:

$$\beta_4(a, x) = s + \frac{1 - h}{1 + s + h}$$

whence $1 + s - h$ divides $1 - h$. Considering that $s = \alpha h$, it is easily seen that $h = s$ and $h = 1$ are the only possibilities. $\square$

Lemma 1 is a trivial consequence of the proposition above.

We can also obtain a well known result on linear space as a consequence of Proposition 32.

Corollary 33 (Doyen and Hubaut 1971). *Let Γ be a geometry belonging to the following diagram*

$$(\text{L.A}_m) \qquad \overset{L}{\underset{r}{\circ}\!\!-\!\!-\!\!\underset{s}{\circ}}\!\!-\!\!-\!\!\underset{s}{\circ}\cdots\underset{s}{\circ}\!\!-\!\!-\!\!\underset{s}{\circ}$$

where r, s are finite parameters and $m \geq 2$. Then either Γ is a projective geometry or $r + 1$ divides s. Furthermore, if $m \geq 3$, then Γ is either a projective geometry or an affine geometry.

If $s > 1$, apply Proposition 32 to the dual of Γ (note that Int holds in Γ, see Proposition 1 of Lunardon and Pasini (to appear) and use Theorems 2 and 7 to conclude. If $s = 1$, then the above diagram becomes A_{m+1} and the conclusion follows from Theorem 2.

Remark 34. Corollary 33 could also be proved directly, by some easy counting arguments on the flags of Γ. On the other hand, the main part of the result of Doyen and Hubaut (1971) is concerned with the rank 3 case ($m = 2$), which is not so trivial. They prove that $s = (r + 1)^2$ and $s = (r + 1)^3 + (r + 1)$ are the only possibilities when $m = 2$, apart from $s = r$ and $s = r + 1$, of course (corresponding to projective and affine geometries respectively). Actually only one example is known with these exceptional parameters, namely the Steiner system $S(22, 6, 3)$ for M_{22} (we have $r = 1$ and $s = (r + 1)^2 = 4$ in this case).

Corollary 33 has some trivial, but relevant consequences. For instance, most of the theorems of §2 can be stated in a more general form for the following diagrams instead of the more special ones considered in §2, provided that the finite parameters are assumed and that the diagram contains at least two 'projective' strokes $\circ\!\!-\!\!-\!\!\circ$;

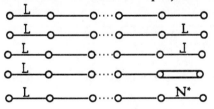

4. The main theorem

We are now ready to prove the following

Theorem 35. *Let Γ be a geometry satisfying Int and belonging to the diagram* $pG.A_m.pG$ *of §3 with $m \geq 3$ and r, s, t as in §1. Then either Γ is one of the geometries of the list of §2.1 or it belongs to one of the following diagrams;*
(i) Af.A_m.N* (§2.5) *with $s = 2$ or 3;*
(ii) N.A_m.N* (§2.5);
(iii) N.C_{m+1} (§2.7).

Indeed, we have just to apply Lemma 1 (namely Proposition 32) and to consider the possibilities allowed by it. Each of the cases that arise is covered by some of the theorems in §2, except for the three cases (i), (ii) and (iii) above.

5. Other examples

5.1   Remarks on the rank 4 case

As we noticed in Remark 34, only one example is known in L.A_2 which is not a projective or affine geometry, namely the Steiner system $S(22, 6, 3)$, belonging to

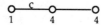

where $\underset{}{\circ}\!\!\overset{c}{\rule{1cm}{0.4pt}}\!\!\underset{}{\circ}$ denotes the class of finite linear spaces with lines of size 2. It has been proved by Hughes (1965) that $S(22, 6, 3)$ and $AG(n, 2)$ are indeed the only possibilities arising in the following diagram

(c.A_{n-1})

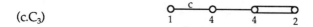

$(n \geq 3, s > 1)$

Two $pG.A_2.pG$ geometries arise in connection with $S(22, 6, 3)$, namely a geometry for the group Fi_{22} belonging to the following diagram (Buekenhout 1985)

(c.C_3)

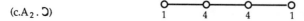

and a geometry for the group HS belonging to the following diagram (Buekenhout 1985)

(c.A_2.Ɔ)

where o——$^{\supset}$——o is the dual of o——^{C}——o, of course.   Both these geometries satisfy Int.   It may be that these are indeed the only exceptional cases that can arise in pG.A$_2$.pG when Int is assumed and $s > 1$.   Anyway, the geometry for HS is in fact the only exceptional case arising in c.A$_2$.$\supset$ when Int is assumed, the 'regular' cases being the two biaffine spaces obtained from $PG(4, 2)$ (Hughes this volume).

5.2   A few more examples related to some of the geometries of §2, but not in pG.A$_m$.pG

Let us turn back to the construction described at the end of §2.7, starting from $P_1 = PG(2, q)$ and either a partial geometry Π realised inside P_1 or a maximal arc K of P_1.   If we embed P_1 in a projective geometry $P = PG(n, q)$ of arbitrary dimension (instead of $n = 3$ or 4 as we did there) and we also add all the subspaces of P that do not meet P_1, then we obtain geometries with arbitrarily long diagrams, as follows:

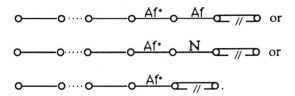

Clearly, the ways of specializing the last stroke ⊂⎯//⎯⊃ depend on what Π or K are.

We have already met an example obtained from generalized Fischer spaces in §2.7.   We notice that many of the generalized Fischer spaces (Cuypers 1989a) are in fact affine polar spaces or minimal standard quotients of affine polar spaces.   However, other interesting examples arise in that context.   Here is one of them ($\overline{Sp(V, f)}$ in Cuypers 1989a).   Start from a non-trivial symplectic form f on a finite dimensional vector space V and let $P(V)$ be the projective geometry of the linear subspaces of V.   Take the points of $P(V)$ and all the proper subspaces of $P(V)$ on which f is non-trivial and drop the rest.   The geometry obtained in this way belongs to the following diagram

o——$\overset{Af^*}{}$——o——$\overset{Af}{}$——o————o····o————o

Let us turn back to the construction of §2.5.   In some sense, the basic idea of that construction is nothing but selecting those subspaces of $PG(m, q)$ that have 'maximal distance' from both X and Y.   We can apply the same idea in other contexts too. For instance, start from a polar space Γ, fix a point p of it and take all the singular subspaces of Γ that have maximal distance from p, compatible with their rank. What we obtain in this way is the affine polar space $\Gamma = \Gamma - H$, where $H = p^\perp$.

If we do the same in building of type F_4 starting from a point p of it, then we obtain

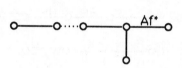

If we work with a point p and a symp S in a F_4 building, taking the elements that, compatible with their type, have maximal distance from both p and S, then we obtain the following

If we start from a building of type D_n, choose a maximal singular subspace M and select those singular subspaces that have maximal distance from M (compatibly with their type), then what we obtain belongs to

Is it possible to collect all the above into one precise and general construction? Also, would such a construction give us the attenuated polar spaces Γ/X where $X = Y^\perp$ for some singular subspace Y of Γ?

5.3   Examples with different kinds of residues

We have mentioned an example in Remark 27, described in Hobart and Hughes (1990) and Del Fra et al. (to appear). It belongs to pG.A$_2$ and the stroke ⊄⫽⊐ is realized in two ways in it; namely as ⊂⊐ and as o—o. The parameters are as follows:

This example arises as J-shadow space from a geometry belonging to the following disconnected diagram, where J is the pair of types as marked in the picture (the reader is referred to Buekenhout 1981; or Tits 1974, Ch. 12 for the definition of shadow spaces).

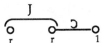

When $r = 1$, this example also appears as top residue of rank 3 inside geometries of longer diagram constructed as follows. Start from a Coxeter complex belonging to a diagram of the following form and take the $\underline{1}$ -shadow space

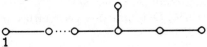

in particular

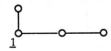

The following is obtained

where all parameters are equal to 1 and is realized as C_2 and A_2. Clearly, we can apply a similar trick to thin geometries instead of Coxeter complexes, if we like.

The case of $r = 2$ is met as top rank 3 residue in geometries constructed by the same trick as above, starting from the following

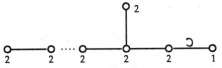

the resulting geometry has parameters as follows

and is realized as and as .

The example discussed above can also be realized as the dual of a tessellation of a sphere in 6 triangular faces (this is indeed the description given in Hobart and Hughes (1990)). Clearly, a lot of examples of pG.pG geometries with different kinds of residues in the same position of the diagram can be obtained from tessellations of closed surfaces in triangles and squares such that every vertex belongs to either 3 or 4 faces. All examples of this kind are thin, of course.

References

M. Biliotti and A. Pasini (1982), Intersection properties in geometries, *Geom. Dedicata* **13**, 257-275.

R. Bose (1963), Strongly regular graphs, partial geometries and partial balanced designs, *Pacific J. Math.* **13**, 389-419.

F. Buekenhout (1969), Une charactérization des espaces affins basée sur la notion de droite, *Math. Z.* **3**, 367-371.

F. Buekenhout (1979), Diagrams for geometries and groups, *J. Combin. Theory Ser. A* **27**, 121-151.

F. Buekenhout (1981), The basic diagram of a geometry, *Geometries and Groups*, Lecture Notes in Math. **893**, 1-25, Springer-Verlag, Berlin.

F. Buekenhout (1985), Diagram geometries for sporadic groups, *Finite Groups — Coming of Age*, Contemp. Math. **45**, 1-32, American Mathematical Society, Providence.

P. Cameron, D. Hughes and A. Pasini (1990), Extended generalized quadrangles, *Geom. Dedicata* **35**, 193-228.

A. Cohen and E. Shult (1990), Affine polar spaces, *Geom. Dedicata* **35**, 43-76.

H. Cuypers (1989a), Geometries and permutation groups of small rank, Part II (generalized Fischer spaces), Thesis, State Univ. Utrecht.

H. Cuypers (1989b), Geometries and permutation groups of small rank, Part III (the dual of Pasch axiom), Thesis, State Univ. Utrecht.

H. Cuypers (to appear), The dual of Pasch axiom, *European J. Combin.*

H. Cuypers (submitted), Locally generalized quadrangles with affine planes.

H. Cuypers and A. Pasini (submitted), Locally polar geometries with affine planes.

F. De Clerck (to appear), Some classes of rank 2 geometries, in *Handbook of Incidence Geometry*, (ed. F. Buekenhout), North-Holland, Amsterdam.

F. De Clerck and J.A. Thas (1978), Partial geometries in finite projective spaces, *Arch. Math.* **30**, 537-540.

A. Del Fra, D. Ghinelli and A. Pasini (to appear), On locally partial geometries with different types of residues, *Proc. Combinatorics '90, Gaeta* .

A. Del Fra, D. Ghinelli and A. Pasini (submitted), Affine attenuated spaces.

A. Del Fra, D. Ghinelli, T. Meixner and A. Pasini (to appear), Flag-transitive extensions of C_n geometries, *Geom. Dedicata*.

P. Dembowski (1968), *Finite Geometries*, Springer-Verlag, Berlin.

M. Deza and M. Laurent, Bouquets of matroids, d-injection geometries and diagrams, *J. Geom.* **29**, 13-35.

J. Doyen and X. Hubaut (1971), Finite regular locally projective spaces, *Math. Z.* **119**, 83-88.

S. Hobart and D.R. Hughes (1990), Extended partial geometries: nets and dual nets, *European J. Combin.* **11**, 357-372.

D.R. Hughes (1965), Extensions of designs and groups: projective, symplectic and certain affine groups, *Math. Z.* **89**, 199-205.

D.R. Hughes (1990), Extended partial geometries: dual 2-designs, *European J. Combin.* **11**, 459-471.

D.R Hughes (to appear a), On partial geometries of rank n, *Proc. Combinatorics '90, Gaeta.*

D.R. Hughes (to appear b), On $C_2.c$ geometries.

D.R. Hughes (this volume), On some rank 3 partial geometries.

D.R. Hughes and F. Piper (1985), *Design Theory*, Cambridge University Press, Cambridge.

W.M. Kantor (1974), Dimension and embedding theorems for geometric lattices, *J. Combin. Theory Ser. A* **17**, 173-195.

W.M. Kantor (1979), Locally polar lattices, *J. Combin. Theory Ser. A* **26**, 90-95.

C. Lefévre-Percsy (1990), Infinite (Af, Af*)-geometries, *J. Combin. Theory Ser. A* **55**, 133-139.

C. Lefévre-Percsy and L. Van Nypelseer (to appear), Finite rank 3 geometries with affine planes and dual affine point residues, *Discrete Math.*

G. Lunardon and A. Pasini (to appear), Finite C_n-geometries (a survey), *Note Mat.*

D.V. Pasechnick (to appear), Dual linear extensions of generalized quadrangles, *European J. Combin.*

A. Pasini (1988), Geometric and algebraic methods in the classification of geometries belonging to Lie diagrams, *Ann. Discrete Math.* **37**, 315-356.

S. Payne and J.A. Thas (1984), *Finite Generalized Quadrangles*, Pitman, London.

M. Ronan (1982), Locally truncated buildings and M_{24}, *Math. Z.* **180**, 489-501.

M. Ronan (1986), extending locally truncated buildings and chamber systems, *Proc. London Math. Soc.* **53**, 385-406.

A. Sprague (1981), Incidence structures whose planes are nets, *European J. Combin.* **2**, 193-204.

A. Sprague (1985), Rank 3 incidence structures admitting dual-linear, linear diagram, *J. Combin. Theory Ser. A* **38**, 254-259.

J. Thas (1973), Construction of partial geometries, *Simon Stevin* **46**, 95-98.

J. Thas (1974), Construction of maximal arcs and partial geometries, *Geom. Dedicata* **3**, 61-64.

J. Thas (1978), Partial geometries in finite affine spaces, *Math. Z.* **158**, 1-13.

J. Thas (1980), Partial three-spaces in finite projective spaces, *Discrete Math.* **32**, 299-322.

J. Thas and F. De Clerck (1977), Partial geometries satisfying the axiom of Pasch, *Simon Stevin* **51**, 123-137.

J. Tits (1974), *Buildings of Spherical Type and Finite BN-pairs*, Lecture Notes in Math. **386**, Springer-Verlag, Berlin.

J. Tits (1981), A local approach to buildings, *The Geometric Vein*, 519-547, Springer-Verlag, Berlin.

J. Tits (1986), Immeubles de type affine, in *Buildings and the Geometry of Diagrams*, Lecture Notes in Math. **1181**, 159-190, Springer-Verlag, Berlin.

L. Van Nypelseer (to appear), Rank m geometries with affine hyperplanes and dual affine point residues, *European J. Combin.*

On an exceptional semi-translation plane

M.J. de Resmini
University of Rome I

1. Introduction

In 1970 N.L. Johnson (1970) constructed an exceptional semi-translation plane of order 16, say π, which is of Lenz-Barlotti class 1-5a and can be derived from a dual translation plane, namely the semifield plane of order 16 with kern $GF(4)$.

We investigate the combinatorial structure of π; more precisely, its subplanes and complete arcs. Since π is a semi-translation plane and a derived plane, it has a distinguished line, its line at infinity and a special set of points on this line, the derivation set Δ.

We show that π is generated by quadrangles and a generating quadrangle can have up to two vertices on the line at infinity. If a quadrangle does not generate π then it generates a Fano subplane of π. The Fano subplanes can be classified by looking at their behaviour with respect to the line at infinity and the three possible cases occur. However, we can also classify the Fano subplanes by looking at the number of Baer subplanes to which they may be completed. Such an approach yields two types of subplanes of order 2: the maximal ones and those which complete to a unique Baer subplane. All Baer subplanes of π have the same points at infinity, namely the points of Δ.

As to complete arcs, π contains small complete k-arcs of the usual sizes in planes of order 16, namely $k = 10, 11, 12, 13$ (de Resmini (1990) and its references). Far more interesting are the complete 14-arcs, all of which are hyperbolic. They occur in all non-Desarguesian translation planes of order 16; however, in π they form some nice configurations consisting of pairs, triples and quadruples of arcs, which share eight finite points and possibly some point at infinity. Also, there exist distinct projectively inequivalent types of complete 14-arcs (§6).

Finally, the hyperovals in π are formed by glueing together hyperovals belonging to distinct Baer subplanes and sharing their points at infinity.

2. Notation and terminology

The description of π in Johnson (1970) allows us to write down its lines as sets of their points. The points of π are denoted by $A0$, Aj, Bj, Cj, Dj, Fj, Hj, Kj, Lj, Mj, Nj, Pj, Rj, Sj, Tj, Wj, Xj, Zj, $(j = 1, 2, ..., 16)$ and the lines by the corresponding lower case letters and same subscripts. The line at infinity is $a0$ and its points are $A0$, Aj. The remaining lines on $A0$ are the lines aj, where $a1$ contains the points Bj, $a2$ the Cj, ..., $a16$ the Zj. Similarly, the lines on $A1$ are the lines bj and the finite points on bj are all points with subscript j. Next, the lines on $A2$ are the cj, those on $A3$ the dj, ..., those on $A16$ the zj. To describe the lines we take into account that π admits an involution $\wp$ which acts on finite points and lines by keeping the letters fixed and pairing off the subscripts as follows: $(1, 2)$, $(3, 6)$, $(4, 10)$, $(5, 16)$, $(7, 12)$, $(8, 15)$, $(9, 11)$, $(13, 14)$. Therefore, in Table 1, we write down the subscripts of the finite points of half the lines cj, ..., zj. The letters used are to be inserted in alphabetic order. For instance, from the Table, we read

f3:　　A4　B4　C1　D13　F11　H3　K10　L9　M6　N8　P7　R2　S14　T12　W15　X5　Z16

and by applying $\wp$ we get

f6:　　A4　B10　C2　D14　F9　H6　K4　L11　M3　N15　P12　R1　S13　T7　W8　X16　Z5.

With our notation, the derivation set Δ consists of the points $A0$, $A1$, $A2$, $A7$, $A12$.

We observe that π admits another involution $\mathfrak{S}$, which acts on finite points by keeping the subscripts fixed and pairing off the letters as follows: (B, C), (D, K), (F, P), (H, Z), (L, S), (M, X), (N, R), (T, W). Both $\wp$ and $\mathfrak{S}$ fix $a0$ pointwise. The involutions $\wp$ and $\mathfrak{S}$ play major roles in the construction of hyperovals.

Table 1.

c1:	1	2	3	4	5	6	7	8	9	10	11	12	13	14	15	16
c3:	3	6	1	7	11	2	4	13	16	12	5	10	8	15	14	9
c4:	4	10	7	1	8	12	3	5	14	2	13	6	11	9	16	15
c5:	5	16	11	8	1	9	13	4	6	15	3	14	7	12	10	2
c7:	7	12	4	3	13	10	1	11	15	6	8	2	5	16	9	14
c8:	8	15	13	5	4	14	11	1	12	16	7	9	3	6	2	10
c9:	9	11	16	14	6	5	15	12	1	13	2	8	10	4	7	3
c13:	13	14	8	11	7	15	5	3	10	9	4	16	1	2	6	12

d1:	1	3	7	6	4	15	13	2	16	14	10	8	12	11	9	5
d3:	3	1	15	14	5	7	8	9	10	6	16	13	11	12	2	4
d4;	4	7	9	16	13	1	11	15	6	10	14	5	8	2	12	3
d5:	11	5	16	9	8	10	4	14	12	1	15	7	13	3	6	2
d7:	7	4	1	10	3	9	5	12	14	16	6	11	2	8	15	13
d8:	13	8	6	7	12	14	1	10	9	15	2	3	4	5	16	11
d9:	16	9	4	2	1	5	12	3	8	11	7	10	6	14	13	15
d13:	8	13	14	15	11	6	3	16	2	7	9	1	5	4	10	12
f1:	1	4	10	7	16	13	14	5	2	11	8	9	15	12	6	3
f3:	4	1	13	11	3	10	9	6	8	7	2	14	12	15	5	16
f4:	3	7	5	8	4	6	15	10	11	1	12	16	2	9	13	14
f5:	15	16	8	5	2	1	3	12	13	6	10	7	4	14	11	9
f7:	7	3	6	1	14	5	16	13	12	8	11	15	9	2	10	4
f8:	14	9	7	10	15	11	1	8	6	13	5	4	16	3	2	12
f9:	5	8	2	3	11	15	12	9	4	16	14	6	13	10	7	1
f13:	9	14	11	13	12	7	4	2	5	10	6	1	3	16	8	15
h1:	1	14	5	2	11	8	9	15	12	6	3	4	10	7	16	13
h3:	14	1	8	6	13	5	4	16	3	2	12	9	7	10	15	11
h4:	16	7	11	10	5	13	3	14	4	12	2	15	1	6	9	8
h5:	3	15	10	11	1	12	16	2	9	13	14	7	5	8	4	6
h7:	7	16	13	12	8	11	15	9	2	10	4	3	6	1	14	5
h8:	9	4	2	5	10	6	1	3	16	8	15	14	11	13	12	7
h9:	8	6	7	14	3	4	12	10	13	9	11	5	15	16	1	2
h13:	4	9	6	8	7	2	14	12	15	5	16	1	13	11	3	10
k1:	1	6	9	5	8	7	16	4	15	3	2	11	14	10	13	12
k3:	6	1	7	3	12	9	11	13	2	5	15	16	10	14	4	8
k4:	10	7	1	4	2	15	8	5	12	13	9	14	6	16	3	11
k5:	8	14	4	1	6	13	10	9	3	15	5	7	2	11	12	16
k7:	7	10	15	13	11	1	14	3	9	4	12	8	16	6	5	2
k8:	16	11	5	9	14	3	1	2	13	7	4	6	8	12	15	10
k9:	13	15	14	8	5	10	12	7	16	2	6	4	9	1	11	3
k13:	11	16	3	7	10	5	6	15	4	9	13	1	12	8	2	14
l1:	1	7	8	9	10	11	12	13	14	15	16	2	3	4	5	6
l3:	3	4	13	16	12	5	10	8	15	14	9	6	1	7	11	2
l4:	4	3	5	14	2	13	6	11	9	16	15	10	7	1	8	12
l5:	5	13	4	6	15	3	14	7	12	10	2	16	11	8	1	9
l7:	7	1	11	15	6	8	2	5	16	9	14	12	4	3	13	10
l8:	8	11	1	12	16	7	9	3	6	2	10	15	13	5	4	14
l9:	9	15	12	1	13	2	8	10	4	7	3	11	16	14	6	5
l13:	13	5	3	10	9	4	16	1	2	6	12	14	8	11	7	15

m1:	1	8	12	11	9	5	3	7	6	4	15	13	2	16	14	10
m3:	8	1	5	4	10	12	13	14	15	11	6	3	16	2	7	9
m4:	11	7	13	3	6	2	5	16	9	8	10	4	14	12	1	15
m5:	5	4	3	13	14	8	11	10	1	2	16	7	6	15	9	12
m7:	7	11	2	8	15	13	4	1	10	3	9	5	12	14	16	6
m8:	3	13	11	12	2	4	1	15	14	5	7	8	9	10	6	16
m9:	10	16	15	1	7	6	12	11	5	14	2	9	8	3	4	13
m13:	13	3	4	5	16	11	8	6	7	12	14	1	10	9	15	2
n1:	1	15	6	13	2	9	10	16	8	12	4	5	11	3	7	14
n3:	15	1	9	12	14	6	5	7	4	13	8	10	3	11	16	2
n4:	9	7	15	2	16	10	13	1	3	5	11	6	4	8	14	12
n5:	13	6	2	15	4	5	9	11	14	10	1	7	16	12	3	8
n7:	7	9	10	5	12	15	6	14	11	2	3	13	8	4	1	16
n8:	10	5	13	6	11	12	1	4	7	9	16	15	2	14	8	3
n9:	3	14	16	4	15	1	12	6	2	8	13	11	7	5	9	10
n13:	5	10	12	9	3	13	15	8	16	6	7	1	14	2	4	11
p1:	1	10	16	8	12	4	5	11	3	7	14	15	6	13	2	9
p3:	10	1	4	7	9	16	15	2	14	8	3	5	13	6	11	12
p4:	6	7	3	1	15	14	9	12	16	11	4	13	5	10	8	2
p5:	9	13	1	3	5	11	6	4	8	14	12	7	15	2	16	10
p7:	7	6	14	11	2	3	13	8	4	1	16	9	10	5	12	15
p8:	5	15	8	16	6	7	1	14	2	4	11	10	12	9	3	13
p9:	14	11	9	13	4	2	12	5	7	6	15	3	1	8	10	16
p13:	15	5	7	4	13	8	10	3	11	16	2	1	9	12	14	6
r1:	1	9	15	12	6	3	4	10	7	16	13	14	5	2	11	8
r3:	9	1	3	16	8	15	14	11	13	12	7	4	2	5	10	6
r4:	15	7	4	14	11	9	16	8	5	2	1	3	12	13	6	10
r5:	16	3	14	4	12	2	15	1	6	9	8	7	11	10	5	13
r7:	7	15	9	2	10	4	3	6	1	14	5	16	13	12	8	11
r8:	4	14	12	15	5	16	1	13	11	3	10	9	6	8	7	2
r9:	6	5	1	11	14	13	12	16	15	10	3	8	4	9	2	7
r13:	14	4	16	3	2	12	9	7	10	15	11	1	8	6	13	5
s1:	1	12	13	14	15	16	2	3	4	5	6	7	8	9	10	11
s3:	3	10	8	15	14	9	6	1	7	11	2	4	13	16	12	5
s4:	4	6	11	9	16	15	10	7	1	8	12	3	5	14	2	13
s5:	5	14	7	12	10	2	16	11	8	1	9	13	4	6	15	3
s7:	7	2	5	16	9	14	12	4	3	13	10	1	11	15	6	8
s8:	8	9	3	6	2	10	15	13	5	4	14	11	1	12	16	7
s9:	9	8	10	4	7	3	11	16	14	6	5	15	12	1	13	2
s13:	13	16	1	2	6	12	14	8	11	7	15	5	3	10	9	4

t1:	1	13	2	16	14	10	8	12	11	9	5	3	7	6	4	15
t3:	13	1	10	9	15	2	3	4	5	16	11	8	6	7	12	14
t4:	5	7	6	15	9	12	4	3	13	14	8	11	10	1	2	16
t5:	4	11	15	6	10	14	5	8	2	12	3	7	9	16	13	1
t7:	7	5	12	14	16	6	11	2	8	15	13	4	1	10	3	9
t8:	8	3	16	2	7	9	1	5	4	10	12	13	14	15	11	6
t9:	9	10	13	7	2	8	12	14	6	3	1	16	5	11	15	4
t13:	3	8	9	10	6	16	13	11	12	2	4	1	15	14	5	7

w1:	1	16	4	15	3	2	11	14	10	13	12	6	9	5	8	7
w3:	16	1	2	13	7	4	6	8	12	15	10	11	5	9	14	3
w4:	14	7	12	5	1	3	10	11	2	9	6	8	13	15	16	4
w5:	10	8	5	12	13	9	14	6	16	3	11	7	1	4	2	15
w7:	7	14	3	9	4	12	8	16	6	5	2	10	15	13	11	1
w8:	11	6	15	4	9	13	1	12	8	2	14	16	3	7	10	5
w9:	15	4	11	6	8	16	12	1	9	7	5	13	10	2	3	14
w13:	6	11	13	2	5	15	16	10	14	4	8	1	7	3	12	9

x1:	1	11	14	10	13	12	6	9	5	8	7	16	4	15	3	2
x3:	11	1	12	8	2	14	16	3	7	10	5	6	15	4	9	13
x4:	8	7	2	11	12	16	14	4	1	6	13	10	9	3	15	5
x5:	14	10	11	2	9	6	8	13	15	16	4	7	12	5	1	3
x7:	7	8	16	6	5	2	10	15	13	11	1	14	3	9	4	12
x8:	6	16	10	14	4	8	1	7	3	12	9	11	13	2	5	15
x9:	4	13	3	5	6	9	12	2	10	1	8	15	16	7	14	11
x13:	16	6	8	12	15	10	11	5	9	14	3	1	2	13	7	4

z1:	1	5	11	3	7	14	15	6	13	2	9	10	16	8	12	4
z3:	5	1	14	2	4	11	10	12	9	3	13	15	8	16	6	7
z4:	13	7	16	12	3	8	6	2	15	4	5	9	11	14	10	1
z5:	6	9	12	16	11	4	13	5	10	8	2	7	3	1	15	14
z7:	7	13	8	4	1	16	9	10	5	12	15	6	14	11	2	3
z8:	15	10	3	11	16	2	1	9	12	14	6	5	7	4	13	8
z9:	11	3	10	15	13	7	12	8	1	5	4	14	2	6	16	9
z13:	10	15	2	14	8	3	5	13	6	11	12	1	4	7	9	16

3. Quadrangles

The plane π is generated by quadrangles. Also, since π is not a translation plane, a generating quadrangle can have up to two vertices on $a0$, as well as two of its diagonal points on $a0$. There are also generating quadrangles with two vertices in Δ. Next we list some generating quadrangles:

C3 D7 F16 K1, H3 L9 W2 Z4, C1 D7 K2 P15, C1 D7 P15 S9, L12 M16 R14 X10, H14 L4 D10 F8, H14 L4 F8 R7, A3 D9 H16 F8, A3 A5 F12 P5, A7 A12 F5 L13, K9 Z16 M16 P2 (diagonal points A5 A9), A0 A7 D11 L14, A0 B3 B4 C3, F3 F12 T3 T12, K3 K7 L3 L7, B5 B6 T5 T6, A0 A1 M3 T7, A0 A2 C7 Z3, C1 F14 N12 K10 (diagonal points A3 A14).

Whenever a quadrangle does not generate π, it generates a Fano subplane of π. For completeness, we also list some such quadrangles (see §4):

A3 D9 H16 M7, A0 A1 M9 N2, B1 B2 C1 C2, B3 B6 D3 D6, F7 K2 F5 K14, F3 F6 P3 P6, D3 D6 W3 W6, F1 F2 P1 P2, N3 N6 R3 R6, B3 B7 C1 C4, D4 D10 W4 W10, F4 F10 P4 P10, C3 T15 C15 T3, F2 F8 Z2 Z8, P7 P8 W7 W8, C7 C12 L7 L12, D13 D16 F13 F16.

We observe that the number of quadrangles which generate π is larger than the number of those generating Fano subplanes. In order to have an idea of the distribution of these two types of quadrangles, we started with a given triangle and let the fourth vertex of the quadrangle range over a prescribed line. The following results confirm the previous observation.

Starting triangle: A0 L3 M7, the fourth vertex ranges over f4. Only with A4 do we obtain a quadrangle generating a Fano subplane and this is maximal (§4).

Starting triangle: D3 N3 N9, fourth vertex on m5 (D3 lies on m5). There is a generating quadrangle for a Fano subplane only with K8. This subplane is exterior to a0 (§4).

Starting triangle: A7 L4 S8, fourth vertex on b3. Only L3 yields a generating quadrangle for a Fano subplane which is tangent to a0 and maximal (§4).

Starting triangle: A0 L3 M7, fourth vertex on a5. Generating quadrangles for Fano subplanes are obtained only with H6 and H13. Both subplanes are tangent to a0 and maximal (§4).

Starting triangle: F4 F7 T4, fourth vertex on r2. Fano subplane only with A11; it is tangent to a0 and maximal (§4).

Starting triangle: M5 M9 W5, fourth vertex on h3. The points C1 and Z11 provide generating quadrangles for Fano subplanes which are exterior to a0, whereas P2 yields a generating quadrangle for a Fano subplane which is tangent to a0 and completes to a Baer subplane (§4).

All remaining points of h3 give generating quadrangles for π.

4. Subplanes

Since π is a derived plane, it contains Baer subplanes. Also, the way π is constructed shows that all its Baer subplanes have the same five points on a0, namely the points in Δ, that is A0 A1 A2 A7 A12. As already observed, π contains Fano subplanes and its Baer subplanes can be obtained by extending Fano subplanes.

We can classify the Fano subplanes of π in a natural way by looking at their behaviour with respect to the line at infinity. Thus we obtain secant, tangent and exterior subplanes according to the number, three, one or zero, of points on a0. Such a classification can be refined by taking into account the number of points in Δ. Consequently, a secant Fano subplane can have zero, one, two or three points in Δ. However, no secant Fano subplane was found with no point in Δ. Similarly, there are two types of tangent Fano subplanes.

On the other hand, we can also classify the Fano subplanes of π according to the number of Baer subplanes to which they complete. The situation in π is completely different from that occuring in the non-desarguesian translation planes of order 16 where a Fano subplane can extend to more than one Baer subplane, up to seven in the Johnson-Walker plane and in the derived semifield plane (de Resmini (1990) and its references). More precisely, a Fano subplane of π is either maximal, or extends to a unique Baer subplane. Since the points at infinity of all Baer subplanes are known, some maximal Fano subplanes are easily recognised.

Next, we provide some examples of Fano and Baer subplanes. We just list their points and lines; the incidences can be easily obtained with the help of Table 1.

(i) Fano subplanes which are exterior to a0.

All such subplanes are obviously maximal.

Points: D3 K8 L3 N1 N3 N9 T13;   lines: a9 b3 c1 m5 n12 r7 x8.
Points: C1 H5 L2 M5 M9 M11 W5;   lines: a8 b5 c12 d3 l7 r3 z15.
Points: C7 D5 M5 M9 M10 W5 Z11;   lines: a8 b5 c12 d8 f4 k4 p2.

(ii) Tangent Fano subplanes to a0.

By the previous argument, all those Fano subplanes which are tangent to a0 at a point in a0-Δ are maximal. Those tangent to a0 at a point in Δ are either maximal or complete to a unique Baer subplane. In the latter case, we list the points not in Δ and the lines other than a0 to be added in order to get the Baer subplane.

Points: A3 B4 D9 F4 H16 M7 N4; lines: b4 d4 d6 d12 h10 n3 s4, maximal.

Points: A7 L3 L4 L11 N11 R11 S8;   lines: a7 b11 l6 l10 l15 t3 z6, maximal.

Points:A0 H6 H9 L3 L16 M1 M7;   lines: a5 a7 a8 d12 l13 m1 z10, maximal.

Points: A0 H9 H13 L3 L15 M2 M7;   lines: a5 a7 a8 m1 p3 s6 t2, maximal.

Points: A11 F4 F7 F11 H12 M4 T4;   lines: a4 b4 d8 r2 r5 r9 x4,
maximal.
Points: A2 L5 M3 M5 M9 P2 W5;   lines: a8 b5 c4 c12 c13 l8 s16, Baer
with L2 L3 L9 M2 P3 P5 P9 W2 W3 W9, a7 a10 a14 b2 b3 b9 c15 l7 l10 l14
s1 s6 s11.
Points: A2 B7 C7 L1 L2 L7 S12;   lines a7 b7 c1 c7 c12 l7 s2, Baer
with B1 B2 B12 C1 C2 C12 L12 S1 S2 S7, a1 a2 a12 b1 b2 b12 c2 l1 l2 l12 s1
s7 s12.

(iii) Secant Fano subplanes.

We classify them according to the number, one, two or three, of points
they have in Δ.   Obviously, all those secant Fano subplanes having
less that three points in Δ are maximal.

Points: A0 A4 A8 L3 L6 M7 M12;   lines: a0 a7 a8 f5 f16 m1 m2,
maximal.
Points: A1 A8 A16 D13 K9 N9 R13;   lines: a0 b9 b13 m4 m14 z3 z6,
maximal.
Points: A0 A1 A5 M2 M9 N2 N9;   lines: a0 a8 a9 b2 b9 h5 h7,
maximal.
Points: A0 A1 A10 B3 B6 D3 D6;   lines: a0 a1 a3 b3 b6 p4 p10,
maximal.
Points: A0 A2 A3 F5 F7 K2 K14;   lines: a0 a4 a6 c3 c8 d8 d10,
maximal.
Points: A0 A1 A4 F1 F5 K1 K5;   lines: a0 a4 a6 b1 b5 f5 f7,
maximal.
Points: A0 A2 A5 M8 M9 P3 P10;   lines: a0 a8 a10 c1 c12 h2 h7,
maximal.
Points: A0 A1 A9 D3 D6 W3 W6;   lines: a0 a3 a14 b3 b6 n1 n2,
maximal.
Points: A0 A1 A11 L4 L10 M4 M10;   lines: a0 a7 a8 b4 b10 r1 r2,
maximal.
Points: A0 A3 A7 B3 B7 C1 C4;   lines: a0 a1 a2 d3 d7 l3 l7,
maximal.
Points: A0 A1 A9 D4 D10 W4 W10;   lines: a0 a3 a14 b4 b10 n7 n12,
maximal.

The next Fano subplanes all have three points in Δ and extend to a
Baer subplane.   Again, we give the finite points and lines to be added
to obtain the Baer subplane.

Fano subplane: A0 A1 A2 B1 B2 C1 C2, a0 a1 a2 b1 b2 c1 c2; Baer with
B7 B12 C7 C12 Lj Sj; a7 a12 b7 b12 c7 c12 lj sj, j = 1, 2, 7, 12.
Fano subplane: A0 A1 A2 F3 F6 P3 P6, a0 a4 a10 b3 b6 c7 c12; Baer
with F4 F10 P4 P10 D3 D6 D4 D10 K3 K4 K6 K10, a3 a6 b4 b10 c1 c2 l5 l16
l13 l14 s8 s15 s9 s11.

Fano subplane: A0 A1 A2 F1 F2 P1 P2, a0 a4 a10 b1 b2 c4 c10; Baer with F7 F12 P7 P12 D1 D2 D7 D12 K1 K2 K7 K12, a3 a6 b7 b12 c3 c6 l8 l15 l9 l11 s5 s16 s13 s14.

Fano subplane: A0 A1 A2 N3 N6 R3 R6, a0 a9 a11 b3 b6 c5 c16; Baer with N4 N10 R4 R10 M3 M6 M4 M10 X3 X6 X4 X10, a8 a15 b4 b10 c13 c14 l8 l15 l9 l11 s1 s2 s7 s12.

Fano subplane: A0 A1 A2 T9 T11 W9 W11, a0 a13 a14 b9 b11 c4 c10; Baer with T8 T15 W8 W15 H8 H15 H9 H11 Z8 Z15 Z9 Z11, a5 a16 b8 b15 c3 c6 l5 l16 l13 l14 sj.

Fano subplane: A0 A1 A2 H1 H4 M1 M4, a0 a5 a8 b1 b4 c5 c8; Baer with H9 H14 M9 M14 K1 K4 K9 K14 S1 S4 S9 S14, a6 a12 b9 b14 c6 c12 l2 l10 l11 l13 s3 s7 s15 s16.

Fano subplane: A0 A2 A7 B4 B14 W1 W9, a0 a1 a14 c4 c14 l4 l14; Baer with B1 B9 W4 W14 F1 F4 F9 F14 N1 N4 N9 N14, a4 a9 b1 b4 b9 b14 c1 c9 l1 l9 s1 s4 s9 s14.

Fano subplane: A0 A1 A2 C3 C15 T3 T15, a0 a2 a13 b3 b15 c6 c8; Baer with C7 C16 T7 T16 P3 P7 P15 P16 R3 R7 R15 R16, a10 a11 b7 b16 c5 c12 l1 l4 l9 l14 s2 s10 s11 s13.

Fano subplane: A0 A1 A2 F2 F8 Z2 Z8, a0 a4 a16 b2 b8 c5 c10; Baer with F4 F16 Z4 Z16 M2 M4 M8 M16 C2 C4 C8 C16, a2 a8 b4 b16 c1 c15 ß l11 l12 l14 s2 s7 s9 s13.

Fano subplane: A0 A1 A2 P7 P8 W7 W8, a0 a10 a14 b7 b8 c6 c16; Baer with P10 P14 W10 W14 L7 L8 L10 L14 M7 M8 M10 M14, a7 a8 b10 b14 c1 c11 l2 l3 l5 l9 s4 s12 s13 s15.

Fano subplane: A0 A1 A7 D13 D16 F13 F16, a0 a3 a4 b13 b16 ß l10; Baer with D5 D14 F5 F14 K5 K16 K13 K14 P5 P16 P13 P14, a6 a10 b5 b14 c8 c15 c9 c11 sj l4 l6.

Fano subplane: A0 A1 A7 C1 C11 D1 D11, a0 a2 a3 b1 b11 l7 l8; Baer with C6 C16 D6 D16 H1 H6 H11 H16 N1 N6 N11 N16, a5 a9 b6 b16 c2 c3 c5 c9 l10 l14 s4 s12 s13 s15.

5. Small complete arcs

The complete k-arcs for $k = 10, 11, 12, 13$ will be referred to as small complete arcs. Such arcs occur in all planes of order 16 which up to now have been investigated for complete arcs (de Resmini (1990) and its references). Also, 13 is the largest possible size for a complete arc other than a hyperoval in the Desarguesian plane of order 16 (Hirschfeld (1979); Fisher et al. (1986)). Since classifying the small complete arcs is a large and not particularly interesting problem, a few examples are listed for each possible size. We observe that no complete k-arc has been found in a plane of order 16 with $k < 10$.

Complete 10-arcs:

B1 B2 C1 C2 D4 D10 K4 K10 Ai Aj, i,j = 3, 5, 6, 7, 8, 9, 11, 12, 15, 16; B1 B2 S1 S2 M3 M6 N3 N6 A2 Aj, j = 3, 4, 5, 6, 9, 11, 13, 14, 15; B4 B10 D4 D10 M13 M14 Z13 Z14 N7 N12; F3 F6 K3 K6 H5 H16 W5 W16 Ai Aj, i,j = 3, 4, 5, 6, 8, 9, 11, 12, 14, 16; B3 B6 P4 P10 H7 H12 Z13 Z14 D3 D6; A7 A12 L1 L2 M7 M12 N13 N14 J Y, J, Y = B13 B14, D3 D6, X7 X12; A0 A1 B1 C2 D4 K10 F6 P3 J Y, J, Y = H9 Z11 M16 X5; B1 B2 C1 C2 H9 H11 Z9 Z11 Ai Aj, i, j = 3, 4, 6, 7, 9, 11, 13, 14, 15, 16; B1 B2 C1 C2 T8 T15 W8 W15 Ai Aj, i, j = 3, 5, 6, 7, 8, 9, 10, 11, 13, 14; F4 F10 P4 P10 H9 H11 Z9 Z11 Ai Aj, i, j = 3, 4, 5, 9. 10, 11, 12, 13, 15, 16; D3 F4 K10 P6 H15 Z11 T9 W8 Ai Aj, i, j = 0, 1, 3, 4, 5, 8, 11, 13, 14, 16; F7 F12 P7 P12 D5 D16 K5 K16 Ai Aj, i, j = 3, 4, 5, 7, 8, 9, 11, 12, 13, 15.   The hyperoval B1 B2 C1 C2 A7 A12 contained in the Baer subplane whose finite points are Bj Cj Lj Sj completes to a 10-arc with the following quadruples of points: F3 F6 P3 P6, H13 H14 Z13 Z14, M9 M11 X9 X11, N8 N15 R8 R15, T5 T16 W5 W16 (cf. §7), j = 1, 2, 7, 12.

Complete 11-arcs:

C4 C10 K4 K10 F7 F12 Z7 Z12 B13 B14 A14; B3 C6 D4 K10 F2 P1 H5 Z16 B5 C16 A5; B4 B10 D4 D10 M13 M14 Z13 Z14 H1 H2 A4; B3 B6 P4 P10 H7 H12 Z13 Z14 D4 D10 A3; D1 D5 H1 H5 T3 T4 P3 P4 B16 N9 Y, Y = S16 or Z13; C4 C10 F4 F10 B8 B15 P8 P15 M5 M16 A7; C7 C12 H3 H6 K5 K16 M8 M15 J U Y, J U Y = T5 T16 A5 or W5 W16 A4.

Complete 12-arcs:

C4 C10 K4 K10 F7 F12 Z7 Z12 H3 H6 Ai Aj, i, j = 2, 6, 7; B3 B6 D4 K10 F2 P1 H5 Z16 + any of the quadruples: D9 K11 A1 A5, D9 K11 T3 W6, M2 X1 A0 A13; B4 B10 D4 D10 M13 M14 Z13 Z14 + any of the quadruples: K9 K11 A4 A5, K9 K11 W1 W2, P7 P12 Ai Aj, i, j = 3, 6, 7, R5 R16 A3 A4, S7 S12 A4 A7, T1 T2 W1 W2; D8 D15 N8 N15 C4 C10 T4 T10 + one of: B1 B2 Ai Aj, i, j = 2, 3, 8, 13, 16, H3 H6 Ai Aj, i, j = 5, 8, 13, L9 L11 S9 S11; B3 B6 P4 P10 H7 H12 Z13 Z14 + one of: D4 D10 F7 F12, L3 L6 Ai Aj, i, j = 2, 3, 5, R4 R10 A3 A12, T5 T16 A1 A3; A7 A12 L1 L2 M7 M12 N13 N14 P13 P14 F1 F2; B1 B2 C3 C6 F7 F12 P8 P15 + H1 H2 M3 M6, H1 H2 S7 S12, or X8 X15 A5 A9; C4 C10 F4 F10 B8 B15 P8 P15 + H7 H12 A2 A4 or H9 H11 Ai Aj, i, j = 2, 4, 12, 13; C7 C12 H3 H6 K5 K16 M8 M15 + one of the quadruples: D8 D15 Z5 Z16, D8 D15 A4 A14, F8 F15 A5 A10, L7 L12 A4 A10, N13 N14 X13 X14, N13 N14 A1 A4, S13 S14 A1 A5.

Complete 13-arcs:

B3 B6 P4 P10 H7 H12 Z13 M7 M12 R13 R14 A2 Z14; C3 C6 F7 F12 P8 P15 M8 M15 N3 B1 B2 N6 A9; C7 C12 H3 H6 K5 K16 M8 M15 B7 B12 Z5 Z16 A14.

6. Complete 14-arcs

The existence of complete 14-arcs is a common feature of non-desarguesian translation planes of order 16 and they occur also in the dual Lorimer plane (de Resmini (1990) and its references; de Resmini and Puccio (1987)). Also π contains such arcs. In order to provide a first classification of the complete 14-arcs in π, we look at the numbers and distributions of their j-points, a j-point for a complete arc being a point on j tangents to the arc. There are two inequivalent types of complete 14-arcs in π. We call arcs of type (i) those for which j takes the values 0, 2, 4, 6, and arcs of type (ii) those for which $j = 0, 2, 4, 6, 10$. Denote by g_j the number of j-points. For all arcs of type (i), $g_0 = 27$, $g_2 = 52$, $g_4 = 158$, $g_6 = 36$, whereas for all arcs of type (ii), $g_0 = 19$, $g_2 = 64$, $g_4 = 166$, $g_6 = 20$ and $g_{10} = 4$.

All arcs of both types have two points on a0, an even intersection with Δ, and three 0-points (interior points) in Δ, whereas the remaining points on the line at infinity are 4-points. Furthermore, for the arcs of type (ii), the four 10-points form a Fano subplane together with the three 0-points on a0.

A quite interesting fact about the 14-arcs in π is that they usually come in pairs, triples or quadruples. More precisely, there are 8-arcs which complete in different ways to 14-arcs. Consequently, the arcs obtained in this way share at least eight finite points. They can also share some point on a0. However, different situations occur with respect to the numbers of distinct points at infinity which a pair, triple, or quadruple can use. Finally, arcs in the same pair, triple, or quadruple may not be of the same type. But, all arcs in the same pair, triple, or quadruple have the same 0-point on a0.

Therefore, it seems more interesting to list the examples of complete 14-arcs in π by dividing them into examples of pairs, triples and quadruples of arcs. Relevant comments will be provided together with the examples. We also observe that an arc can belong to distinct pairs, triples or quadruples.

Our last examples consist of a hexad of complete 14-arcs, all of which extend the same hyperoval in a Baer subplane, and of a quadruple of arcs all of which extend a hyperoval in a Baer subplane.

Together with each arc we also list its 0-points, 10-points if any, and 6-points, and they are listed in the same order as the arcs.

Pairs of complete 14-arcs:

(1) M4 M10 N5 N16 X3X6 R13 R14 B3 B6 C4 C10 A3 A6

M4 M10 N5 N16 X3 X6 R13 R14 B4 B10 C3 C6 A4 A10

0-points: A0 A7 A12 B1 B2 B7 B12 C1 C2 C7 C12 H8 H15 M8 M15 N9 N11 R8 R15 T9 T11 W8 W15 X9 X11 Z9 Z11

6-points: B5 B16 B13 B14 C5 C16 C13 C14 H9 H11 L1 L2 L7 L12 L13 L14 M9 M11 N8 N15 R9 R11 S1 S2 S5 S16 S7 S12 T8 T15 W9 W11 X8 X15 Z8 Z15;

0-points: A0 A7 A12 B1 B2 B7 B12 C1 C2 C7 C12 H9 H11 M9 M11 N8 N15 R9 R11 T8 T15 W9 W11 X8 X15 Z8 Z15

6-points: B5 B16 B13 B14 C5 C16 C13 C14 H8 H15 L1 L2 L5 L16 L7 L12 M8 M15 N9 N11 R8 R15 S1 S2 S7 S12 S13 S14 T9 T11 W8 W15 X9 X11 Z9 Z11.

(2)　　B7 B12 H7 H12 C1 C2 Z1 Z2 M9 M11 X8 X15 A13 A16

　　　　B7 B12 H7 H12 C1 C2 Z1 Z2 T13 T14 W5 W16 A8 A15

10-points: F8 F15 P9 P11

0-points: A0 A7 A12 B3 B6 C4 C10 L3 L6 M5 M16 N5 N16 R13 R14 S4 S10 X13 X14

6-points: B4 B10 C3 C6 F9 F11 L4 L10 M13 M14 N13 N14 P8 P15 R5 R16 S3 S6 X5 X16;

0-points: A0 A7 A12 B8 B15 B9 B11 C8 C15 C9 C11 H4 H10 H9 H11 T4 T10 T8 T15 W3 W6 W9 W11 Z3 Z6 Z8 Z15

6-points: B5 B16 B13 B14 C5 C16 C13 C14 H3 H6 H8 H15 L5 L16 L8 L15 L9 L11 S8 S15 S9 S11 S13 S14 T3 T6 T9 T11 W4 W10 W8 W15 Z4 Z10 Z9 Z11.

Notice that the two arcs are of different types.

(3)　　B4 B10 C4 C10 D1 D2 K1 K2 F3 F6 P3 P6 A7 A12

　　　　B4 B10 C4 C10 D1 D2 K1 K2 L7 L12 S7 S12 A7 A12

10-points: B1 B2 C1 C2

0-points: A0 A1 A2 B8 B15 B9 B11 C8 C15 C9 C11 H1 H2 T1 T2 W1 W2 Z1 Z2

6-points: H7 H12 L7 L12 L8 L15 L9 L11 S7 S12 S8 S15 S9 S11 T7 T12 W7 W12 Z7 Z12;

10-points: D4 D10 K4 K10

0-points: A0 A1 A2 D8 D15 D9 D11 H4 H10 K8 K15 K9 K11 T4 T10 W4 W10 Z4 Z10

6-points: F3 F6 F8 F15 F9 F11 H3 H6 P3 P6 P8 P15 P9 P11 T3 T6 W3 W6 Z3 Z6.

Notice that the two arcs share ten points, two of which are in Δ.

(4)　　B1 B2 C1 C2 M13 M14 X13 X14 L8 L15 S8 S15 A9 A11

　　　　B1 B2 C1 C2 M13 M14 X13 X14 T7 T12 W7 W12 A9 A11

10-points: M3 M6 X3 X6

0-points: A0 A1 A2 H7 H12 M1 M2 M7 M12 T1 T2 W1 W2 X1 X2 X7 X12 Z7 Z12

6-points: H1 H2 N1 N2 N3 N6 N7 N12 R1 R2 R3 R6 R7 R12 T7 T12 W7 W12 Z1 Z2;

10-points: D13 D14 K13 K14

0-points: A0 A1 A2 B8 B15 B13 B14 C8 C15 C13 C14 L9 L11 L13 L14 S9 S11 S13 S14

6-points: B5 B16 B9 B11 C5 C16 C9 C11 D5 D16 K5 K16 L5 L16 L8 L15 S5 S16 S8 S15.

Also these arcs share ten points, but the common points at infinity are not in Δ. They also share the tangents on A3, A12, A13 and A15.

(5)   F4 F10 P4 P10 B9 B11 C9 C11 D1 D2 K1 K2 A13 A16
    F4 F10 P4 P10 B9 B11 C9 C11 L13 L14 S13 S14 A3 A10

10-points: B5 B16 C5 C16

0-points: A0 A1 A2 B3 B6 B4 B10 C3 C6 C4 C10 M5 M16 N5 N16 R5 R16 X5 X16

6-points: L3 L6 L4 L10 L13 L14 M13 M14 N13 N14 R13 R14 S3 S6 S4 S10 S13 S14 X13 X14;

10-points: F7 F12 P7 P12

0-points: A0 A1 A2 F5 F16 F13 F14 M7 M12 N7 N12 P5 P16 P13 P14 R7 R12 X7 X12

6-points: D1 D2 D5 D16 D13 D14 K1 K2 K5 K 16 K13 K14 M1 M2 N1 N2 R1 R2 X1 X2.

(6)   F4 F10 P4 P10 L13 L14 S13 S14 N5 N16 R5 R16 A9 A11
    F4 F10 P4 P10 L13 L14 S13 S14 B9 B11 C9 C11 A3 A10

Notice that the second arc of this pair is also the second one of pair (5); thus, this arc belongs to two distinct pairs. For the first arc of the pair:

10-points: T3 T6 W3 W6

0-points: A0 A1 A2 M1 M2 N7 N12 R7 R12 T5 T16 T13 T14 W5 W16 W13 W14 X1 X2

6-points: H3 H6 H13 H14 M7 M12 N1 N2 R1 R2 X7 X12 Z3 Z6 Z5 Z16 Z13 Z14 H5 H16.

(7)   F4 F10 P4 P10 H1 H2 Z1 Z2 B3 B6 C3 C6 A3 A4
    F4 F10 P4 P10 H1 H2 Z1 Z2 M7 M12 X7 X12 A5 A13

10-points: N1 N2 R1 R2

0-points: A0 A1 A2 D1 D2 F1 F2 K1 K2 N5 N16 N13 N14 P1 P2 R5 R16 R13 R14

6-points: D7 D12 F7 F12 K7 K12 M5 M16 M7 M12 M13 M14 P7 P12 X5 X16 X7 X12 X13 X14;

10-points: L4 L10 S4 S10

0-points: A0 A1 A2 L5 L16 L13 L14 M4 M10 N4 N10 R4 R10 S5 S16 S13 S14 X4 X10

6-points: B3 B6 B5 B16 B13 B14 C3 C6 C5 C16 C13 C14 M3 M6 N3 N6 R3 R6 X3 X6.

(8)   B4 B10 C4 C10 L9 L11 S9 S11 T5 T16 W5 W16 A5 A16
    B4 B10 C4 C10 L9 L11 S9 S11 T13 T14 W13 W14 A13 A14

10-points: T1 T2 W1 W2

0-points: A0 A1 A2 D1 D2 F1 F2 K1 K2 P1 P2 T8 T15 T9 T11 W8 W15 W9 W11

6-points: D7 D12 F7 F12 H7 H12 H8 H15 H9 H11 K7 K12 P7 P12 Z7 Z12 Z8 Z15 Z9 Z11;

10-points: T7 T12 W7 W12

0-points: D7 D12 F7 F12 K7 K12 P7 P12 T8 T15 T9 T11 W8 W15 W9 W11 A0 A1 A2
6-points: H1 H2 H8 H15 H9 H11 K1 K2 D1 D2 F1 F2 P1 P2 Z1 Z2 Z8 Z15 Z9 Z11.

We observe that for a given complete 14-arc it is easy to construct Baer subplanes, whose finite points split into eight 0-points and eight 6-points for the arc.   Also, both the 0-points and the 6-points can be partitioned into quadrangles which generate Fano subplanes.

Triples of complete 14-arcs:
(1)   B4 B10 C7 C12 L7 L12 S4 S10 H8 H15 W8 W15 A5 A13
      B4 B10 C7 C12 L7 L12 S4 S10 H9 H11 W9 W11 A14 A16
      B4 B10 C7 C12 L7 L12 S4 S10 T1 T2 Z1 Z2 A2 A12
10-points: H13 H14 W13 W14
0-points: A0 A1 A7 D7 D12 F1 F2 H1 H2 H7 H12 K1 K2 P7 P12 W1 W2 W7 W12
6-points: D1 D2 F7 F12 K7 K12 P1 P2 T1 T2 T7 T12 T13 T14 Z1 Z2 Z7 Z12 Z13 Z14;
10-points: H5 H16 W5 W16
0-points: A0 A1 A7 D1 D2 F7 F12 H1 H2 H7 H12 K7 K12 P1 P2 W1 W2 W7 W12
6-points: D7 D12 F1 F2 K1 K2 P7 P12 T1 T2 T7 T12 T5 T16 Z1 Z2 Z5 Z16 Z7 Z12;
10-points: T4 T10 Z4 Z10
0-points: A0 A1 A7 M4 M10 N4 N10 R4 R10 T8 T15 T9 T11 X4 X10 Z8 Z15 Z9 Z11
6-points: H3 H6 H8 H15 H9 H11 M3 M6 N3 N6 R3 R6 W3 W6 W8 W15 W9 W11 X3 X6.
      The three arcs are of type (ii).
(2)   L8 L15 S9 S11 H5 H16 Z13 Z14 M8 M15 X9 X11 A13 A16
      L8 L15 S9 S11 H5 H16 Z13 Z14 N8 N15 R9 R11 A5 A14
      L8 L15 S9 S11 H5 H16 Z13 Z14 T7 T12 W1 W2 A1 A2
0-points: A0 A7 A12 F8 F15 F9 F11 H7 H12 M3 M6 N4 N10 P8 P15 P9 P11 R3 R6 T1 T2 W7 W12 X4 X10 Z1 Z2
6-points: D8 D15 D9 D11 D13 D14 F5 F16 F13 F14 H1 H2 K5 K16 K8 K15 K9 K11 M4 M10 N3 N6 P5 P16 P13 P14 R4 R10 T7 T12 W1 W2 X3 X6 Z7 Z12;
0-points: A0 A7 A12 D8 D15 D9 D11 H7 H12 K8 K15 K9 K11 M4 M10 N3 N6 R4 R10 T1 T2 W7 W12 X3 X6 Z1 Z2
6-points: D5 D16 D13 D14 F5 F16 F8 F15 F9 F11 H1 H2 K5 K16 K13 K14 M3 M6 N4 N10 P8 P15 P9 P11 P13 P14 R3 R6 T7 T12 W1 W2 X4 X10 Z7 Z12;
0-points: A0 A7 A12 L1 L2 L7 L12 M7 M12 M9 M11 N1 N2 N9 N11 R7 R12 R8 R15 S1 S2 S7 S12 X1 X2 X8 X15

6-points: B1 B2 B4 B10 B7 B12 C1 C2 C3 C6 C7 C12 L3 L6 L4 L10 M1 M2 M8 M15 N7 N12 N8 N15 R1 R2 R9 R11 S3 S6 S4 S10 X7 X12 X9 X11.

All three arcs are of type (i).

(3)   D7 D12 K1 K2 L13 L14 S5 S16 F4 F10 P3 P6 A1 A2

D7 D12 K1 K2 L13 L14 S5 S16 H1 H2 Z7 Z12 A8 A11

D7 D12 K1 K2 L13 L14 S5 S16 T1 T2 W7 W12 A9 A15

0-points: A0 A7 A12 H7 H12 H13 H14 L1 L2 L7 L12 S1 S2 S7 S12 T5 T16 T7 T12 W1 W2 W13 W14 Z1 Z2 Z5 Z16

6-points: B1 B2 B7 B12 B9 B11 C1 C2 C7 C12 C8 C15 H1 H2 H5 H16 L8 L15 L9 L11 S8 S15 S9 S11 T1 T2 T13 T14 W5 W16 W7 W12 Z7 Z12 Z13 Z14;

10-points: N5 N16 R13 R14

0-points: A0 A7 A12 D3 D6 F3 F6 K4 K10 N3 N6 N4 N10 P4 P10 R3 R6 R4 R10

6-points: D4 D10 F4 F10 K3 K6 M3 M6 M4 M10 M5 M16 P3 P6 X3 X6 X4 X10 X13 X14;

10-points: M13 M14 X5 X16

0-points: A0 A7 A12 D3 D6 F3 F6 K4 K10 M3 M6 M4 M10 P4 P10 X3 X6 X4 X10

6-points: D4 D10 F4 F10 K3 K6 N3 N6 N4 N10 N13 N14 P3 P6 R3 R6 R4 R10 R5 R16.

The first arc is of type (i), the other two are of type (ii).

(4)   F13 F14 R13 R14 K5 K16 X5 X16 C1 C2 L7 L12 A8 A11

F13 F14 R13 R14 K5 K16 X5 X16 C3 C6 L4 L10 A3 A10

F13 F14 R13 R14 K5 K16 X5 X16 T3 T6 Z4 Z10 A5 A16

10-points: H1 H2 W7 W12

0-points: A0 A2 A12 B5 B16 B8 B15 C5 C16 C9 C11 L8 L15 L13 L14 S9 S11 S13 S14

6-points: B9 B11 B13 B14 C8 C15 C13 C14 H7 H12 L5 L16 L9 L11 S5 S16 S8 S15 W1 W2;

0-points: A0 A2 A12 B9 B11 B13 B14 C5 C16 C9 C11 L8 L15 L13 L14 S5 S16 S8 S15 T8 T15 T9 T11 Z8 Z15 Z9 Z11

6-points: B5 B16 B8 B15 C8 C15 C13 C14 H3 H6 H8 H15 H9 H11 L5 L16 L9 L11 S9 S11 S13 S14 T3 T6 T4 T10 W4 W10 W8 W15 W9 W11 Z3 Z6 Z4 Z10;

0-points: A0 A2 A12 C5 C16 C13 C14 D7 D12 F1 F2 K7 K12 L5 L16 L13 L14 M7 M12 N1 N2 P1 P2 R7 R12 X1 X2

6-points: B4 B10 B5 B16 B13 B14 C3 C6 C4 C10 D1 D2 F7 F12 K1 K2 L3 L6 L4 L10 M1 M2 N7 N12 P7 P12 R1 R2 S3 S6 S5 S16 S13 S14 X7 X12.

The first arc is of type (ii), the other ones are of type (i). None of the arcs has points in Δ.

(5)   B1 B2 C1 C2 L3 L6 S3 S6 D8 D15 K8 K15 A6 A10

B1 B2 C1 C2 L3 L6 S3 S6 D9 D11 K9 K11 A3 A4

B1 B2 C1 C2 L3 L6 S3 S6 F7 F12 P7 P12 A7 A12

10-points: D13 D14 K13 K14
0-points: A0 A1 A2 D1 D2 D7 D12 K1 K2 K7 K12 M1 M2 N7 N12 R7
R12 X1 X2
6-points: F1 F2 F7 F12 F13 F14 M7 M12 N1 N2 P1 P2 P7 P12 P13 P14 R1
R2 X7 X12;
10-points: D5 D16 K5 K16
0-points: A0 A1 A2 D1 D2 D7 D12 K1 K2 K7 K12 M7 M12 N1 N2 R1 R2
X7 X12
6-points: F1 F2 F5 F16 F7 F12 M1 M2 N7 N12 P1 P2 P5 P16 P7 P12 R7
R12 X1 X2;
10-points: F3 F6 P3 P6
0-points: A0 A1 A2 F8 F15 F9 F11 H3 H6 P8 P15 P9 P11 T3 T6 W3 W6
Z3 Z6
6-points: D4 D10 D8 D15 D9 D11 H4 H10 K4 K10 K8 K15 K9 K11 T4 T10
W4 W10 Z4 Z10.
(6)   B1 B2 C1 C2 F7 F12 P7 P12 H4 H10 Z4 Z10 A5 A14
    B1 B2 C1 C2 F7 F12 P7 P12 T4 T10 W4 W10 A13 A16
    B1 B2 C1 C2 F7 F12 P7 P12 L3 L6 S3 S6 A7 A12
The third arc of this triple is the same as the third arc of triple (5). So
an arc can belong to different triples.   For the first and second arc:
10-points: M4 M10 X4 X10
0-points: A0 A1 A2 B4 B10 B5 B16 C4 C10 C5 C16 L4 L10 L13 L14 S4 S10
S13 S14
6-points: B3 B6 B13 B14 C3 C6 C13 C14 L3 L6 L5 L16 M3 M6 S3 S6 S5
S16 X3 X6;
10-points: N4 N10 R4 R10
0-points: A0 A1 A2 B4 B10 B13 B14 C4 C10 C13 C14 L4 L10 L5 L16 S4
S10 S5 S16
6-points: B3 B6 B5 B16 C3 C6 C5 C16 L3 L6 L13 L14 N3 N6 R3 R6 S3 S6
S13 S14.
(7)   L4 L10 S3 S6 M4 M10 X3 X6 D1 D2 K7 K12 A5 A14
    L4 L10 S3 S6 M4 M10 X3 X6 F8 F15 P9 P11 A6 A10
    L4 L10 S3 S6 M4 M10 X3 X6 N13 N14 R5 R16 A4 A10
10-points: T1 T2 W7 W12
0-points: A0 A7 A12 D13 D14 F5 F16 K5 K16 P13 P14 T5 T16 T13 T14
W5 W16 W13 W14
6-points: D5 D16 F13 F14 H1 H2 H5 H16 H13 H14 K13 K14 P5 P16 Z5
Z16 Z7 Z12 Z13 Z14;
0-points: A0 A7 A12 D5 D16 F5 F16 H3 H6 H4 H10 K13 K14 M7 M12
N7 N12 P13 P14 R1 R2 X1 X2 Z3 Z6 Z4 Z10
6-points: D13 D14 F13 F14 H8 H15 H9 H11 K5 K16 M1 M2 N1 N2 P5
P16 R7 R12 T3 T6 T4 T10 T9 T11 W3 W6 W4 W10 W8 W15 X7 X12 Z8
Z15 Z9 Z11;

0-points: A0 A7 A12 H9 H11 L1 L2 L7 L12 M8 M15 N9 N11 R8 R15 S1
S2 S7 S12 T8 T15 W9 W11 X9 X11 Z8 Z15
6-points: B1 B2 B7 B12 B13 B14 C1 C2 C5 C16 C7 C12 H8 H15 L5 L16
L13 L14 M9 M11 N8 N15 R9 R11 S5 S16 S13 S14 T9 T11 W8 W15 X8 X15
Z9 Z11.

Notice that the first arc is of type (ii), whereas the other two are of
type (i);   the second and third arc also share one point at infinity.

(8)   H1 H2 Z1 Z2 T13 T14 W13 W14 L5 L16 S5 S16 A7 A12

      H1 H2 Z1 Z2 T13 T14 W13 W14 B3 B6 C3 C6 A9 A11

      H1 H2 Z1 Z2 T13 T14 W13 W14 B4 B10 C4 C10 A8 A15

10-points: L1 L2 S1 S2
0-points: A0 A1 A2 L3 L6 L4 L10 M1 M2 N1 N2 R1 R2 S3 S6 S4 S10 X1
X2
6-points: B3 B6 B4 B10 B7 B12 C3 C6 C4 C10 C7 C12 M7 M12 N7 N12
R7 R12 X7 X12;
10-points: B9 B11 C9 C11
0-points: A0 A1 A2 B5 B16 B13 B14 C5 C16 C13 C14 D9 D11 F9 F11 K9
K11 P9 P11
6-points: D8 D15 F8 F15 K8 K15 L5 L16 L8 L15 L13 L14 P8 P15 S5 S16 S8
S15 S13 S14;
10-points: B8 B15 C8 C15
0-points: A0 A1 A2 B5 B16 B13 B14 C5 C16 C13 C14 D8 D15 F8 F15 K8
K15 P8 P15
6-points: D9 D11 F9 F11 K9 K11 L5 L16 L9 L11 L13 L14 P9 P11 S5 S16 S9
S11 S13 S14.

Observe that the second arc of this triple forms a pair with the first one
of pair (7), whereas the third arc forms a pair with the second one of
pair (8).

Quadruples of complete 14-arcs:

(1)   F9 F11 W9 W11 P8 P15 T8 T15 D7 D12 K1 K2 A4 A10

      F9 F11 W9 W11 P8 P15 T8 T15 L13 L14 S5 S16 A8 A9

      F9 F11 W9 W11 P8 P15 T8 T15 M4 M10 X3 X6 A8 A9

      F9 F11 W9 W11 P8 P15 T8 T15 N5 N16 R13 R14 A4 A10

10-points: H7 H12 Z1 Z2
0-points: A0 A7 A12 M5 M16 N8 N15 N13 N14 R5 R16 R9 R11 X9 X11
X13 X14
6-points: H1 H2 M9 M11 M13 M14 N5 N16 N9 N11 R8 R15 R13 R14
X5 X16 X8 X15 Z7 Z12;
0-points: A0 A7 A12 B1 B2 B8 B15 C7 C12 C9 C11 L7 L12 L8 L15 N3 N6
N4 N10 R3 R6 R4 R10 S1 S2 S9 S11
6-points: B7 B12 B9 B11 C1 C2 C8 C15 L1 L2 L9 L11 M3 M6 M4 M10
M13 M14 N5 N16 N13 N14 R5 R16 R13 R14 S7 S12 S8 S15 X3 X6 X4 X10
X5 X16;

0-points: A0 A7 A12 B5 B16 B13 B14 C5 C16 C13 C14 H7 H12 M7 M12
N1 N2 R7 R12 T1 T2 W7 W12 X1 X2 Z1 Z2
6-points: B3 B6 B4 B10 C3 C6 C4 C10 H1 H2 L4 L10 L5 L16 L13 L14 M1
M2 N7 N12 R1 R2 S3 S6 S5 S16 S13 S14 T7 T12 W1 W2 X7 X12 Z7 Z12;
0-points: A0 A7 A12 D1 D2 F1 F2 K7 K12 L8 L15 L9 L11 M1 M2 N1 N2
P7 P12 R7 R12 S8 S15 S9 S11 X7 X12
6-points: B8 B15 B9 B11 B13 B14 C5 C16 C8 C15 C9 C11 D7 D12 F7 F12
K1 K2 L5 L16 L13 L14 M7 M12 N7 N12 P1 P2 R1 R2 S5 S16 S13 S14 X1
X2.

We observe that this quadruple splits into two pairs such that the
arcs in the same pair also share their points at infinity.   Also, only the
first arc of the quadruple is of type (ii).
(2)   D7 D12 K7 K12 T4 T10 W4 W10 B8 B15 C8 C15 A9 A11
         D7 D12 K7 K12 T4 T10 W4 W10 F9 F11 P9 P11 A3 A4
         D7 D12 K7 K12 T4 T10 W4 W10 L1 L2 S1 S2 A5 A14
         D7 D12 K7 K12 T4 T10 W4 W10 N3 N6 R3 R6 A3 A10
0-points: A0 A1 A2 B4 B10 B5 B16 C4 C10 C5 C16 L3 L6 L5 L16 N8 N15
N9 N11 R8 R15 R9 R11 S3 S6 S5 S16
6-points: B3 B6 B13 B14 C3 C6 C13 C14 L4 L10 L13 L14 M5 M16 M8
M15 M9 M11 N5 N16 N13 N14 R5 R16 R13 R14 S4 S10 S13 S14 X5 X16
X8 X15 X9 X11;
10-points: H13 H14 Z13 Z14
0-points: A0 A1 A2 D13 D14 F13 F14 K13 K14 M9 M11 N8 N15 R8 R15
P13 P14 X9 X11
6-points: D5 D16 F5 F16 H5 H16 K5 K16 M8 M15 N9 N11 P5 P16 R9
R11 X8 X15 Z5 Z16;
10-points: N4 N10 R4 R10
0-points: A0 A1 A2 B4 B10 B13 B14 C4 C10 C13 C14 L4 L10 L5 L16 S4
S10 S5 S16
6-points: B3 B6 B5 B16 C3 C6 C5 C16 L3 L6 L13 L14 N3 N6 R3 R6 S3 S6
S13 S14;
10-points: L7 L12 S7 S12
0-points: A0 A1 A2 D3 D6 F4 F10 H7 H12 K3 K6 P4 P10 T7 T12 W7
W12 Z7 Z12
6-points: D4 D10 F3 F6 H1 H2 K4 K10 L1 L2 P3 P6 S1 S2 T1 T2 W1 W2
Z1 Z2.

Notice that the first arc of this quadruple is of type (i);   the second
and fourth also share one point at infinity.
(3)   F4 F6 T4 T6 P3 P10 W3 W10 B8 B9 C11 C15 A3 A10
         F4 F6 T4 T6 P3 P10 W3 W10 D2 D7 K1 K12 A13 A14
         F4 F6 T4 T6 P3 P10 W3 W10 M11 M15 X8 X9 A13 A14
         F4 F6 T4 T6 P3 P10 W3 W10 N5 N14 R13 R16 A3 A10
0-points: A0 A2 A7 B1 B4 B6 B12 C2 C3 C7 C10 L2 L4 L6 L7 M5 M13
M14 M16 S1 S3 S10 S12 X5 X13 X14 X16

6-points: B2 B3 B7 B10 C1 C4 C6 C12 L1 L3 L10 L12 M8 M9 M11 M15 N5 N8 N9 N13 N14 N16 R5 R11 R13 R14 R15 R16 S2 S4 S6 S7 X8 X9 X11 X15;

10-points: H1 H12 Z2 Z7

0-points: A0 A2 A7 M4 M6 M8 M9 N4 N6 N11 N15 R3 R8 R9 R10 X3 X10 X11 X15

6-points: H2 H7 M3 M10 M11 M15 N3 N8 N9 N10 R4 R6 R11 R15 X4 X6 X8 X9 Z1 Z12;

0-points: A0 A2 A7 B3 B4 B6 B10 C3 C4 C6 C10 D1 D12 F1 F12 K2 K7 M2 M7 N2 N7 P2 P7 R1 R12 X1 X12

6-points: B8 B15 B9 B11 C8 C15 C9 C11 D2 D7 F2 F7 K1 K12 L3 L4 L6 L8 L9 L10 M1 M12 N1 N12 P1 P12 R2 R7 S3 S4 S6 S10 S11 S15 X2 X7;

0-points: A0 A2 A7 H1 H12 L8 L9 L11 L15 M2 M7 N1 N12 R2 R7 S8 S9 S11 S15 T2 T7 W1 W12 X1 X12 Z2 Z7

6-points: B5 B8 B9 B11 B14 B15 C8 C9 C11 C13 C15 C16 H2 H7 L5 L13 L14 L16 M1 M12 N2 N7 R1 R12 S5 S13 S14 S16 T1 T12 W2 W7 X2 X7 Z1 Z12.

Observe that only the second arc of this quadruple is of type (ii). Furthermore, the quadruple splits into two pairs and arcs of the same pair also share their points at infinity. One such pair contains both arcs of the same type, the other one does not.

Complete 14-arcs on hyperovals in Baer subplanes:

The points B1 B2 C1 C2 A7 A12 form a hyperoval in the Baer subplane on the points: A0 Aj Bj Cj Lj Sj, j = 1, 2, 7, 12. This hyperoval completes to the following six complete 14-arcs:

    B1 B2 C1 C2 A7 A12 L3 L6 S3 S6 F7 F12 P7 P12

    B1 B2 C1 C2 A7 A12 D7 D12 K7 K12 L4 L10 S4 S10

    B1 B2 C1 C2 A7 A12 H7 H12 Z7 Z12 L13 L14 S13 S14

    B1 B2 C1 C2 A7 A12 L5 L16 S5 S16 T7 T12 W7 W12

    B1 B2 C1 C2 A7 A12 L8 L15 S8 S15 N7 N12 R7 R12

    B1 B2 C1 C2 A7 A12 L9 L11 S9 S11 M7 M12 X7 X12.

The first arc of this hexad is the third arc of both triple (5) and triple (6). The second arc of the hexad forms a triple with the following arcs:

B1 B2 C1 C2 L4 L10 S4 S10 F8 F15 P8 P15 A3 A4 (10-points: F5 F16 P5 P16)

B1 B2 C1 C2 L4 L10 S4 S10 F9 F11 P9 P11 A6 A10 (10-points: F13 F14 P13 P14).

For the arcs apart from the first one in the hexad:

10-points: D4 D10 K4 K10

0-points: A0 A1 A2 D8 D15 D9 D11 H4 H10 K8 K15 K9 K11 T4 T10 W4 W10 Z4 Z10

6-points: F3 F6 F8 F15 F9 F11 H3 H6 P3 P6 P8 P15 P9 P11 T3 T6 W3 W6 Z3 Z6;

10-points: H13 H14 Z13 Z14
0-points: A0 A1 A2 H3 H6 H4 H10 M13 M14 N13 N14 R13 R14 X13 X14 Z3 Z6 Z4 Z10
6-points: M5 M16 N5 N16 R5 R16 T3 T6 T4 T10 T5 T16 W3 W6 W4 W10 W5 W16 X5 X16;
10-points: T5 T16 W5 W16
0-points: A0 A1 A2 M5 M16 N5 N16 R5 R16 T3 T6 T4 T10 W3 W6 W4 W10 X5 X16
6-points: H3 H6 H4 H10 H13 H14 M13 M14 N13 N14 R13 R14 X13 X14 Z3 Z6 Z4 Z10 Z13 Z14;
10-points: N8 N15 R8 R15
0-points: A0 A1 A2 D8 D15 F8 F15 K8 K15 N5 N16 N13 N14 P8 P15 R5 R16 R13 R14
6-points: D9 D11 F9 F11 K9 K11 M5 M16 M9 M11 M13 M14 P9 P11 X5 X16 X9 X11 X13 X14;
10-points: M9 M11 X9 X11
0-points: A0 A1 A2 D9 D11 F9 F11 K9 K11 M5 M16 M13 M14 P9 P11 X5 X16 X13 X14
6-points: D8 D15 F8 F15 K8 K15 N5 N16 N8 N15 N13 N14 P8 P15 R5 R16 R8 R15 R13 R14.

Notice that all six 14-arcs are of type (ii) and have the same 0-points on a0.

The points A0 A1 D3 F4 K10 P6 form a hyperoval in the Baer subplane on the points A0 A1 A2 A7 A12 Dj Fj Kj Pj, j = 3, 6, 4, 10. This hyperoval extends to four complete 14-arcs which come in two pairs, and arcs in the same pair share ten points, two of which are on a0. The quadruple follows:

  D3 F4 K10 P6 M15 N8 R9 X11 B16 L14 C13 S5 A0 A1
  D3 F4 K10 P6 M15 N8 R9 X11 H1 T7 W2 Z12 A0 A1
  D3 F4 K10 P6 M9 N11 R15 X8 B5 L13 C14 S16 A0 A1
  D3 F4 K10 P6 M9 N11 R15 X8 H12 T2 W7 Z1 A0 A1

10-points: H7 Z2 T1 W12
0-points: A2 A7 A12 F5 F16 K5 K16 D13 D14 M13 M16 N5 N14 P13 P14 R5 R14 X13 X16
6-points: D5 D16 F13 F14 H1 K13 K14 M5 M14 N13 N16 P5 P16 R13 R16 T7 W2 X5 X14 Z12;
10-points: B14 C5 L16 S13
0-points: A2 A7 A12 H3 H9 H10 H11 T4 T6 T8 T15 W4 W6 W9 W11 Z3 Z8 Z10 Z15
6-points: B16 C13 H4 H6 H8 H15 L14 S5 T3 T9 T10 T11 W3 W8 W10 W15 Z4 Z6 Z9 Z11;
10-points: H2 T12 W1 Z7
0-points: A2 A7 A12 D5 D16 F13 F14 K13 K14 M13 M16 N5 N14 P5 P16 R5 R14 X13 X16

6-points: D13 D14 F5 F16 H12 K5 K16 M5 M14 N13 N16 P13 P14 R13 R16 T2 W7 X5 X14 Z1;
10-points: B13 C16 L5 S14
0-points: A2 A7 A12 H4 H6 H9 H11 T3 T8 T10 T15 W3 W9 W10 W11 Z4 Z6 Z8 Z15
6-points: B5 C14 H3 H8 H10 H15 L13 S16 T4 T6 T9 T11 W4 W6 W8 W15 Z3 Z9 Z10 Z11.

Our classification of complete 14-arcs could be refined and a natural refinement is provided by the arc having two or no points in Δ.

Therefore, there are two inequivalent types of arcs of type (i) and two inequivalent types of those of type (ii).

7. Hyperovals

The hyperovals in π have a nice structure, in the sense that they are obtained by glueing together four suitable hyperovals belonging to convenient disjoint Baer subplanes, that is, which have no common finite point, and sharing their points at infinity.   Moreover, the same "good" Baer hyperoval lies on distinct hyperovals in π.

Next, we list some examples of hyperovals in π and show their composing Baer hyperovals.   We also give the finite points of the involved Baer subplanes.
Example 1.
   D4 D10 K4 K10 H3 H6 Z3 Z6 F5 F16 P5 P16 T13 T14 W13 W14 A7 A12
   D4 D10 K4 K10 T3 T6 W3 W6 F13 F14 P13 P14 H5 H16 Z5 Z16 A7 A12
   D4 D10 K4 K10 M3 M6 X3 X6 F8 F15 P8 P15 N9 N11 R9 R11 A7 A12
   D4 D10 K4 K10 N3 N6 R3 R6 F9 F11 P9 P11 M8 M15 X8 X15 A7 A12
The Baer subplanes and the Baer hyperovals associated with these four hyperovals are the following ones.

Baer subplanes	Baer hyperovals
Dj Kj Fj Pj, j = 3, 6, 4, 10	D4 D10 K4 K10 A7 A12
Dj Kj Fj Pj, j = 5, 16, 13, 14	F5 F16 P5 P16 A7 A12
	F13 F14 P13 P14 A7 A12
Hj Zj Tj Wj, j = 3, 6, 4, 10	H3 H6 Z3 Z6 A7 A12
	T3 T6 W3 W6 A7 A12
Hj Zj Tj Wj, j = 5, 16, 13, 14	T13 T14 W13 W14 A7 A12
	H5 H16 Z5 Z16 A7 A12
Mj Xj Nj Rj, j = 3, 6, 4, 10	M3 M6 X6 A7 A12
	N3 N6 R3 R6 A7 A12
Dj Kj Fj Pj, j = 8, 15, 9, 11	F8 F15 P8 P15 A7 A12
	F9 F11 P9 P11 A7 A12

Mj Xj Nj Rj, j = 8, 15, 9, 11   N9 N11 R9 A7 A12
                                       M8 M15 X8 X15 A7 A12

Example 2.

F4 F10 P4 P10 D9 D11 K9 K11 M3 M6 X3 X6 N8 N15 R8 R15 A7 A12
F4 F10 P4 P10 M9 M11 X9 X11 N3 N6 R3 R6 D8 D15 K8 K15 A7 A12
F4 F10 P4 P10 D13 D14 K13 K14 H3 H6 Z3 Z6 T5 T16 W5 W16 A7 A12
F4 F10 P4 P10 D5 D16 K5 K16 H13 H14 Z13 Z14 T3 T6 W3 W6 A7 A12

These four hyperovals use the same Baer subplanes as those in Example 1, but with a different, not disjoint, collection of Baer hyperovals which are easily found.

Example 3.

F3 F6 P3 P6 D8 D15 K8 K15 M4 M10 X4 X10 N9 N11 R9 R11 A7 A12
F3 F6 P3 P6 D9 D11 K9 K11 N4 N10 R4 R10 M8 M15 X8 X15 A7 A12
F3 F6 P3 P6 D5 D16 K5 K16 H4 H10 Z4 Z10 T13 T14 W13 W14 A7 A12
F3 F6 P3 P6 D13 D14 K13 K14 H5 H16 Z5 Z16 T4 T10 W4 W10 A7 A12

Also this quadruple of hyperovals is associated with the Baer subplanes in Example 1. Again, it is easy to find the composing Baer hyperovals.

We observe that the hyperovals listed use the existence of a partition of the affine points of π by the affine points of a suitable family of Baer subplanes.

Other examples of hyperovals are:

H9 H11 Z9 Z11 M5 M16 X5 X16 N8 N15 R8 R15 T13 T14 W13 W14 A7 A12

H9 H11 Z9 Z11 M8 M15 X8 X15 N13 N14 R13 R14 T5 T16 W5 W16 A7 A12.

Even though no exhaustive search was carried out, but only a thorough one, it seems that the only possible points at infinity for a hyperoval in π are A7 and A12 (cf. §6).

Acknowledgement. The author wishes to thank William Cherowitzo for helpful discussions.

References

M.J. de Resmini (1990), On the derived semifield plane of order 16, *Ars Combin.* **29B**, 97-109.

M.J. de Resmini and L. Puccio (1987), Some combinatorial properties of the dual Lorimer plane, *Ars Combin.* **24A**, 131-148.

J.C. Fisher, J.W.P. Hirschfeld and J.A. Thas (1986), Complete arcs in planes of square order, *Ann. Discrete Math.* **30**, 243-250.

J.W.P. Hirschfeld (1979), *Projective Geometries over Finite Fields,* Oxford University Press, Oxford.

N.L. Johnson (1970), A note on semi-translation planes of class 1-5a, *Arch. Math.* **21**, 528-532.

Strongly regular graphs induced by polarities of symmetric designs

**W.H. Haemers[1], D.G. Higman[2]
and S.A. Hobart[3]**
Tilburg University[1], University of Michigan[2]
and University of Wyoming[3].

1. Introduction

Throughout, we shall assume that A is the $(0, 1)$ incidence matrix of a symmetric 2-(v, k, λ) design D with a polarity with v_1 ($\neq 0$) absolute points and $v_2 = v - v_1$ ($\neq 0$) non-absolute points. So A takes the form:

$$A = \begin{bmatrix} A_1 & C \\ C^T & A_2 \end{bmatrix},$$

where A_1 and A_2 are symmetric matrices, A_1 has 'ones' on the diagonal and A_2 has zeros on the diagonal. Thus $A_1 - I$ and A_2 are adjacency matrices of graphs Γ_1 and Γ_2, say. We call Γ_1 and Γ_2 the graphs *induced* by the polarity of D. The above decomposition of A is called the *polar decomposition*. A polar decomposition is *regular* if Γ_1 and Γ_2 are regular and *strongly regular* if Γ_1 and Γ_2 are strongly regular, complete or empty. The incidence structure of absolute points and non-absolute blocks, given by the matrix C, is called the *polar structure* of the design D.

We assume that D is non-trivial; that is, $0 < \lambda < k - 1 < v - 2$. Polar decompositions of trivial designs are obvious and not very interesting.

In the present paper, we study strongly regular polar decompositions. It is, in a certain sense, an extension to a paper on strongly regular graphs with strongly regular decompositions by Haemers and Higman (1989). The subject is motivated by the following example. Let D be the design of points and hyperplanes of $PG(n, 4)$ ($n \geq 2$). Then the graphs induced by the unitary polarity are strongly regular (see for instance Hubaut 1975). The parameters for this example are:

$$v = \frac{4^{n+1} - 1}{3}, \ k = \frac{4^n - 1}{3}, \ \lambda = \frac{4^{n-1} - 1}{3}, \ v_1 = \frac{2^{2n+1} - (-2)^n - 1}{3}.$$

We shall prove that all parameters of a strongly regular polar decomposition can be expressed in terms of only one parameter a. If

a is a power of -2 we find the values above, but unfortunately, other values of a for which no examples are known, remain feasible.

For strongly regular Γ_i (i= 1, 2) the parameters are denoted by v_i, k_i', λ_i and μ_i, and the eigenvalues of A_i by k_i, ρ_i and σ_i, such that $|\rho_i| \geq |\sigma_i|$. Note that $k_1 = k_1' + 1$ and $k_2 = k_2'$. The multiplicity of σ_i is denoted by φ_i. If Γ_i is complete or empty we put $\varphi_i = v_i - 1$ (ρ_i disappears).

2. Preliminaries

The following three Lemmas are the analogues of 2.2, 2.4 and 2.5 of Haemers and Higman (1989). Since the proofs are the same, we omit them here.

Lemma 1. *Suppose Γ_1 is regular of degree $k_1' = k_1 - 1$. Then*

$$\left| \frac{k_1 v - v_1 k}{v - v_1} \right| \leq \sqrt{(k - \lambda)}.$$

Equality holds if and only if the polar decomposition is regular.

Note that for the polar decomposition to be regular $k - \lambda$ (the order of D) must be square.

For a regular polar decomposition we define k_2 to be the degree of Γ_2 and

$$a = \frac{k_1 v - v_1 k}{v - v_1}. \tag{1}$$

Then we easily have

$$a = k_1 + k_2 - k = \pm\sqrt{(k - \lambda)}, \ |a| \geq 2. \tag{2}$$

So, a, $-a$ and k are the eigenvalues of A with multiplicities φ (say), $v - 1 - \varphi$ and 1 respectively.

Lemma 2. *Suppose A_1 has eigenvalues k_1, σ_1 and ρ_1 ($|\sigma_1| \leq |\rho_1|$) with multiplicities 1, φ_1 and $v_1 - 1 - \varphi_1$ ($1 \leq \varphi_1 \leq v_1 - 1$), respectively. Let the polar decomposition be regular. Then A_2 has eigenvalues k_2, $-\sigma_1$, $-\rho_1$, a, $-a$ with multiplicities 1, φ_1, $v_1 - 1 - \varphi_1$, $\varphi - v_1$ and $v_2 - \varphi$, respectively. Also, $|\sigma_1| \leq |\rho_1| \leq |a|$ and $|\sigma_1| \neq |a|$.*

Lemma 3. *With the hypotheses of Lemma 2, the polar decomposition is strongly regular if and only if one of the following occurs:*

(i) $v_1 = v_2 = \varphi$;

(ii) $\rho_1 = a$, $v_1 = \varphi$;

(iii) $\rho_1 = -a$, $v_2 = \varphi$.

It is easy to see that if (ii) occurs for a matrix A, then (i) occurs for the complementary case; that is, for the design with incidence matrix (J denotes the all-one matrix),

$$J - \begin{bmatrix} A_2 & C^T \\ C & A_1 \end{bmatrix}.$$

Note that Lemma 3 also applies if Γ_1 is complete or empty (the conditions $\rho_1 = \pm a$ are meaningless in this case). Next we show that case (i) does not occur.

Lemma 4. *A strongly regular polar decomposition with $v_1 = v_2 = \varphi$ does not exist.*

Proof. Assume $v \geq 2k$ (otherwise consider the complement). Trace $A = v_1 = \varphi = k + \varphi a - (v - \varphi - 1)a$ implies $k = v_1 - a$. Hence, by (1) and (2), $k_1 = k_2 = v_1/2 = v_2/2$. Therefore, since Γ_2 is strongly regular, $k_2(k_2 - \lambda_2 - 1) = \mu_2(v_2 - k_2 - 1) = \mu_2(k_2 - 1)$. This implies $\mu_2 \equiv 0 \mod k_2$, hence Γ_2 is complete bipartite, so $\rho_2 = -k_2$. This is impossible, since then by Lemma 2, $|a| \geq |\rho_2| = k_2 = v_2/2 = v/4$, which is a contradiction to $a^2 < k \leq v/2$ and $|a| \geq 2$. □

3. Main theorem

Theorem 5. *If D has a strongly regular polar decomposition then $a \equiv 4 \mod 6$, and, up to taking complements, all other parameters are expressible in terms of a as follows:*

$v = (16a^2 - 1)/3,$ $\qquad k = (4a^2 - 1)/3,$ $\qquad \lambda = (a^2 - 1)/3,$

$v_1 = \varphi = (2a + 1)(4a - 1)/3,$ $\qquad v_2 = 2a(4a - 1)/3,$

$k_1' = k_1 - 1 = 2(a - 1)(a + 2)/3,$ $\qquad k_2' = k_2 = a(2a - 1)/3,$

$\rho_1 = -\rho_2 = a,$ $\qquad \sigma_2 = -\sigma_1 = a/2,$

$\varphi_1 = \varphi_2 = 8(a - 1)(2a + 1)/9,$

$\mu_1 = (a - 1)(a + 2)/6,$ $\qquad \lambda_1 = (a^2 + 4a - 2)/6,$

$\mu_2 = a(a + 2)/6,$ $\qquad \lambda_2 = a(a - 1)/6.$

Proof. We may restrict ourselves to case (ii) of Lemma 3. By this result we can express all parameters in terms of k and a only. We have $\lambda = k - a^2$ and hence, using $\lambda(v - 1) = k(k - 1)$:

$$v = \frac{k^2 - a^2}{k - a^2}.$$

Trace $A = v_1 = \varphi = k + \varphi a - (v - 1 - \varphi)a$ yields

$$v_1 = \frac{k(k + a)(a - 1)}{(k - a^2)(2a - 1)}, \qquad v_2 = \frac{a(k + a)(k + 1 - 2a)}{(k - a^2)(2a - 1)}.$$

Next, using (1) and (2), we find

$$k_2 = \frac{ak}{2a - 1}, \quad v_2 - k_2 - 1 = \frac{k(a + 1)(a - 1)^2}{(k - a^2)(2a - 1)}.$$

Since $k - a^2 = \lambda > 0$ and $|a| \geq 2$ it follows that $0 < k_2 < v_2 - 1$, so Γ_2 is not the empty graph or the complete graph. Moreover, $2a - 1$ divides k. Define $x = k/(2a - 1)$, then $k_2 = ax$ and a and x have the same sign. By use of $\rho_2 = -a$ (by Lemma 2 and 3) and $(k_2 - \rho_2)(k_2 - \sigma_2) = (k_2 + \rho_2\sigma_2)v_2$ (because Γ_2 is strongly regular) we find after some computation:

$$\sigma_2 = \frac{a(ax + a - 2x)}{2ax - x - 1}.$$

Define

$$d = 4\sigma_2 - 2a + 3 = \frac{4a^2 + 2a - 3x - 3}{2ax - x - 1},$$

then $d \geq 0$, since $d \leq 1$ would imply $(2a - 1)(a + 1) \leq x(1 - a) < 0$, which contradicts $|a| \geq 2$. Moreover, using $2ax - x - 1 = k - 1 \geq a^2$, we have

$$d \leq 4 + \frac{2}{a} - \frac{3(x + 1)}{2ax - x - 1} < \begin{cases} 4 + \dfrac{2}{a} \leq 5 & \text{if } a \text{ and } x \text{ are positive} \\[2mm] 4 - \dfrac{3}{2a - 1} < 5 & \text{if } a \text{ and } x \text{ are negative.} \end{cases}$$

Since d is odd we can conclude that $d = 1$ or $d = 3$. If $d = 1$, then $x = 2a - 1$ and $v_1 = (2a - 1)(4a^2 - 3a + 1)/(3a - 1)$, which is no integer. Hence $d = 3$, so

$$x = \frac{2a+1}{3}, \quad \sigma_2 = \frac{a}{2}, \quad a \equiv 4 \bmod 6.$$

All other parameters now follow in a straightforward manner. □

If $a = (-2)^{n-1}$ we find the parameters of the example given in the introduction. No other examples are known. This leaves $a = 10$ (D is a 2-(533, 133, 33) design) as the smallest unsolved case. By Theorem 5, only for $a = -2$ is one of the graphs induced by the polarity complete or empty. For $a = -2$, however, the structure is unique. So $PG(2, 4)$ with its unitary polarity and the complement provide the only strongly regular polar decomposition for which one of the induced graphs is empty of complete. In this case, the polar structure is $AG(2, 3)$, the affine plane of order 3. If $|a| > 2$, it can be shown (by similar, but less complicated arguments as for Theorem 2.8 of Haemers and Higman (1989)) that the polar structure is a strongly regular design (see Higman (1988) for definition and notation) with parameters

$$(n_1, S_1, a_1, b_1, n_2, S_2, a_2, b_2) = (v_1, k-k_1, \lambda-2-\lambda_1, \lambda-\mu_1, v_2\, k-k_2, \lambda-\lambda_2, \lambda-\mu_2) \quad (3)$$

and that, conversely, a strongly regular design whose parameters satisfy (3) with v_1, k, k_1 etc. as given in Theorem 5, is the polar structure of some symmetric design with a strongly regular polar decomposition.

A symmetric design with a polarity can be seen as a strongly regular graph for which loops are admitted. More precisely, if we allow loops in graphs, a strongly regular graph is a simple strongly regular graph or a symmetric design with a polarity. In this more general setting, a strongly regular graph with strongly regular decomposition either belongs to the case without loops treated in Haemers and Higman (1989), or is a design with a strongly regular polar decomposition. This can be proved as follows. For a strongly regular decomposition of a symmetric design with a polarity it follows by eigenvalue arguments as in Lemma 3, that if Γ_1 is a symmetric design with a polarity, then so is Γ_2. This, however, is impossible, by Rahilly (1988). Thus both Γ_1 and Γ_2 must be strongly regular graphs without loops, so we have a strongly regular polar decomposition.

References

W.H. Haemers and D.G. Higman (1989), Strongly regular graphs with strongly regular decomposition, *Linear Algebra Appl.* **114/115**, 379-398.

D.G. Higman (1988), Strongly regular designs and coherent configurations of type ($^3_3$), *European J. Combin.* **9**, 411-422.

X.L. Hubaut (1975), Strongly regular graphs, *Discrete Math.* **13**, 357-381.
A. Rahilly (1988), Divisions of symmetric designs into two parts, *Graphs and Combin.* **4**, 67-73.

A problem on squares in a finite field and its application to geometry

J.W.P. Hirschfeld and T. Szönyi
University of Sussex and Eötvös Loránd
University Budapest

1. Introduction

The starting point of this work is the following lemma in Hirschfeld and Szönyi (submitted):

For q an odd prime power, let α be an element of $GF(q^2) \backslash GF(q)$. Then there are precisely $(q-1)/2$ elements u in $GF(q)$ for which $\alpha - u$ is a non-zero square in $GF(q^2)$.

It should be observed that, if α is allowed to be any element of $GF(q)$, then the number of elements u is $q-1$ when $\alpha \in GF(q)$, since every element of $GF(q)$ is a square in $GF(q^2)$.

A natural generalization of the lemma replaces $GF(q^2)$ by $GF(q^h)$ as follows:

For q an odd prime power, let α be an element of $GF(q^h)$ not in $GF(q^{h/2})$ if h is even. If N denotes the number of elements u in $GF(q)$ for which $\alpha - u$ is a non-zero square in $GF(q^h)$, is N close to $q/2$ (as might be expected)?

For $h = 1$, the number of u is just the number of non-zero squares in $GF(q)$, that is, $N = (q-1)/2$. The case $h = 2$ is the lemma above. So the problem is interesting when $h \geq 3$. In Theorem 2 it will be shown that

$$|N - q/2| \leq (h-1)\sqrt{q}/2. \qquad (1)$$

This supports the intuitive notion that being a square is a "random event" with probability $1/2$. So the answer to the question posed is "yes" if the exponent h is small compared to q.

The first application of Theorem 2 is to the existence of blocking sets in affine and projective spaces. Improving a bound of Beutelspacher and Eugeni (1987), we show that there exist blocking sets with respect to lines of $AG(h, q)$ when $q \geq h^2 + 1$. For q even, the proof of this result requires a slight modification of the proof of Theorem 2.

The second application of Theorem 2 is to the existence of relatively large (k, n)-arcs in $PG(2, q)$ for q odd. Using the general construction of the authors (submitted) we construct these sets from a pencil of touching conics.

2. Solution of the algebraic problem

Throughout this section, let q be odd and ω be the quadratic character of $GF(q)$; that is,

$$\omega(x) = \begin{cases} 0 & \text{for } x = 0 \\ 1 & \text{for } x \text{ a non-zero square} \\ -1 & \text{for } x \text{ a non-square.} \end{cases}$$

The key result on character sums with polynomial arguments, called Weil sums, is the following due to Burgess (see Lidl and Niederreiter 1983, Theorem 5.41).

Proposition 1. *Let f in $GF(q)[X]$ be a monic polynomial of positive degree that is not the square of a polynomial. Let d be the number of distinct roots of f in its splitting field over $GF(q)$. Then*

$$\left| \sum_{x \in GF(q)} \omega(f(x)) \right| \leq (d - 1)\sqrt{q}. \quad \square$$

The solution of the algebraic problem follows readily from this result.

Theorem 2. *For q odd and $h > 1$ let $\alpha \in GF(q^h)$ and suppose for h even that $\alpha \notin GF(q^{h/2})$. If*

$$N = \left| \{u \in GF(q) \mid \alpha - u \text{ is a non-zero square in } GF(q^h)\} \right|, \quad (2)$$

then

$$|N - q/2| \leq (h - 1)\sqrt{q}/2. \quad (1)$$

Proof. Let $v_h: GF(q^h) \rightarrow GF(q)$ be the usual norm; that is,

$$v_h(x) = x \cdot x^q \cdot x^{q^2} \dots \dots x^{q^{h-1}}. \quad (3)$$

Now, x is a non-zero square in $GF(q^h)$

$$\Leftrightarrow x^{(q^h - 1)/2} = 1$$
$$\Leftrightarrow v_h(x)^{(q - 1)/2} = 1$$
$$\Leftrightarrow v_h(x) \text{ is a non-zero square in } GF(q).$$

However,

$$\begin{aligned}
v_h(\alpha - u) &= (\alpha - u)(\alpha - u)^q \ldots (\alpha - u)^{q^{h-1}} \\
&= (\alpha - u)(\alpha^q - u) \ldots (\alpha^{q^{h-1}} - u) \\
&= f_h(u),
\end{aligned}$$

where the polynomial f_h given by

$$f_h(X) = (\alpha - X)(\alpha^q - X) \ldots (\alpha^{q^{h-1}} - X)$$

and so is a polynomial of degree h in X defined over $GF(q)$. As $\alpha \notin GF(q^{h/2})$, the roots of f_h, namely $\alpha, \alpha^q, \ldots, \alpha^{q^{h-1}}$, have odd multiplicity. Typically, if α is not contained in a proper subfield of $GF(q^h)$, all these roots are distinct. So, f_h is not a constant multiple of a square of a polynomial and Proposition 1 can be applied to f_h. Hence

$$\left| \sum_{u \in GF(q)} \omega(v_h(\alpha - u)) \right| = \left| \sum_{u \in GF(q)} \omega(f_h(u)) \right| \le (h - 1)\sqrt{q}. \tag{4}$$

When $\alpha \notin GF(q)$, the norm $v_h(\alpha - u) \ne 0$ and so f_h has no zeros in $GF(q)$. Hence every element in the sum of the left-hand side of (4) is $+1$ or -1. By definition N is the number of 1's, whence $q - N$ is the number of -1's. Thus (4) becomes

$$|q - 2N| \le (h - 1)\sqrt{q},$$

which in turn gives (1). When $\alpha \in GF(q)$, then $N = (q - 1)/2$, which also satisfies (1). $\square$

Remark. When $h = 3$, the smallest q for which Theorem 2 applies is $q = 5$; here

$$|N - 5/2| \le \sqrt{5}$$

and $N \ne 0$.

3. Application 1: Blocking sets in affine and projective spaces

3.1 General results on blocking sets

A *blocking set* in a space (affine or projective) is a set of points which meets every line but contains no line entirely. The fundamental question about blocking sets is that of existence.

First, fix q, the order of the space. Using a geometric version of Ramsey's theorem due to Graham, Leeb and Rothschild (1972), the following non-existence result was proved by Mazzocca and Tallini (1985).

Theorem 3. *There exists $h_0 = h_0(q)$ such that there are no blocking sets in $AG(h, q)$ for $h \geq h_0$.* □

On the other hand, when the dimension is small, blocking sets exist. For example, in $PG(2, q)$, there are blocking sets when $q > 2$; see Hirschfeld (1979, Chapter 13). In $PG(3, q)$, Rajola (1988) showed that blocking sets exist for $q > 4$. Blokhuis and Fisher (private communication) gave the following elegant example. Consider $AG(3, q)$ as $GF(q^3)$ and let S be the set of squares in $GF(q^3)$; then S is a blocking set in $AG(3, q)$ for $q \geq 5$. Theorem 4 gives a simple proof of this fact. In contrast to Theorem 3, Beutelspacher and Eugeni (1986) proved by a recursive construction that, for $q \geq 2^h$, there do exist blocking sets in $AG(h, q)$. Here we improve this bound.

3.2   Blocking sets for q odd

Theorem 4. *Let q and h be odd with $q \geq (h - 1)^2 + 1$. Then there is a blocking set in $AG(h, q)$.*

Proof. Identify $AG(h, q)$ with $GF(q^h)$ and consider $S = \{s^2 \mid s \in GF(q^h)\}$. The lines of $AG(h, q)$ in this representation are the sets $\{a + tb \mid t \in GF(q)\}$ with $b \neq 0$. So to show that S is a blocking set, it must be shown that the equation

$$s^2 = a + tb \qquad (5)$$

has at least one and less than q solutions in s as t varies in $GF(q)$. Since

$$s^2 = a + tb = -b(-a/b - t),$$

it follows that when $-b$ is a square we must find the number N of t such that $-a/b - t$ is a square; similarly when $-b$ is a non-square. From Theorem 2, N satisfies

$$|N - q/2| \leq (h - 1)\sqrt{q}/2.$$

This gives the result that for $q > (h - 1)^2$ we have $1 \leq N \leq q - 1$. □

Theorem 5. *Let q be odd and let $q \geq h^2 + 1$. Then there is a blocking set in $AG(h, q)$.*

Proof. For h odd, Theorem 4 is a stronger result. For h even, embed $AG(h, q)$ in $AG(h + 1, q)$. By Theorem 4 there is a blocking set S in $AG(h + 1, q)$. Then $S \cap AG(h, q)$ is a blocking set. $\square$

Theorem 6. *Let q be odd and let $q \geq h^2 - 1$. Then there is a blocking set in $PG(h, q)$.*

Proof. As $PG(h, q)$ is the disjoint union of $AG(h, q)$ and $PG(h - 1, q)$, induction on h and Theorem 5 give the result. $\square$

3.3 Blocking sets for q even

To obtain a result similar to the case of q odd, it is necessary to consider an additive character for $GF(q)$. Let

$$\chi(x) = 1 \quad \text{for } x = u^2 + u$$
$$\chi(x) = -1 \quad \text{for } x \neq u^2 + u.$$

The following result is due to Carlitz and Uchiyama (1957), and is a variation of Weil's theorem as given in Theorem 5.38 by Lidl and Niederreiter (1983).

Proposition 7. *Suppose that f in $GF(q)[X]$ of degree d cannot be written in the form $f(X) = g(X)^2 + g(X) + c$. Then*

$$\left| \sum_{x \in GF(q)} \chi(f(x)) \right| \leq (d - 1)\sqrt{q}. \quad \square \tag{6}$$

Theorem 8. *In $AG(h, q)$ with q even, h odd and $q > (h - 1)^2$, there exists a blocking set.*

Proof. Identify $AG(h, q)$ with $GF(q^h)$ and consider

$$S = \{s \in GF(q^h) \mid v_h(s) = u^2 + u \text{ for some } u \text{ in } GF(q)\},$$

where v_h is the usual norm as in (3). The lines of $AG(h, q)$ are $\{a + tb \mid t \in GF(q)\}$, where $b \neq 0$. To show that S is a blocking set, it must be shown that the number N of solutions in t of the equation

$$v_h(a + tb) = u^2 + u \tag{7}$$

satisfies $1 \leq N < q$.   Since h is the degree in T of $v_h(a + Tb)$ and h is odd, so $v_h(a + Tb) \neq g(T)^2 + g(T) + c$, the latter having even degree in T. Thus, by Proposition 7,

$$\left| \sum_{x \in GF(q)} \chi(v_h(a + tb)) \right| \leq (h - 1)\sqrt{q}. \tag{8}$$

Since $q \geq (h-1)^2 + 1$, the sum in (8) contains both 1 and -1.   So S is a blocking set.   □

Remark. For $h = 3$, Theorem 8 gives the bound $q \geq 5$, whence there are blocking sets in $PG(3, q)$, q even, for $q \geq 8$.

Theorem 9.   *In $AG(h, q)$ with q even and $q \geq h^2 + 1$, there exists a blocking set.*

Proof. See the proof of Theorem 5.   □

Theorem 10. *In $PG(h, q)$ with q even and $q \geq h^2 + 1$, there exists a blocking set.*

Proof. See the proof of Theorem 6.   □

4. Application 2: Large (k, n)-arcs with small n in $PG(2, q)$

A (k, n)-arc in a projective plane of order q is a set of k points with some n but no $n + 1$ collinear.   The maximum value of k for which a (k, n)-arc exists in $PG(2, q)$ is denoted by $m_n(2, q)$.   For the foundations of this topic, see Hirschfeld (1979, Chapter 12).   For other results, see Hill and Mason (1981), Hirschfeld (1983), Mason (1984).

Next we survey some results on lower bounds for $m_n(2, q)$.   By taking a union of disjoint conics, we have

$$m_n(2, q) \geq [n/2](q + 1). \tag{9}$$

Similarly, when q is a square, the example of a disjoint union of Baer subplanes $PG(2, \sqrt{q})$, shows that

$$m_n(2, q) \geq (n - \sqrt{q})(q - \sqrt{q} + 1). \tag{10}$$

If n is large compared to q, that is $n > q^{(1/2)+\delta}$, then

$$m_n(2, q) \geq (1 - \varepsilon)nq \tag{11}$$

for $q \geq q_0(\varepsilon, \delta)$; see Hirschfeld and Szönyi (submitted).   The previous lower bounds other than (9) give no information when $n < \sqrt{q}$.   The bound in (11) remains valid for smaller n with $n \sim q^\lambda$ and $1/4 < \lambda \leq 1/2$ in planes of square order.   Now using Theorem 2, we construct various (k, n)-arcs in $PG(2, q)$, where $q = s^h$ and n is essentially smaller than $\sqrt{q}$.   The construction is based on the general scheme of the authors' paper (submitted).   The key lemma there is the following result.

Proposition 11. *For a subset B of $GF(q)$, q odd, consider the set of points in $PG(2, q)$ given by*

$$\mathcal{K}(B) = \cup_{b \in B}\{(x, x^2 + b, 1) \mid x \in GF(q)\}.$$

If $|l \cap \mathcal{K}(B)| \in \{j_1, \ldots, j_m\}$ is satisfied for $l = l_c$, where l_c has equation $y = cz$, then it is satisfied for all q^2 lines l not through $(0, 1, 0)$. $\square$

In fact, $|l_c \cap \mathcal{K}(B)| = 2N(c) + Z(c)$, where

$$N(c) = |\{b \in B \mid c - b \text{ is a non-zero square}\}|$$

$$Z(c) = \begin{cases} 0 & \text{when } c \notin B \\ 1 & \text{when } c \in B. \end{cases}$$

From this observation, Theorem 2 implies the following result.

Theorem 12. *Let $q = s^h$ and let $B = GF(s) \subset GF(q)$. Then $\mathcal{K}(B)$ is a (qs, n)-arc in $AG(2, q)$ for some $n \leq s + (h - 1)\sqrt{s}$. When h is essentially smaller than $\sqrt{s}$, the number n is asymptotically s, whence $\mathcal{K}(B)$ is a (k, n)-arc with $k \sim nq$.* $\square$

Finally, as a further application, we construct some more "almost regular" large (k, n)-arcs in planes of order s^3.   The result can be generalized to s^h with h odd.

Theorem 13. *With $q = s^3$, let $\{b_1, \ldots, b_{s^2}\}$ be a set of coset representatives in $GF(s^3)/GF(s)$ and let $B = \cup^t_{i=1}(b_i + GF(s))$. Then $\mathcal{K}(B)$ is a (k, n)-arc with $k = tsq$ and $n \leq ts + 2t\sqrt{s}$. When $t \leq \sqrt{s}$, we have $n \sim ts$ and $k \sim nq$; that is, this construction gives "almost regular" large (k, n)-arcs with $\sqrt[3]{q} < n < q$.* $\square$

Acknowledgement.   The second author thanks L. Rònyai for fruitful discussions.

References

A. Beutelspacher and F. Eugeni (1986), On blocking sets in projective and affine spaces of large order, *Rend. Mat.* **6**, 587-595.

L. Carlitz and S. Uchiyama (1957), Bounds for exponential sums, *Duke Math. J.* **24**, 37-41.

R.L. Graham, K. Leeb and B.L. Rothschild (1972), Ramsey's theorem for a class of categories, *Adv. in Math.* **8**, 417-433.

R. Hill and J.R.M. Mason (1981), On (k, n)-arcs and the falsity of the Lunelli-Sce conjecture, *Finite Geometries and Designs*, L.M.S. Lecture Note Series **49**, 153-168, Cambridge University Press, Cambridge.

J.W.P. Hirschfeld (1979) *Projective Geometries over Finite Fields*, Oxford University Press, Oxford.

J.W.P. Hirschfeld (1983), Maximum sets in finite projective spaces, *Surveys in Combinatorics*, L.M.S. Lecture Note Series **82**, 55-76, Cambridge University Press, Cambridge.

J.W.P. Hirschfeld and T. Szönyi (submitted), Constructions of large arcs and blocking sets in finite planes.

R. Lidl and H. Niederreiter (1983), *Finite Fields*, Addison-Wesley, Reading, Mass.

J.R.M. Mason (1984), A class of $((p^n - p^m)(p^n - 1), p^n - p^m)$-arcs in $PG(2, q)$, *Geom. Dedicata* **15**, 355-361.

F. Mazzocca and G. Tallini (1985), On the non-existence of blocking sets in $PG(n, q)$ and $AG(n, q)$ for all large enough n, *Simon Stevin* **1**, 43-50.

S. Rajola (1988), A blocking set in $PG(3, q)$, $q \geq 5$, *Combinatorics '86*, Ann. Discrete Math. **37**, 391-394, North-Holland, Amsterdam.

On totally irregular simple collineation groups

C.Y. Ho and A. Gonçalves

University of Florida
and Federal University of Rio de Janeiro.

1. Introduction

A collineation group of a finite projective plane is called totally irregular if the stabilizer of any point of the plane is non-trivial. This concept is developed on one hand from Hering's strong irreducibility (1979) (that is, no fixed point, line, triangle or proper subplane) and on the other hand from the result that the order of a projective plane is bounded by a function of the order of the collineation group and the number of its regular point orbits (Ho (1987)). Extensive research has been done on strongly irreducible collineation groups containing perspectivities, and the corresponding geometries. Non-abelian simple strongly irreducible collineation groups containing a perspectivity have been characterized (Reifart and Stroth (1982)). They are $PSL(2, q)$, q odd, $PSL(3, q)$, $PSU(3, q)$, A_7, or J_2. The only family of the rank one simple groups of Lie type not included in this list is $SZ(2^{2k+1})$, which are collineation groups of Lüneberg planes. Planes obtained in this way are Desarguesian (Hering and Walker (1977, 1979)). On the other hand, planes admitting totally irregular collineation groups include Desarguesian, Hughes, Figueroa, and all translation planes with non-trivial complement (Kantor shows that there is a translation plane with trivial complement).

The geometry of planes admitting small strongly irreducible collineation groups seems very difficult to determine. One evidence of this difficulty can be seen from the fact that $PSL(2, 5)$, $PSL(2, 7)$ and $PSL(2, 9)$ act on infinitely many Desarguesian planes. On the other hand totally irregular collineation groups appear to control the geometry of the planes better (see Gonçalves and Ho (1987)).

Classifications of Desarguesian planes by doubly transitive collineation groups (Ostrom-Wagner) or by the property that either each point is a centre or each line is an axis of a perspectivity (Cofman-Piper) are examples of totally irregular collineation groups with few orbits. Cyclic planes (in general planar difference sets) also admit totally irregular collineation groups, by using Hall's multiplier theorem (Hughes and Piper (1973)). In this paper the following two results determine the structure of a totally irregular non-abelian simple collineation group containing an involutory perspectivity.

Theorem 1. *Let* G *be a totally irregular non-abelian simple collineation group of a finite projective plane* Π *containing an involutory homology. Then one of the following occurs:*
(i) G *acts strongly irreducibly on the subplane generated by all centres and axes of involutory homologies of* G *and* G *is isomorphic to* $PSL(2, q)$, q *odd,* $PSL(3, q)$, $PSU(3, q)$, A_7, *or* J_2.
(ii) G *is isomorphic to* $PSU(3, 2^t)$ *and* G *fixes a point or a line which is not a centre or an axis of any perspectivity of* G. *Further, commuting involutions of* G *have a common centre and a common axis.*

In particular, if G *is not isomorphic to* $PSU(3, 2^t)$, *then* G *does not fix any point or line.*

Theorem 2. *Let* G *be a non abelian simple totally irregular collineation group of a finite projective plane containing an involutory elation. Then one of the following occurs:*
(i) G *acts strongly irreducibly on the subplane generated by all centres and axes of involutory elations of* G *and* G *is isomorphic to* $PSL(3, 2^t)$ *or* $PSU(3, 2^t)$.
(ii) G *is isomorphic to* $PSL(2, 2^t)$, $PSU(3, 2^t)$ *or* $SZ(2^{2k+1})$ *and* G *fixes a point or a line which is not a centre or an axis of any perspectivity of* G.

Note that in Theorems 1 and 2 the subplane generated by all centres and axes of involutory perspectivities of G coincide with the subplanes generated by all centres and axes of perspectivities of G. The following theorem analyzes the totally irregular action of $PSU(3, q)$, q odd.

Theorem 3. *Let* $G \cong PSU(3, q)$, q *odd, be a totally irregular collineation group of a finite projective plane* Π *of order* n, *containing an involutory homology. Let* $\hat{G}$ *be the full collineation group of* Π. *Then the number of subgroups of prime order of* G *is larger than* n *and one of the following occurs:*
(i) Π *is a Desarguesian plane of order* q^2 *or* q^4.
(ii) Π *is a non Desarguesian plane of order* q^4 *containing a* $\hat{G}$-*invariant Desarguesian subplane of order* q^2.
(iii) $G \trianglelefteq \hat{G}$, $n < q^8$, *and* $\hat{G}$ *leaves invariant a unital order* q.

More information concerning $PSU(3, q)$ can be found in Lemma 29 and Moorhouse (1987).

2   Defininitions, notations and known results

Let $\Pi = (\mathcal{P}, \mathcal{L})$ be a projective plane of order n, and let $G \leq \hat{G} = Aut(\Pi)$ be a collineation group of Π.   For prime p dividing $|G|$, $Syl_p(G)$ denotes the set of all Sylow p-subgroups of G.   For a subset K of G, $\mathcal{P}(K)$ (respectively $\mathcal{L}(K)$) denotes the set of fixed points (respectively lines) of K, $Fix(K) = (\mathcal{P}(K), \mathcal{L}(K))$, $K^\# = \{k \in K \mid k \neq 1\}$, and $I(K)$ is the set of involutions of K.

For $A \in \mathcal{P}, l \in L$, let $[A] = \{b \in L \mid A \in b\}$, $(l) = \{P \in \mathcal{P} \mid P \in l\}$.   For $a, b \in L$, we use $a \cap b$ to denote $(a) \cap (b)$.

We call a collineation $1 \neq \sigma$ a *generalized perspectivity* (respectively *perspectivity*) if $\mathcal{P}(\sigma) \subseteq$ (respectively $=$) $(a) \cup \{A\}$ and $L(\sigma) \subseteq$ (respectively $=$) $[A] \cup \{a\}$ for some $A \in \mathcal{P}$ and $a \in L$, and say that $A = C_\sigma$ is the *centre* and $a = a_\sigma$ the *axis* of σ.   A perspectivity is called an *elation* (respectively *homology*) if its centre is (respectively not) on its axis.   We call a collineation σ *planar*, *triangular*, *flag*, *anti-flag* respectively according to $Fix(\sigma)$ is a subplane, triangle, an incident point line pair, or a non incident point line pair.   The same terminology applies to collineation groups also.

For $A \in \mathcal{P}$ and $l \in L$, let $G(A, l)$ be the subgroup of G consists of perspectivities with centre A and axis l.   Let $G(A)$ (respectively, $G(a)$) be the subgroup of all perspectivities with centre A, (respectively, axis a) of G.   The substructure generated by centres and axes of involutory perspectivities of G is denoted by $\Pi(G)$.   We call a collineation group G *totally irregular* if each G-orbit of points is irregular (that is, the stabilizer of any point in G is not 1).   A collineation group is strongly irreducible if it does not leave invariant any point, line, triangle or proper subplane.

Here a simple group means a non-abelian simple group.   All objects considered in this paper are of finite cardinality.   Other notation and terminology concerning groups (respectively projective planes) can be found in Gorenstein (1968) or Huppert (1967) (respectively Dembowski (1968) or Hughes and Piper (1973)).   For the convenience of the reader, we record some known results used in this paper.

Theorem 4   (Hering (1979)).   *Let G be a strongly irreducible collineation group containing perspectivities.   Then G contains a unique minimal subgroup M and one of the following holds:*
(i) Each element of M is regular or planar.
(ii) $G \cong Z_3 \times Z_3$ and $CG(M) = M$.   Moreover either each subgroup of M is triangular, or M contains two triangular and two planar subgroups of order 3.
(iii) M is non-abelian simple.

Lemma 5   (Hering (1979)).   *A perspectivity leaving invariant a subplane has its centre and axis belonging to this subplane.*

Lemma 6   (Hering (1979)).   *Let α and β be two perspectivities of G and $\delta = \alpha\beta$.   Then δ is a generalized perspectivity.   If $C_\alpha \notin a_\beta$ and $C_\beta \notin a_\alpha$, then $\mathcal{P}(\alpha) \subseteq \{a_\alpha \cap a_\beta\} \cup (C_\alpha C_\beta)$.*

Theorem 7   (Hughes and Piper (1973)).   *Let Π_0 be a subplane of order m of Π.   Then either $n = m^2$ and Π_0 is a Baer subplane or else $m^2 + m \leq n$.*

Proposition 8   (Hughes and Piper (1973)).   *If $|G(A, l)| > 1$ for at least two choices of points A on l, then the subgroup of elations with common axis l is elementary abelian.*

Proposition 9   (Hughes and Piper (1973)).   *If α and β are two homologies with distinct centres A and B but common axis, then there exists an elation $\sigma \in \langle \alpha, \beta \rangle$ such that $A^\sigma = B$.*

Proposition 10   (Hughes and Piper (1973)).   *Let G be a permutation group on a set Ω, and let p be a prime.   If $\Omega_0 \subseteq \Omega$ is such that, for any X $\in \Omega_0$, G contains an element σ of order p fixing X and no other point of Ω, then Ω_0 is contained in an orbit of G.*

Theorem 11   (Gorenstein (1968)).   *Sylow 2-subgroups of a Frobenius complement are either cyclic or quaternion.*

Theorem 12   (Z^*-Theorem, Glaubermann (1966)).   *In a simple group a conjugacy class of involutions intersects any Sylow 2-subgroup in more than one element.*

Theorem 13. (Hering (1968)). *Let (Ω, G) be a transitive G-set where* $|\Omega| > 1$. *Assume that for some* $\alpha \in \Omega$, *the stabilizer* G_α *contains the normal subgroup Q of even order such that* $Q_\beta = 1$ *for all* $\beta \in \Omega \setminus \{\alpha\}$. *If S is the normal closure of Q in G, then one of the following occurs:*
(i) $S \cong SL(2, q)$, $SZ(q)$, $SU(3, q)$ *or* $PSU(3, q)$, *where q is a power of 2 and* $|\Omega| = q + 1, q^2 + 1, q^3 + 1, q^3 + 1$, *respectively.*
(ii) $S = QO(S)$ *and Q is a Frobenius complement.*

Theorem 14 (Reifart and Stroth (1982)). *A strongly irreducible simple collineation group containing perspectivities is isomorphic to one of the following:* $PSL(2, q)$, q *odd,* $PSL(3, q)$, $PSU(3, q)$, A_7, *or* J_2. *Moreover if $G \cong J_2$, then the order of the plane is m^4, where m is odd.*

Lemma 15 (Ho (1990)). *Supppose G is totally irregular.   Then $n <$* $|G| + \sqrt{|G|}$.

3. Totally irregular simple collineation groups containing perspectivities

In this section we assume that G is a totally irregular collineation group containing a perspectivity.

Lemma 16. *G does not fix the axis or centre of any perspectivity of G.*

Proof. Let α be an axis of a perspectivity of G and assume that G fix a. As G is a simple, this implies that a is a common axis for G.

Suppose G contains an elation.   Hence, as G is simple, $G = G(a, a)$, the elation group of G with axis a, which cannot be a totally irregular collineation group.   Therefore G does not contain any elation.   By Propositon 9, $G = G(A, a)$ with $A \notin a$.   However the only fixed point outside a for any non trivial element of G is A.   This contradicts the total irregularity of G.

Assume now that G fix a centre A of perspectivity of G.   The simplicity of G implies that $G = G(A)$, the subgroup of perspectivities with centre A. Suppose G contains an elation. Hence, as G is simple, $G = G(A, A)$ the subgroup of elations of G with centre A.   The total irregularity of G forces the existence of two elations with distinct axes. By dual result of Proposition 8, G is elementary abelian.   This contradiction proves that G does contain any elation. By the dual of Proposition 9, $G = G(A, a)$ with $A \notin a$.   This is a contradiction as before.   This proves Lemma 16.

Lemma 17. *Let S be a 2-group of collineations of* Π. *Suppose any two commuting involutory homologies of S have the same centre and axis. Then every homology of S has a common centre A and axis a. In particular, the set* $\{1\} \cup \{homologies\ of\ S\} = S(A, a)$ *is a characteristic subgroup of S.*

Proof. Let i be an involutory homology of S with centre A and axis a. Let j be any involutory homology of S. If $ij = ji$ then $C_j = A$, $a_j = a$, by our hypothesis.

Assume now that $ij \neq ji$. Then $y = ij$ is a 2-element of order $4m$. Let $k = y^{2m}$. Then k is the central involution of $\langle i, j \rangle$, a dihedral group of order $8m$. From $k = i(ik) = iy^m$, k is an involutory homology by Lemma 6. Our hypothesis now implies $C_i = C_k = C_j$, and $a_i = a_k = a_j$. Hence any involutory homology is in $S(A, a)$. Let h be any homology in S. The involution t of $\langle h \rangle$ is a homology with centre C_h and axis a_h. But $t \in S(A, a)$. This implies $h \in S(A, a)$. This proves Lemma 17.

Corollary 18. *Let S be as in Lemma 17. If S contains a homology, then S contains a central involutory homology.*

Proof. This follows from $1 \neq S(A, a) \trianglelefteq S$.

A similar argument applying to the case of elations yields the following.

Lemma 19. *Let S be a 2-group of collineations of* Π . *Suppose any two commuting involutory elations have the same centre (respectively axis). Then* $\{1\} \cup \{elations\ of\ S\}$ *is a characteristic subgroup of S with a common centre (respectively axis). If S contains an elation, then S contains a central involutory elation.*

4. Proof of Theorem 1

In this section we assume that G is a totally irregular simple collineation group containing an involutory homology. We divide the proof of the first part of Theorem 1 (that is, (i) or (ii) occurs) into three cases.

Case 1. G does not fix any point or line.

As G is simple, the substructure $\Pi(G)$ is a G-invariant subplane. By Lemma 5, G acts strongly irreducibly on $\Pi(G)$. By Theorem 14, Theorem 1(i) holds.

Case 2. G fixes a line l.

By Lemma 16, l is not an axis of any perspectivity of G. Hence centres of all perspectivities of G lie on l. In particular, any two commuting involutory homologies have the same centre and axis.   Let S be a Sylow 2-subgroup of G.   By Corollory 18, S contains a central involutory homology i. Let $C_i = A$ and $a_i = a$.   By Lemma 17, 1 $\neq S(A, a) \leq G(A, a)$. By Theorem 12, Glaubermann's Z^* theorem, $S(A, a)$ contains $j \in iG$ with $j \neq i$.   By Theorem 11, $G(A, a)$ cannot be a Frobenius complement.   However, as $A \notin a$, an elation in $G(A, A)$ does not fix any line not incident with A.   Thus $G(A, A)G(A, a)$ is a Frobenius group with kernel $G(A, A)$. So the subgroup of elations $G(A, A) = 1$. In a similar way $G(a, a) = 1$. From this, Propositon 9 and its dual imply that $G_A = G_a = G_{A,a}$. So $Q = G(A, a) \trianglelefteq G_A$.

Let $\Omega = \{$centres of involutory homologies of $G\}$. By Propositon 10, G is transitive on Ω. As $|Q|$ is even and commuting involutory homologies have the same centre and axis, Q acts semiregularly on $\Omega \backslash \{A\}$.   Since Q is not a Frobenius complement, Theorem 13 implies that G is 2-transitive on Ω and one of the following occurs, where q is an appropriate power of 2:

(i) $G \cong PSL(2, q)$, $|\Omega| = q + 1$.

(ii) $G \cong SZ(q)$, $|\Omega| = q^2 + 1$.

(iii) $G \cong PSU(3, q)$, $|\Omega| = q^3 + 1$.

We will show that both (i) and (ii) cannot occur in the present case.

Assume (i) occurs.   As G is generated by two Sylow 2-subgroups, which are elementary abelian, G fixes the point of intersection L of two involutory axes.   Hence all $q + 1$ involutory axes pass through L. Since odd order elements of G are inverted by involutions, by Lemma 6, fixed points of any non trivial element of G belong to l or the $q + 1$ involutory axes.   If $L \notin l$, then total irregularity of G forces $n = q$. However n is odd as there is an involutory homology.   Therefore $L \in l$.   This time total irregularity of G forces $n = q + 1$. Let H be a subgroup of order $q - 1$ normalizing a Sylow 2-subgroup T of G. Then H leaves invariant the common involutory axis t of T.   Since the fixed points of non-trivial elements of H lie on l, H acts fixed-point-freely on the points of t distinct from L. This implies $q - 1$ divides $q + 1$. As q is even, $q = 2$.   Then $G \cong PSL(2, 2)$ is not simple.   This contradiction proves (i) cannot occur.

Assume (ii) occurs. This time G is generated by the centres, which are elementary abelian, of two Sylow 2-subgroups.   As before, all $q^2 + 1$ involutory axes are concurrent with a point L. The fact that odd order elements of G are inverted by involutions and total irregularity of G

imply $L \in l$ as in (i).   Using the subgroup of order $q - 1$ in the normalizer of a Sylow 2-subgroup, a similar argument as in (i) yields $q - 1$ divides $q^2 + 1$.   This implies $q = 2$, a contradiction.   This proves (ii) cannot occur and Thorem 1(ii) holds in this case.

Case 3. G fixes a point.

Let $\Lambda = \{$axes of involutory homologies of $G\}$.   Using Λ in place of Ω, argue dually as in case (2), we obtain that commuting involutory homologies in G have the same centre and axis, and we reach the same three possibilities for G.   Since $PSL(2, q)$ and $SZ(q)$ are generated by two elementary abelian 2-subgroups, all involutory centres are collinear.   This is a line fixed by G in both situations.   We can now apply case (2) to eliminate these two possibilities.   This completes the proof of the first part of Theorem 1.

We now prove the second part of Theorem 1.   Suppose $G \ncong PSU(3, 2^t)$.   Hence (i) of Theorem 1 occurs.   Assume G fixes a point (respectively line).   By Lemma 16, all involutory axes (respectively centres) are incident with this point (respectively line).   Hence this point (respectively line) is in $\Pi(G)$.   This contradiction proves Theorem 1.

5 The group A_7

Proposition 20.   *A collineation group $G \cong A_7$ contains a perspectivity if and only if it contains an involutory perspectivity.*

Proof.   It suffices to prove that if G contains a perspectivity, then G contains an involutory perspectivity.   This follows from the fact that an involution of G can be expressed as a product of two elements of the same order 3, 5, 7 (one could see this by using a subgroup isomorphic to A_4, A_5, $PSL(2, 7)$ respectively).

In the rest of this section, let $G \cong A_7$ containing an involutory elation.   Since G has only one conjugacy class of involutions, all involutions of G are elations. G contains two conjugacy classes of Klein 4-groups, I and II.   A Sylow 2-subgroup of G is a dihedral group of order 8.   It has exactly two Klein 4-subgroups.   Hence one of these subgroups belongs to I and the other to II.

Lemma 21.   *If commuting involutions of G have a common axis (respectively centre) then G fixes a point (respectively line).*

Proof.   Let $H \cong PSL(3, 2)$ be a subgroup of G. Assume that commuting involutions of G have a common axis.   Given any two Sylow 2-

subgroups P and Q of H, there are $P = T_1, T_2, \ldots T_e = Q$, Sylow 2-subgroups such that $T_r \cap T_s \neq 1$ for $1 \leq r, s \leq e$. This implies that H have a common axis, say, a_H. As $G = \langle J, K \rangle$, where $J \cong PSL(3, 2) \cong K$, G fixes all points in $a_J \cap a_K$. The case where commuting involution have a common centre can be treated similarly.

Proposition 22. *If G contains an involutory elation, then G fixes a point or a line.*

Proof. Assume that G does not fix any point or line. Let $A \in I$ and $B \in II$ such that $\langle A, B \rangle \in Syl_2(G)$. Suppose the involutions of A have a common axis. If the involutions of B also have a common axis, then $\langle A, B \rangle$ will have a common axis, namely $a_{A \cap B}$. This is a contradiction by Lemma 21. Therefore involutions of B do not have a common axis. As commuting elations either have a common axis or a common centre, involutions of B have a common centre. If the involutions of A also have a common centre in the present case, then $\langle A, B \rangle$ will have a common centre, namely $C_{A \cap B}$. This contradicts Lemma 21. Hence involutions of A have distinct centres. Using a similar argument to treat the case in which involutions of A have a common centre, we may assume, without loss of generality, that the following holds.

(5.1) Involutions of a subgroup in I have a common axis but distinct centres, and involutions of a subgroup in II have a common centre but distinct axes.

We now prove Proposition 22 in the following steps.

(5.2) Suppose $E, F \in I$ and $\langle E, F \rangle \cong S_4$. Let $K = O_2(\langle E, F \rangle)$. Then $K \in II$ and $a_E \cap a_F = C_K$.

Proof. The fact that $K \in II$ is clear. This implies that $a_E \neq a_F$ by (5.1). As $K \in II$, K has a common centre C_K. Let $E \cap K = \langle e \rangle$ and $F \cap K = \langle f \rangle$. Then $C_K = C_e \in a_E$ as $E \in I$. Also $C_K = C_f \in a_F$ as $F \in I$. Therefore $C_K \in a_E \cap a_F$ and so $C_K = a_E \cap a_F$.

(5.3) For any involution e, $a_e = a_E$ for some $E \in I$ such that $e \in E$.
Proof. This follows from the fact that an involution always belongs to a subgroup in I.

(5.4) Let $E \neq F \in I$. Then $a_E \cap a_F \in \{$ centres of involutory elations of $G \}$.

Proof. Let $H = \langle E, F \rangle$ and let S be a Sylow 2-subgroup of H containing E.

Suppose $|S| = 8$. Then S contains a subgroup $A \in II$ and $S = \langle E, A \rangle$. Since S contains only one subgroup in I, $S \neq \langle E, F \rangle$. If $a_E = a_F$, then H has a common axis. This implies that S has a common axis, which contradicts (5.1). Hence $a_E \cap a_F = P$ is a point. Since involutions are elations and $P^i = P$ for any involution i of H, $P \in a_i$. By (5.1) A has a common centre but distinct axes. So C_A is the intersection of the distinct axes of the three involutions of A. But this forces $P = C_A$ and the proof in this case is complete.

Suppose $|S| = 4$. Again $S \neq H$. Then $H \cong A_5$. If $a_E = a_F$, then H has a common axis. Let $K \in HG$ such that $\langle H, K \rangle = G$. This will imply that G fixes a point of $a_H \cap a_K$, a contradiction. Therefore $a_E \neq a_F$. Let $P = a_E \cap a_F$. As before, all axes of involutions of H are incident with P. Let $N = NG(H)$. Then $N \cong S_5$, and $P^N = P$. Hence the axes of all involutions of N are incident with P. Let Q be a Sylow 2-subgroup of N containing E. Then $Q = \langle E, B \rangle$ with $B \in II$. By (5.1), C_B is the intersection of the distinct axes of the three involutions of B. This forces $P = C_B$. The proof of (5.4) is now complete.

Let Γ (respectively Λ) be the set of centres (respectively axes) of all involutions of G. By (5.3) and (5.4), $\{\Gamma, \Lambda\}$ is a closed substructure. Hence it is a subplane. Since the stabilizer of an involutory axis contains the normalizer of a Klein 4-group in I, $|\Gamma|$ divides $3 \cdot 5 \cdot 7$. However, there is no projective plane having 3, 5, 15, 35 or 105 points and a projective plane of order 2 or 4 does not admit A_7 as a collineation group. This contradiction proves Proposition 22.

A consequence of this is that if $G \cong A_7$ is totally irregular containing a perspectivity, then its involutions are homologies. Instead of giving a direct proof now, we present the proof in 6.1 of the next section.

6. Proof of Theorem 2

In this section we assume that G is a totally irregular simple collineation group containing an involutory elation. We divide the proof of Theorem 2 into three cases.

Case 1. G does not fix any point or line.

As G is simple, the substructure $\Pi(G)$, generated by centres and axes of all involutory elations of G, is a G-invariant subplane. By Lemma 5, G acts strongly irreducibly on $\Pi(G)$. By Theorem 14, G is isomorphic to one of the following: $PSL(2, q)$, q odd, $PSL(3, q)$,

$PSU(3, q)$, A_7 or J_2. Moreover, if $G \cong J_2$, then $n = m^4$ is odd. This implies that J_2 cannot occur. If $G \cong A_7$, then by Proposition 22, G cannot be strongly irreducible. By Hering and Walker (1979), $PSL(2, q)$ occurs as a strongly irreducible collineation group containing an elation only when $q = 7$. However $PSL(2, 7) \cong PSL(3, 2)$. Thus Theorem 2(i) holds in this case.

Case 2. G fixes a line l.

By Lemma 16, l is not an axis of any perspectivity of G. Hence centres of all perspectivities lie on l. In particular, any two commuting involutory elations have the same centre. Let S be a Sylow 2-subgroup of G. By Lemma 19, S contains a central involutory elation i. Let $C_i = A$. Thus $1 \neq S(A) \leq G(A)$, and $S(A)$ is a characteristic subgroup of S. As G is simple, by Theorem 12, $S(A)$ contains $j \in iG$, with $j \neq i$. By Theorem 11, $G(A) = Q$ cannot be a Frobenius complement.

Let $\Omega = \{$centres of involutory elations of $G\}$. By Proposition 10, G is transitive on Ω. Also $Q = G(A) \trianglelefteq G_A$ is even and acts semi-regularly on $\Omega \setminus \{A\}$. Since Q is not a Frobenius complement, Theorem 13 implies that Theorem 2(ii) holds.

Case 3. G fixes a point L.

By Lemma 16, L is not a centre of any perspectivity of G. Hence axes of all perspectivities of G pass through L. In particular, any two commuting involutory elations have the same axis. Let $\Lambda = \{$axes of involutory elations of $G\}$. Using Λ in place Ω and arguing dually we have again by Theorem 13 that Theorem 2(ii) holds. This completes the proof of Theorem 2.

Corollary 23. *If A_7 is a totally irregular collineation group containing a perspectivity, then an involution is a homology.*

Proof. This follows from Theorem 2, Propositions 20 and 22.

7. Proof of Theorem 3.

In this section we assume that $G \cong PSU(3, q)$, $q = p^r$ odd, is a totally irregular collineation group of Π containing involutory perspectivities. As q is odd, involutions of G are homologies by Theorem 2.

In addition to the notations in §2, we let $\Pi(G) = (\overline{\mathcal{P}}, \overline{L})$ be the substructure generated by centres and axes of all involutory homologies of G, and $\Pi(\hat{G}) = (\hat{\mathcal{P}}, \hat{L})$ be the substructure generated by centres and axes of all involutory homologies of $\hat{G} = Aut(\Pi)$. Also let $P \in Syl_p(G)$, $Z(P)$ be the centre of P, and $H = NG(P)$. For any subplane $\Gamma, o(\Gamma)$ denotes the order of Γ. Let $\Omega = \{X \in \mathcal{P} \mid X \notin b, \forall b \in \overline{L}\}$.

Lemma 24. $n < q^8$.

Proof. Since $|G| = q^3(q^3 + 1)\dfrac{(q^2 - 1)}{d}$, where $d = (3, q - 1)$, so $|G| \leq q^8 - (q^6 + q^3 - q^5)$. Hence $\sqrt{|G|} < q^4$. Therefore $n < |G| + \sqrt{|G|} < q^8$ by Lemma 15.

Lemma 25. If $\Pi(G)$ is a subplane of order m, then $|\Omega| = (n - m^2)(n - m^4)$.

Proof. For an integer r, let $v(r) = r^2 + r + 1$. Then $|\Omega| = v(n) - v(m) - v(m)[n + 1 - (m + 1)] = [v(n) - v(m)n] + v(m)(m - 1) = [n^2 + 1 - (m^2 + m)n] + m^3 - 1 = (n - m^2)(n - m)$.

Lemma 26. If Π is Desarguesian, then $\Pi(G) \cong PG(2, q^2)$ and $n = q^2$ or q^4.

Proof. By Theorem 1, G acts strongly irreducibly on $\Pi(G)$. Since Π is Desarguesian, $n = q^{2m}$ for some m by Mitchell (1911). By Lemma 24, $n = q^2, q^4$, or q^6.

Assume $n = q^6$. A subgroup L of order $(q^2 - q + 1)/d$, where $d = 1$ or 3 is cyclic. Note that 3 divides $q^2 - q + 1$ only when $q + 1 \equiv 0 \pmod 3$. In this case the element of order 3 normalizing L is diagonalizable over $GF(q^2)$. Hence it has three non-collinear fixed points in $\Pi(G)$. Since this element is induced from a linear transformation, it is not planar. Hence this element cannot fix any point in Ω.

Let $A \in \Omega$. If $g \in G_A$, then $g \in H \in LG$. By Mitchell (1911), $|\mathcal{P}(L)| = 3$ and $NG(L)$ has an element of order 3 permuting these three points. This implies that $\mathcal{P}(h) = \mathcal{P}(L)$ for all $1 \neq h \in L$. As L acts semi-regularly on $\Pi(G)$, $\mathcal{P}(L) \subseteq \Omega$. For $J \neq K \in LG, \langle J, K \rangle$ contains elements of order not dividing $q^2 - q + 1$. Hence $\mathcal{P}(J) \cap \mathcal{P}(K) = \phi$. From this and total irregularity of G we obtain $|\Omega| = |\mathcal{P}(L)| \cdot |G: NG(L)|$. As $|NG(L)| = 3|L|$, this implies $|\Omega| = |G| / |L|$. However $|\Omega| = (q^6 - q^4)(q^6 - q^2)$. This contradiction proves the lemma.

Lemma 27. Assume $q \neq 3$ and $\Pi(G) \ncong PG(2, q)$. Then $o(\Pi(G)) > q^3$.

Proof. By Gonçalves and Ho (1987), $H = NG(P)$ fixes a point line pair (P^*, l^*), and $P^* \notin l^*$ as $\Pi(G) \not\leq PG(2, q)$, and $o(\Pi(G)) \geq 2q^2 - 1$.

Assume, by contradiction, $o(\Pi(G)) < q^3$. As $Z(P)$ fixes at least $2q^2$ points on l^* and $\Pi(G) \not\leq \mathcal{P}(Z(P))$, $Z(P)$ is a generalized homology on $\Pi(G)$. In particular, $o(\Pi(G)) = \bar{n} \equiv 1 \pmod{q}$. By Gonçalves and Ho (1987), P acts fixed-point-freely on $\overline{\mathcal{P}} \backslash \{P^*\} \cap (l^*)$. This implies that $q^3 \mid \bar{n} - 1$. As q is odd and $q \mid \bar{n} - 1$, we obtain $q^3 \mid \bar{n} - 1$. Thus $\bar{n} > q^3$. This contradiction proves the lemma.

Lemma 28. *The number of subgroups of prime order of G is larger than n.*

Proof. We count the fixed points of non-trivial elements of G in the following way.

Let $I = \{$points fixed by involutions$\}$. Using the properties that involutions are homologies; distinct involutions have distinct centres and axes; and each involution commutes with $q^2 - q$ other involutions, we obtain

$$|I| \leq (\text{number of involutions})(n + 2 - q^2 - q) \tag{1}$$

Let Σ be the set of subgroups of G of odd prime order such that a generator is either inverted or centralized by an involution. Let $\langle \sigma \rangle \in \Sigma$ such that σ is either inverted or centralized by an involution i, and let $\Gamma(\sigma) = \mathcal{P}(\sigma) \backslash I$. If σ is a generalized perspectivity, then $|G(\sigma)| < n - 1$. Suppose that σ is planar, then i induces an homology of $Fix(\sigma)$. This implies that $|G(\sigma)| \leq o(Fix(\sigma))^2 - 1 \leq n - 1$. Let $II = \{$ points outside I fixed by a subgroup in $\Sigma \}$. Then

$$|II| \leq |\Sigma|(n - 1). \tag{2}$$

Let Ψ be the set of subgroups of G of odd prime order such that no generator is inverted or centralized by an involution. Let $\langle v \rangle \in \Psi$ and $\Lambda(v) = \mathcal{P}(v) \backslash \{I \cap II\}$. If $T = Fix(v)$ is not a Baer subplane, then $|\Lambda| \leq n - 1$. Suppose T is a Baer subplane. Assume $q = 3$. Then v has order 7. Let a be the axis of an involution i. Since T is a Baer subplane, a carries at least one point A of T. Thus $|G_A|$ has order divisible by 2 and 7. This implies that $G_A \cong PSL(2, 7)$ as $G_A \neq G$. But then A is fixed by at least three commuting involutions. This contradiction enables us to assume that $q \neq 3$. Let H be the normalizer of a Sylow p-

subgroup of G, where q is a power of p. By Gonçalves and Ho (1987), the q^2 distinct centres of involutions of H are incident with a line l^*. Note that the points of l^* are in $I \cap II$. Since T is a Baer subplane, l^* carries a point D of T. Since distinct involutions have distinct centres, D cannot be a centre of an involution. Since H is a maximal subgroup, D cannot be the point P^* in Gonçalves and Ho (1987) fixed by H. So D must be of type B in Gonçalves and Ho (1987). By Gonçalves and Ho (1987) table I, $G_D \subseteq G_{l^*} = H$. This implies that $v \in H$. Thus v commutes with an involution. This contradiction proves that $|\Lambda| \leq n - 1$. Let $III = \{$points outside $I \cup II$ fixed by a subgroup in $\Psi\}$. Then

$$|III| \leq |\Psi|(n - 1) \tag{3}$$

Let the number of subgroups of prime order of G be c. From the total irregularity of G, (1), (2) and (3) we obtain $n^2 + n + 1 < c(n - 1)$. This implies that $n < c$ as desired.

Lemma 29. *Assume $q = 3$. Then the following hold.*
(i) *If $Z(P)$ is not planar, then $\Pi(G) \cong PG(2, 9)$ and $n = 9$ or 81.*
(ii) *If $Z(P)$ is planar, then $\Pi = \Pi(G)$ and G acts strongly irreducibly on Π.*

Proof. (i) The present condition implies that $\Pi \cong PG(2, 9)$ and $9 = q^2 \mid n$ by Gonçalves and Ho (1987).

Assume $\Omega \neq \phi$. As $q + 1 = 4$, $q - 1 = 2$, the only possible order of an element fixing a point in Ω is 7. Let L be a subgroup of order 7. Then $NG(L)$ is a Frobenius group of order 21, in which an element τ of order 3 is inverted by an involution of G. So $|\mathcal{P}(L)| \equiv 0 \pmod 3$. As τ has no fixed point or line in Ω, $Fix(L)$ is a triangle or a subplane. This implies that $|\mathcal{P}(L)| \equiv 3 \pmod 9$. Let $n = 9s$. By Lemma 25, $|\Omega| = (n - 3^4)(n - 3^2) = 3^4(s - 9)(s - 1)$. On the other hand, $|\Omega| = |\mathcal{P}(L)||G|/21$. Hence 9 divides $|\mathcal{P}(L)|$. This contradiction establishes (i).

(ii) Assume $Z(P)$ is planar. By Gonçalves and Ho (1987), $\Pi(H)$ is a subplane of the subplane $Fix(Z(P))$, and $H/Z(P) \cong (Z_3 \times Z_3).Z_8$ acts strongly irreducibly on $\Pi(H)$.

Let $\sigma \in P \backslash Z(P)$. As σ is inverted by an involution in H, by Hering (1979) σ is triangular. Thus σ fixes a non-incident point line pair such that the point (respectively line) is incident with three axes (respectively centres) of involutory homologies.   From this we obtain $o(\Pi(H)) \geq 7$ as $o(\Pi(H)) \equiv 1 \pmod 3$.   As $\Pi(G) \not\subseteq Fix(Z(P))$, so $\Pi(H) \subset \Pi(G)$.   Thus $m = o(\Pi(G)) \geq 7^2$.

We now count the fixed points of subgroups of prime order as follows: the points of $\Pi(G)$; the fixed points outside $\Pi(G)$ of an involution or a subgroup generated by a non central 3-element;   the fixed points of a subgroup generated by a central 3-element or 7-element.   Using the total irregularity of G and $q = 3$, we obtain: $n^2 + n + 1 \leq (m^2 + m + 1) + (q^2(q^2 - q + 1) + q(q + 1)(q^3 + 1))(n - m) + ((q^3 + 1) + (q^3(q^2 - 1)(q + 1)/3))(n + \sqrt{n} + 1)$.   After dividing by $n + \sqrt{n} + 1$, a direct computation shows that $n < 7^4$.   However, $7^2 \leq m \leq \sqrt{n}$.   This contradiction establishes (ii).

(iii)   If J_2 acts strongly irreducibly on Π, then by Theorem 14 and Lemma 25, $n = 3^4, 5^4, 7^4$.   By Prince (1988), there exist three J_2 orbits of sizes: 315, 1008 and 8400.   This eliminates the possibility $n = 3^4$.   The possibility $n = 7^4$ is eliminated by Lemma 28.   Suppose $n = 5^4$.   By Gonçalves and Ho (1987) and the action of the central 3-element we see that a non central 3-element is triangular.   Hence the fixed points of the non-central 3-elements belong to the union of 2-axes.   Thus in the proof of Lemma 28, we could drop the 336 subgroups generated by non central 3-elements and obtain that the order of the plane n is less than 379 which is the number of subgroups corresponding to 63 involutions, 28 subgroups generated by central 3-elements, and 288 subgroups of order 7.   However this implies that $n = 5^4 < 379$.   This contradiction proves the lemma.

Lemma 30.   *Assume that G is not strongly irreducible, then the following holds.*
(i)   *If $q = 3$, then $Z(P)$ is not planar, $\Pi(G) \cong PG(2, 9)$, and $n = 9$ or 81.*
(ii)   *If $q \neq 3$ and $n \leq q^6$, then $\Pi(G) \cong PG(2, q^2)$.*

Proof.   As G is not strongly irreducible, $\Pi(G) \neq \Pi$.   If $q = 3$, (i) follows from Lemma 29.   If $q \neq 3$, (ii) follows from Lemma 27.

We now prove Theorem 3.   Lemma 28 proves that the number of subgroups of prime order of G is larger than n.   As q is odd, Theorem 1 implies that G (respectively $\Pi(\hat{G})$) acts strongly irreducibly on the subplane $\Pi(G)$ (respectively $\Pi(\hat{G})$).   Let $\hat{K}$ be the kernel of the action of $\hat{G}$ on $\Pi(\hat{G})$.   Since $\Pi(G) \subseteq \Pi(\hat{G})$, $G \cap \hat{K} = 1$.   So $G \cong G\hat{K}/\hat{K} \leq \hat{G}/\hat{K}$.   Let $1 \neq k \in \hat{K}$ and i be an involution of G.   If $[i, k] \neq 1$, then as a product of two involutions, it cannot be planar. However $[i, k] \in \hat{K}$. This implies that $[i, k] = 1$. Hence $G\hat{K} = G \times \hat{K}$.

We look at $\Pi(\hat{G})$ for a moment. The restriction of G on $\Pi(\hat{G})$ is isomorphic to G and we use G to denote this restriction. By Theorem 4, $\hat{G}/\hat{K}$ contains a unique minimal normal subgroup M, which is simple. Since $\hat{G}/\hat{K}$ is solvable, $G \leq M$. As M is a normal subgroup containing perspectivities, M acts strongly irreducibly on $\Pi(\hat{G})$. As $G \cong PSU(3, q)$, Theorem 14 implies that M is isomorphic to one of the following groups: $PSL(3, \hat{q}) J_2, PSU(3, \hat{q})$.

Assume $M \cong PSL(3, \hat{q})$. By Bloom (1967), $\hat{q} = q^{2t}$ for some $t \geq 1$. By Hering and Walker (1979), $\Pi(\hat{G}) \cong PG(2, \hat{q})$. As G acts totally irregularly on $\Pi(\hat{G})$, by Lemma 26 we obtain that $\hat{q} = q^2$ or q^4 and $\Pi(G) \cong PG(2, q^2)$. By Lemma 24, $\Pi = \Pi(\hat{G})$ is Desarguesian if $o(\Pi(\hat{G})) = q^4$. If $o(\Pi(\hat{G})) = q^2$, then it is a $\hat{G}$ invariant Desarguesian subplane.

By Lemma 29 we have $M \not\cong J_2$.

Assume $M \cong PSU(3, \hat{q})$. By Mitchell (1911), $\hat{q} = q^t$, with t odd. Suppose $t = 1$. Then $G \trianglelefteq \hat{G}$, $\Pi(G) = \Pi(\hat{G})$ and, by Gonçalves and Ho (1987), $\hat{G}$ leaves invariant a unital of order q in $\Pi(G)$. Suppose $t > 1$. So $\hat{q} \geq q^3$. By Lemma 25, $G \not\trianglelefteq \hat{G}$, $G \subset M$. By Lemma 24 and Lemma 27, as $\hat{q} \neq 3$, $\Pi(\hat{G}) \cong PG(2, (\hat{q})^2)$ is a Desarguesian subplane of $(\hat{q})^2 = q^{2t} \geq q^6$. As $\Pi(G) \subset \Pi(\hat{G})$, $\Pi(G) \cong PG(2, q^2)$. By Lemma 24, $\Pi = \Pi(\hat{G}) \cong PG(2, q^6)$. This contradicts Lemma 26. This completes the proof of Theorem 3.

As a consequence of the lemmas of this section and Theorem 3, we have the following.

Theorem 31. *If G does not act strongly irreducibly on Π and $G \not\trianglelefteq \hat{G}$, then $\Pi(G) \cong PG(2, q^2)$ and $n = q^4$. Moreover Π is either Desarguesian or contains $\Pi(G)$ as a $\hat{G}$ proper invariant subplane.*

Acknowledgement. The first author was partially supported by a NSA grant MDA 904-90-H-1013. The second author wishes to express his gratitude to CNPq of Brasil and to the University of Florida.

References.

D. Bloom (1967), The subgroups of $PSL(3, q)$ for q odd, *Trans. Amer. Math. Soc.* **127**, 150-178.
P. Dembowski (1970), *Finite Geometries*, Springer-Verlag, Berlin.

A. Gonçalves and C.Y. Ho (1987), On *PSU*(3, *q*) as collineation groups, *J. Algebra* **111**, 1-13.

G. Glaubermann (1966), Central elements in core-free groups, *J. Algebra* **4**, 403-420.

D. Gorenstein (1968), *Finite Groups*, Harper and Row, New York.

C. Hering (1968), Zweifach transitive Permutationsgruppen, in denen zwei die maximale Anzahl von Fixpunkten von involutionen ist, *Math. Z.* **105**, 150-147

C. Hering (1979), On the structure of finite collineation groups of projective planes, *Abh. Math. Sem. Univ. Hamburg* **49**, 155-182.

C. Hering and M. Walker (1977), Perspectivities in irreducible collineation groups of projective planes I, *Math. Z.* **155**, 95-101.

C. Hering and M. Walker (1979), Perspectivities in irreducible collineation groups of projective planes II, *J. Statist. Planning Inference* **3**, 151-177.

C.Y. Ho (1987), On the order of a projective plane with a totally irregular collineation group, *Proc. Sympos. Pure Math.* **47**, Part 2, 423-429, American Mathematical Society, Providence.

C.Y. Ho (1990), Totally irregular collineation groups and finite Desarguesian planes, *Coding Theory and Design Theory, Part II*, IMA Volumes in Mathematics and its Applications vol **21**, 127-131.

D. Hughes and F. Piper (1973), *Projective Planes*, Springer-Verlag, Berlin.

B. Huppert (1967), *Endliche Gruppen I*, Springer-Verlag, Berlin.

H. Mitchell (1911), Determination of the ordinary and modular ternary linear groups, *Trans. Amer. Math. Soc.* **12**, 207-242.

E. Moorhouse (1987), Unitary and other linear groups acting on finite projective planes, PhD thesis, *Univ. of Toronto*.

A. Reifart and G. Stroth (1982), On finite groups containing perspectivities, *Geom. Dedicata* **13**, 7-46.

A. Prince (1988), On certain permutation representations of the Hall-Janko group, *Proc. Roy. Soc. Edinburgh Sect A* **110**, 295-303.

On some rank 3 partial geometries

D.R. Hughes
Queen Mary and Westfield College, London

1. Introduction

If we let $\overset{\alpha}{\circ\!=\!=\!=\!\circ}$ represent the family of partial α-geometries, then in the natural way,

$$\underset{\alpha_1}{\circ\!=\!=\!\circ}\underset{\alpha_2}{=\!\circ} \cdots \underset{\alpha_{n-1}}{\circ\!=\!=\!\circ} \tag{1}$$

with n nodes, represents a family of geometries which we call *rank n partial geometries*. (In this paper we also assume residual connectedness and the Intersection Property.) We suppose that the rank 2 residues have *order* (sometimes these numbers have been called *parameters* or *indices*): a pair (s, t) for a diagram $\underset{s}{\circ\!=\!=}\overset{\alpha}{=\!=}\underset{t}{\circ}$, indicating that there are $s + 1$ points on a line and $t + 1$ lines on a point. But there are geometries for the rank n diagram above, without the parameters α_i, where the α_i are not constant, and we avoid this situation by including the α_i in the definitions. Then each stroke above will have three parameters, hence the whole picture has the form:

$$\underset{s_0}{\circ\!=}\underset{\alpha_1}{=}\underset{s_1}{\circ}\underset{\alpha_2}{=}\underset{s_2}{\circ} \cdots \underset{s_{n-2}}{\circ}\underset{\alpha_{n-1}}{=}\underset{s_{n-1}}{\circ} \tag{2}$$

A geometry for (2) will be said to have *order* $(s_0, s_1, s_2, ..., s_{n-1})$.

Such geometries have been studied in a large number of special cases, and we wish to draw together some of these different results in this paper. In section 2 we give a survey of many known results; some of the material in this section was covered long ago in Buekenhout (1979), and we include it for completeness, and to make this paper easier to read. In section 3 we give some constructions of rank 3 pGs. In section 4 we introduce the important function φ and the concept of *triangularity*, which play central roles in these geometries. In section 5 we consider a special class called L.pG geometries, and section 6 determines bounds on the diameter for L.pG geometries, giving "weak finiteness theorems" (Cameron (to appear)). Then in section 7 we consider the sub-family of A_2.pG geometries, and give characterisations in some cases; here we get "strong finiteness theorems". Section 8 introduces another class of rank 3 pGs, the pG.L geometries, and section 9 gives some partial diameter bounds in this family.   In sections 5, 7 and 8 we pay particular attention to the triangular case.   Note that arbitrary rank 3 pGs do not satisfy even weak finiteness conditions, as shown by the geometries for

$$\underset{s}{\circ}\!=\!=\!=\!\underset{t}{\circ}\!=\!=\!\underset{u}{\circ}:$$ there are infinite geometries even for the diagram $\underset{1}{\circ}\!=\!=\!\underset{1}{\circ}\!=\!=\!\underset{1}{\circ}$.

For background about diagrams, see Buekenhout (1981, 1985).   We have chosen the diagram $\circ\!=\!=\!=\!\circ$ (rather than, say, $\circ\!\!\overset{Pg}{-\!-}\!\!\circ$) for pGs since (among other things) it is often useful, or necessary to indicate the parameter "α".   (And also a pG is "almost" a GQ and not far from linear space.)   It is sometimes convenient (eg., on a blackboard) to write $\circ\!=\!\!=\!\!\circ$ as $\circ\!=\!\!=\!\circ$, and to write $\circ\!=\!\!\underset{\alpha}{}\!=\!\circ$ as $\circ\!=\!\!\underset{\alpha}{}\!=\!\circ$.   Unless otherwise stated, all geometries in this paper are finite.

The author would like to express his gratitude to several colleagues for their helpful comments and mathematical observations: especially Andries Brouwer, Francis Buekenhout, Frank De Clerck, Anne Delandtsheer, Sylvia Hobart, Dimitrii Pasechnik and Joseph Thas.   And above all, we would like to express our appreciation of the many useful interventions of Antonio Pasini, whose contributions are found everywhere in this paper.

There have been other investigations of rank n pGs, sometimes, as in Laskar and Dunbar (1978), with a number of strong extra conditions, and a somewhat different aim;   and sometimes, as in Del Fra et. al. (this volume), with a large rank (as well as slightly different restrictions on the rank 2 residues which occur).   In any case, it appears that much of the difficulty in a (hypothetical, and indeed unlikely) complete analysis of rank n pGs will lie in the rank 3 case.

2. Diagrams and geometries

A *geometry* or *structure S of rank n* is an incidence structure with n *types* of objects, called {0, 1, 2, ..., n - 1}, or *points, lines, planes, ..., hyperplanes,* with an incidence relation such that two objects of the same type are not incident (so we could think of an n-partite graph instead).   A *flag* is a collection of objects $X_1, X_2, ..., X_i$ of S such that any two are incident, and a *maximal* flag, or a *chamber*, is a flag with n types.   If $F = \{X_1, X_2, ..., X_i\}$ is a flag, then the *residue* S_F is the set of all objects of S which are incident with every object of F.   We say that a geometry is *connected* if given any two types i, j, the incidence graph of the objects of S of types i and j is a connected graph;   then S is *residually connected* if S and all residues S_F, where F consists of fewer than $n - 1$ objects, are connected. S has the *intersection property*, abbreviated IP, if, for any two flags F_1 and F_2, the point set of the intersection $S_{F_1} \cap S_{F_2}$ is either empty or is the point set of the residue of a flag F which is incident to both F_1 and F_2.   For more detailed information, see Buekenhout (1981, 1985).   Often the objects of type 0 are called *points* and those of type 1, *lines*; we call S a *semilinear geometry*, or *space*, if any two points are on at most one common line: this is a weak form of IP. ("Semilinear" is sometimes called "partial linear", which might be confusing in this paper, and is grammatically suspect in any event.)

We shall use the term *antiflag* to refer to a pair (P, X) where P is a point and X is an object of some fixed type j ($\neq$ points) such that P is not on X. The type j chosen will depend on the context.

In the rank two case, semilinear geometries are especially important.   Such a geometry is *linear* if two points are always on exactly one line; a *dual* linear space is a semilinear space in which every pair of lines have exactly one point in common.   Here *antiflag* will mean a pair (P, y) where y is a line not on the point P, and two points are *collinear* if they lie on a common line.

Definition 1.   The semilinear space S is called a *partial α-geometry of order* (s, t), where α, s, t are positive integers, if
(1)   every line of S has $s + 1$ points;
(2)   every point of S is on $t + 1$ lines;
(3)   if (P, y) is an antiflag, then there are exactly α points on y collinear with P.

We say that such an S is a $pG_\alpha(s, t)$, or a pG_α, or merely a pG.

Definition 2.   The family of pGs is classified into subfamilies as follows:
(1)   A $pG_{s+1}(s, t)$ is a *linear space*, more precisely a 2-design for $(s(t + 1) + 1, s + 1, 1)$.
(2)   A $pG_{t+1}(s, t)$ is a *dual linear space*, more precisely the dual of a 2-design for $(t(s + 1) + 1, t + 1, 1)$.
(3)   A $pG_t(s, t)$ is a *net*, more precisely a net of type $(s + 1, t + 1)$.
(4)   A $pG_s(s, t)$ is a *dual net*, or sometimes a 2-*transversal design*.
(5)   A $pG_1(s, t)$ is a *generalised quadrangle* of order (s, t), or a $GQ(s, t)$.
(6)   A $pG_\alpha(s, t)$, with $1 < \alpha < \min\{s, t\}$ is a *proper pG*.

Now we describe somewhat intuitively the meaning of the diagram (Δ), with $n > 2$;   in what follows, "indices" are written above a node to indicate the type associated with that node.

$$\text{(}\Delta\text{)}$$

First, a $pG_\alpha(s, t)$ *belongs to* $\underset{s}{\text{o}}\!\!=\!\!=\!\!\overset{\alpha}{\rule{1.5cm}{0pt}}\!\!=\!\!=\!\!\underset{t}{\text{o}}$.   Then a geometry S of rank n, with types $\{0, 1, 2, ..., n - 1\}$, *belongs to* (Δ) if, whenever X is an object of type i then the residue S_X belongs to (Δ_i), which is (Δ) with node i deleted, and the edges on that node deleted.   If this results in (Δ_i) being disconnected, eg.:

$$(\Delta_i)$$

Then in the residue S_X , every object of type $0, 1, ... , i - 1$ is incident with every object of type $i+ 1, i + 2, ..., n - 1$.

Definition 3.   A residually connected geometry S of rank n which belongs to (Δ) is a *rank n partial geometry*, or *rank n pG*.

As a rule, we shall want our geometries to have IP as well, *and except where the contrary is stated, it is always implicitly assumed.*   To avoid confusion, we always mean "rank 2" when we refer to a pG, and write rank n pG when the rank is higher.   There are "special" diagrams for many subfamilies of pG.

Definition 4.   The following diagrams are used for the subfamilies given:

(1) $\underset{}{\text{o}}\!\!\overset{L}{\rule{1cm}{0pt}}\!\!\text{o}$ for linear space, in general;

$\text{o}\rule{1cm}{0pt}\text{o} = \underset{s}{\text{o}}\!\!=\!\!=\!\!\overset{s+1}{\rule{1cm}{0pt}}\!\!=\!\!=\!\!\underset{s}{\text{o}}$ for projective plane, in this case of order s, (that is, both a linear space and a dual linear space);

$\underset{}{\text{o}}\!\!\overset{C}{\rule{1cm}{0pt}}\!\!\text{o} = \underset{1}{\text{o}}\!\!=\!\!=\!\!\overset{2}{\rule{1cm}{0pt}}\!\!=\!\!=\!\!\underset{t}{\text{o}}$, or equivalently, a 2-design with block-size two, in this case with $t + 2$ points;

$\underset{}{\text{o}}\!\!\overset{Af}{\rule{1cm}{0pt}}\!\!\text{o} = \underset{q-1}{\text{o}}\!\!=\!\!=\!\!\overset{q}{\rule{1cm}{0pt}}\!\!=\!\!=\!\!\underset{q}{\text{o}}$ for affine plane, in this case of order q.

(2) $\underset{}{\text{o}}\!\!\overset{\text{J}}{\rule{1cm}{0pt}}\!\!\text{o}$ for dual linear space, with appropriate adaption from (1) above, eg., $\text{o}\!\!\overset{\supset}{\rule{1cm}{0pt}}\!\!\text{o}$, and either $\underset{}{\text{o}}\!\!\overset{fA}{\rule{1cm}{0pt}}\!\!\text{o}$ or $\underset{}{\text{o}}\!\!\overset{Af*}{\rule{1cm}{0pt}}\!\!\text{o}$ for dual affine plane.

(3) $\underset{}{\text{o}}\!\!\overset{N}{\rule{1cm}{0pt}}\!\!\text{o} = \underset{s}{\text{o}}\!\!=\!\!=\!\!\overset{t}{\rule{1cm}{0pt}}\!\!=\!\!=\!\!\underset{t}{\text{o}}$ for net; in this case with $s + 1$ points on a line, and $t + 1$ parallel classes;

$\text{o}\!\!\overset{*}{\rule{1cm}{0pt}}\!\!\text{o} = \underset{q-1}{\text{o}}\!\!=\!\!=\!\!\overset{q-1}{\rule{1cm}{0pt}}\!\!=\!\!=\!\!\underset{q-1}{\text{o}}$ for "doubly affine plane" with q points on a line. (That is, both a net and a dual net).

(4) $\underset{s}{\circ}\overset{\mathcal{N}}{\quad\quad}\circ = \underset{s}{\overset{}{\circ}}\overset{}{\underset{t}{\rule{3em}{0.4pt}}}\overset{}{\underset{t}{\circ}}$ for dual net, or 2-transversal design

(sometimes written $\circ\overset{N^*}{\quad\quad}\circ$); in this case with $(s + 1)(t + 1)$ points in $s + 1$ equivalence classes of $t + 1$ points each.

(5) $\circ\!\!=\!\!=\!\!=\!\!\circ = \underset{s}{\circ}\overset{}{\underset{1}{\rule{3em}{0.4pt}}}\overset{}{\underset{t}{\circ}}$ for GQ.

We give a few construction techniques for pGs of rank n below; we have chosen some which are useful to us, or are non-standard. (We thank Antonio Pasini for pointing out Example 7.) To simplify these constructions, we anticipate Theorem 10 by pointing out that

$$\underset{q}{\overset{0}{\circ}}\rule{2em}{0.4pt}\underset{q}{\overset{1}{\circ}}\rule{2em}{0.4pt}\underset{q}{\overset{2}{\circ}}\cdots\underset{q}{\overset{n\text{-}2}{\circ}}\rule{2em}{0.4pt}\underset{q}{\overset{n\text{-}1}{\circ}}$$

is the diagram for a rank n projective geometry $PG(n, q)$. (The objects of type i in such a $PG(n, q)$ might be said to have *rank* i; this may cause no confusion, but we avoid it.)

Definition 5. If P is a geometry of rank n, and we choose a subset T of the set $\{0, 1, \ldots, n - 1\}$ of types, then the T-*truncation* of P is the geometry whose objects are the objects of P with types in T, and incidence inherited from P.

Example 6. Let $\Omega = \{W_1, W_2, \ldots, W_k\}$ be a set of subspaces in the rank n projective geometry $\mathcal{P} = PG(n, q)$, of respective types $n_1, n_2, \ldots, n_k$, and satisfying $n_i - n_{i+1} \geq 2$ for $i = 1, 2, \ldots, k - 1$. Let $\mathcal{P}(\Omega)$ be the geometry consisting of all the elements X of $PG(n, q)$ such that for each i, $X \cap W_i$ is minimal. (This last condition is equivalent to: for each i, $X \cap W_i = 0$ or $X + W_i = \mathcal{P}$.) Then $\mathcal{P}(\Omega)$ is a geometry for a diagram of the form

$$\underset{q}{\circ}\rule{1.5em}{0.4pt}\underset{q}{\circ}\cdots\underset{q}{\circ}\rule{1em}{0.4pt}\underset{q}{\circ}\overset{\text{fA}}{\underset{q\text{-}1}{\rule{1.5em}{0.4pt}}}\underset{q}{\circ}\overset{\text{Af}}{\underset{q}{\rule{1.5em}{0.4pt}}}\underset{q}{\circ}\rule{1em}{0.4pt}\underset{q}{\circ}\cdots\underset{q}{\circ}\overset{\text{fA}}{\underset{q\text{-}1}{\rule{1.5em}{0.4pt}}}\underset{q}{\circ}\overset{\text{Af}}{\underset{q}{\rule{1.5em}{0.4pt}}}\underset{q}{\circ}\cdots$$

with $n\text{-}n_1$ and $n\text{-}n_2$ labels above,

that is, with a pair $\circ\overset{\text{fA}}{\rule{1.5em}{0.4pt}}\circ\overset{\text{Af}}{\rule{1.5em}{0.4pt}}\circ$ at each node $n - n_i$ of the diagram for $\mathcal{P}$. Furthermore, $\mathcal{P}(\Omega)$ does not satisfy IP in general. In particular, if $n = 3$ and $\Omega = \{W\}$, where W is a line, then $\mathcal{P}(\Omega)$ is a geometry for

$\circ\overset{\text{fA}}{\rule{1.5em}{0.4pt}}\circ\overset{\text{Af}}{\rule{1.5em}{0.4pt}}\circ$ without IP.

Example 7. Let $Q = Q^+(3, q)$ be a ruled quadric in $PG(3, q)$, q even. Let S be the structure of points of $PG(3, q)$ not on Q, of lines not meeting Q

and of planes which meet Q in a non-degenerate conic.   The incidence rules are "natural", except that if the plane h meets Q in the conic $h \cap Q$, with nucleus X, then X is *not* on the plane h. Then S is a geometry for

$$\underset{q}{\circ}\overset{J}{\rule{1cm}{0.4pt}}\underset{\frac{q}{2}-1}{\circ}\overset{L}{\rule{1cm}{0.4pt}}\underset{q}{\circ} \, ,$$

and also lacks IP.

Example 8. Let Q be a non-singular quadric in $PG(n, q)$, and h a hyperplane of $PG(n, q)$. Let S be the geometry whose points, lines, planes, ... are the points, lines, planes, ... of Q which are not contained in h. Then S is a geometry for

with m nodes, where m depends on n, the type of Q, and h, and t depends on the type of Q.   These are called *affine polar spaces*, and have been studied (but not always with this name) in Buekenhout (1977, 1979), Buekenhout and Hubaut (1977) and Cohen and Shult (1990).

Example 9.   Let $\mathcal{P}$ be a projective geometry $PG(n, q)$, and W a subspace of type n - d. Let $\mathcal{B} = \mathcal{B}(n, d, q)$ be the set of all subspaces X of type $i \leq d$ such that $X \cap W = 0$; call the subspaces in $\mathcal{B}$ of type d *points*, those of type d - 1 *lines*, etc.   Then $\mathcal{B}$ is a geometry for

$$\circ \overset{N}{\rule{1cm}{0.4pt}} \circ \rule{1cm}{0.4pt} \circ \cdots \circ \rule{1cm}{0.4pt} \circ \, ,$$

of rank d.   These are called *attenuated spaces* in Sprague (1981);   the truncation on points and lines gives a *d-net*.

Much additional information about pGs will be found in De Clerck (1984, to appear).   A good deal is already known about rank n pGs as well, and will be found in particular in Buekenhout (1979).   We give next a long list of characterisations of rank n pGs, where we permit ourselves a little informality in the description.

Theorem 10. *If S is a finite residually connected rank n pG with IP belonging to the given diagram, then S can be characterised as follows.*

(i) $\underset{s}{\circ}\rule{1cm}{0.4pt}\underset{s}{\circ}\rule{1cm}{0.4pt}\underset{s}{\circ}\cdots\underset{s}{\circ}\rule{1cm}{0.4pt}\underset{s}{\circ}$ *with $n \geq 2$ nodes: S is a projective geometry of rank n and order $s \geq 1$, and conversely.*   (Tits 1956).

(ii) $\underset{s}{\circ}\rule{1cm}{0.4pt}\underset{s}{\circ}\overset{L}{\rule{1cm}{0.4pt}}\underset{t}{\circ}$: *S is a projective geometry of order* $s \geq 1$, *rank* n, *and* $t = s^{n-2} + s^{n-3} + \dots + s$, *truncated on* $\{0, 1, 2\}$, *and conversely.* (Buekenhout (1979)).

(iii) $\circ\overset{L}{\rule{1cm}{0.4pt}}\underset{s}{\circ}\rule{1cm}{0.4pt}\underset{s}{\circ}\cdots\underset{s}{\circ}\rule{1cm}{0.4pt}\underset{s}{\circ}$ *with* n *nodes: if* $n \geq 4$, *then S is an affine geometry* $AG(n, s)$ *or a projective geometry* $PG(n, s)$; *if* $n = 3$, *then S is either an affine geometry* $AG(3, s)$, *a projective geometry* $PG(3, s)$, *or one of an unknown family of Lobachevsky spaces (of which only one example is known to exist: the Mathieu 3-design for* $(22, 6, 1)$). See Doyen and Hubaut (1971) for more.

(iv) $\underset{s}{\circ}\rule{1cm}{0.4pt}\underset{s}{\circ}\rule{1cm}{0.4pt}\underset{s}{\circ}\cdots\underset{s}{\circ}\rule{1cm}{0.4pt}\underset{s}{\circ}\overset{}{=\!=\!=}\underset{t}{\circ}$ *with* $n \geq 3$ *nodes: S is a polar space of rank* n *and order* (s, t), *and conversely.*   (Tits (1981)).

(v) $\circ\overset{Af}{\rule{1cm}{0.4pt}}\underset{}{\circ}\rule{1cm}{0.4pt}\circ\cdots\circ\rule{1cm}{0.4pt}\circ$ *with* $n \geq 2$ *nodes: S is an affine geometry of rank* n, *and conversely.*   (Buekenhout (1979)).

(vi) $\underset{s-1}{\circ}\overset{Af}{\rule{1cm}{0.4pt}}\underset{s}{\circ}\overset{L}{\rule{1cm}{0.4pt}}\circ$: *if* $s > 3$, *then S is an affine geometry of rank* $n \geq 3$, *truncated on* $\{0, 1, 2\}$, *and conversely;   for* $s \leq 3$, *there are other examples.*   (See Buekenhout (1979)).

(vii) $\circ\overset{I}{\rule{1cm}{0.4pt}}\circ\overset{L}{\rule{1cm}{0.4pt}}\circ$: *here S is a projective geometry of rank* $n \geq 3$, *truncated at* $\{i - 1, i, i + 1\}$; *but notice also Example (6).*   (Sprague (1985)).

(viii) $\circ\overset{C}{\rule{1cm}{0.4pt}}\underset{s}{\circ}\rule{1cm}{0.4pt}\underset{s}{\circ}\cdots\underset{s}{\circ}\rule{1cm}{0.4pt}\underset{s}{\circ}$ *with* $n \geq 3$ *nodes: if* $n = 3$, *then S is a 3-design for* $(4, 3, 1)$ *with* $s = 1$, *for* $(8, 4, 1)$ *with* $s = 2$, *or for* $(22, 6, 1)$ *with* $s = 4$, *and conversely;   if* $n > 3$, *then S is either an* n-*design for* $(n + 1, n, 1)$ *with* $s = 1$, *or is an affine geometry* $AG(n, 2)$ *with* $s = 2$, *and conversely.*   (Dembowski (1968); Hughes (1965); Hughes and Piper (1973)).

(ix) $\circ\overset{C}{\rule{1cm}{0.4pt}}\circ\overset{C}{\rule{1cm}{0.4pt}}\circ\rule{1cm}{0.4pt}\underset{s}{\circ}$: *S is a 4-design for* $(5, 4, 1)$ *with* $s = 1$, *or for* $(23, 7, 1)$ *with* $s = 4$, *and conversely.*   (Dembowski (1968); Hughes (1965); Hughes and Piper (1973))

(x) $\circ\overset{C}{\rule{1cm}{0.4pt}}\circ\overset{C}{\rule{1cm}{0.4pt}}\circ\overset{C}{\rule{1cm}{0.4pt}}\underset{s}{\circ}$ *is a 5-design for* $(6, 5, 1)$ *with* $s = 1$, *or for* $(24, 8, 1)$ *with* $s = 4$, *and conversely.*   (Hughes (1965)).

(xi) $\circ\overset{C}{\rule{1cm}{0.4pt}}\circ\overset{\supset}{\rule{1cm}{0.4pt}}\circ$: *S is a semibiplane, and conversely.*

(xii) $\circ\overset{C}{\rule{1cm}{0.4pt}}\underset{s}{\circ}\rule{1cm}{0.4pt}\underset{s}{\circ}\overset{\supset}{\rule{1cm}{0.4pt}}\circ$: *S is either a 4-design for* $(5, 4, 1)$ *with* $s = 1$; *the geometry of points and hyperplane complements from* $PG(3, 2)$, *with 15 points and 8 points on a block, with* $s = 2$; *the doubly affine geometry of rank 4 over* $GF(2)$ *with 16 points and 8 points on a block, also with* $s = 2$; *or the "Higman–Sims" geometry with 100 points and 22 points on a block, with* $s = 4$.   *(Here the types of the geometry are called points, lines, planes and blocks respectively from left to right; in each case the number of blocks on a point equals the number of points on a block.)*   (Hughes (1982, 1983)).

(xiii) $\underset{q\text{-}1}{\circ}\!\!\overset{Af}{\rule{1.2cm}{0.4pt}}\!\!\underset{q}{\circ}\rule{0.8cm}{0.4pt}\underset{q}{\circ}\cdots\underset{q}{\circ}\rule{0.8cm}{0.4pt}\underset{q}{\circ}\!\!\overset{fA}{\rule{1.2cm}{0.4pt}}\!\!\underset{q\text{-}1}{\circ}$: *with n > 2 nodes: S is the "doubly affine" geometry obtained from PG(n, q) by deleting a point X and a hyperplane h, and all the objects of PG(n, q) containing X or contained in h.   Here there are two cases: X on h, and X not on h. Example (xii) with s = 2 is a special case of this with n = 4 and q= 2. (Lefèvre-Percsy and Van Nypelseer (to appear); Van Nypelseer (to appear)).*

(xiv) $\circ\!\!\overset{\subset}{\rule{1.2cm}{0.4pt}}\!\!\circ\!\!\overset{Af}{\rule{1.2cm}{0.4pt}}\!\!\circ$: *S is an inversive plane (a Möbius plane), and conversely.   (Dembowski (1968); Hughes (1965)).*

(xv) $\circ\!\!\overset{\subset}{\rule{1.2cm}{0.4pt}}\!\!\circ\!\!\underset{s\text{-}1}{\overset{\subset}{\rule{1.2cm}{0.4pt}}}\!\!\circ\!\!\underset{s}{\overset{Af}{\rule{1.2cm}{0.4pt}}}\!\!\circ$: *S is either a 4-design for (6, 4, 1) with s = 2, or (11, 5, 1) with s = 3, or possibly an unknown 4-design for (171, 15, 1) with s = 13 (extending a non-egglike, hence unknown, inversive plane), and conversely.   (Dembowski (1968); Hughes (1965); Hughes and Piper (1973); Kantor (1974); Penttila (to appear)).*

(xvi) $\circ\!\!\overset{\subset}{\rule{1.2cm}{0.4pt}}\!\!\circ\!\!\overset{\subset}{\rule{1.2cm}{0.4pt}}\!\!\circ\!\!\underset{s\text{-}1}{\overset{\subset}{\rule{1.2cm}{0.4pt}}}\!\!\circ\!\!\underset{s}{\overset{Af}{\rule{1.2cm}{0.4pt}}}\!\!\circ$: *S is either a 5-design for (7, 5, 1) with s = 2, or (12, 6, 1) with s = 3, and conversely.   (Dembowski (1968); Hughes (1965); Hughes and Piper (1973)).*

(xvii) $\underset{1}{\circ}\!\!\overset{\subset}{\rule{1.2cm}{0.4pt}}\!\!\underset{s}{\circ}\!\!\overset{L}{\rule{1.2cm}{0.4pt}}\!\!\underset{t}{\circ}$: *S is a 3-design for (s(t + 1) + 2, s + 2, 1), and conversely; (this gives exactly all 3-designs with λ = 1).*

(xviii) $\circ\!\!\overset{N}{\rule{1.2cm}{0.4pt}}\!\!\circ\rule{1.2cm}{0.4pt}\circ\cdots\circ\rule{1.2cm}{0.4pt}\circ$, *with d nodes: these are exactly the attenuated spaces of Example 9.   (Sprague (1981))   Note that we consider an affine plane to be a net.   (So some of the geometries of Example 15 are included here.)*

(xix) $\circ\!\!\overset{Af}{\rule{1.2cm}{0.4pt}}\!\!\circ\rule{1.2cm}{0.4pt}\circ\cdots\circ\rule{1.2cm}{0.4pt}\circ\!\!\overset{}{=\!\!=}\!\!\circ$, *with n > 2 nodes: these are exactly the affine polar spaces of Example 8.   (Cohen and Shult (1990)).*

   (NB.   We do not pretend that the references are exhaustive or complete;   some of these results are almost folk-lore and are quite easy to prove: others are far more elaborate.   There are other characterisations known beside those in this theorem, and a good many of the characterisations above do *not* use IP, nor even finiteness. But (vii) for instance certainly needs IP, as Examples 6 and 7 show.)

   The 3-(22, 6, 1), 4-(23, 7, 1) and 5-(24, 8, 1) designs arising in (iii), (viii), (ix) and (x) of Theorem 10 are, of course, the "large" Mathieu designs;   similarly, the 4-(11, 5, 1) and 5-(12, 6, 1) designs of (xv) and (xvi) are the "little" Mathieu designs.   These have the appropriate Mathieu groups involved in their automorphism groups.   The

Higman-Sims geometry of (xii) has the group HS in its automorphism group.   (And in each of these cases, the cited group has index ≤ 2 in the full automorphism group.)   These observations are among the interesting reasons for studying rank n pGs.

After constructing some rank 3 pGs in section 3 and studying some special properties of those geometries in section 4, we shall pass on to two special families of such geometries in the rest of the paper.   These are:

Definition 11. (i) If S is a residually connected geometry with IP for

$$\circ \overset{L}{\rule{2.5cm}{0.4pt}} \circ\!\!=\!\!\rule{1.5cm}{0.4pt}\!\!=\!\!\circ$$

then we call S an L.pG *geometry*.   If the diagram is:

$$\underset{s}{\circ} \overset{L}{\rule{2.5cm}{0.4pt}} \underset{t}{\circ}\!\!=\!\!\underset{\alpha}{\rule{1.5cm}{0.4pt}}\!\!=\!\!\underset{u}{\circ}$$

then we say that S is an $L.pG_\alpha$ geometry of *order* (s, t, u).   The subfamily of L.pGs for the diagram $\circ\!\!-\!\!-\!\!\circ\!\!=\!\!-\!\!=\!\!\circ$ are called $A_2.pG$ *geometries*.

(ii)   If S is a residually connected geometry with IP for

$$\circ\!\!=\!\!\rule{1.5cm}{0.4pt}\!\!=\!\!\circ \overset{L}{\rule{2.5cm}{0.4pt}} \circ$$ then we say that S is a pG.L *geometry*.   If the diagram is:

$$\underset{s}{\circ}\!\!=\!\!\underset{\alpha}{\rule{1.5cm}{0.4pt}}\!\!=\!\!\underset{t}{\circ} \overset{L}{\rule{2.5cm}{0.4pt}} \underset{u}{\circ}$$

then S is a $pG_\alpha.L$ geometry of *order* (s, t, u).

(In both diagrams, the types are *points*, *lines*, and *planes*, from left to right.)

An important special case of the above are the *extended partial geometries*, EpGs: an $EpG_\alpha(s, t)$ is an $L.pG_\alpha$ geometry of order $(1, s, t)$. Many examples of EpGs are known, as well as characterisations and non-existance theorems.   The extensive literature is reviewed in Hughes (to appear), but see also Buekenhout (1977); Buekenhout and Hubaut (1977); Cameron et. al. (1990); Del Fra et al. (this volume).

3. Some rank 3 pGs

In this section we construct some rank 3 pGs.   We recall some geometries of section 2, and add some new ones.

Example 12. (i) If $\mathcal{P}$ is the projective geometry $PG(n, K)$ of rank $n \geq 3$, and $T = \{i - 1, i, i + 1\}$, with $i < n$, then the T-truncation of $\mathcal{P}$ is a

geometry for o—$\underset{}{\overset{J}{\quad}}$—o—$\overset{L}{\quad}$—o , and is a geometry for o———o—$\overset{L}{\quad}$—o
if $i = 1$.   (Compare Theorem 10 (vii).)

(ii)   If $\mathcal{P}$ is a polar space of rank $n \geq 3$, and $T = \{n - 3, n - 2, n - 1\}$, then

the T-truncation of $\mathcal{P}$ is a geometry for o—$\overset{J}{\quad}$—o═══o (so its dual is

a geometry for o═══o—$\overset{L}{\quad}$—o).

(iii)   If A is an affine geometry $AG(n, K)$ of rank $n \geq 3$, and $T = \{0, 1, 2\}$,

then the T-truncation of A is a geometry for o—$\overset{Af}{\quad}$—o—$\overset{L}{\quad}$—o.
(Theorem 10 (vi).)

Example 13.   Let $\mathcal{P}$ be the projective plane of order 2, and construct a geometry S as follows: the 14 *points* of S are the points and lines of $\mathcal{P}$; the 28 *lines* of S are the antiflags of $\mathcal{P}$; the 21 planes of S are the 4-tuples (A, B, x, y) where A and B are distinct points, and x and y are the unique pair of distinct lines of $\mathcal{P}$ which do not contain A, or B. Incidence is "natural".

Then it can be shown (quite easily) that S is a geometry for

$$\underset{1}{o}═══\underset{1}{o}—\overset{L}{\quad}—\underset{2}{o} ,$$

and that a point-residue S_P is a 2-(4, 2, 1), which can be considered to be the point-line truncation of the projective geometry PG(3, 1).

Example 14.   Let M be the Mathieu 5-design for (24, 8, 1).   Then M has 759 blocks and satisfies:

(∗)   given any set of four points, the remaining 20 points are divided uniquely into five sets of 4 points in such a way that the union of any two of these six 4-sets is a block.

We construct S as follows: the *points* of S are the blocks of M; the *lines* of S are the set of three blocks which cover M (from (∗) it is clear that such sets exist); the *planes* of S are the sets of six 4-sets which cover M, as in (∗).   Incidence is "natural": a line contains the 3 blocks which form the line, and is in any plane whose six 4-sets can be "pasted together" into the three blocks of the line, and so on.   (See Hughes and Piper (1985), Chapter 8, for more details on M.)

It can be shown (again quite easily) that S is a geometry for

$$\underset{2}{o}═══\underset{2}{o}—\overset{L}{\quad}—\underset{6}{o} ,$$

and that the point-residue S_P is the point-line truncation of the projective geometry PG(3, 2).

These geometries of the last two examples will be found in Ronan (1982) and Ronan and Smith (1980).   In Ronan (1982) it is shown that together with the dual of the {1, 2, 3}-truncation of a polar space of rank 4, they are the only possible geometries for $\circ\!=\!\!=\!\!=\!\!\circ\overset{L}{\rule{1cm}{0.4pt}}\circ$ whose point-residues are isomorphic to the {0, 1}-truncation of a PG(3, q) (here as usual we allow $q = 1$) and which satisfy a property called there "Int".   It would be good to know the exact status of that property "Int", (eg. its relationship to IP) and indeed, what can be said in general about geometries for $\circ\!=\!\!=\!\!=\!\!\circ\overset{L}{\rule{1cm}{0.4pt}}\circ$ whose point-residues are projective geometries.

Example 15. Let $\mathcal{D}$ be a $pG_\alpha(t, u)$ which is contained in the projective geometry PG(n, q), and in no sub-PG(n_1, q) for $n_1 < n$; consider this PG(n, q) as the hyperplane at infinity of an affine geometry AG($n + 1$, q).   Construct the geometry $S = S(\mathcal{D})$ as follows: the *points* of S are the points of   AG($n + 1$, q); the *lines* of S are the lines of AG($n + 1$, q) which contain (exactly) one point of $\mathcal{D}$; the *planes* of S are the planes of AG($n + 1$, q) which contain (exactly) one line of $\mathcal{D}$.   Then S is a rank 3 pG for $\underset{q\text{-}1}{\circ}\overset{N}{\rule{1cm}{0.4pt}}\underset{t}{\circ}\!=\!\!=\!\!=\!\underset{\alpha}{}\overset{}{\rule{1cm}{0.4pt}}\underset{u}{\circ}$, possibly without IP.   Special cases are:

(A) For $n = 2$:

(i) If $\mathcal{D}$ is a projective plane PG(2, p), with $q = p^m$, then S is a geometry for $\underset{q\text{-}1}{\circ}\overset{N}{\rule{1cm}{0.4pt}}\underset{p}{\circ}\overset{}{\rule{1cm}{0.4pt}}\underset{p}{\circ}$, so its dual S^T is a geometry for $\circ\!\!\rule{1cm}{0.4pt}\!\!\circ\overset{N}{\rule{1cm}{0.4pt}}\circ$. (Compare Theorem 10 (xviii): S is isomorphic to $\mathcal{B}(m + 2, 3, p)$ of Example 9.)

(ii) If $\mathcal{D}$ is an affine plane of order r, then we have $\underset{q\text{-}1}{\circ}\overset{N}{\rule{1cm}{0.4pt}}\underset{r\text{-}1}{\circ}\overset{Af}{\rule{1cm}{0.4pt}}\underset{r}{\circ}$; note that q need not equal r, and we can even have $q = 4^m$ and $r = 3$; if $q = r$, then the diagram is $\underset{q\text{-}1}{\circ}\overset{*}{\rule{1cm}{0.4pt}}\underset{q\text{-}1}{\circ}\overset{Af}{\rule{1cm}{0.4pt}}\underset{q}{\circ}$.

(iii) If $\mathcal{D}$ is a dual affine plane of order r, then the dual S^T is $\underset{r\text{-}1}{\circ}\overset{Af}{\rule{1cm}{0.4pt}}\underset{r}{\circ}\overset{N}{\rule{1cm}{0.4pt}}\underset{q\text{-}1}{\circ}$; if $r = q$, this gives $\circ\overset{Af}{\rule{1cm}{0.4pt}}\circ\overset{fA}{\rule{1cm}{0.4pt}}\circ$.

(iv) If $\mathcal{D}$ is a 2-design, or a dual 2-design, embedded in PG(2, q) (eg. a unital, or dual unital of various possible orders), this gives other examples for $\circ\overset{N}{\rule{1cm}{0.4pt}}\circ\overset{L}{\rule{1cm}{0.4pt}}\circ$, and equally for $\circ\overset{L}{\rule{1cm}{0.4pt}}\circ\overset{N}{\rule{1cm}{0.4pt}}\circ$.

(v)   We may choose $\mathcal{D}$ to be a net in $PG(2, q)$, with p points on a line and $k \leq p$ parallel classes, where $q = p^m$, with possibly $m = 1$.   Then we have a geometry for $\circ \underset{}{\overset{N}{\text{——}}} \circ \underset{}{\overset{N}{\text{——}}} \circ$.

(vi)   Dualising (v), we can construct geometries for $\circ \underset{}{\overset{И}{\text{——}}} \circ \underset{}{\overset{N}{\text{——}}} \circ$.

(vii)   If $\mathcal{D}$ is a dual net, then S is the geometry for $\circ \underset{}{\overset{N}{\text{——}}} \circ \underset{}{\overset{И}{\text{——}}} \circ$.

(viii)   For q even, there are certain well-known proper pGs in $PG(2, q)$, coming from "hyperovals" and "maximal k-arcs".   (See De Clerck, to appear, for more.)   In particular, this will yield geometries for

$$\underset{4^m - 1}{\circ} \overset{N}{\underset{2}{\text{——}}} \circ \underset{2}{=\!=\!=} \circ \text{ , for every } m \geq 1.$$

(B).   For $n = 3, 4$, or 5, we can take $\mathcal{D}$ to be a GQ thus giving geometries for $\underset{q-1}{\circ} \overset{N}{\underset{t}{\text{——}}} \circ \underset{u}{=\!=\!=} \circ$ where the point-residue   can be an arbitrary classical GQ (and some non-classical ones as well).

(i)   This will give, in particular, examples for $t = q$, hence for $\underset{q-1}{\circ} \overset{Af}{\underset{q}{\text{——}}} \circ \underset{u}{=\!=\!=} \circ$ (these are affine polar spaces: see Example 8).

(ii)   If $\mathcal{D}$ is a $GQ(q - 1, q + 1)$ embedded in $PG(3, q)$ (see Payne and Thas (1984)), then with $n = 3$, we have geometries for $\underset{q-1}{\circ} \overset{*}{\underset{q-1}{\text{——}}} \circ \underset{q+1}{=\!=\!=} \circ$.

(C).   For any n, we can let $\mathcal{D}$ be an $AG(n, r)$ or a dual $AG(n, r)$, in general for $q = r^m$, to construct geometries

$$\underset{q-1}{\circ} \overset{N}{\underset{r-1}{\text{——}}} \circ \overset{L}{\underset{u}{\text{——}}} \circ \text{ and } \underset{q-1}{\circ} \overset{N}{\underset{u}{\text{——}}} \circ \overset{J}{\underset{r-1'}{\text{——}}} \circ$$

as well as their duals.   Similarly, if $\mathcal{D}$ is a $PG(n, r)$ or a dual $PG(n, r)$, we have $\underset{q-1}{\circ} \overset{N}{\underset{r}{\text{——}}} \circ \overset{L}{\underset{u}{\text{——}}} \circ$ and $\underset{q-1}{\circ} \overset{N}{\underset{u}{\text{——}}} \circ \overset{J}{\underset{r}{\text{——}}} \circ$, plus their duals. (here u depends on r, n and the type of $\mathcal{D}$; in each case, we understand that we are embedding a truncation, which has only points and lines.) And, of course, this construction could be carried out with other pGs embedded minimally in $PG(n, q)$.

(NB.   The condition that $\mathcal{D}$ not be contained in a $PG(n_1, q)$ for smaller n_1 is to ensure that S is connected.   Example 15 is a special case of a more general construction, where we embed a rank m geometry $\mathcal{D}$ in a $PG(n, q)$, again with n minimal, which is in turn the hyperplane at infinity of $AG(n + 1, q)$, and then construct a rank $(m + 1)$ geometry S whose first "stroke" is a net $\underset{q-1}{\circ} \overset{N}{\underset{t}{\text{——}}} \circ$, the rest given by $\mathcal{D}$.   If $\mathcal{D}$

has $q + 1$ points on a line, then the first stroke of S will be $\circ \!\!\overset{Af}{\rule{2em}{0.4pt}}\!\! \circ$, while if $\mathcal{D}$ has q points on a line, then the first stroke of S will be $\circ \!\!\overset{*}{\rule{2em}{0.4pt}}\!\! \circ$, as above.

Many of the geometries constructed above will have *quotients*, and for some of these cases, little is known about this situation.)

Theorem 16. *Let $S = S(\mathcal{D})$ as in Example 15. Then S satisfies IP if and only if whenever two lines of $\mathcal{D}$ meet in $PG(n, q)$, then their point of intersection is in $\mathcal{D}$.*

Proof. It is only necessary to verify that if two planes h_1 and h_2 of S meet in a line in $AG(n + 1, q)$, then that line is in S; this is easily seen to be equivalent to demanding that if two lines of $\mathcal{D}$ meet in $PG(n, q)$, the hyperplane at infinity, then their point of intersection is in $\mathcal{D}$. $\square$

Hence in Example 15 (A) the geometries of (i), (iii) and those from the dual 2-designs of (iv), will have IP. In case (B), the geometries of (i), but not of (ii), will have IP.

4. Triangularity

In Theorem 10 and Examples 12, 13, 14 and 15 we have seen several rank 3 pGs. Now we develop some important general ideas about rank 3 pGs; by an *antiflag* in such a geometry we shall mean a non-incident point-plane pair (if we want to refer to a point-line antiflag, we shall make that clear).

Definition 17. (i) If S is a rank 3 pG, then $\Gamma = \Gamma(S)$ is the graph whose vertices are the points of S, with two adjacent when they lie on a common line.

(ii) In analogy with the situation for points, two planes of S are *collinear* if they have a line in common.

(iii) If (P, h) is an antiflag in S then $\mathcal{A}(P, h)$ is the set of points of h which are collinear with P and of the planes on P which are collinear with h. By $\mathcal{A}_P(P, h)$ we mean the set of lines PX, for points $X \in \mathcal{A}(P, h)$, and of planes in $\mathcal{A}(P, h)$, while $\mathcal{A}_h(P, h)$ means the set of points of $\mathcal{A}(P, h)$ and of the lines $g \cap h$, for planes g in $\mathcal{A}(P, h)$.

(iv) For an antiflag (P, h), $\varphi(P, h)$ is the number of points in $\mathcal{A}(P, h)$ and $\varphi^*(P, h)$ is the number of planes in $\mathcal{A}(P, h)$.

Theorem 18. *The point-plane geometry $\mathcal{A}(P, h)$ is isomorphic with the point-line geometry $\mathcal{A}_h(P, h)$ and with the line-plane geometry $\mathcal{A}_P(P, h)$.*

Proof. The mapping $X \rightarrow X$ and $g \rightarrow g \cap h$ maps the points and planes of $\mathcal{A}(P, h)$ onto $\mathcal{A}_h(P, h)$, preserving the incidence, and similarly mapping $X \rightarrow PX, g \rightarrow g$ maps the points and planes of $\mathcal{A}(P, h)$ onto $\mathcal{A}_P(P, h)$, also preserving incidence. $\square$

In view of Theorem 18, we shall often speak about $\mathcal{A}(P, h)$ when we mean either $\mathcal{A}_P(P, h)$ or $\mathcal{A}_h(P, h)$; the context will make it clear what we mean. It is useful sometimes to know that both S_P and S_h (ie. h) contain isomorphic subgeometries.

Theorem 19. *If S is a rank 3 pG with diagram*

and we define $\beta = (\alpha_1 - 1)\alpha_2 + 1$, $\beta^* = (\alpha_2 - 1)\alpha_1 + 1$, then we have
(a) $\alpha_2 \cdot \varphi (P, h) = \alpha_1 \cdot \varphi^*(P, h)$;
(b) $\varphi(P, h) \geq max \{\beta, \alpha_1\beta^*/\alpha_2\}$ and $\varphi^*(P, h) \geq max \{\beta^*, \alpha_1\beta/\alpha_2\}$;
(c) $\alpha_2\beta \gtreqless \alpha_1\beta^*$ according as $(\alpha_2 - \alpha_1)(\alpha_1 - 1)(\alpha_2 - 1) \gtreqless 0$;
(d) $\varphi(P, h) \leq min \{(t + 1)(st + \alpha_1)/\alpha_2, \alpha_1(u + 1)(tu + \alpha_2)/\alpha_2, (s + 1)(st + \alpha_1)/\alpha_1, (t + 1)(tu + \alpha_2)/\alpha_2, (u + 1)(tu + \alpha_2)/\alpha_1\}$.

Proof. Suppose X in h is collinear with P, and y is the line on P and X. Then in the residue S_X, y is a line not on the plane h, so there are α_2 planes on y which meet h in a line. If g is one of these planes, and z is the line $g \cap h$, then P is collinear with α_1 points on z, one of which is X; so P is collinear with at least $\alpha_2(\alpha_1 - 1) + 1$ points in h. So $\varphi(P, h) \geq \beta$.

Now let us count flags (Q, w) where Q is a point and w a line of $\mathcal{A}_h(P, h)$. We have $\alpha_2 \cdot \varphi(P, h) = \alpha_1 \cdot \varphi^*(P, h)$ which gives (a).
The dual geometry S^T goes with the diagram

and it is immediate that if P^T and h^T are the plane and point, respectively, coming from P and h, then in S^T, $\varphi(h^T, P^T) = \varphi^*(P, h)$. So $\varphi^*(P, h) \geq \alpha_1(\alpha_2 - 1) + 1$, from which we have both inequalities of (b).

Now $\varphi^*(P, h)$ is bounded above by the number of lines in h, as well as the number of planes in the residue S_P, and this gives us two more inequalities for $\varphi(P, h)$. And finally, $\varphi(P, h)$ is bounded above by the number of points in h and the number of lines in S_P. This gives us the upper bounds for $\varphi(P, h)$, and (c) is merely simple computation. □

(In fact, all these ideas can be extended to rank n pGs by considering point-maximal subspace antiflags, and defining $\mathcal{A}(P, h)$ to be a geometry of rank $n - 1$ inside of either h (= S_h) or S_P.) Now we shall assume that S is a rank 3 pG for $\underset{s}{\circ}\!=\!\!\underset{\alpha_1}{}\!\!=\!\!\underset{t}{\circ}\!=\!\!\underset{\alpha_2}{}\!\!=\!\!\underset{u}{\circ}$, and study the case that the function φ takes on the minimal value β.

Definition 20. We say that the antiflag (P, h) is *triangular* if $\varphi(P, h) = (\alpha_1 - 1)\alpha_2 + 1$, and that P is a *triangular point* if $\varphi(P, h) = 0$, or $(\alpha_1 - 1)\alpha_2 + 1$ for all antiflags (P, h). Finally, S is *triangular* if all its points are.

Theorem 21. (a) *The antiflag (P, h) is triangular if and only if $\mathcal{A}(P, h)$ is a 2-design (including the possibility that $\mathcal{A}(P, h)$ is the trivial 2-design consisting of only one point and some of the lines of h through that point, or of only one line together with some points on that line.)* (b) *P is a triangular point if and only if whenever P is collinear with two points on the line y, then there is a plane on P and y.*

Proof. (a) $\varphi(P, h) = (\alpha_1 - 1)\alpha_2 + 1$ if and only if every point in $\mathcal{A}_h(P, h)$ is collinear with every other. (If $\alpha_1 = 1$ then $\mathcal{A}_h(P, h)$ will consist of one point and α_2 lines of h through that point.)
(b) Suppose P is a triangular point, that X and Y are points collinear with P and that there is a line on X and Y. There is a plane on X and Y, and then from (a), there is a plane on P, X, and Y, and in any plane h containing points collinear with P, the structure $\mathcal{A}_h(P, h)$ is a 2-design. □

Corollary 22. *If S is triangular, then either $\alpha_1 = 1$, or $\alpha_2 = 1$, or $\alpha_2 \geq \alpha_1$. If in addition S^T is triangular, then $\alpha_1 = \alpha_2$, or $\alpha_1 = 1$ or $\alpha_2 = 1$.*

Proof. If S is triangular, then $\beta \geq \beta^*$; thus see Theorem 19 (c). □

Theorem 23. *If S is a rank 3 pG, both triangular and semilinear, then S satisfies IP.*

Proof. If S is such a geometry, and if a line y meets a plane h in two points P and Q, then there is a plane g on y which meets h in P and Q; thus the entire IP for this case will be settled if we prove: if two planes h and g meet in two points, then h and g meet in a line.

Suppose $h \cap g$ includes the two points P and Q. In h there is a point X collinear with both P and Q, so the 2-design $\mathcal{A}_g(X, g)$ includes the points P and Q, hence there is a line $z = h \cap g$ on P and Q. $\quad\square$

Triangularity is a strong property, and seems to be possessed by many examples which come from classical geometries, by various constructions. This is particularly evident in the study of EpGs and EGQs, and in the A_2.pG geometries below (Theorem 44). The following comes easily from Theorem 21:

Corollary 24. *If S is a triangular rank 3 pG, and if (P, y) is a point-line antiflag, then P is collinear with 0, 1, or α_1 points of y.*

Corollary 25. *If (P, h) is a triangular antiflag, then $\mathcal{A}(P, h)$ is a 2-design for $((\alpha_1 - 1)\alpha_2 + 1, \alpha_1, 1)$, and so both the pG h and the pG S_P contain a substructure isomorphic to such a 2-design.*

Proof. Straightforward from Theorems 21 and 18. $\quad\square$

Before proceeding into a more detailed analysis of some special cases, we examine the triangularity, or lack of the same, in the examples already given. Notice that $\varphi^*(P, h)$ is the value of φ for the point-plane antiflag (h^T, P^T) in the dual S^T, and that $\beta^* = (\alpha_2 - 1)\alpha_1 + 1$ is the corresponding value of β. Further, S is triangular at (P, h) *and* S^T is triangular at (h^T, P^T) if and only if both $\mathcal{A}(P, h)$ and its dual $\mathcal{A}^*(P, h)$ are 2-designs. But since we include trivial structures with only one line, as well as the structure of one point and some lines as a 2-design, this is a slightly unexpected definition. So in (b) below, $\mathcal{A}_h$ has one point, and either two or three lines, all on that point. Hence both S and S^* are triangular at the "same" antiflag if and only if $\mathcal{A}_h$ is either a projective plane, consists of a single point and some lines on that point, or consists of a single line and some points on that line.

Theorem 26. *The geometries of Examples 12 to 15, and their duals where appropriate, are triangular as follows:*
(a) *The geometries of Example 12 are all triangular, as are their duals. (Here any $\mathcal{A}(P, h)$ is a projective plane.)*
(b) *The geometries of Examples 13 and 14 are triangular, as are their duals. (Here any $\mathcal{A}(P, h)$ consists of one point, plus certain lines on that point.)*
(c) *The geometries of Example 15 (A) with the diagram*

o—N—o⇒———⇒o *are triangular as follows:*

(i) *when $\mathcal{D}$ is a 2-design: if and only if $\mathcal{D}$ is a projective plane (for then $\mathcal{A}(P, h)$ will be an affine plane);*
(ii) *when $\mathcal{D}$ is a dual 2-design: if and only if $\mathcal{D}$ is either a projective plane or is the dual of a (trivial) 2-(4, 2, 1) (in this second case, $\mathcal{A}(P, h)$ will be a 2-(3, 2, 1); that is, a projective plane of order one);*
(iii) *when $\mathcal{D}$ is a net: never;*
(iv) *when $\mathcal{D}$ is a dual net: if and only if $\mathcal{D}$ is the dual of a (trivial) 2-(4, 2, 1), as in (ii);*
(v) *when $\mathcal{D}$ is the structure of points not on a complete k-arc, and the lines of that k-arc: never.*
(d) *The dual geometries of Example 15 (A), with diagram*

o⇒———⇒o—И—o, *are never triangular except in the dual in (c)(ii) above, when $\mathcal{D}$ is the dual of a 2-(4, 2, 1).*
(e) *The geometries of Example 15 (B) are never triangular, except when $\mathcal{D}$ is a grid (1, 1).*
(f) *The geometries of Example 15 (C) are never triangular, if $n \geq 2$.*

Proof. The proofs are all straightforward, involving only computing β and $\varphi(P, h)$, or β^* and $\varphi^*(P, h)$. In Example 15, $\varphi(P, h)$ is always the number of points of $\mathcal{D}$ minus the number of points (in $\mathcal{D}$) on a line of $\mathcal{D}$. □

Observe that the triangular geometries constructed in Example 15 with $\mathcal{D}$ the dual of a 2-(4, 2, 1) are themselves duals of extended dual nets, and that the extended dual nets are triangular too.

5. On L.pG geometries

In this section we will study L.pG geometries in more detail and then in the next section find bounds on their diameters.

Lemma 27. *Let S be an $L.pG_\alpha$ geometry of order (s, t, u).*

(a) *Two points lie on a common line if and only if they lie in a common plane.*

(b) *Each line contains $s + 1$ points and lies in $u + 1$ planes.*

(c) *The points and lines of each plane form a 2-design for $(s(t + 1) + 1, s + 1, 1)$, hence a plane contains $s(t + 1)$ points and $(t + 1)(s(t + 1) + 1)/(s + 1)$ lines.*

(d) *The lines and planes on a point form a $pG_\alpha(t, u)$, hence there are $(t + 1)(tu + \alpha)/\alpha$ lines and $(u + 1)(tu + \alpha)/\alpha$ planes on a point.*

(e) *Two collinear points are in exactly one line and $u + 1$ common planes.*

Proof. It is all quite straightforward. □

Corollary 28. *If S is an $L.pG_\alpha(s, t, u)$, and if $\varphi(P, h) \neq 0$, then $\alpha s + 1 \leq \varphi(P, h) \leq st + s + 1$. Furthermore, $s + 1$ divides $\alpha \cdot \varphi(P, h)$.*

Proof. Everything is immediate from Theorem 19, where we choose the upper bound $s(t + 1) + 1$ given by the number of points in h. □

Corollary 29. *S has a triangular point P if and only if $\varphi(P, h) = 0$ or $\alpha s + 1$, for all planes h not on P.*

Proof. This follows from Theorem 21. □

Now we examine some special values of α. The following is easy:

Theorem 30. *A geometry for* $\underset{\circ}{}\,\underline{\quad L\quad}\,\underset{\circ}{}\,\underline{\quad L\quad}\,\underset{\circ}{}$ *is always triangular.*

Proof. In this case we have $\alpha = t + 1$, and substituting in Corollary 28, we see that $\varphi(P, h) = st + s + 1 = \alpha s + 1$. □

Theorem 31. *If S is an $L.pG_t$ geometry of order (s, t, u) (that is, the point-residues of S are dual nets), then $\varphi(P, h) \geq st + s$ if $\varphi(P, h) \neq 0$, and the diameter of S is two.   If S is triangular at some point, then $s = 1$ and S is an extended dual net.*

Proof. Here we have $\alpha = t$, and if $\varphi(P, h) \neq 0$, then $\varphi(P, h) \geq st + 1$. The number of points in the plane h is $st + s + 1$, so $\mathcal{A}'$, the set of points not in $\mathcal{A} = \mathcal{A}_h(P, h)$ has at most s points.   On each point X of $\mathcal{A}$ there are t lines of $\mathcal{A}$ and hence just one line y_X not in $\mathcal{A}$.   Since the lines of $\mathcal{A}$ have all their $s + 1$ points in $\mathcal{A}$, the line y_X must join X to all the points of $\mathcal{A}'$. Now if $|\mathcal{A}'| \geq 1$, and there are at least two distinct lines y_X and y_Z, then they will intersect in more than one point, which is not possible. So there is only one line y_X, but it contains at most $s - 1$ points of $\mathcal{A}$, which is all the points of $\mathcal{A}$, which violates $\varphi(P, h) \geq st + 1$.

Hence $|\mathcal{A}'| \leq 1$ and $st + s \leq \varphi(P, h) \leq st + s + 1$. Now if S is triangular at P, then $s = 1$, so S is an extended dual net. In any case, $\varphi(P, h) > 0$ implies $\varphi(P, h) \geq st + s$. For the diameter statement, see Corollary 39. □

(NB. The diameter assertion can also be proved directly without Corollary 39, but it hardly seems worth the trouble. The structure of $\mathcal{A}$ inside of S_P is interesting, and possibly could be used to characterise those geometries more sharply, since $\mathcal{A}$ possesses much more of the structure to be a sub-geometry of a dual net.)

See Hobart and Hughes (1990) for more information about extended dual nets, in particular those which are locally triangular; see section 7 for more about the case $s = t$, and Example 15 (A)(i)(iii)(iv) for geometries of this class.

In Higman (1982) a Γ-*geometry* is defined to be a semilinear space in which a point is collinear with 0, 1 or all points of any line. This gives us the following property of triangularity, from Theorem 21 and Corollary 24 (the converse presumably holds under certain extra hypotheses).

Corollary 32. *If S is a triangular L.pG, then the point-line truncation of S is a Γ-geometry.*

6. Diameter bounds for L.pG geometries

Here we develop some diameter bounds for L.pG geometries. These are "weak finiteness theorems", in that they depend on the orders. Many much stronger diameter bounds for EpGs have been found (see for instance Cameron et al. (1990)), especially by Del Fra, Ghinelli and Pasini, and references to these results will be found in Hughes (to appear): it seems reasonable to expect that similarly stronger bounds exist for L.pG geometries. Notice, by the way, that the rank 3 pGs of Example 15 all have diamater $\delta \leq 3$, and usually have $\delta = 3$.

Definition 33. If S is an L.pG geometry, and P is a fixed point of S, then for any plane h in S, the *distance* $d(P, h)$ is the minimum distance from P to any point of h and $f_P(h)$ is the number of points on h at a distance $d(P, h)$ from P.

Lemma 34. *Let S be an L.pG$_\alpha$. Suppose g and h are planes on the point P, and $\mathcal{E}$ is the set of lines on P all lying on g. If $\alpha = 1$, assume in addition that g and h intersect only in the point P. Define $\mathcal{E}_1$ as follows: a line y of h is in $\mathcal{E}_1$ if y lies on P and there is a plane of S which contains y and some line of $\mathcal{E}$. Then $|\mathcal{E}_1| \geq |\mathcal{E}|$.*

Proof. In the residue S_P, the "points" of $\mathcal{E}$ lie on the "line" g, and h is another "line" in the pG$_\alpha$ S_P. Now see Lemma 2.9 of Hobart and Hughes (1990), or simply count. ☐

Lemma 35. *If S is an L.pG$_\alpha$(s, t, u), and h is not on P, and if $\varphi_0(P)$ is the minimum non-zero value of $\varphi(P, h)$, then $f_P(h) \geq \varphi_0(P) + d(P, h) - 1$.*

Proof. If $d(P, h) = 1$, then this is the definition of $\varphi_0(P)$. Suppose the lemma has been proved for planes at a distance less than i from P, and that $d(P, h) = i$. Then let X be a point in h at distance i from P and on a plane g which is at a distance $i - 1$ from P; if $\alpha = 1$, we can choose g so that $g \cap h = P$. We know that X is collinear with all the $f_P(g)$ points of g at a distance $i - 1$ from P; call this set of points $\mathcal{D}$. The lines from X to $\mathcal{D}$ each contain at most s points of $\mathcal{D}$, so there must be at least $f_P(g)/s$ lines joining X to the points of $\mathcal{D}$. Call this set of lines $\mathcal{E}$, and apply Lemma 34: there are at least $f_P(g)/s$ lines on P, and lying in h, which lie in planes that contain points of $\mathcal{D}$. All points on each of these lines are collinear with points at distance $i - 1$ from P; there are s points on each such line, other than P, and P is also at a distance i from P, so $f_P(h) \geq f_P(g) + 1$, which proves the lemma. ☐

In the proof above, a slight "improvement" can be made as follows: if h is at distance two, then X is joined to $\mathcal{D}$ by at least $(\alpha s + 1)/s = \alpha + 1/s$ lines, so by at least $\alpha + 1$ lines, so $f_P(h) \geq (\alpha + 1)s + 1$. Proceeding in this fashion, we can show:

Lemma 36. *If S is an L.pG$_\alpha$(s, t, u), and $d(P, h) \geq 2$, then*

$$f_P(h) \geq (\alpha - 1) + d(P, h))s + 1.$$

Theorem 37. *Suppose that S is an L.pG$_\alpha$(s, t, u), and that φ_0 is the minimum value of all $\varphi_0(P)$. Then the diameter δ of S satisfies:*

$$\delta \le max\{2, min(s(t + 1) - 2\varphi_0 + 4, t - 2\alpha + 3)\}.$$

Proof. Suppose that $\delta \ge 3$, and consider this configuration: a plane g at distance $\delta - 2$ from P, a plane h at distance $\delta - 1$ from P, a point X at distance $\delta - 1$ from P, in $g \cap h$, and a point Y at distance δ from P, in h. (Such a set of objects must exist.) We know that g contains $f_P(g)$ points at distance $\delta - 2$ from P. But Y is collinear with some points on g, hence with at least $\varphi(Y, g) \ge \alpha s + 1$ points of g, and these points must be at distance $\delta - 1$ from P (otherwise Y would not be at distance δ). Now g contains $s(t + 1) + 1$ points, so $f_P(g) + \varphi(Y, g) \le s(t + 1) + 1$. Substituting $f_P(g) \ge \varphi_0 + \delta - 3$ and $\varphi(Y, g) \ge \varphi_0$, we have $\delta \le s(t + 1) - 2\varphi_0 + 4$.

Now suppose that $s(t + 1) - 2\varphi_0 + 3 \le 2$. Then if there could be a point Y at distance three from P, it is collinear with points on a plane g, and g contains points at distance one. But g must contain at least φ_0 points collinear with P, and at least φ_0 collinear with Y and hence at distance two. So $2\varphi_0 \le s(t + 1) + 1$, the number of points on a plane. This contradicts $s(t + 1) - 2\varphi_0 + 4 \le 2$. So we have $\delta \le max\{2, s(t + 1) - 2\varphi_0 + 4\}$.

For the other bound, if $\delta \ge 4$, then $d(P, g) \ge 2$, and we can use Lemma 36. Substituting $f_P(g) \ge (\alpha + \delta - 3)s + 1 > (\alpha + \delta - 3)s$, and $\varphi(Y, g) \ge \alpha s + 1$ we have $\delta < t - 2\alpha + 4$, so $\delta \le t - 2\alpha + 3$.

Suppose $t - 2\alpha + 3 = 3$, so $t = 2\alpha$. Then, more or less as above, there is a plane g containing points at distance two, and containing points collinear with some point Y at distance four, and this leads to too many points on g. Finally, if $t - 2\alpha + 3 = 2$, then there is a plane g at distance one, but containing points collinear with some point Y at distance three, which also leads to too many points on g. Thus we have the rest of the inequality. $\square$

In the case of an EpG_α, the second bound above, $t - 2\alpha + 3$, is never better than the first, but if $s > 1$, it seems quite possible that this second bound can improve the first one (indeed, by a lot). For instance, if S is triangular, then the second bound is better precisely when $s > 1$ and $t > 2\alpha - 1$.

Corollary 38. *If S is an $EpG_\alpha(s, t)$ and φ_0 is the minimum value of all non-zero $\varphi(P, h)$ in S, then the diameter δ of S satisfies $\delta \ge max\{2, s - 2\varphi_0 + 5\}$.*

Corollary 39. (a) *Suppose S is an* $L.pG_{t+1}(s, t, u)$ *(so the point-residues are linear spaces); then the diameter of S is one.*
(b)   *Suppose S is an* $L.pG_t(s, t, u)$ *(so the point-residues are dual nets); then the diameter of S is at most two.*

Proof.   In case (a), S is a geometry for ○——L——○——L——○, and it is easy to see directly that S is a 2-design (using either the lines, or the planes as blocks).

In the case (b), just substitute $\alpha = t$ in Theorem 37.   □

7. On $A_2.pG$ geometries

A special case of L.pG geometries are the $A_2.pG$ *geometries*: those going with the diagram ○————○═══α═══○ .   Here the respective three orders must be s, s, t, from left to right.   There are two especially interesting families of such geometries known: (truncated) projective geometries of rank ≥ 3, and polar spaces of rank 3, as in Theorem 10 (ii) and (iii).   In Theorem 44 we give a characterisation of these two cases: they are the triangular examples.   But there are other such geometries, as in Example 9, and we characterise in Theorem 41 all $A_2.pG$ geometries whose point-residues are non-proper.   In general, the results are much stronger in this $A_2.pG$ case.   Thus in Theorem 42 we prove "strong finiteness": the diameter of an $A_2.pG$ geometry is bounded by two.

Lemma 40.   *If S is an* $A_2.pG_\alpha$ *geometry of order* (s, s, t) *and if* $\varphi(P, h) \neq 0$, *then the dual of* $\mathcal{A}(P, h)$ *is a 2-design with parameters* $(\alpha(s + 1) - s, \alpha, 1)$. *Thus* $\varphi(P, h) = (\alpha(s + 1) - s)(s + 1)/\alpha$.

Proof.   If $X \in \mathcal{A}(P, h)$, then by the proof of Theorem 19, X lies on α planes which both contain P and intersect h in a line.   So the lines of $\mathcal{A}_h(P, h)$ contain $s + 1$ points, and there are α of them on each point of $\mathcal{A}_h(P, h)$.   Any two of these lines meet, since h is a projective plane.   If there are m lines in $\mathcal{A}_h(P, h)$, then the dual of $\mathcal{A}_h(P, h)$ is a 2-$(m, \alpha, 1)$, with its number of " blocks on a point" equal to $s + 1$.   Hence $m = \alpha(s + 1) - s$, and $\varphi(P, h)$ is the number of "blocks" in this dual structure. Thus

$$\varphi(P, h) = \frac{m(m-1)}{\alpha(\alpha - 1)},$$

which finishes the proof. ☐

The $A_2.pG$ geometries are "known" whenever the point-residues are not proper (and we conjecture that the proper pGs will not yield many more examples):

Theorem 41. (a) *A geometry for* $\circ\!\!-\!\!-\!\!-\!\!\circ\overset{L}{-\!\!-}\!\circ$ *is a truncated projective geometry.*

(b) *A geometry for* $\circ\!\!-\!\!-\!\!-\!\!\circ\overset{J}{-\!\!-}\!\circ$ *is a "dual matroid".*

(c) *A geometry for* $\circ\!\!-\!\!-\!\!-\!\!\circ\overset{N}{-\!\!-}\!\circ$ *is the inversive plane of order 2 (a 3-design for* (5, 3, 1), *or a* $PG(4, 1)$), *or is a polar space of order* (s, 1) *and rank 3 (a geometry for the diagram* $\underset{s}{\circ}\!\!-\!\!-\!\!\underset{s}{\circ}\!\!=\!\!=\!\!\underset{1}{\circ}$).

(d) *A geometry for* $\circ\!\!-\!\!-\!\!-\!\!\circ\overset{и}{-\!\!-}\!\circ$ *is a dual attenuated space.*

(e) *A geometry for* $\circ\!\!-\!\!-\!\!-\!\!\circ\!\!=\!\!=\!\!\circ$ *is a polar space.*

Proof. (a), (d) and (e) are immediate consequences of Theorem 10, while (b) (and a discussion of matroids) can be found in Buekenhout (1979).

For (c), we note that the net S_P, considered as a point-line geometry for convenience, contains a structure with $t(s + 1) - s$ lines, any two of which meet. But the number of lines cannot exceed the number of parallel classes of lines, since two parallel lines do not meet. Hence $t(s + 1) - s \le t + 1$, which yields $s(t - 1) \le 1$. So $t = 1$ (= α), which gives us an (arbitrary) polar space for $\underset{s}{\circ}\!\!-\!\!-\!\!\underset{s}{\circ}\!\!=\!\!=\!\!\underset{1}{\circ}$, or $t = 2$, $s = 1$, which leads immediately to $\underset{1}{\circ}\overset{C}{-\!\!-}\underset{1}{\circ}\overset{Af}{-\!\!-}\underset{2}{\circ}$, the inversive plane of order 2. ☐

(NB. Of course, Theorem 41 implies a classification - up to "matroids" - of geometries for

$$\circ\!\!-\!\!-\!\!-\!\!\circ\!\!-\!\!-\!\!-\!\!\circ\cdots\circ\!\!-\!\!-\!\!\circ\!\!=\!\!=\!\!\circ$$

where the pG at the end is improper.

The proof of (c) is equivalent to a special case of a theorem due to De Clerck (2.1.6 in Haemers (1979)) concerning a sub-$pG_\alpha(s_1, t_1)$ in a $pG_\alpha(s, t)$: we must have $s = s_1$ or $s \ge s_1 t_1 + \alpha - 1$. Using that theorem one can show, for instance, that a point-residue in an $A_2.pG$ geometry cannot be one of the structures coming from one of the known complete k-arcs in a projective plane, excepting trivially.)

Theorem 42. *If* (P, h) *is an antiflag in the* $A_2.pG_\alpha$ *geometry S, then* $\varphi(P, h) \neq 0$. *Furthermore, the diameter of S is at most two.*

Proof. If there could exist a pair (P, h) such that $\varphi(P, h) = 0$, then by connectivity we can find two planes g_1 and g_2, and two points A and B on g_2 such that B is collinear with points on g_1 and A is not. Let h be a plane on B which meets g_1 in a line z (as in the proof of Lemma 40 or Theorem 19). Let w be the line on A and B.

There are $s + 1$ lines on B, in the plane h, which meet z, and considering the residue S_B, we see that the line w is in a plane with α of these $s + 1$ lines. Let k be such a plane; k meets h in a line u, and u meets z, since u and z are lines in the projective plane h. Thus the plane k contains points of g_1, and so A is collinear to points of g_1, contrary to hypothesis. This proves the first statement of the theorem.

The second statement follows immediately: for if P and Q are two points, and h is a plane on Q, then P is collinear with some point of h, and that point is collinear with Q □

Now, from Theorem 18 the dual 2-design (or linear dual space) $\mathcal{A}(P, h)$ is a sub-design of both the projective plane h and the pG_α S_P. Note that if $\alpha = 1$, then $\mathcal{A}(P, h)$ is trivial: it has one line and $s + 1$ points. But if $\alpha > 1$, then the structure $\mathcal{A}(P, h)$ can be more interesting. If $\alpha = s + 1$, then S is a truncated projective geometry, and $\mathcal{A}_h(P, h) = h$, for all antiflags. But for $\alpha \le s$, we have:

Theorem 43. *If S is an A_2. pG_α geometry of order (s, s, t), with $\alpha \le s$, then:*

(a) α *divides s;*

(b) *if (P, y) is a point-line antiflag, then P is collinear with $s + 1 - s/\alpha$ points of y, or with all points of y;*

(c) *if (P, y) is a point-line antiflag, then P is collinear with all points of y if and only if P and y lie in a common plane.*

Proof. If there is a plane on P and y, then clearly P is collinear with all points of y. Suppose no such plane exists, and let h be a plane on y. Then y is not one of the lines of the dual 2-design $\mathcal{A}_h(P, h)$. Let us suppose y contains m points of $\mathcal{A}_h(P, h)$; at each of these m points y meets α lines of $\mathcal{A}_h(P, h)$, and since it meets all the lines of $\mathcal{A}_h(P, h)$, it follows that $m\alpha = \alpha(s + 1) - s - s/\alpha$. The value of m does not depend on y, thus proving the rest of the theorem. □

Theorem 44. *An $A_2.pG$ geometry is triangular if and only if it is a polar space of rank 3, or a projective geometry of rank ≥ 3 truncated on {points, lines, planes}.*

Proof. S is triangular if and only if $\varphi(P, h) = \alpha s + 1$; using the value of φ from Lemma 40, this simplifies to $\alpha^2 - \alpha(s + 2) + (s + 1) = 0$, or $(\alpha - 1)(\alpha - (s + 1)) = 0$. Thus S is triangular if and only if $\alpha = 1$, or $s + 1$. But in the first case S is an $A_2.GQ$ geometry, that is, a geometry for ○────○════○, and is a polar space of rank 3 (see Theorem 10 (iv)). In the second case, S is a geometry for ○────○─$\overset{L}{}$─○, which is a truncated projective geometry (see Theorem 10 (ii)).   □

The polar spaces (of any rank) are characterised by the property that for every point-line antiflag (P, y), P is collinear with one or all the points of y. This is a larger class that $A_2.GQ$ geometries, in any case, but it leads to the natural problem of characterising the point-line structures where "one" is replaced by another constant γ.

8. On pG.L geometries

In this section we consider pG.L geometries, which are rather like the last ones, but turned around. Here if S is a triangular pG.L, then S necessarily satisfies IP, from Theorem 23. Note that Examples 12, 13 and 15 (A)(i)(ii)(iv) include a number of pG.L geometries. We follow this in the next section with results on diameters. First we have an analogue to Lemma 27:

Lemma 45. *Let S be a $pG_\alpha.L$ geometry of order (s, t, u).*
(a) *Two points lie on a common plane if they lie in a common line. (Note that this is not "if and only if" in general, unlike Lemma 27.*
(b) *Each line contains $s + 1$ points and lies in $u + 1$ planes.*
(c) *The points and lines of each plane form a $pG_\alpha(s, t)$, hence there are $(s + 1)(st + \alpha)/\alpha$ points and $(t + 1)(st + \alpha)/\alpha$ lines in a plane.*
(d) *The lines and planes on a point form a 2-design for $(t(u + 1) + 1, t + 1, 1)$, so there are $t(u + 1) + 1$ lines and $(u + 1)(t(u + 1) + 1)/(t + 1)$ planes on a point.*
(e) *If two points are collinear, then they lie on exactly one line and $u + 1$ common planes.*

Proof. Completely straightforward.   □

Corollary 46. *If (P, h) is an antiflag, and $\varphi(P, h) \neq 0$, then $\alpha(t + 1) - t \leq \varphi(P, h) \leq st + \alpha$, and α divides $(t + 1) \cdot \varphi(P, h)$.*

Proof. Both inequalities come from Theorem 19, using the first of the upper-bounds for $\varphi(P, h)$ given there. ◻

Corollary 47. *(P, h) is a triangular antiflag if and only if $\varphi(P, h) = \alpha(t + 1) - t$, and this is equivalent to the following:*
(a) when $\alpha = 1$, if and only if $\mathcal{A}(P, h)$ consists of a single point;
(b) when $\alpha > 1$, if and only if $\mathcal{A}(P, h)$ is a sub-$pG_\alpha(\alpha - 1, t)$, that is, a 2-design for $(\alpha t + \alpha - t, \alpha, 1)$.

Proof. Simply a rephrasing of Corollary 25. ◻

Corollary 47, together with the divisibility condition of Corollary 46 gives strong restrictions on these triangular pG.L geometries in specific cases:

Corollary 48. *Suppose S is a triangular pG.L geometry and a plane is a $pG_s(s, t)$ (that is, a dual net).*
(a) If $s > 1$, then S is a $\{1, 2, 3\}$-truncation of a geometry for

which is in turn a truncated projective space of order 1.

(b) If $s = 1$, then S is a (triangular) geometry for ⊶⚬—ᴸ—⚬ .

Proof. Note that in a dual net $pG_s(s, t)$, there are $s + 1$ equivalence classes of $t + 1$ points each, under the relationship of not being collinear. But then the sub-$pG_s(s - 1, t)$ $\mathcal{A}_h(P, h)$ will be a 2-design with $s(t + 1) - t$ points. Any two points of $\mathcal{A}_h(P, h)$ are collinear; hence there is at most one point from each equivalence class in $\mathcal{A}_h(P, h)$. So $s(t + 1) - t \leq s + 1$, which leads to $s \leq 1 + 1/t$. If $t > 1$ this yields $s = 1$ and if $t = 1$ this gives $s \leq 2$.

Suppose $s = 2, t = 1$. The diagram for S is ⚬⚬—ᴸ—⚬ in fact, and since S has IP, the conclusion of (a) follows from Sprague's result (Theorem 10 (vii)). If on the other hand $s = 1$, then (b) follows immediately. ◻

It would be very useful if the following conjecture were true, as we see in the conjecture which follows it.

Conjecture 49. *If S is a finite linear space with s + 1 points on a line, such that every triangle lies in a sub-geometry isomorphic to an affine plane with s points on a line, then S is a projective geometry PG(n, s).*

Conjecture 50. *If S is a triangular pG.L geometry in which a plane is a net, then S is an attenuated space (Example 9).*

Proof. (*Depending on Conjecture 49*) Here $\mathcal{A}_P(P, h)$ will be an affine plane of order t, and in particular, embedded in the linear space S_P, which has $t + 1$ points on a line. If we choose any three lines x, y, z on P, then it is easy to see that for any point $Q \neq P$ on x, there are points $R \neq P$ and $S \neq P$ on y and z respectively such that Q is collinear with R and S; hence there is a plane h on Q, R, S. Since P is collinear with all three of these points, they all lie in $\mathcal{A}_h(P, h)$, which is an affine plane of order t. Thus (projecting into S_P, as in Theorem 18), any triangle in S_P is contained in an affine plane of order t. Now, using Conjecture 49, and Theorem 10 (xviii), the result follows. $\square$

In a pG.L geometry, minimal φ plays a powerful role, as it does in L.pG geometries. That is, it is associated with the most "geometric" cases, and it might not be unreasonable to hope for some kind of converse of Theorem 26 in so far as it applies to pG.L geometries.

The maximum value of φ gives:

Theorem 51. *Let $\mathcal{A} = \mathcal{A}_h(P, h) \neq 0$. Then $\varphi(P, h)$ takes on its maximum value $st + \alpha$ as follows:*
(a) *when $\alpha = 1$, if and only if $\mathcal{A}$ is an ovoid in the GQ h;*
(b) *when $\alpha > 1$, then:*
  (i) *if $\alpha = s + 1$, then if and only if $\mathcal{A} = h$*
  (ii) *if $\alpha = s$, then $\mathcal{A}$ consists of all points of the dual net h, except exactly one equivalence class of $t + 1$ points, and of all lines of h.*
  (iii) *if $\alpha = t$, then $\mathcal{A}$ consists of $t(s + 1)$ points in the net h such that every line contains t points of $\mathcal{A}$ (and every line of h is in $\mathcal{A}$).*
(*If $\alpha = t + 1$, then S is of course a truncated projective geometry, and can be shown to be truncated at {points, lines, planes}.*)

Proof. In each case, $\mathcal{A}$ has the property that every line of the $pG_\alpha(s, t)$ meets $\mathcal{A}$ in α points. So (a) is just the definition of an ovoid in a GQ.

For (b) we first compute in each case. So in case (i), $\varphi(P, h) = st + s + 1$ is the number of points in a $pG_{s+1}(s, t)$. In case (ii), $\mathcal{A}$ certainly contains all but $t + 1$ points of h, and every line meets $\mathcal{A}$ in s points, but this means that every point of h not in the equivalence class of X must be in $\mathcal{A}$, proving the statement.

(iii) is just a translation of the facts about $\mathcal{A}$. □

Probably a more detailed analysis of these various cases in Theorem 57 would allow us to say a lot more, perhaps even characterise some of the cases. (The pG_α analogue of an ovoid in a GQ thus might be a set of $st + \alpha$ points such that every line of the pG meets the set in α points; Thas has already called these α-*ovoids*.)

9. Diameter bounds for pG.L geometries

Now we pass on to the diameter bounds for these pG.L geometries. The situation here is less clear, and we have results for $\alpha > 1$ only. (Dimitrii Pasechnik pointed out an error in an earlier version of this paper for the case $\alpha = 1$, for which the author expresses his thanks; Pasechnik (to appear) has in fact studied the diameters of these GQ.L geometries, and has some partial results).

Let us assume that S is a pG_α.L with $\alpha > 1$.

Lemma 52. *If X is at distance $i > 1$ from P, then every line X either contains $\alpha + i - 2$ points at distance $i - 1$ from P, or contains $\alpha + i - 1$ points at distance i from P.*

Proof. If X is at distance one, then there is a line x through X and P; if y is any other line on X, then the plane determined by y and x is a pG_α, so P is collinear to α points on y. If Y is at distance two from P, then any line that joins Y to a point at distance one will be like a line y just discussed and will contain α points at distance one from P. Any other line, w say, will form with a line y a plane, and in that plane we have two lines y and w, meeting in a point X, and with a set $\mathcal{B}$ of α points on y, none of them X; so the α points of $\mathcal{B}$ are joined to at least α points on w by lines (Lemma 2.9 of Hobart and Hughes (1990), or see Lemma 34 above), so w contains at least $\alpha + 1$ points at distance two from P.

The proof proceeds like this for each i; we omit the easy details. □

Theorem 53. *If $\alpha > 1$, then the diameter δ of a pG_α.L of order (s, t, u) satisfies $\delta \leq max\{2, s - 2\alpha + 4\}$.*

Proof. Suppose $\delta \geq 3$. If there is a point X at distance δ from P, then it is collinear with a point Y at distance $\delta - 1$, which is on a line y which contains a point at distance $\delta - 2$. From Lemma 52, y contains at least $\alpha + \delta - 3$ points at distance $\delta - 2$ from P. On the other hand, since y contains points collinear with X, it contains α points collinear with X, and these points must be at distance $\delta - 1$ from P. So $(\alpha + \delta - 3) + \alpha \leq s + 1$, so $\delta \leq s - 2\alpha + 4$.

Now suppose that $s - 2\alpha + 4 \leq 2$, whence $s + 2 \leq 2\alpha$. Let X be at distance three from P, if possible. Then there is a point Y at distance two from P, and a line XY, and similarly a point Z collinear with P and a line $y = ZY$. Since y contains at least one point collinear with X, it contains exactly α; since y contains at least one point collinear with P, it contains exactly α also. So $2\alpha \leq s + 1$, a contradiction. $\square$

(NB. Notice that here we do not use IP).

Corollary 54. If S is a $pG_s.L$ of order (s, t, u) with $s > 1$, then the diameter of S is two.

Proof. From Theorem 53, the diameter is at most $\max\{2, 4 - s\}$ which is at most 2, since $s > 1$. If P is a point and h is a plane not on P (such planes must exist), then by Corollary 46 P is collinear with $\varphi(P, h) \leq st + \alpha = s(t + 1)$ points of h, so there at least $t + 1$ points of h not collinear with P, hence at distance two from P. $\square$

Thus the result in Theorem 53 is best possible in some cases: eg., when $\alpha = s$. The case $\alpha = s + 1$ also leads to $\delta = 1$ simply by applying Lemma 52, but this could, of course, be seen more directly.

References

F. Buekenhout (1977), Extensions of polar spaces and the doubly transitive symplectic groups, *Geom. Dedicata* **6**, 13-21.

F. Buekenhout (1979), Diagrams for geometries and groups, *J. Combin. Theory Ser. A* **27**, 121-151.

F. Buekenhout (1981), The basic diagram of a geometry, *Geometries and Groups*, Lecture Notes in Math. 893, 1-29, Springer-Verlag, Berlin.

F. Buekenhout (1985), Diagram geometries for sporadic groups, *Finite Groups - Coming of Age*, **45**, 1-32, American Mathematical Society, Providence.

F. Buekenhout and X. Hubaut (1977), Locally polar spaces and related rank 3 groups, *J. Algebra* **45**, 391-434.

P.J. Cameron (to appear), Some Buekenhout geometries, *Proc. L.M.S. symposium on Groups and Combinatorics*, 1990.

P.J. Cameron, D.R. Hughes and A. Pasini (1990), Extended generalised quadrangles, *Geom. Dedicata* **35**, 193-228.

A.M. Cohen and E.E. Shult (1990), Affine polar spaces, *Geom. Dedicata* **35**, 43-76.

F. De Clerck (1984), Substructure of partial geometries, *Quaderno del seminario di geometrie combinatorie* **5**, Istituto di Matematica Applicata della Facoltà d'Ingegneria, l'Aquila.

F. De Clerck (to appear), Some classes of rank 2 geometries, *Handbook of Incidence Geometry*, (ed. F. Buekenhout), North Holland, Amsterdam.

P. Dembowski (1968), *Finite Geometries*, Springer-Verlag, Berlin.

J. Doyen and X. Hubaut (1971), Finite regular locally projective spaces, *Math. Z.* **119**, 83-88.

W.H. Haemers (1979), *Eigenvalue techniques in design and graph theory*, Thesis, Mathematisch Centrum, Amsterdam.

D.G. Higman (1982), Admissible graphs, *Finite Geometries*, Lecture Notes in Pure and Applied Math. **82**, 211-222, (ed. N.L. Johnson, M.L. Kallaher and C.T. Long), Dekker, New York.

S.A. Hobart and D.R. Hughes (1990), Extended partial geometries: nets and dual nets, *European J. Combin.* **11**, 357-372.

D.R. Hughes (1965), Extensions of designs and groups: projective, symplectic and certain affine groups, *Math. Z.* **89**, 199-205.

D.R. Hughes (1982), Semi-symmetric 3-designs, *Finite Geometries*, Lecture Notes in Pure and Applied Math. **82**, 223-235, (ed. N.L. Johnson, M.L. Kallaher and C.T. Long), Dekker, New York.

D.R. Hughes (1983), On the non-existence of a semi-symmetric 3-design with 78 points, *Ann. Discrete Math.* **18**, 473-480.

D.R. Hughes (to appear), On partial geometries of rank n, *Combinatorics '90*.

D.R. Hughes and F.C. Piper (1973), *Projective Planes*, Springer-Verlag, Berlin.

D.R. Hughes and F.C. Piper (1985), *Design Theory*, Cambridge University Press, Cambridge.

W.M. Kantor (1974), Dimension and embedding theorems for geometric lattices, *J. Combin. Theory Ser. A* **17**, 173-195.

R. Laskar and J. Dunbar (1978), Partial geometry of dimension three, *J. Combin. Theory Ser. A* **24**, 187-201.

C. Lefèvre-Percsy and L. Van Nypelseer (to appear), Finite rank 3 geometries with affine plane and dual affine point residues, *Discrete Math.*

D.V. Pasechnik (to apppear), Dual linear extensions of generalized quadrangles.

T. Penttila (to appear), One-point extensions of finite inversive planes.

M.A. Ronan (1982), Locally truncated buildings and M_{24}. *Math. Z.* **180**, 489-501.

M.A. Ronan and S.D. Smith (1980), 2-local geometries for some sproadic groups, *Proc. Symp. Pure Math.* **37**, 283-289, American Mathematical Society, Providence.

A.P. Sprague (1981), Incidence structures whose planes are nets, *European J. Combin.* **2**, 193-204.

A.P. Sprague (1985), Rank 3 incidence structures admitting dual linear-linear diagram, *J. Combin. Theory Ser. A* **38**, 254-259.

J. Tits (1956), Les groupes de Lie exceptionels et leur interprétation géométrique, *Bull. Soc. Math. Belg.* **8**, 48-81.

J. Tits (1981), A local approach to buildings, *The Geometric Vein*, 519-547, Springer-Verlag, Berlin.

L. Van Nypelseer (to appear), Rank n geometries with affine hyperplanes and dual affine point residues, *European J. Combin.*

Intersection numbers of quasi-multiples
of symmetric designs

D. Jungnickel and V.D. Tonchev
University of Giessen

Dedicated to Professor Hanfried Lenz on the occasion of his 75th birthday.

1. Introduction

We assume that the reader is familiar with the basic notation and facts from design theory. Our notation follows that from Beth et al. (1985), Cameron and van Lint (1980), Hall (1986), Hughes and Piper (1985), and Tonchev (1988).

Throughout, let D be an s-fold quasi-multiple of a symmetric (v, k, λ) design; that is, the parameters of D are

$$v' = v, \ k' = k, \ r' = sk, \ \lambda' = s\lambda, \ b' = sv \qquad (1)$$

where v, k, λ satisfy

$$k(k - 1) = \lambda(v - 1). \qquad (2)$$

Since a design with parameters (1) may exist regardless of the existence or non-existence of a symmetric (v, k, λ) design, we shall consider arbitrary triples of integers (v, k, λ) which satisfy (2), even when they do not satisfy the Bruch-Ryser-Chowla condition. Our goal will be directed to answering the following

Question 1. What block intersection numbers can a quasi-multiple design D have?

According to their block intersection numbers, quasi-multiples of symmetric designs can be divided into two classes.

Class (a). D is a multiple; that is, a union of identical copies of a symmetric (v, k, λ) design. Then D is clearly a quasi-symmetric design with intersection numbers $x = \lambda, y = k$. Recall that a 2-(v, k, λ) design is quasi-symmetric with intersection numbers x, y $(x < y)$ if the cardinality of intersection of pairs of blocks takes two distinct values x and y. Quasi-symmetric designs have received considerable attention because of their fruitful relations with other combinatorial structures such as strongly regular graphs and self-dual codes (see for example the recent surveys Shrikhande (1990) and Tonchev (to appear)).

Class (b). *D* is not a multiple; that is a proper quasi-multiple design. In this case we will be interested in the following

Question 2. Under what conditions can *D* be proper and still quasi-symmetric?

A proper quasi-multiple quasi-symmetric design is "exceptional" in the sense of Neumaier (1982). The extended version of Neumaier's table for parameter sets of exceptional quasi-symmetric designs (1982), Tonchev (to appear), contains 6 sets for proper quasi-multiple designs:

(a)   #1: $4\times(19, 9, 4)$, $x = 3, y = 5$;

(b)   #10: $8\times(22, 7, 2)$, $x = 1, y = 3$;

(c)   #22: $4\times(37, 9, 2)$, $x = 1, y = 3$;

(d)   #38: $9\times(49, 16, 5)$, $x = 4, y = 7$;

(e)   #61: $16\times(61, 25, 10)$, $x = 9, y = 13$;

(f)   #62: $3\times(61, 21, 7)$, $x = 6, y = 9$.

The non-existence of a quasi-symmetric design with parameters (a) was shown by Calderbank (1987).

A design with parameters (b) can be constructed as residual of the unique 4-(23, 7, 1) design (Goethals and Seidel 1970), hence it is, in fact, a 3-(22, 7, 4) design and as shown by Tonchev (1986), this is the unique quasi-symmetric design with the given parameters. Both the 4-(23, 7, 1) and 3-(22, 7, 4) designs have been recently characterized as the only quasi-symmetric 3-designs with $x = 1$ (Calderbank and Norton (1990); Pawale and Sane (to appear)).

The existence of quasi-symmetric designs in the remaining cases is still unsettled.

The unique example (b) shows that proper quasi-multiple quasi-symmetric designs may exist in general and the rest of the table hints that admissable parameters seem not to be that scarce. These examples, however, suggest the following

Conjecture 3. Assume $(k, \lambda(s - 1)) = 1$. Then any quasi-symmetric *s*-fold quasi-multiple design is a multiple.

In the next section we give some sufficient conditions for a quasi-symmetric design to be a multiple, which support this conjecture. In particular, the conjecture is true for $k = p$ a prime, and a quasi-multiple of a projective plane is quasi-symmetric if and only if it is a multiple. A characterization of the unique quasi-symmetric 2-

(22, 7, 16) design as the only proper quasi-multiple quasi-symmetric design with $x = 1$ and $s = k + 1$ is also proved.   Finally, we consider some examples of quasi-doubles provided by cyclic difference sets.

We mention that a characterization of the doubles of symmetric designs among the quasi-symmetric designs in terms of extremal intersection numbers has been given by Beutelspacher (1982).

2. Preliminaries

Proposition 4. *The largest intersection number of D is at least* $\lambda + 1$ *provided that* $s > 1$.

Proof.   Assume the contrary and count in two ways all pairs (p, B) with $p \in B \cap B_0$, where B_0 is a fixed block distinct from B.   We have

$$k(sk - 1) = |\{(p, B): p \in B \cap B_0\}| \leq \lambda(sv - 1),$$

whence

$$s(k^2 - \lambda v) = s(k - \lambda) \leq k - \lambda,$$

that is, $s \leq 1$, a contradiction. $\square$

Proposition 5. *Let D be quasi-symmetric.   Then the intersection numbers* x, y *satisfy the following:*
(i) $k(sk - 1)(x + y - 1) - (sv - 1)xy = k(k - 1)(s\lambda - 1)$;
(ii) $y - x \mid k - x$;
(iii)   *D is a multiple if and only if* $x = \lambda$ *or* $y = k$;
(iv)   *if D is not a multiple then* $x \leq \lambda - 1$;
(v)   *if* $s \geq (v - 1)/2$ *then D is a multiple.*

Proof. The assertions (i), (ii) follow directly from Lemmas 2.1, 2.2 of Sane and Shrikhande (1987) respectively.   To prove (iii), substitute $y = k$ in (i).   This gives

$$(k - 1)(sk - s\lambda) = x(sv - sk),$$

whence, using (2), one obtains $x = (k - 1)(k - \lambda)/(v - k) = \lambda$. Conversely, putting $x = \lambda$ in (i) implies $y = k$.
  (iv) By Proposition 2.2 from Sane and Shrikhande (1987)

$$x < k^2/v < s\lambda.$$

Using (2), we have

$$k^2/v = (k^2 - \lambda v + \lambda v)/v = (k - \lambda)/v + \lambda < \lambda + 1$$

and since D is not a multiple, $x \le \lambda - 1$ by (iii).
Finally,

$$b \le v(v - 1)/2$$

for every quasi-symmetric design without repeated blocks and up to
the complementation $b = v(v - 1)/2$ only for a trivial 2-$(v, 2, 1)$ design
or the unique 4-$(23, 7, 1)$ design (Cameron and van Lint 1980), which
implies (v). $\square$

3. Quasi-symmetric quasi-multiple designs

Theorem 6. *Any quasi-symmetric quasi-multiple of a projective
plane is a multiple.*

Proof. Assume the contrary.   Then $x = 0$ by Proposition 5(iv).   Now
since $\lambda = 1$ Proposition 5(i) gives

$$(sk - 1)(y - 1) = (k - 1)(s - 1).$$

By Proposition 4, $y \ge 2$.   Hence

$$(sk - 1)(y - 1) \ge sk - 1.$$

On the other side, $k + s > s \ge 2$ implies $sk - 1 > (k - 1)(s - 1)$, a
contradiction.
   The next theorem proves our conjecture for k a prime.

Theorem 7. *Let D be a quasi-symmetric quasi-multiple design and
assume that k is a prime.   If k does not divide $s - 1$, then D is a
multiple.*

Proof.   Since k is a prime and $k > \lambda$, we have $k \mid \lambda(s - 1)$.   Therefore
$k \mid sv - 1 = (sk(k - 1) + \lambda(s - 1))/\lambda$, whence by Proposition 5,(i) $k \mid xy$.
Since $x < y \le k$, the last implies $y = k$.   Thus the assertion now follows
by Proposition 5(iii). $\square$

Remarks. (i) Example (b) shows that k may be a prime dividing $s - 1$
for a proper quasi-multiple design; in fact, we have $s = k + 1$ in this
case.
(ii) The last theorem thus holds in case that $s \le k$.   In particular, this
gives the following

Corollary 8. *If $2n + 1$ is a prime then a quasi-symmetric 2-$(4n + 3, 2n + 1, sn)$ design is an s-fold multiple Hadamard 2-$(4n + 3, 2n + 1, n)$ design.*

Proof.   Here we have $(v - 1)/2 = k = 2n + 1$.   Thus if $s > 2n + 1$ the statement follows by Proposition 5(v).   On the other hand, if $s \leq 2n + 1 \leq k$, the assertion follows from Theorem 7.   ☐

The next proposition can be viewed as an improvement of Theorem 6.

Theorem 9. *Let D be quasi-symmetric.   Then $x > 0$.*

Proof. Suppose that $x = 0$. Proposition 5(i) gives

$$(y - 1)(sk - 1) = (k - 1)(s\lambda - 1).$$

Thus $sk - 1 \mid (k - 1)(s\lambda - 1)$.   Put $d = (sk - 1, k - 1)$, $e = (sk - 1, s\lambda - 1)$. Then $d \mid s - 1$ and $e \mid k - \lambda$.   So $sk - 1 \mid de \mid (s - 1)(k - \lambda)$, which is absurd, since $sk - 1 > s(k - 1) > (s - 1)(k - \lambda)$.   ☐

4. Quasi-multiple quasi-symmetric designs with $x = 1$

In view of Theorem 7 and Example (b), it is natural to consider the case $s = k + 1$.   Adding the assumption that $x = 1$, we can characterize the only known non-trivial example;   that is, the unique quasi-symmetric 2-$(22, 7, 16)$ design with $x = 1$, $y = 3$, being an 8-fold quasi-multiple of the non-existing symmetric 2-$(22, 7, 2)$ design.

Theorem 10. *Let D be a quasi-symmetric design with $x = 1$ which is a $(k + 1)$-fold quasi-multiple of a symmetric (v, k, λ) design.   Then either D is a $(k + 1)$-fold multiple of a projective plane of order $k - 1$ or D is the unique quasi-symmetric 2-$(22, 7, 16)$ design.*

Proof. If D is a $(k + 1)$-fold multiple then $x = \lambda = 1$.   We may thus assume that D is a proper quasi-symmetric design from now on. Substituting $x = 1$ and $s = k + 1$ in Proposition 5(i) and using

$$sv - 1 = (sk(k - 1) + \lambda(s - 1))/\lambda$$

gives

$$(k^2 + k - 1)y - [((k + 1)(k - 1)/\lambda) + 1]y = (k - 1)((k + 1)\lambda - 1),$$
$$\lambda y(k^2 + k - 2) - (k - 1)(k + 1)y = \lambda(k - 1)((k + 1)\lambda - 1),$$

$$\lambda y(k + 2) - (k + 1)y = (k + 1)\lambda^2 - \lambda. \tag{3}$$

Now (3) shows that $\lambda \mid (k + 1)y$, thus $\lambda \mid (k + 1)(k - 2)y = (k^2 - k - 2)y$. From (2) we also have $\lambda \mid k(k - 1)$, thus $\lambda \mid (k^2 - k)\,y$. Hence

$$\lambda \mid 2y. \tag{4}$$

From (3) we also get

$$k + 1 \mid \lambda(y + 1). \tag{5}$$

We now claim that $(\lambda, k + 1) = 2$. Note first that $(\lambda, k + 1) > 1$: otherwise (5) gives $k + 1 = y + 1$, or $k = y$; this is, D a multiple, a contradiction. Now let $p \mid (\lambda, k + 1)$ for some prime $p > 2$ and let $p^a \parallel \lambda$ (the last notation means that p^a divides λ, but p^{a+1} does not). Then (3) gives the contradiction $\lambda \equiv 0 \bmod p^{a+1}$, as we have $p^a \mid y$ by (4). Finally, assume that $4 \mid (\lambda, k + 1)$. Then let $2^a \parallel \lambda$. This time (3) gives the contradiction $\lambda \equiv 0 \bmod 2^{a+1}$, as (4) now shows that $2^{a-1} \mid y$. Therefore we have

$$(\lambda, k + 1) = 2 \tag{6}$$

But then (5) shows that $(k + 1)/2 \mid y + 1$; as $y < k$ this implies $y + 1 = (k + 1)/2$, that is,

$$k = 2y + 1. \tag{7}$$

Substituting (7) in (3), we obtain

$$\lambda y(2y + 3) + \lambda = 2(y + 1)(y + \lambda^2),$$

or $2(y + 1)(y + \lambda^2 - \lambda y) = \lambda(y + 1)$; that is, $2(y + \lambda^2 - \lambda y) = \lambda$, whence,

$$\lambda(2\lambda - 1) = 2y(\lambda - 1). \tag{8}$$

By (4) we can write $2y = \lambda z$ for some integer z. Then (8) becomes

$$2\lambda - 1 = z(\lambda - 1),$$

whence $z = (2\lambda - 1)/(\lambda - 1)$, and thus $\lambda = 2, z = 3$, whence

$$y = 3, \quad x = 1, \quad k = 7, \quad v = 22, \quad s = 8. \tag{9}$$

As noticed above, the quasi-symmetric designs with parameters (9) is unique (Tonchev 1986). □

5. Quasi-double designs from difference sets

The quasi-symmetric 2-(22, 7, 16) design is the only known example of a proper quasi-multiple design which is also quasi-symmetric.

We now give examples of quasi-double designs with three intersection numbers showing that Proposition 4, Theorem 6 and Corollary 8 are best possible.   We use the following construction method. Let S be a cyclic (v, k, λ) difference set in G and consider the quasi-double design $D = \text{dev}(S, -S)$ of the symmetric (v, k, λ) design $\text{dev}(S)$. Here $\text{dev}(S, -S)$ is the design with point set G and block set $\{S + x : x \in G\} \cup \{-S + x : x \in G\}$.

Example 11. If S is a planar difference set with parameters $(n^2 + n + 1, n + 1, 1)$ then D is a design with block intersection numbers 0, 1, 2 (Jungnickel and Vedder 1984).

Example 12. Let S be a Paley Hadamard difference set in $GF(q)$, $q = 4n - 1$; that is, let S be the set of all non-zero squares in $GF(q)$. Then D is a 2-$(4n - 1, 2n - 1, 2n - 2)$ design with intersection numbers 0, $n - 1$, n. This is easily seen using the fact that the complementary difference set of S is just $-S \cup \{0\}$.

A computer investigation of cyclic difference sets with $k \leq 100$ as listed in Baumart (1971) suggests the following

Conjecture 13.   Let S be a cyclic (v, k, λ) difference set such that $D = \text{dev}(S, -S)$ has only three distinct intersection numbers 0, $\lambda, \lambda + 1$. Then S is either a planar difference set (that is $\lambda = 1$), or a Hadamard difference set of Paley type.

In fact, the computer search suggests an even stronger conjecture. All cyclic difference sets with $k \leq 100$ which are neither planar nor Paley-Hadamard, in fact, lead to at least 4 intersection numbers. Moreover, only 4 examples have just 4 intersection numbers, namely those with parameters (15, 7, 3), (37, 9, 2), (101, 25, 6) and (197, 49, 12). Observing that the last three of these are biquadratic residue difference sets (see Hall 1986), suggests the following result

Theorem 14. *Let t be an odd number for which $q = 4t^2 + 1$ is a prime power and let S be the difference set of biquadratic residues in $GF(q)$. Then each block of $D = \text{dev}(S, -S)$ has intersection numbers*

$$0, \frac{(t-1)^2}{4}, \lambda = \frac{(t^2-1)}{4}, \frac{(t+1)^2}{4},$$

with multiplicities 1, t^2, $6t^2$, t^2 *respectively.*

Proof. It is sufficient to show that $-S$ has intersection numbers 0, $\frac{(t-1)^2}{4}$, $\lambda = \frac{(t^2-1)}{4}$, $\frac{(t+1)^2}{4}$ with multiplicities 1, t^2, $6t^2$, t^2 respectively, with the translates of S. We use group ring notation. Thus, let G be the elementary abelian group of order q, written multiplicatively and consider the group ring ZG of G over the integers. By abuse of language, denote the formal sum of the elements of G in the difference set S again by S. Observe that the cardinality of $Sx \cap S^{(-1)}$ (where $S^{(-1)} = \{s^{-1}, s \in S\}$ now corresponds to $-S$) is just the coefficient of x^{-1} in the product $S^2 \in ZG$. We thus have to compute S^2. We will now use the results given in 11.6 of Hall (1986). What we call S here, is called C_0 by Hall. One has (by Hall op. cit. 11.6.23)

$$S^2 = a_{00} \cdot 1 + \Sigma c_{00k} C_k,$$

where the elements C_0, C_1, C_2, C_3 of ZG correspond to the multiplicative cosets of the biquadratic residues in $GF(q)$. Also, by Hall (op. cit., 11.6.37) $a_{00} = 0$ and (by 11.6.31), $c_{00k} = (0, k)$ (these are the cyclotomic numbers). Since $q = 1 + 4t^2$, we have $x = 1, y = t$ in (11.6.94) and thus (11.6.95) gives $(0, 0) = (0, 2) = \frac{t^2-1}{4}, (0, 1) = \frac{(t-1)^2}{4}$ and $(0, 3)$

$= \frac{(t+1)^2}{4}$.   Hence $S^2 = \frac{(t^2-1)(C_0 + C_2)}{4} + \frac{(t-1)^2 C_1}{4} + \frac{(t+1)^2 C_3}{4}$ which proves the assertion. $\square$

The case $(15, 7, 3)$ is exceptional. Here S is a twin prime difference set and just accidently leads to 4 intersection numbers only. In fact, one has

Theorem 15. *Let q and $q + 2$ be two prime powers and let S be the Hadamard difference set of twin prime power type in $GF(q) \times GF(q + 2)$. Then any block of $D = \text{dev}(S, -S)$ has intersection numbers*

$$q, \frac{(q^2 + 2q - 7)}{4} = \lambda - 1, \frac{(q^2 + 2q - 3)}{4} = \lambda, \frac{(q^2 + 4q - 5)}{4} \text{ and } \frac{(q^2 + 4q - 1)}{4}$$

with multiplicities 1, $(q^2 - 1)/2$, $(3q^2 + 4q - 3)/2$, $q + 1$ and $q - 1$ *respectively.*

The proof uses straightforward, but rather lengthy arguments, again involving the related group algebra, so we omit it.

Altogether, it seems that the construction of infinite series of quasi-doubles of symmetric designs having only a few intersection numbers is a difficult problem.

Acknowledgement. The second author was a Research Fellow of the Alexander von Humbolt Foundation at the University of Giessen and on leave from the University of Sofia, Bulgaria whilst writing this paper.

References

L.D. Baumert (1971), *Cyclic Difference Sets* LMN **182**, Springer, Berlin.

T.H. Beth, D. Jungnickel and H. Lenz (1985), *Design Theory*, B.I. Wissenschaftsverlag, Mannheim, and (1986) Cambridge University Press, Cambridge.

A. Beutelspacher (1982), On extremal intersection numbers of a block design, *Discrete Math.* **42**, 37-49.

A.R. Calderbank (1987), The application of invariant theory to the existence of quasi-symmetric designs, *J. Combin. Theory Ser. A* **44**, 94-109.

A.R. Calderbank and P. Norton (1990), Quasi-symmetric 3-designs and elliptic curves, *SIAM J. Discrete Math.* **3**, 178-196.

P.J. Cameron and J.H. van Lint (1980), *Graphs, Codes and Designs*, Cambridge University Press, Cambridge.

J-M Goethals and J.J. Seidel, Strongly regular graphs derived from combinatorial designs, *Canad. J. Math.* **22**, 597-614.

M. Hall (1986), *Combinatorial Theory*, 2nd ed., Wiley, New York.

D.R. Hughes and F.C. Piper (1985), *Design Theory*, Cambridge University Press, Cambridge.

D. Jungnickel and K Vedder (1984), On the geometry of planar difference sets, *European J. Combin.* **5**, 143-148.

A. Neumaier (1982), Regular sets and quasi-symmetric designs, *Combinatorial Theory* (ed. D. Jungnickel and K. Vedder), Lecture Notes in Math. **969**, 258-275, Springer-Verlag, Berlin.

R.M. Pawale and S.S. Sane (to appear), A short proof of a conjecture on quasi-symmetric 3-designs, *Discrete Math.*

S.S. Sane and M.S. Shrikhande (1987), Quasi-symmetric 2, 3, 4-designs, *Combinatorica* **7**, 291-301.

M.S. Shrikhande (1990), Designs, intersection numbers and codes, *Coding Theory and Design Theory*, Part II, 304-318, (ed. D. Ray-Chaudhuri), Springer-Verlag, New York.

V.D. Tonchev (1986), Quasi-symmetric designs and self-dual codes, *European J. Combin.* **7**, 67-73.

V.D. Tonchev (1988), *Combinatorial Configurations*, Longman Scientific and Technical, J. Wiley and Sons, New York.

V.D. Tonchev (to appear), Self-orthogonal designs, *Contemporary Math.*

The discriminant of a cubic curve

H. Kaneta and T. Maruta
Okayama University

1. Introduction

Let K be an algebraically closed field of any characteristic. Denote by $PG(n, K)$ the n-dimensional projective space over K. A cubic form

$$f = X^3a + Y^3b + Z^3c + XYZm + X^2Ya_2 + X^2Za_3 + Y^2Xb_1 + Y^2Zb_3 + Z^2Xc_1 + Z^2Yc_2$$

in $K[X, Y, Z]$ defines a cubic curve $V(f)$ in $PG(2, K)$. The cubic form f or the cubic curve $V(f)$ is said to be singular if the intersection $V(f) \cap V(f_X) \cap V(f_Y) \cap V(f_Z)$ is not empty. An irreducible homogeneous polynomial $D \in K[A, ..., C_2]$ is called a discriminant of the cubic form f, or of the cubic curve $V(f)$ if f is singular if and only if $D(a, ..., c_2) = 0$. We shall denote $D(a, ..., c_2)$ by $D(f)$. It is known that the discriminant exists when char $K \neq 3$ (van der Waerden 1950). An element g of $GL(3, K)$ gives rise to a cubic form $f_g(X, Y, Z) = f(g(X, Y, Z)) = X^3a^* + ... + Z^2Yc_2^*$ and to an element g^* of $GL(10, K)$ such that $g^*(a, ..., c_2) = (a^*, ..., c_2^*)$. Note that $f_{gg'} = (f_g)_{g'}$ or, equivalently, $(gg')^* = g'^*g^*$ for g, g' in $GL(3, K)$. An $H \in K[A, ..., C_2]$ is called invariant if $H_g(A, ..., C_2) = H(g^*(A, ..., C_2))$ is equal to H for any $g \in SL(3, K)$. Recall that $H_{gg'} = (H_{g'})_g$ for any $g, g' \in GL(3, K)$. Denote by I the invariant polynomial ring consisting of all invariant polynomials. Recall that $H_g = (\det g)^d H$ if $H \in I$ is homogeneous and $g \in GL(3, K)$ with deg $H = d$. When char $K = 0$, or ≥ 5, there exist S and T in I with deg $S = 4$ and deg $T = 6$, (Elliott 1855, Salmon 1879). Our main theorems are as follows:

Theorem 1. *If char $K = 2$, then $I = K[M, D]$ where D is the discriminant.*

Theorem 2. *If char $K = 2$, then non-singular cubics $V(f)$ and $V(f^*)$ are projectively equivalent if and only if $J(f) = J(f^*)$, where $J(f) = M(f)^{12}/D(f)$.*

Theorem 3. *If char $K = 3$, there exists a discriminant D of degree 12.*

Theorem 4. *If char $K = 3$, then $I = K[I_2, D]$, where $I_2 = B_1C_1 + A_2C_2 + A_3B_3 - M^2$.*

Theorem 5. *If char K = 3, then non-singular cubics V(f) and V(f*) are projectively equivalent if and only if J(f)= J(f*), where J(f) = $I_2(f)^6/D(f)$.*

We shall frequently use the following result:

Lemma 6. *Assume that P, Q, and R $\in$ $K[X_1, ..., X_n]$ are homogeneous and that P is prime and not a factor of Q.   If P(x) = 0 and Q(x) $\neq$ 0 implies R(x) = 0, then P divides R.*

2. Char K = 2

Let *f* be a cubic form

$$X^2(Xa_1 + Yb_1 + Zc_1) + Y^2(Xa_2 + Yb_2 + Zc_2) + Z^2(Xa_3 + Yb_3 + Zc_3) + XYZm.$$

Throughout this section $x_{ij} = a_ib_j + a_jb_i$, $y_{ij} = a_ic_j + a_jc_i$ and $z_{ij} = b_ic_j + b_jc_i$.

Lemma 7. *There exists an irreducible homogeneous polynomial $D \in K[A_1, ..., M]$ such that a cubic form f is singular if and only if D(f) = 0.*

Proof.   Since *f* is singular if and only if there exists a nontrivial $(x, y, z) \in K^3$ such that $f_X = f_Y = f_Z = 0$, we can argue as in van der Waerden (1950, Sections 81 and 82).

Such a D is called the discriminant of the cubic form *f*, or of the cubic curve V(f).   We can obtain its explicit form as follows.   Assume that $a_1(a_1b_2 + a_2b_1)m \neq 0$. If f_X, f_Y and f_Z vanish for some (x, y, z), then z $\neq 0$ and $b_1f_X + a_1f_Y = 0$ so that

$$x = (x_{12}y^2 + x_{13}z^2 + b_1myz)/a_1mz. \tag{1}$$

Substituting the right-hand side of (1) for x in $f_X = 0$ and $f_Z = 0$ we get

$$y^4x_{12}^2 + y^2z^2(a_1a_2 + b_1^2)m^2 + yz^3a_1m^3 + z^4(x_{13}^2 + a_1a_3m^2) = 0, \tag{2}$$

$$y^4c_1x_{12}^2 + y^3za_1x_{12}m^2 + y^2z^2(a_1^2c_2 + b_1^2c_1 + a_1b_1m)m^2 + \\ yz^3a_1x_{13}m^2 + z^4(c_1x_{13}^2 + a_1^2c_3m^2) = 0. \tag{3}$$

Hence the resultant R of (2) and (3) vanishes.   Conversely, if there exists a nontrivial zero (y, z) common to (2) and (3), that is, if R vanishes, then z $\neq 0$ and (x, y, z), x being defined by (1), is a solution to $f_X = f_Y = f_Z = 0$.   Thus, if $a_1x_{12}mD \neq 0$, then R $\neq 0$.   Now we have R = const $a_1^{n_1}x_{12}^{n_1}m^{n_3}D^{n_4}$ for some non-negative integers n_i.   Using

Lemma 9 (below), R turns out to be of the form $a_1^{12}x_4^{12}m^8D^*$, where D^* runs as follows:

$$m^9a_1b_2c_3$$
$$+ m^8(a_1a_3b_2b_3 + b_1b_2c_1c_3 + a_1a_2c_2c_3)$$
$$+ m^7(a_1x_{23}y_{23} + b_2x_{13}z_{13} + c_3y_{12}z_{12})$$
$$+m^6(a_1a_3x_{23}^2 + a_1a_2y_{23}^2 + b_2b_3x_{13}^2 + b_1b_2z_{13}^2 + c_2c_3y_{12}^2 + c_1c_3z_{12}^2 +$$
$$a_1^2b_2b_3c_2c_3 + a_2^2b_1b_3c_1c_3 + a_3^2b_1b_2c_1c_2 + a_2a_3b_1^2c_2c_3 + a_1a_3b_2^2c_1c_3 +$$
$$a_1a_2b_3^2c_1c_2 + a_2a_3b_3b_2c_1^2 + a_1a_3b_1b_3c_2^2 + a_1a_2b_1b_2c_3^2 + a_1a_2b_2b_3c_3c_1$$
$$+ a_1a_2c_2c_3b_3b_1 + b_1b_2a_2a_3c_3c_1 + c_1c_2b_2b_3a_3a_1 + b_1b_2c_2c_3a_3a_1$$
$$+c_1c_2a_2a_3b_3b_1 + a_1^2b_2^2c_3^2)$$
$$+m^5(a_2z_{12}z_{13}^2 + a_3z_{13}z_{12}^2 + b_1y_{12}y_{23}^2 + b_3y_{23}y_{12}^2 + c_1x_{13}x_{23}^2 + c_2x_{23}x_{13}^2 +$$
$$a_3b_2c_3y_{12}^2 + a_1c_1b_2y_{23}^2 + a_1b_2c_2z_{13}^2 + a_1b_3c_3z_{12}^2 + a_2b_2c_3x_{13}^2 + a_1b_1c_3x_{23}^2$$
$$+ a_1b_1c_1z_{23}^2 + a_2b_2c_2y_{13}^2 + a_3b_3c_3x_{12}^2)$$
$$+ m^4(y_{12}^2y_{23}^2 + x_{13}^2x_{23}^2 + z_{12}^2z_{13}^2 + (a_2a_3 + b_1b_3 + c_1c_2)d^2) + m^3d^3 + d^4,$$

where $d = c_1x_{23} + c_2x_{13} + c_3x_{23}$.

Lemma 8. *D coincides with D^* above.*

Proof. It can be easily seen that none of a_1, $a_1b_2 + a_2b_1$ and m divides D^*, Hence $D^* = \text{const } D^n$ for an integer n, which must be equal to one.

Lemma 9. *The resultant R_4 of two quartics $ay^4 + by^3z + cy^2z^2 + dyz^3 + ez^4$ and $a'y^4 + b'y^3 + c'y^2z^2 + d'yz^3 + e'z^4$ is equal to the following determinant:*

$$\begin{vmatrix} (ab') & (ac') & (ad') & (ae') \\ (ac') & (ad')+(bc') & (ae')+(bd') & (be') \\ (ad') & (ae')+(bd') & (be')+(cd') & (ce') \\ (ae') & (be') & (ce') & (de') \end{vmatrix}$$

where $(ab') = ab' - ba'$ and so on.

Proof. Denote by D_4 the above determinant. Both R_4 and D_4 are homogeneous polynomials of degree 8. Recall that R_4 is the discriminant of the two quartics so that R_4 is prime. Since $R_4 = 0$ implies $D_4 = 0$ (Salmon 1855, section 84), R_4 divides D_4. If $a = 1, b = c = d = e = 0, a' = b' = c' = d' = 0$ and $e' = 1$, then $R_4 = D_4 = 1$.

Let $f_m = X^3 + Y^3 + Z^3 + XYZm$. It can be easily seen that f_m is singular if and only if $m^3 = 1$. The set of inflexions of the non-singular cubic $V(f_m)$ coincides with $V(X^3 + Y^3 + Z^3) \cap V(XYZ)$, which consists of 9 points. Also, there exist exactly 4 triangles each of which contains 9 inflexions on its sides, (Hirschfeld 1979, chap. 11). The four triangles are given as:

$V(XYZ)$

$V((X + Y + Z)(X + \omega Y + \omega^2 Z)(X + \omega^2 Y + \omega Z)) = V(X^3 + Y^3 + Z^3 + XYZ)$

$V((\omega X + Y + Z)(X + \omega Y + Z)(X + Y + \omega Z)) = V(\omega(X^3 + Y^3 + Z^3 + \omega XYZ))$

$V((\omega^2 X + Y + Z)(X + \omega^2 Y + Z)(X + Y + \omega^2 Z)) = V(\omega^2(X^3 + Y^3 + Z^3 + \omega^2 XYZ))$

where $\omega^2 + \omega + 1 = 0$. Denote them by Δ_0, Δ_1, Δ_2 and Δ_3 respectively. It is clear that a permutation matrix $[\sigma]$ in $GL(3, K)$ maps Δ_i onto itself.

Lemma 10 (Bretagnolle-Nathan 1958). *A non-singular cubic $V(f)$ is projectively equivalent to some $V(f_m)$, where $f_m = X^3 + Y^3 + Z^3 + XYZm$ with $m^3 \neq 1$.*

In the proof of Lemma 10 it is crucial that there are three collinear inflexions, which can be verified directly as in Campbell (1926): there exists at least one line touching $V(f)$ only at one point, from which it follows that there exists another line with the same property.

Lemma 11. *Let $g \in GL(3, K)$ and $m^3 \neq 1$. Then g is a projective automorphism of $V(f_m)$ if and only if*
(1) $g = \lambda(diag\ [1, \varepsilon, \varepsilon^2])[\sigma]$ *when $m \neq 0$,*
(2) $g = \lambda(diag\ [1, \varepsilon_2, \varepsilon_3])[\sigma]\ \Omega_j\ (j = 0, 1, 2)$ *when $m = 0$,*
where $\lambda \in K$ with $\lambda \neq 0$, $\varepsilon^3 = \varepsilon_2^3 = \varepsilon_3^3 = 1$, $[\sigma]$ is a permutation matrix and

$$\Omega_0 = \begin{bmatrix} 1 & 1 & 1 \\ 1 & \omega^2 & \omega \\ 1 & \omega & \omega^2 \end{bmatrix}, \quad \Omega_1 = \begin{bmatrix} \omega & 1 & 1 \\ 1 & \omega & 1 \\ 1 & 1 & \omega \end{bmatrix}, \quad \Omega_2 = \begin{bmatrix} \omega^2 & 1 & 1 \\ 1 & \omega^2 & 1 \\ 1 & 1 & \omega^2 \end{bmatrix}.$$

Proof. Only the case $m \neq 0$ will be proved. Assume that g maps $V(f_m)$ onto itself. Denote X, Y, and Z by X_1, X_2 and X_3 respectively. Assume first, that $g(\Delta_0) = \Delta_0$. There exists a permutation $\sigma \in S_3$ such that $g(V(X_i)) = V(X_{\sigma(i)})$ $(i = 1, 2, 3)$. Now g^{-1} takes the form $(diag[\mu_1, \mu_2, \mu_3])\ [\sigma]$, where $[\sigma]$ denotes the permutation matrix corresponding to σ. Since $f_m(g^{-1}(x_1, x_2, x_3))$ is proportional to f_m, we get $\mu_1^3 = \mu_2^3 = \mu_3^3 = \mu_1\mu_2\mu_3$.

Thus setting $\mu_i = \varepsilon_i\mu_1$ $(i = 2, 3)$, we have $\varepsilon_2{}^3 = \varepsilon_3{}^3 = \varepsilon_2\varepsilon_3 = 1$. Assume that $g(\Delta_0) = \Delta_1$. Then g^{-1} takes the form $(\mathrm{diag}[\mu_1, \mu_2, \mu_3])$ $[\sigma]$ Ω_0. Since $f(g^{-1}(x_1, x_2, x_3) = f_{m'}$, we can show that $\mu_1{}^3 = \mu_2{}^3 = \mu_3{}^3$ and that $\varepsilon m/(1 + \varepsilon m) = \varepsilon m$, where $\mu_i = \varepsilon_i\mu_1$ and $\varepsilon = \varepsilon_2\varepsilon_3$. This implies that $1 + \varepsilon m = 1$, a contradiction. Similarly it cannot happen that $g(\Delta_0) = \Delta_i$ $(i = 2, \text{ or } 3)$.

Theorem 12 (Hirschfeld 1979, Lemma 11.5.2). *Non-singular cubics $V(f_m)$ and $V(f_{m'})$ are projectively equivalent if and only if, for some ε_1, ε_2 and ε_3 with $\varepsilon_1{}^3 = \varepsilon_2{}^3 = \varepsilon_3{}^3 = 1$, we have*

$$m' = \varepsilon_1 m \quad \text{or} \quad m' = \varepsilon_2\varepsilon_3 m/(1 + \varepsilon_3 m) \tag{4}$$

Proof. Let $g \in GL(3, K)$ map $V(f_{m'})$ onto $V(f_m)$. When $g(\Delta_0) = \Delta_0$, g^{-1} takes the form $(\mathrm{diag}[\mu_1, \mu_2, \mu_3])$ $[\sigma]$, where $[\sigma]$ is a permutation matrix. Since $f_m(g^{-1}(X, Y, Z))$ is proportional to $f_{m'}$, we have $\mu_1{}^3 = \mu_2{}^3 = \mu_3{}^3$ and $m\varepsilon_2\varepsilon_3 = m'$ with $\mu_i = \varepsilon_i\mu_1$ $(i = 2, 3)$. In case $g(\Delta_0) = \Delta_i$ for some $i \neq 0$, there exists $g' = \mathrm{diag}[\delta^{-1}, 1, 1]$ with $\delta^3 = 1$ such that $g^* = g'g$ maps $V(f_{m'})$ and Δ_0 onto $V(f_{\delta m})$ and Δ_1 respectively. Thus g^{*-1} takes the form $(\mathrm{diag}[\mu_1, \mu_2, \mu_3])$ $[\sigma]$ Ω_0 $(\Omega_0$ is the same as in Lemma 11). Since $f_{\delta m}$ $(g^{-1}(X, Y, Z)$ is proportional to $f_{m'}$, it follows that $\mu_i = \delta_i\mu_1$ $(i = 2, 3)$ with $\delta_i{}^3 = 1$ and that $m' = \delta m/(1 + \delta_2\delta_3\delta m)$. Putting $\varepsilon_3 = \delta_1\delta_2\delta$ and $\varepsilon_2 = (\delta_1\delta_2)^2$, we get (4). Conversely, if (4) holds, we can show that $V(f_m)$ and $V(f_{m'})$ are equivalent.

Denote by d_ε and t_ε the linear fractional transformations defined by the following matrices:

$$D = \begin{bmatrix} \varepsilon & 0 \\ 0 & 1 \end{bmatrix} \quad \text{and} \quad T = \begin{bmatrix} \varepsilon & 0 \\ \varepsilon & 1 \end{bmatrix} \quad (\varepsilon^3 = 1)$$

respectively. We note that $t_\varepsilon t_\delta = d_\delta$ or $d_{1/\varepsilon}t_{\delta/\varepsilon}$ according as $\varepsilon = 1$ or $\varepsilon \neq 1$, and that $t_\varepsilon d_\delta = t_{\varepsilon\delta}$ and that $(t_\delta)^{-1} = d_{1/\delta}t_1$.

Lemma 13. *If $m^3 \neq 0,1$ and $\varepsilon_1{}^3 = \varepsilon_2{}^3 = \varepsilon_3{}^3 = 1$, then the 12 values $\varepsilon_1 m$ and $\varepsilon_2\varepsilon_3/(1 + \varepsilon_3 m)$ are mutually distinct.*

Proof. Note first that $\varepsilon_1 m \neq \varepsilon_2\varepsilon_3 m/(1 + \varepsilon_3 m)$. Indeed, otherwise we have $1 = (1 + \varepsilon_3 m)^3$, which implies $(\varepsilon_3 m)^3 = 1$, a contradiction. If $\varepsilon_1 m = \delta_1 m$, clearly $\varepsilon_1 = \delta_1$. If $\varepsilon_2\varepsilon_3 m/(1 + \varepsilon_3 m) = \delta_2\delta_3 m/(1 + \delta_3 m)$, then m

$= (t_{\varepsilon_3})^{-1} (d_{\varepsilon_2})^{-1} d_{\delta_2} t_{\delta_3} m = d_{1/\varepsilon_3} t_{\delta_2/\varepsilon_2} t_{\delta_3} m$ is equal to $d_{1/\varepsilon_3} d_{\delta_3} m$ or $d_{1/\varepsilon_3} d_{\varepsilon_2/\delta_2} t_{(\delta_2/\varepsilon_2)^2} \delta_3 m$ according as $\varepsilon_2 = \delta_2$ or not. Thus, unless $(\varepsilon_2, \varepsilon_3) = (\delta_2, \delta_3)$, we have $(1 + \varepsilon m)^3 = 1$ for some ε with $\varepsilon^3 = 1$, again a contradiction.

The following result is now immediate.

Lemma 14. *The order of the subgroup* $G = \left\{ d_{\varepsilon_1} d_{\varepsilon_2} t_{\varepsilon_3} ; \varepsilon_1{}^3 = \varepsilon_2{}^3 = \varepsilon_3{}^3 = 1 \right\}$ *is equal to 12.*

Proof of Theorem 2. The rational function $j(m) = \Pi_{a \in G}\, am = m^{12}/(1 + m^3)^3$ is clearly G-invariant. The equation $j(m) = 0$ has only one solution $m = 0$ in $K \backslash \{m^3 = 1\}$, while the equation $j(m) = \lambda$ for a given $\lambda \neq 0$ has exactly 12 solutions $\{am_1; a \in G\}$, where $j(m_1) = \lambda$. Let $V(f_m)$ and $V(f_{m'})$ be non-singular cubics. They are projectively equivalent if and only if $m' = am$ for some $a \in G$ which is equivalent to $j(m) = j(m')$. Since $J = M^{12}/D$ is an absolute invariant and since $J(f_m) = j(m)$, Theorem 2 now follows.

Proof of Theorem 1. We can now proceed as in the case char $K = 0$ (Elliott 1895, section 295). Let $I_1 = M$ and $I_2 = D$. It suffices to show that I belongs to $K[I_1, I_2]$ for a homogeneous I in I.
(a) I is algebraically dependant on I_1 and I_2. Indeed $J_1 = I^{d_2}/I_2^d$ and $J_2 = I_1^{d_2}/I_2$ with $d = \deg I > 0$ and $d_2 = \deg I_2$ are absolute invariants. Let $f_m = X^3 + Y^3 + Z^3 + XYZm$. Then $j_k(m) = j_k(f_m)$ $(k = 1, 2)$ are algebraically dependent in the field $K(m)$ of the rational functions of m. There exists $b \in K \backslash \{0\}$ such that $\Sigma b_{n_1 n_2} j_1{}^{n_1} j_2{}^{nj} = 0$ in $K(m)$. Let $N_k = \max n_k$ and multiply by $i_2{}^{N_1 + dN_2}$ to obtain

$$\Sigma b_{n_1 n_2}\, i_2{}^{N_1 + dN_2 - n_1 - dn_2}\, i_1{}^{n_1}\, i^{n_2} = 0 \ \text{ in } K[m], \tag{5}$$

where $i_1 = I_1(f_m)$, $i_2 = I_2(f_m)$ and $i = I\,(f_m)$. This implies

$$\Sigma b_{n_1 n_2}\, I_2{}^{N_1 + dN_2 - n_1 - dN_2}\, I_1{}^{n_1}\, I^{n_2} = 0 \ \text{ in } K[a_1, ..., m], \tag{6}$$

To see this, denote by I' the right-hand side of (6), which is an invariant homogenous polynomial. For a cubic form f such that $I_2(f) \neq 0$ there exists a $g \in GL(3, K)$ such that $f_g = f(g(X, Y, Z))$ is proportional to f_m with $m^3 \neq 1$. Now (5) implies that $I'(f_g) = 0$. Thus if $I_2(f) \neq 0$, then $I'(f) = 0$, which implies that I' vanishes identically.

(b)   Let $f_{abcm} = X^3a + Y^3b + Z^3c + XYZm$.   Then $I(f_{abcm}) \in K[abc, m]$. Indeed, we have $i_1 = I_1(f_{abcm}) = m$ and $i_2 = I_2(f_{abcm}) = abc(abc + m^3)^3$.   By (6) $i = I(f_{zbcm})$ are algebraically dependent on abc and m.   Since i_1 and i_2 belong to the field $K(abc, m)$ and since i belongs to the field $K(a, b, c, m) = (K(abc, m))(b, c)$, it follows that $i \in K(abc, m)$, and hence $i \in K[abc, m]$.

(c) I belongs to $K[I_1, I_2]$.   Using the notation in (b), i takes the form

$$i = (abc)^s i_1{}^t c_0 + (abc)^{s-1} i_1{}^{t+3} c_1 + \ldots + i_1{}^{t+3s} c_s.$$

Substituting $i_2 + (abc)i_1{}^9 + (abc)^2 i_1{}^6 + (abc)^3 i_1{}^3$ for $(abc)^4$, we get

$$i = \sum\nolimits_{k=0}^{3} (abc)^k h_k(i_1, i_2). \tag{7}$$

We can show by induction that $h_k(i_1, i_2)$ takes the form $c_{n_1 n_2}{}^{(k)} i_1{}^{n_1} i{}^{n_2}$ with $n_1 + dn_2 = d - 3k$.   In particular, $h_k(I_1, I_2) \in I$ is homogeneous and of degree $d - 3k$ $(k = 0, 1, 2, 3)$.   To complete the proof we shall show that $h_k = 0$ for $k = 1, 2$ and 3.   Let $f_m = X^3 + Y^3 + Z^3 + XYZm$ with $m^3 \neq 1$.   There exists a $g \in GL(3, K)$ such that $\det g = \varepsilon$ with $\varepsilon^3 = 1$ and that $f_m(g(X, Y, Z)) = (1 + \varepsilon m)(X^3 + Y^3 + Z^3) + XYZm$ (see the proof of Theorem 12).   By (7) we have

$$\varepsilon^d I(f_m) = \sum_{k=0}^{3}(1 + \varepsilon m)^{3k} \varepsilon^{d-3k} h_k(I_1(f_m), I_2(f_m)) .$$

Hence we have

$$I(f_m) = \sum_{k=0}^{3}(1 + \varepsilon'm)^{3k} h_k(I_1(f_m), I_2(f_m))$$

for $\varepsilon' = 0, 1, \omega, \omega^2$, from which it follows that $h_k(I_1(f_m), I_2(f_m)) = h_k(I_1, I_2)(f_m) = 0$ for $k = 1, 2, 3$.   Therefore $h_k(I_1, I_2)(f)$ vanishes for any cubic form f.   Thus $I = h_0(I_1, I_2) \in K[I_1, I_2]$.

3.   char $K = 3$

In this section, the coefficients of a cubic form f are to be parameterized as in Section 1.   We start with an easy

Lemma 15.   *A cubic form* $f = Y^3b + X^2Za_3 + YZ^2c_2$ *is non-singular if and only if* $ba_3 c_2(b_3{}^2 - bc_2) \neq 0$.

Lemma 16   (Bretagnolle-Nathan 1958, De Groote 1973).   *A nonsingular cubic curve is projectively equivalent to some $V(f_n)$, where $f_n = Y^3 + X^2Z + Y^2Zb_3 + YZ^2c_2$ with $c_2(b_3{}^2 - c_2) \neq 0$.   A non-singular cubic curve $V(f_n)$ has either 3 inflexions, or only one inflexion, according to whether $b_3 \neq 0$, or $b_3 = 0$.*

Lemma 17.   *The polynomial $I_2(A, \ldots, C_2) = B_1C_1 + A_2C_2 + A_3B_3 - M^2$ belongs to I. Thus, a non-singular cubic $V(f)$ has three inflexions, or one according as $I_2(f) = 0$ or not.*

Proof.   It can be easily verified that $I_2(f_g) = (\det g)^2 I_2(f)$, (recall that $SL(3, K)$ is generated by $\lambda_{ij} = E_3 + \lambda E_{ij}$, where $\lambda \in K$, $i \neq j$, E_{ij} is a matrix whose (i, j) component is equal to one).

Let $f = Y^3b + X^2Za_3 + Y^2Zb_3 + YZ^2c_2$ be non-singular with $I_2(f) = a_3b_3 \neq 0$. If $V(\alpha X + \beta Y - Z)$ is the tangent through an inflexion of $V(f)$, $(\alpha, \beta) \in K^2$ must satisfy $a_3\beta + c_2\alpha^2 = 0$ and $b_3\alpha - c_2\alpha\beta = 0$, namely

$$\beta = -c_2\alpha^2/a_3, \quad b_3\alpha + c_2{}^2\alpha^3/a_3 = 0. \tag{8}$$

Conversely for a solution (α, β) of (8), we have $f(X, Y, \alpha X + \beta Y) = X^3a_3\alpha + Y^3(b + b_3\beta + c_2\beta^2)$.   Hence $V(\alpha X + \beta Y - Z)$ is a tangent through an inflexion of $V(f)$.   Since (8) has three solutions (α_i, β_i) $(i = 1, 2, 3)$, every tangent at an inflexion takes the form $V(\alpha_i X + \beta_i Y - Z)$.   Let $l_i = V(-a_3\alpha_i X + c_2\alpha_i{}^2 Y + a_3 Z)$ $(i = 1, 2, 3)$.   Define $g \in GL(3, K)$ by

$$\begin{bmatrix} 1 & \alpha_1 & \alpha_1^2 \\ 1 & \alpha_2 & \alpha_2^2 \\ 1 & \alpha_3 & \alpha_3^2 \end{bmatrix} \begin{bmatrix} 0 & 0 & a_3 \\ -a_3 & 0 & 0 \\ 0 & c_2 & 0 \end{bmatrix} g^{-1} = E_3.$$

Then $g(l_1)$, $g(l_2)$ and $g(l_3)$ are $V(X)$, $V(Y)$ and $V(Z)$ respectively.   Since the three tangents at inflexions of the cubic $V(f_{g^{-1}})$ are $V(X)$, $V(Y)$ and $V(Z)$, so $f_{g^{-1}} = X^3\gamma_1 + Y^3\gamma_2 + Z^3\gamma_3 + XYZ\gamma_4$ with $\gamma_1\gamma_2\gamma_3\gamma_4 \neq 0$.   Actually $f_{g^{-1}}$ is proportional to $X^3b + Y^3b + Z^3(b - b_3{}^2/c_2) + XYZ(-b_3{}^2/c_2)$.   Thus $V(f)$ is projectively equivalent to $V(f_m)$, where $f_m = X^3 + Y^3 + Z^3 + XYZm$ with $m = b_3{}^2/b^{2/3}(b_3{}^2c_2{}^2 - bc_2{}^3)^{1/3}$.   It is immediate that $V(f_m)$ and $V(f_{m'})$ are projectively equivalent if and only if $m = m'$, that is $m^3 = m'^3$.   This gives the following result.

Lemma 18 (Bretagnolle-Nathan 1958). *Let f_n and $f_{n'}$ be non-singular cubics as in Lemma 16. Then $V(f_n)$ and $V(f_{n'})$ are projectively equivalent if and only if $j(f_n) = j(f_{n'})$, where $j(f_n) = b_3{}^6/c_2{}^2(b_3{}^2 - c_2)$.*

In view of Lemmas 16 and 18, we can define an absolute invariant $J(f)$ for non-singular cubic forms f by $J(f) = j(f_n)$, where $f_g = f_n$ for some $g \in GL(3, K)$. We shall show that there exists an irreducible homogeneous polynomial $Q \in K[A, \ldots, C_2]$ such that $J(f) = I_2(f)^6/Q(f_n)$ with $Q(f) \neq 0$ for a non-singular cubic form f. By elimination theory (van der Waerden 1950) there exist a finite number of homogeneous polynomials $D_1, \ldots, D_f \in K[A, \ldots, C_2]$ such that a cubic f is singular if and only if $D_1(f) = \ldots = D_f(f) = 0$. Let f be a non-singular cubic form with $I_2(f) \neq 0$. Assume that the tangents at three inflexions of $V(f)$ take the form $V(\alpha X + \beta Y - Z)$. Since the coefficients of X^2Y and XY^2 of $f(X, Y, \alpha X + \beta Y)$ vanish, we have

$$a_2 + a_3\beta - c_1\alpha\beta + c_2\alpha^2 + m\alpha = 0, \tag{9}$$

$$b_1 + b_3\alpha + c_1\beta^2 - c_2\alpha\beta + m\beta = 0. \tag{10}$$

If $-a_3 + c_1\alpha \neq 0$, then (9) and (10) can be written as

$$\beta = (a_2 + \alpha m + \alpha^2 c_2)/(-a_3 + \alpha c_1), \tag{11}$$

$$a_2{}^2c_1 + a_3{}^2b_1 - a_2a_3m + \alpha I_2a_3 + \alpha^2 I_2c_1 + \alpha^3(a_3c_2{}^2 + b_3c_1{}^2 \\ - c_1c_2m) = 0. \tag{12}$$

A sufficient condition under which $-a_3 + c_1\alpha \neq 0$ for any root of (12) is that $a_3(a_2c_1{}^2 + a_3c_1m + a_3{}^2c_2) \neq 0$. We also note that the discriminant of (12) is equal to $-I_2(f)^3(a_2c_1{}^2 + a_3{}^2c_2 + a_3c_1m)^2/(a_3c_2{}^2 + b_3c_1{}^2 - c_1c_2m)^4$. Recall the calculation in Lemma 17. Regard $(b + b_3\beta + c_2\beta^2 + c_1\beta^3)(-a_3 + c_1\alpha)^3 = 0$ as an equation for α by substituting (11) for β, and denote by R the resultant of this equation and (12). Also, denote by $D_0(f)$ the product $I_2(f) a_3(a_2c_1{}^2 + a_3c_1m + a_3{}^2c_2)(a_3c_2{}^2 + b_3c_1{}^2 - c_1c_2m)R$. We note that $D_0(f) \neq 0$ for a non-singular cubic. $f = Y^3b + X^2Za_3 + Y^2Zb_3 + YZ^2c_2$ with $I_2(f) = a_3b_3 \neq 0$. This gives the following result.

Lemma 19. *Let f be a non-singular cubic form such that $D_0(f) \neq 0$. Then the three tangents at inflexions of $V(f)$ are of the form $V(\alpha X + \beta Y - Z)$, where α and β are given by (12) and (11).*

For a non-singular cubic f satisfying the condition of Lemma 19, let $\alpha_i(i = 1, 2, 3)$ be the roots of (12) and define $[\alpha]$ and h in $GL(3, K)$ by

$$[\alpha] = \begin{bmatrix} 1 & \alpha_1 & \alpha_1{}^2 \\ 1 & \alpha_2 & \alpha_2{}^2 \\ 1 & \alpha_3 & \alpha_3{}^2 \end{bmatrix}, \quad h = \begin{bmatrix} 0 & a_2 & a_3 \\ -a_3 & m & -c_1 \\ c_1 & c_2 & 0 \end{bmatrix}. \tag{13}$$

Since the tangent l_i at the inflexion of $V(f)$ is of the form $V((-a_3 + c_1\alpha_i)\alpha_i X + (a_2 + m\alpha_i + c_2\alpha_i{}^2)Y + (a_3 - c_1\alpha_i)Z)$, the matrix $g = [\alpha]h$ transforms the tangents l_1, l_2 and l_3 to $V(X)$, $V(Y)$ and $V(Z)$. Consequently $f_{g^{-1}} = f(g^{-1}(X, Y, Z))$ takes the form $C_1 X^3 + C_2 Y^3 \ C_3 Z^3 + C_4 XYZ$ with $C_i = C_i(\alpha) \neq 0$. Thus $J(f) = C_4{}^3/C_1 C_2 C_3$.

Lemma 20. *There exist homogeneous polynomials P, $Q \in K[A, ..., C_2]$ of the same degree such that P and Q are mutually prime and that $J(f) = P(f)/Q(f)$ with $Q(f) \neq 0$ for a non-singular cubic form f satisfying the condition of Lemma 19.*

Proof. For a permutation matrix σ of $\{1, 2, 3\}$ denote by $[\sigma]$ the permutation matrix such that $[\sigma](X_1, X_2, X_3) = (X_{\sigma(1)}, X_{\sigma(2)}, X_{\sigma(3)})$. Put $[\alpha_\sigma] = [\sigma][\alpha]$ and $g = g(\alpha_\sigma) = [\alpha_\sigma]h$, where $[\alpha]$ and h are given by (13). Since $g(\alpha_\sigma)^{-1} = g(\alpha)^{-1}[\sigma]^{-1}$, we have $f(g(\alpha_\sigma)^{-1}(X_1, X_2, X_3)) = f(g(\alpha)^{-1}(X_{\sigma^{-1}(1)}, X_{\sigma^{-1}(2)}, X_{\sigma^{-1}(3)}))$. The left-hand side is equal to $C_1(\alpha_\sigma)X_1{}^3 + C_2(\alpha_\sigma)X_2{}^3 + C_3(\alpha_\sigma)X_3{}^3 + C_4(\alpha_\sigma)X_1 X_2 X_3$. Thus

$$C_1 C_2 C_3(\alpha_\sigma) = C_1 C_2 C_3(\alpha), \ C_4(\alpha_\sigma) = C_4(\alpha). \tag{14}$$

It is immediate that there exist polynomials $\gamma_i(\alpha)$ of $\alpha_1, \alpha_2, \alpha_3$ such that $C_i(\alpha) = \gamma_i(\alpha)/\det[\alpha]^3 \det h^3$ $(i = 1, 2, 3, 4)$. Since $\det[\alpha_\sigma] = \text{sgn } \sigma \det[\alpha]$, it follows that $\gamma_1\gamma_2\gamma_3(\alpha)$ and $\gamma_4(\alpha)$ are anti-symmetric functions of α_i, hence divisible by $\det[\alpha]$. Thus $J(f) = (\gamma_4(\alpha)/\det[\alpha])^3 \det[\alpha]^2/\{\gamma_1\gamma_2\gamma_3(\alpha)/\det[\alpha]\}$. Since coefficients of both the numerator and denominator are homogeneous polynomials of degree 6 in $K[a, ..., c_2]$, the numerator and denominator, multiplied by suitable power of $a_3 c_2{}^2 + b_3 c_1{}^2 - c_1 c_2 m$, give rise to homogeneous polynomials P^* and Q^* of the same degree in $K[a, ..., c_2]$. Thus $J(f) = P^*(f)/Q^*(f)$ with $Q^*(f) \neq 0$. Deleting common factors, if any, we are done.

Let $D^*, P_1, Q_1 \in K[A, ..., C_2]$ be homogeneous polynomials such that P_1 and Q_1 are mutually prime, having the same degree, and that if $D^*(f) \neq 0$, f is non-singular and $J(f) = P_1(f)/Q_1(f)$ with $Q_1(f) \neq 0$. For example, $D^* = D_0 D_1$, $P_1 = P$ and $Q_1 = Q$ satisfy the condition (see Lemmas

19 and 20 for the definition of D_0, P and Q). We shall show that $P/Q = P_1/Q_1$ in $K(A, ..., C_2)$. In fact, $PQ_1(f) - P_1Q(f) = 0$ provided $D_0D_1D^*(f) \neq 0$. Thus $PQ_1 = P_1Q$. Since P divides P_1 and vice versa, we have $P_1 = cP$. Similarly, $Q_1 = c'Q$. It is clear that $c = c'$.

Lemma 21. *The polynomials P and Q in Lemma 20 are invariant.*

Proof. Put $D^* = D_0D_1$. If $D^*(f) \neq 0$, then $J(f) = P(f)/Q(f)$ by Lemma 20. If $D^*(f)D^*_g(f) \neq 0$, then $P(f)/Q(f) = J(f)$, which is equal to $J(f_g) = P(f_g)/Q(f_g)$ (see the introduction for the definition of f_g and D^*_g). By the argument above, $P_g = c_gP$ for some c_g in $K \setminus \{0\}$. Since $P_{gg'} = (P_{g'})_g$, we have $c_{gg'} = c_{g'}c_g$ for any $g, g' \in GK(3, K)$. Thus $c_g = (\det g)^d$ for a non-negative integer. Indeed, c_g is a polynomial function on $M(3, K)$ such that $\det g \neq 0$ implies $c_g \neq 0$. If c_g is a constant, then $c_g = 1$. Otherwise, $c_g = 0$ implies $\det g = 0$. Since $\det g$ is prime, $\det g$ divides c_g. Now consider $c_g{}' = c_g / \det g$. Repeating this process, we arrive at $c_g = (\det g)^d$. In our case, $d = \deg P$.

Lemma 22. *Let Q_0 be an invariant homogeneous polynomial such that $Q_0(X^3 + Y^3 + XYZ) = 0$. Then $Q_0(f) = 0$ for any singular cubic f.*

Proof. Since $(X - Y)^3 + Y^3 + (X - Y)YZ = X^3 - Y^2Z + XYZ$, Q_0 vanishes at this form. Let $f = X^3a + Y^2Zb_3 + XYZm$. Then $Q_0(f) = 0$ provided $ab_3m \neq 0$. Thus $Q_0(f) = 0$ in $K[a, b_3, m]$. In particular $Q_0(X^3 + Y^2Z) = 0$. A singular cubic form can be transformed into one of the following forms: $f_1 = X^3 + Y^3 + XYZ$, $f_2 = X^3 + Y^2Z$, $f_3 = X^3 - XYZ$, $f_4 = X^2Z - YZ^2$, $f_5 = XYZ$, $f_6 = X^2Z$, $f_7 = Y^3$ and $f_8 = Y^2Z + YZ^2$. We have already seen that $Q_0(f) = 0$ except for $i = 4$ and $i = 8$. Since $Q_0((X^2 - YZ)(X\alpha + Y\beta + Z\gamma)) = 0$ unless $V(X\alpha + Y\beta + Z\gamma)$ is tangent to $V(X^2 - YZ)$, it vanishes even when $\alpha = \beta = 0$. Thus $Q_0(f_4) = 0$. Similarly, using $Q_0(f_5) = 0$, we get $Q_0(f_8) = 0$.

Lemma 23. *Let R be an invariant homogeneous polynomial. If R decomposes into prime factors as $R_1{}^{n_1} ... R_i{}^{n_r}$, then all R_i are invariant homogeneous polynomials.*

Proof. It is well known that R_i are homogeneous. For g in $SL(3, K)$ we have $R_g = R_{1g}{}^{n_r} ... R_{rg}{}^{n_r}$. By the uniqueness of the decomposition R_1 is proportional to one of R_{ig}. Let $g = E_3 + \lambda E_{ij}$ ($i \neq j$, $\lambda \in K$) and let k be the number such that R_1 is proportional to R_{kg} for infinite many λ. Then $R_1 = R_{kg}$ for any λ, which implies $k = 1$. Since $SL(3, K)$ is generated by g, $R_1 = R_{1l}$ for any $g \in SL(3, K)$. As the proof of Lemma 21 shows, $R_{1g} = (\det g)^{d_1} R_1$ for $g \in GL(3, K)$ with $d_1 = \deg R_1$.

Lemma 24. *Let P and Q be as in Lemma* 20.
(1) $Q(f) = 0$ *if and only if f is a singular cubic.*
(2) Q *is irreducible.*
(3) $J(f) = P(f)/Q(f)$ *for a non-singular cubic f.*
(4) $P = I_2^6$ *and* $Q(Y^3b + X^2Za_3 + Y^2Zb_3 + YZ^2c_2) = a_3^6 b^2 c_2^2 (b_3^2 - bc_2)$.

Proof. Let $f = Y^3b + X^2Za_3 + Y^2Zb_3 + YZ^2c_2$ be a non-singular cubic so that $ba_3c_2(b_3^2 - bc_2) \neq 0$. Assume that $V(f)$ has three inflexions so that $b_3 \neq 0$. Then $D_0(f) = (a_3b_3)a_3(a_3^2c_2)(a_3c_2^2)(b_3^2 - bc_2) \neq 0$, and $J(f) = b_3^6/b_2c_2^3$ $(b_3^2 - bc_2) = P(f)/Q(f)$ with $Q(f) \neq 0$ by Lemmas 18 and 20. In particular $P(f)Q(f) \neq 0$ if $ba_3b_3c_2(b_3^2 - bc_2) \neq 0$. Therefore $Q(f) = b^{n_1}a_3^{n_2}b_3^{n_3}c_2^{n_4}(b_3^2 - bc_2)^{n_5}$ for some non-negative integers n_i and arbitrary f. We claim that $n_3 = 0$. Let $g = \mathrm{diag}[1, b^{-1/3}, 1]$ with $b \neq 0$. Then $f_g = Y^3 + X^2Za_3 + Y^2Zb_3b^{-2/3} + YZ^2c_2b^{-1/3}$. The equality $Q(f) = b^{d/3}Q(f_g)$ yields $n_2 = 2n_1 + n_3 + 2n_5$; hence $n_2 \geq 6$. Assume that $n_3 \geq 1$. Then a_3b_3 divides $P(f)$ and $Q(f)$. Hence $P(f') = Q(f') = 0$ if f' is non-singular and $I_2(f') = 0$. Recall that f' is singular if and only if $D_1(f') = \ldots = D_4(f') = 0$. Since there exists a non-singular cubic form f' with $I_2(f') = 0$, one of D_i, say D_1, cannot contain I_2 as a factor. Now $P(f') = Q(f') = 0$ provided $I_2(f') = 0$ and $D_1(f') \neq 0$. Thus I_2 divides P and Q, a contradiction. Thus $n_3 = 0$. Now $Q(f) \neq 0$ if f is non-singular. Since Q is invariant, $Q(f') \neq 0$ if f' is non-singular. Since P/Q is an absolute invariant, $J(f') = P(f')/Q(f')$ for any non-singular f'. Let $f_{abcm} = X^3a + Y^3b + Z^3c + XYZm$. This form is non-singular if and only if $abcm \neq 0$. Thus $m^3/abc = J(f_{abcm}) = P(f_{abcm})/Q(f_{abcm})$ provided that $abcm \neq 0$. Since abc is a factor of $Q(f_{abcm})$ it follows that $Q(X^3 + Y^3 + XY^2) = 0$. In view of Lemma 22, $Q(f')$ vanishes if and only if f' is singular. Assume that Q decomposes into prime factors as $Q_1^{q_1} \ldots Q_n^{q_n}$. We may assume that $Q_1(X^3 + Y^3 + XY^2) = 0$. By Lemma 22, $Q_1(f')$ vanishes if and only if f' is singular. Since Q_1 does not vanish unless Q_i vanishes, we have $n = 1$. Since $P(f) = J(f)$ $Q(f) \neq 0$ if f is non-singular with $I_2(f) \neq 0$, so $P(f') \neq 0$ if $I_2Q_1(f') \neq 0$. Thus $P = \mathrm{const}\, I_2^{p_1}Q_1^{p_2}$ for some non-negative integers p_i. Since P is prime to Q, $p_2 = 0$. We claim that $p_1 = 6$. This follows from the fact that $J(f) = b_3^6/b^2c_2^2(b_3^2 - bc_2) = P(f)/Q(f)$ and that b_3 is not a factor of $Q(f)$. At the same time, we get $q_1 = 1$.

Proof of Theorems 3 and 5 These theorems follow from Lemma 24.

Proof of Theorem 4. Let $f = X^3a + Y^3b + Z^3c + XYZm$. since $J(f) = m^3/abc$ and $I_2(f) = -m^2$ when $abcm \neq 0$, we have $D(f) = abcm^9$. Let I be a homogeneous invariant polynomial of positive degree. As in the case of characteristic 2, I is algebraically dependant on I_2 and D. Hence, so is $I(f)$ on $abcm^9$ and m^2, therefore, on abc and m. Thus $I(f) \in K[abc, m]$, because $K(a, b, c, m) = (K(abc, m))(b, c)$. The degree of d is even. Indeed, let g be a permutation matrix such that $g(X, Y, Z) = (Y, X, Z)$. Then we have $(-1)^d I(f) = (\det g)^d I(f) = I(f_g)$, which is equal to $I(f)$, because $I(f) \in K[abc, m]$. We will show that $I \in K[I_2, D]$ by induction on d. If $d = 2$, then $I(f) = \text{const } m^2$, hence $I = -\text{const } I_2$. Assume the assertion holds up to $d - 2 \geq 2$. $I(f)$ takes the form $(abc)^i m^j c_i + (abc)^{i-1} m^{j+3} c_{i-1} + \ldots + (abc) m^{j+3(i-1)} c_1 + m^{j+3i} c_0$, where $c_i \in K$. Put $I' = I - (-1)^{d/2} c_0 I_2^{d/2}$. Then $I'(f) = 0$ if $abc = 0$. By Lemma 22, $I'(f) = 0$ if $D(f) = 0$, which implies that the irreducible polynomial D divides I'. By the induction hypothesis, we get $I \in K[I_2, D]$.

Acknowledgement. The authors would like to thank Prof. H. Morikawa and Prof. Y. Hirano for their valuable suggestions. The first author was partially supported by The Grant-in-aid for General Scientific Research (C 01540182), The Ministry of Education, Science and Culture, Japan.

References

J. Bretagnolle-Nathan (1958), Cubiques définies sur un corps de caractéristique quelconque, *Ann. Fac. Sci. Univ. Toulouse* **22**, 175-234.

A.D. Campbell (1926), Plane cubic curves in Galois fields of order 2^n, *Ann. of Math.* **27**, 395-406.

R. De Groote (1973), Les cubiques dans un plan projectif sur un corps de caracteristique trois, *Acad. Roy. Belg. Bull. A. Sci.* **59**, 1140-1155.

E.A. Elliott (1895), *An Introduction to the Algebra of Quantics*, Clarendon Press, Oxford.

J.W.P. Hirschfeld (1979), *Projective Geometries over Finite Fields*, Clarendon Press, Oxford.

G. Salmon (1855), *Modern Higher Algebra*, Chelsea, New York.

G. Salmon (1879), *Higher Plane Algebraic Curves*, Chelsea, New York.

A. Seidenberg (1968), *Elements of the Theory of Algebraic Curves*, Addison-Wesley, Reading.

B.L. van der Waerden (1950), *Modern Algebra, Vol II*, Fredrick Unger, New York.

Automorphism groups of some generalized quadrangles

W.M. Kantor
University of Oregon

Dedicated to James Hirschfeld on the Occasion of his 50th Birthday.

The following construction was first given in Kantor (1980). Let Q be a finite group and $\mathcal{F}$ a family of subgroups; with each $A \in \mathcal{F}$ is associated another subgroup A^*. These subgroups satisfy the following conditions for some integers $s, t > 1$ and all distinct $A, B, C \in \mathcal{F}$:

$$|Q| = st^2, \quad |\mathcal{F}| = s + 1, \quad |A| = t, \quad |A^*| = st,$$
$$A < A^*, Q = A^*B, A^* \cap B = 1, AB \cap C = 1.$$

A generalized quadrangle $Q(\mathcal{F})$ is constructed by using cosets together with symbols $[A]$ and $[\mathcal{F}]$ as in the following picture (for all $A \in \mathcal{F}$ and $g \in Q$):

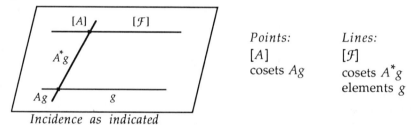

Points: | Lines:
$[A]$ | $[\mathcal{F}]$
cosets Ag | cosets A^*g
 | elements g

Incidence as indicated

"Most" of the known families of finite generalized quadrangles arise by this procedure (see Kantor (1980); Payne (1990)). It seems likely that Q must be a p-group for some prime p. While this remains open, there has been significant progress (Frohardt (1988); Chen and Frohardt (submitted)).

Proposition. Let $G = \text{Aut}Q(\mathcal{F})$. Then one of the following holds:
(i) G fixes a line, namely $[\mathcal{F}]$.
(ii) G fixes a point ∞ but no line. Then ∞ is on $[\mathcal{F}]$, G is transitive on the points collinear with ∞ as well as those not, and G is 2-transitive on the lines on ∞.

(iii) G is transitive on both points and the lines of $Q(\mathcal{F})$, having rank 3 on each of these sets. G is flag-transitive.   The stabilizer of a line is 2-transitive on the points of the line and dually.

Proof. **(i)** Q moves each line $\neq [\mathcal{F}]$.

(ii) Q moves each point not on $[\mathcal{F}]$, so that the fixed point ∞ must be on $[\mathcal{F}]$, say $\infty = [A]$.   Then G must have an element g moving $[\mathcal{F}]$ to some other line on $[A]$.   Write $H = \langle Q, Q^g \rangle$.

Note that Q fixes $[A]$ and is transitive on the lines $\neq [\mathcal{F}]$ on $[A]$, so that H is 2-transitive on the lines through ∞.   Since A^* is transitive on the set of points $\neq [A]$ of the line A^*1, it follows that G is transitive on the set of points collinear with ∞.

Since Q is transitive on the lines missing $[\mathcal{F}]$, H is transitive on the lines not on ∞.   It remains to consider the action of the points not collinear with ∞.   Recall that B^* is transitive on the set of points $\neq [B]$ of the line B^*1; by transitivity on the lines not on ∞, the same is true for each such line.   Thus, if x is any point not collinear with ∞, then x^H contains every point collinear with x but not ∞.   Moreover, every such line must contain at least s points of x^H, so there can be only one such point-orbit x^H.

(iii)   Now without loss of generality G moves every point and line. Then $[\mathcal{F}]^G$ contains two lines with no common points, and hence contains every line since Q is transitive on the lines missing $[\mathcal{F}]$.   If $g \in G$ moves $[\mathcal{F}]$ to a line meeting $[\mathcal{F}]$ at a point x, then $\langle Q, Q^g \rangle$ fixes x and is transitive on the points collinear with x as well as those not collinear with x.   Thus, G has rank 3 on points.   Moreover, $\langle Q, Q^g \rangle$ is 2-transitive on the lines through x.   In particular, G is flag-transitive.

Moreover, it follows that the stabilizer of $[\mathcal{F}]$ is transitive on the lines meeting $[\mathcal{F}]$.   Thus, G has rank 3 on lines. $\square$

Since all rank 3 groups are essentially known (using the classification of finite simple groups), it is clear that one can determine all the possibilities in (iii); and this does, indeed, lead to a characterization of the classical generalized quadrangles.   However, this seems to be an uninteresting and uninformative approach.   Too much information is available and too much is ignored in the above argument (especially the *regularity* of Q on the set of lines missing $[\mathcal{F}]$).   In other words, a more geometric approach is needed.

As a side remark, it is also clear that all generalized quadrangles whose automorphism groups have rank 3 on the set of points can be determined.   Again, this is a straightforward question.   (Note, however, that there is one nonclassical example: the quadrangle with

$s = 3, t = 5$.)   A more interesting question is the determination of all flag-transitive quadrangles — using the aforementioned classification! Besides the classical ones, presumably there are only two others, with $s = 3, t = 5$ and $s = 15, t = 17$.

In the only known nonclassical examples for which (ii) holds, Q is elementary abelian of order q^4 (Kantor (1986); Payne (1989)).

In order to make further progress in the context of the Proposition, it seems necessary to have a way to recover Q from $Q(\mathcal{F})$. In general this is entirely open. However, under "reasonable" assumptions this can be accomplished:

For intersecting lines L and M, let $U_{L, M}$ denote the group of all automorphisms fixing every point of L, every point of M and every line on $L \cap M$.

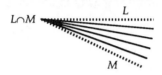

Then $U_{L, M}$ is semiregular on the t lines $\neq L$ through each point $\neq L \cap M$ of L, so that $|U_{L, M}| \leq t$ with inequality if and only if $U_{L, M}$ is regular on those t lines. The groups $U_{L, M}$, and their duals, are exactly those involved in the Moufang condition for generalized quadrangles (Tits (1976); Payne and Thas (1984)).

It is clear that the collection of groups $U_{L, M}$ is canonically determined by a quadrangle. Thus, if Q is generated by some such subgroups then it can be recovered from the quadrangle.

Remark. $U_{[\mathcal{F}], A^{\cdot 1}} = A \Leftrightarrow (A \lhd A^*$ and $g^{-1}Ag \leq A^*, \forall g \in Q)$.   Hence, if this condition holds for (at least) three members of $\mathcal{F}$ then $Q = \langle U_{[\mathcal{F}], M} \mid M$ meets $[\mathcal{F}] \rangle$.

This is straightforward to check. The stated condition holds for all known examples of the construction given at the beginning of this note.

Let q be a prime power.

Example 1.   Let Q be the set $GF(q)^2 \times GF(q) \times GF(q)^2$ equipped with the multiplication $(u, c, v)(u', c', v') = (u + u', c + c' + v \cdot u', v + v')$, where $v \cdot u'$ is the usual dot-product.   Then Q is a group, whose center is $Z(Q) = 0 \times GF(q) \times 0$.   Moreover, $Q/Z(Q)$ can be viewed as a vector space over $GF(q)$.   Let

$$A(\infty) = 0 \times 0 \times GF(q)^2, A(r) = \{(u, uB_r u^t, uM_r) \mid u \in GF(q)^2\},$$
$$\text{and } A^* = AZ(Q) \text{ for } r \in GF(q)^2 \text{ and each } A = A(\infty) \text{ or } A(r),$$

where B_r and M_r are 2×2 matrices satisfying suitable conditions in order to produce a quadrangle with $s = q$ and $t = q^2$ (Payne (1980, 1989, 1990); Kantor (1986)).

In this situation, *if $Q(\mathcal{F})$ is **not** the $O(5, q)$ quadrangle then $Q \trianglelefteq \mathrm{Aut}Q(\mathcal{F})$*. For, in Payne (1989) it is shown that the line $[\mathcal{F}]$ is the *only* line L with the following property: If x_1, x_2, x_3, x_4 and y_1, y_2, y_3, y_4 are two sets of pairwise noncollinear points such that x_1 and y_1 are on L and x_i and y_j are collinear for all i, j except perhaps for $i = j = 4$, then also x_4 and y_4 are collinear. Thus, $\mathrm{Aut}Q(\mathcal{F})$ fixes $[\mathcal{F}]$, and then the preceding Remark implies that Q is normal in $\mathrm{Aut}Q(\mathcal{F})$.

Example 2. Let $s = q$ and $t = q^2$, and assume that $Q/Z(Q)$ is elementary abelian of order q^4 and $A^* = AZ(Q)$ for all $A \in Z$. Note that quadrangles behaving in this manner, but not as in Example 1, have been constructed in Payne (1989).

Once again, *if $Q(\mathcal{F})$ is **not** the $O(5, q)$ quadrangle then $Q \trianglelefteq \mathrm{Aut}Q(\mathcal{F})$*. This time no purely geometric argument presently is available.

By the Remark, Q can be recovered as the group generated by those $U_{[\mathcal{F}],M}$ with M meeting $[\mathcal{F}]$. Also, $Z(Q)$ is the group $U_{[\mathcal{F}]}$ of all automorphisms of $Q(\mathcal{F})$ fixing every line meeting $[\mathcal{F}]$, and is regular on the set of points not on $[\mathcal{F}]$ of each such line.

Now consider possibilities (ii) and (iii) for $\mathrm{Aut}Q(\mathcal{F})$ in the Proposition. If (iii) holds, then $Q(\mathcal{F})$ is Moufang, and the main theorem in Fong and Seitz (1973) can be applied. However, this will not be needed: possibilities (ii) and (iii) will be handled simultaneously.

Consider the group H generated by Q and Q^g for some $g \in G$ such that $[\mathcal{F}]^g$ meets $[\mathcal{F}]$ at a point $[A]$. Without loss of generality $[\mathcal{F}]^g = A^*1$. The stabilizer $H_{[\mathcal{F}]}$ normalizes Q (since Q is canonically determined by $[\mathcal{F}]$), so that the stabilizer $H_{[\mathcal{F}],1}$ lies in $\mathrm{Out}(Q)$.

Let K be the kernel of the 2-transitive action of H on the set of $t + 1$ lines on $[A]$. Then $K \cap Q = A^*$, while A is just the pointwise stabilizer of A^*1 in Q Moreover, $Q/A^* \cong QK/K$ induces a normal subgroup of $H_{[\mathcal{F}]}/K$ regular on the t lines $\neq [\mathcal{F}]$ through $[A]$.

Without loss of generality g interchanges $[\mathcal{F}]$ and A^*1, and hence normalizes $U_{[\mathcal{F}],A^*1} = A$.

Note that $Z(Q)^g$ is transitive on the points $[B] \neq [A]$ of $[\mathcal{F}]$, while A^* is transitive on the points not on $[\mathcal{F}]$ collinear with such a $[B]$. Thus, $\langle A^*, A^{*g} \rangle$ is a subgroup of K transitive on the set of points not collinear with $[A]$. Let E denote the group generated by all the conjugates of A^* under the action of H. Then E is a p-group (as each such conjugate is a p-group normal in K). *Claim:* $E = \langle A^*, A^{*g} \rangle$. For, it suffices to show that E *is regular on the set of points not collinear with* $[A]$. Let $e \in E$ fix a point x not collinear with $[A]$. Each line on $[A]$ has a unique point collinear with x, and hence has a point $\ne [A]$ fixed by e; and since e is a p-element it must fix yet another point on that line. It follows that the set of fixed points and lines of e is a subquadrangle with $s' \le s = q$ and $t' = t = q^2$. By Higman's inequality (Payne and Thas (1984, p.4)) $t \le s'^2$, so that $s' = s$ and hence $e = 1$.

Thus, $E = \langle A^*, A^{*g} \rangle$ has order q^4. It contains a set Ω of $t + 1 = q^2 + 1$ subgroups conjugate under H to $Z(Q)$ and permuted 2-transitively by H. It follows that E must be elementary abelian, and that H acts irreducibly on it.

At this point, I do not know how to show that E is a $GF(q)[H]$-module without invoking some (preclassification!) group theory. The 2-transitive group $\overline{H}$ induced by H on Ω has the property that the stabilizer of $Z(Q)$ has a normal elementary abelian subgroup of order q^2 regular on the remaining members of Ω. It follows from Shult (1972) and Hering, Kantor and Seitz (1972) that $\overline{H} \ge PSL(2, q^2)$. This group $PSL(2, q^2)$ acts (projectively and) irreducibly on the $GF(p)$-space E. However, up to twisting by field automorphisms, $PSL(2, q^2)$ has exactly two projective irreducible modules of size q^4: the natural one (over $GF(q^2)$) and the $\Omega^-(4, q)$-module (Fong and Seitz (1973 4B, C)). Since there is an orbit Ω of $q^2 + 1$ subgroups of size q, the only possibility is that E can be viewed as the $\Omega^-(4, q)$-module with Ω the associated ovoid of singular points.

Now it is easy to recover the generalized quadrangle from the group E. For, the stabilizer in E of a point $[B]$ of $[\mathcal{F}]$ is known — namely, A^*, which fixes every such point; as is that of a line B^* on $[B]$ but not on $[A]$ — namely, $Z(Q)$, which fixes every such line. Since E is regular on the points not collinear with $[A]$, while H is transitive on the lines on $[A]$, the lines not on $[A]$, and the points collinear with $[A]$, it follows that $Q(\mathcal{F})$ can be described in terms of the $\Omega^-(4, q)$-ovoid Ω consisting of $q^2 + 1$ subgroups of E, together with the tangent planes (such as A^*) to that ovoid. Consequently, $Q(\mathcal{F}) \cong Q(\Omega)$ and $Q(\mathcal{F})$ is the $O(5, q)$-quadrangle. $\square$

Remark. Payne and Thas (1984) came very close to obtaining a geometric proof of the classification of finite Moufang quadrangles. Their obstacle was the same one appearing in Example 2! Thus, the above argument can be inserted into the appropriate part of Payne and

Thas (op. cit., Ch. 9) in order to complete their approach to that classification theorem.

Note, however, that the amount of group theory employed was fairly small, certainly miniscule compared to Fong and Seitz (1973): information was required concerning a relatively restricted type of 2-transitive permutation group, together with a fact about very small degree representations of such a group.

Acknowledgement. Research supported in part by NSF grant DMS 87-01794 and NSA grant MDA 904-88-H-2040.

References

X. Chen and D.E. Frohardt (submitted), Normality in a Kantor family.

P. Fong and G.M. Seitz (1973), Groups with a (B, N)-pair of rank 2, *Invent. Math.* **21**, 1-57.

D.E. Frohardt (1988), Groups which produce generalized quadrangles, *J. Combin. Theory Ser. A* **48**, 139-145.

C. Hering, W.M. Kantor and G.M. Seitz (1972), Finite groups with a split BN-pair of rank 1. I., *J. Algebra* **20**, 435-475.

W.M. Kantor (1980), Generalized quadrangles associated with $G_2(q)$, *J. Combin. Theory Ser. A* **29**, 212-219.

W.M. Kantor (1986), Some generalized quadrangles with parameters q^2, q, *Math. Z.* **192**, 45-50.

S.E. Payne (1980), Generalized quadrangles as group coset geometries, *Congr. Numer.* **29**, 717-734.

S.E. Payne (1989), An essay on skew translation generalized quadrangles, *Geom. Dedicata* **32**, 93-118.

S.E. Payne (1990), A census of finite generalized quadrangles, *Finite Geometries, Buildings and Related Topics*, 29-36, Oxford University Press, Oxford.

S.E. Payne and J.A. Thas (1984), *Finite Generalized Quadrangles*, Pitman, London.

E.E. Shult (1972), On a class of doubly transitive groups, *Ill. J. Math.* **16**, 434-455.

J. Tits (1976), Classification of buildings of spherical type and Moufang polygons: A survey, *Teorie Combinatorie*, Volume 1, 229-256, Accademia Nazionale dei Lincei, Rome.

Collineation groups of $(q + t, t)$-arcs of type $(0, 2, t)$ in a Desarguesian plane of order q

G. Korchmáros
University of Basilicata

1. Introduction

In Korchmáros and Mazzocca (1990), some $(2^r + 2^h, 2^h)$-arcs T of type $(0, 2, 2^h)$ in $PG(2, q)$ have been obtained from translation ovals of $PG(2, 2^{r-h})$ via the trace function over the subfield $GF(2^{r-h})$ of $GF(2^r)$. These examples have the common property of having a t-nucleus in the sense that their t-secants are concurrent at a point. Furthermore, they are translation $(2^r + 2^h, 2^h)$-arcs of type $(0, 2, 2^h)$ with respect to the line l_∞. This means that for any two points $A, B \in T$ not on l_∞ there is an elation with axis l_∞ (that is a translation of the affine plane $AG(2, q)$ $= PG(2, q) \setminus l_\infty$) which sends A to B. The improper line l_∞ turns out to be a t-secant of T, and hence the t-nucleus is an improper point.

These examples are of interest also from Klein's point of view, as they have a large linear collineation group with a doubly transitive action on the set of all t-secants different from l_∞. Our main result is that no other $(q + t, t)$-arc of type $(0, 2, t)$ in $PG(2, q)$ has this property; see the corollary in §4.

2. Preliminary results on $(q + t, t)$-arcs of type $(0, 2, t)$

A (k, t)-arc in a projective plane is a set of k points such that t is the greatest number of collinear points in the set. A line r of the plane is called an *i-secant* of a (k, t)-arc T if $|r \cap T| = i$. T is said to be of type $(m_1, \ldots, m_u, t)$ if $0 \le m_1 < \ldots < m_u < t$ and the only values of i for which T admits an i-secant are just $m_1, \ldots, m_u$ and t.

A standard counting argument shows that $(q + t, t)$-arcs of type $(0, 2, t)$ can exist only if q is even. Another necessary condition for the existence is that t divides q. Furthermore, the $(q + t, t)$-arcs of type $(0, 2,t)$ are just the (k, t)-arcs such that through every point on the (k, t)-arc there passes only one t-secant and such that any other line through the point is a 2-secant.

Assume that the plane is a Desarguesian plane $PG(2, q)$, $q = 2^r$ and $r \ge 2$. Then t is also a power of 2 and thus the $(q + t, t)$-arcs of type $(0, 2, t)$ are exactly the $(2^r + 2^h, 2^h)$-arcs of type $(0, 2, 2^h)$.

The known examples have a common property of having a t-nucleus in the sense that their t-secants are concurrent at a point. This should hold true in general; counterexamples could exist at most in the case $r - h + 1 = gcd\ [r,\ h-\ 1],\ r \geq h - 2$. For the investigation of $(2^r + 2^h,\ 2^h)$-arcs of type $(0,\ 2,\ 2^h)$ with a t-nucleus, we have adopted a convenient canonical frame in $PG(2,\ q)$ such that the t-nucleus is Y_∞, a t-secant is the improper line and the q affine points are $(f(z),\ z)$ for a suitable polynomial $f(z)$ over $GF(q)$. Precisely, any $(q + t,\ t)$-arc of type $(0,\ 2,\ t)$ in $PG(2,\ q)$, q even and $q > 2$, can be written in the form

(a) $\begin{cases} T(f) = \{(f(c),\ c,\ 1) : c \in GF(q)\} \cup \{(1,\ 0,\ 0),\ (1,\ w_1,\ 0),\ ...,\ (1,\ w_{t-1},\ 0)\} \\ f(z) = \Sigma a_i z^i \quad \alpha_i \in GF(q) \end{cases}$,

such that:
(b) $f(z)$ is a monic polynomial and $f(0) = 0,\ f(1) = 0$;

(c) the value set of $f(z)$ in $GF(2,\ q)$ has size q/t; more precisely, the equation $f(z) = c,\ c \in GF(q)$ has either 0, or t solutions in $GF(q)$.

Furthermore, if for each $a \in GF(q)$ we form the polynomial $F_a(z) = (f(z) + f(a))/(z + a)$ in the indeterminate z, then we also have
(d) the restriction of $F_a(z)$ to the set of elements of $GF(q)\setminus\{a\}$ which are not roots of $F(z)$ is injective;

(e) for $a \in GF(q)$, none of the polynomials $w_i^{-1} + F_a(z)$ $(1 \leq i \leq t - 1)$ has a root in $GF(q) - \{a\}$.

Conversely, every such set $T(f)$ is a $(q + t,\ t)$-arc of type $(0,\ 2,\ t)$ in $PG(2,\ q)$.

Remark 1. When, in the rest of this paper, we refer to a $(q + t,\ t)$-arc $T(f)$ of type $(0,\ 2,\ t)$ in $PG(2,\ q)$, then this means a pointset $T(f)$ in $PG(2,\ q)$ with $f(z)$ satisfying (a)-(e).

If $r - h$ is a proper divisor of r, then any oval of $PG(2,\ 2^{r-h})$ gives rise to a $(2^r + 2^h,\ 2^h)$-arc of type $(0,\ 2,\ 2^h)$ as follows.

We put $q' = 2^{r-h}$ and denote by $GF(q')$ the subfield of order q' of $GF(q)$. Then we consider the *trace function* (see Lidl and Niederreiter 1983, Ch. 2, §3)

$$L(z) = z + z^{q'} + z^{q'^2} + ...,\ + z^{q/q'} \tag{1}$$

from $GF(q)$ into $FG(q')$ satisfying (i) $L(z_1 + z_2) = L(z_1) + L(z_2)$, and (ii) $L(\zeta z) = \zeta L(z)$, for all $z_1,\ z_2,\ z \in GF(q)$ and all $\zeta \in GF(q')$. Then, for any monic permutation polynomial $g(\zeta)$ of $GF(q^{r-h})$, the set $T(f)$, where $f(z) = g(L(z))$ and $0,\ w_1,\ ...,\ w_{t-1}$ are the roots of $L(z)$ in $GF(q)$, is a $(q + t,\ t)$-arc of type $(0,\ 2,\ t)$ in $PG(2,\ q)$ if and only if

$$\begin{vmatrix} g(\alpha) & \alpha & 1 \\ g(\beta) & \beta & 1 \\ g(\gamma) & \gamma & 1 \end{vmatrix} \neq 0 \qquad\qquad (2)$$

for all distinct $\alpha, \beta, \gamma \in GF(q')$.

Given a permutation polynomial $g(\zeta)$ of $GF(q')$, let $\Omega(g)$ be the pointset of size $q' + 1$ consisting of $(1, 0, 0)$ and all points $(g(\zeta), \zeta, 1)$, $\zeta \in GF(q')$. The condition (8) means geometrically that $\Omega(g)$ cannot have more than two collinear points; in other words, $\Omega(g)$ is an oval in $PG(2, q')$. Therefore ovals of $PG(2, q')$ can be used to obtain $(q + t, t)$-arcs of type $(0, 2, t)$ in $PG(2, q)$. An extensive theory of ovals was developed by Segre and his school in the 1960's. Recently some new classes of ovals have been found; see the survey paper Cherowitzo (1988). From the known ovals, we obtain a series of monic permutation polynomials $g(\zeta)$ of $GF(q')$, $q' = 2^d$, having property (8).

A (k, n)-arc K of $PG(2, q)$ is said to be a *translation* (k, n)-*arc (with respect to a line l_∞)* if for any two points $A, B \in K$ not on l_∞ there is an elation with axis l_∞ (that is a translation of the affine plane $AG(2, q) = PG(2, q) \backslash l_\infty$) which sends A onto B and maps K onto itself. The line l_∞ is called a *translation line* for T. This notion is a natural generalisation of that of translation oval introduced by Segre (1962) and Tits (1962).

Any translation $(q + t, t)$-arc T of type $(0, 2, t)$ has a t-nucleus, and the translation line l_∞ is a t-secant of it. Choose a canonical frame with the improper line l_∞ and put $AG(2, q) = PG(2, q) \backslash l_\infty$. Then T turns out to be of the form $T(f)$ such that $f(z)$ is additive; that is, $f(a + b) = f(a) + f(b)$ holds identically in $GF(q)$. Conversely, if $F(z)$ is additive, then $T(f)$ is a translation $(q + t, t)$-arc of type $(0, 2, t)$.

3. A class of $(q + t, t)$-arcs of type $(0, 2, t)$ in $PG(2, q)$ admitting a large group of linear collineations

In Korchmáros and Mazzocca (1990), some translation $(2^r + 2^h, 2^h)$-arcs of type $(0, 2, 2^h)$ in $PG(2, q)$ have been obtained from translation ovals of $PG(2, 2^{r-h})$ via the trace function $L(z)$ over the subfield $GF(2^{r-h})$ of $GF(2^r)$, in the following way. For a monic permutation polynomial $g(\zeta)$ of $GF(q^{r-h})$, let $F(z) = g(L(z))$. Then $T(f)$ is a $(q + t, t)$-arc of type $(0, 2, t)$ if and only if $g(\zeta)$ is a permutation polynomial of $GF(2^{r-h})$ satisfying (8). Furthermore, a necessary and sufficient condition that $T(f)$ be of translation is that $g(\zeta)$ be additive. These conditions mean that $\Omega(g) = \{g(\zeta), \zeta, 1) \mid \zeta \in GF(q')\} \cup \{(1, 0, 0)\}$ is a translation oval written in a canonical form.   Notice that all translation ovals are known, (Hirschfeld 1975; Payne (1971), see also Hirschfeld 1979, §8.5). It turns out that $g(\zeta) = \zeta^{2^e}$ where $gcd(e, r - h) = 1$.   For later convenience, we note that then $g(\eta)^{-1} = \eta^{2^n}$ where $n + e = r$.

We see that our $T(f)$ consists of all affine points $(f(z), z)$, $z \in GF(2^r)$, and of all improper points (m) where m runs over all roots of $L(z)$, while its t-secants are the improper line $[\infty]$ and vertical lines $[k]$ with equation $x = k, k \in GF(2^{r-h})$.

Next we investigate the linear collineation group of $PG(2, q)$ which leaves $T(f)$ invariant.   We begin by observing that they translation group T preserving $T(f)$ consists of all collineations $t(u, v)$ with equations $x' = x + u, y' = y + v, v \in GF(2^r), L(v) = u^{2^n}, u \in GF(2^{r-h})$.   To see how T acts on $T(f)$, it is convenient to write T as the direct product of two of its subgroups $T_1 = \{t(0, v) \mid v \in GF(2^r), L(v) = 0\}$ and $T_2 = \{t(u, u^{2^n}) \mid u \in GF(2^{r-h})\}$, since T_1 is a regular permutation group on each affine t-secant of $T(f)$, while T_2 acts on the set of all t-secants of $T(f)$ as the group of all permutations $k' = k + u, u \in GF(2^{r-h})$, on $GF(2^{r-h}) \cup \{\infty\}$.

$T(f)$ is also invariant under a shear group S consisting of all collineations $s(v)$ with equations $x' = x, y' = vx + y$ where $L(v) = 0$, $v \in GF(2^r)$. S acts on each t-secant of $T(f)$ different from $[0]$ as a regular permutation group.   Clearly TS is an elementary abelian 2-group of order 2^{2h+r}, and we show that TS is the elation group of $T(f)$.   Observe that any elation e preserving $T(f)$ is either a translation and hence $e \in T$, or a shear whose axis is a vertical line $x = c$.   In the latter case e has equations $x' = x, y' = xv + y + cv$, where $v \in GF(q), L(v) = 0, L(cv) = 0$.   It turns out that $e = t(0, cv)s(v)$ and thus $e \in TS$.   This proves our assertion.   As a consequence, we have that T_1S is the group of all linear collineations preserving each t-secant of $T(f)$.

A further collineation group W preserving $T(f)$ is that consisting of all collineations with equations $x' = wx, y' = w^{2^n}y$ where $w \in GF(2^{r-h})^*$. W fixes $(0, 0)$ and acts as a regular permutation group on the remaining points of $[0]$ because $r - h$ and n are relative prime.   Since W normalises S, so SW is a Frobenius group with a natural permutation representation on both pointsets $[\infty] \cap (f)$ and $[0] \cap T(f)$. Such a permutation representation is doubly transitive if and only if $r = 2h$.

W normalises T_2, and T_2W acts on the set of all t-secants of $T(f)$ as the group of all permutations $k' = wk + u$ on $GF(2^{r-h}) \cup \{\infty\}$, where u,

$w \in GF(2^{r-h})$ and $W \neq 0$. Since W also normalises T, the collineation group generated by T, S and W coincides with their product TSW.

We are able to show that TSW is the full affine linear collineation group preserving $T(f)$. As we have seen, T_2W acts on the set of all affine t-secants as a doubly transitive permutation group. From this we can infer that for any affine linear collineation a preserving T there is an element $b \in T_2W$ such that ab leaves each vertical line invariant. Thus ab is an elation with centre Y_∞. As we have already seen, this implies $ab \in T_1S$. Thus $a \in T_1ST_2W$. Since $T_1ST_2 = TS$, our assertion follows.

Therefore we can describe the affine linear collineation group of $T(f)$ as follows:

Theorem 1. $T(f)$ *is a translation* $(2^r + 2^h, 2^h)$-*arc of type* $(0, 2, 2^h)$ *in* $PG(2, 2^r)$ *whose translation group is* T. *The affine linear collineation group* A *preserving* $T(f)$ *is the semidirect product of an elementary abelian normal subgroup* T_1S *of order* 2^{2h} *with a subgroup* T_2W *of order* $2^{r-h}(2^{r-h} - 1)$. *Here,* T_1S *consists of elations with centre* Y_∞ *while* T_2W *is isomorphic to* $ASL(2, 2^{r-h})$ *and acts on the set of all affine t-secants of* $T(f)$ *as* $ASL(2, 2^{r-h})$ *in its naturally doubly transitive representation on* $GF(2^{r-h})$. *Also* SW *is a Frobenius group of order* $2^h(2^{r-h} - 1)$ *and each t-tangent of* $T(f)$ *is invariant under a suitable conjugate under* SW *in* A.

In some special cases, $PG(2, 2^r)$ has further (projective) collineations preserving $T(f)$. Let f denote the collineation with equations $\rho x_1 = x_3$, $\rho x_2 = x_2$, $\rho x_3 = x_1$. It is easily checked that f interchanges $[0]$ with $[\infty]$ and takes $[k]$ to the pointset $\overline{k} = \{(k^{-1}, k^{-1}y) \mid L(y) = k^{2^n}, y \in GF(2^r)\}$. Thus f preserves T if and only if $\overline{k} = [k^{-1}]$ for each $k \in GF(2^{r-h})^*$. Since $L(y) = k \setminus s(2^n)$ can be replaced with $L(k^{-1}y) = k \setminus s(2^n - 1)$, the conditions $\overline{k} = [k^{-1}]$ and $k^{2^{n+1} - 1} = 1$ are equivalent. If each $k \in GF(2^{r-h})^*$ satisfies the latter condition, then $r - h$ divides $n + 1$. Observe that the converse is also true. It follows that $(r - h) \mid (n + 1)$ is a necessary and sufficient condition for f to preserve $T(f)$.

Let G denote the full linear collineation group of $PG(2, 2^r)$ preserving $T(f)$. We show that if $(r - h) \mid (n + 1)$ then $G = \langle A, f \rangle$, and $G = A$ otherwise. Let $\overline{G}$ be the permutation group induced by G on the set of all t-secants of $T(f)$. The kernel of this representation is the subgroup consisting of all elations with centre Y_∞. By Theorem 1 this subgroup is T_1S. Let us identify $[k]$ with k and $[\infty]$ with ∞. Then $\overline{G}$ becomes a permutation group on $GF(2^{r-h}) \cup \{\infty\}$, and is a subgroup of

$PSL(2, 2^{r-h})$ because G contains no non-linear collineations. The subgroup $\overline{A}$ of $\overline{G}$ consists of all permutations $k' = ak + b, a, b \in GF(2^{r-h})$, $a \neq 0$. Thus $\overline{A}$ is a maximal subgroup of $PSL(2, 2^{r-h})$. Suppose that $G \neq A$. Since TS is a subgroup of A, we see that $\overline{G}$ contains $\overline{A}$ properly and hence $\overline{G} = PSL(2, 2^{r-h})$. As a consequence, $\overline{G}$ must contain the involutorial permutation $\overline{w}: k' = 1/k$. Let $\overline{f}$ be an involutorial permutation $\overline{G}$ which comes from a translation $t \in T_2$. As $\overline{G}$ has a unique class of involutorial elements, $\overline{w}$ is a conjugate of $\overline{f}$ under a suitable permutation $\overline{s}$ of $\overline{G}$ Take an element $s \in G$ acting on $GF(2^{r-h})$ $\cup \{\infty\}$ as $\overline{s}$. Then $u = sts^{-1}$ is an involutorial element of G such that $\overline{u} = \overline{w}$. Clearly u maps X_∞ onto a point of $[0]$, say $(0, v)$ where $v \in GF(2^r)$ and $L(v) = 0$. Then the collineation $v = t(0, v)ut(0, v)$ sends X_∞ into $(0, 0)$. Since v is still an involutorial element of G such that $\overline{v} = \overline{w}$, v coincides with f. Finally, $f \in G$ implies $(r - h) \mid (n + 1)$, as we have previously shown. This completes the proof of our assertion.

Finally we determine the abstract structure of G. For the case $G = A$, see Theorem 1. Suppose that $G = \langle A, f \rangle$. Then G acts transitively on the set of all t-secants of $T(f)$. This allows us to apply a known result concerning finite G-transitive sets (see Hering 1972, Theorem 2 and Lemma 3). It turns out that the closure H of T_2 in G is isomorphic to $PSL(2, 2^{r-h})$, and hence G is the semidirect product of H with T_1S.

Let us formulate our results as a theorem:

Theorem 2. *The full linear collineation group G preserving $T(f)$ is in general the group A described in Theorem 1. An exception is only possible if $r - 1$ divides $n + 1$ in which case G is the semidirect product of $T_1 S$ with a subgroup $H \cong PSL(2, 2^{r-h})$. H acts on the set of all t-secants of $T(f)$ as $PSL(2, 2^{r-h})$ in its natural triply transitive representation on $GF(2^{r-h}) \cup \{\infty\}$.*

4. Translation $(q + t, t)$-arcs of type $(0, 2, t)$ in $PG(2, q)$ admitting a linear collineation group doubly transitive on the set of all affine t-secants

We prove that the $(q + t, t)$-arcs of type $(0, 2, t)$ indicated in the section title are precisely those constructed in §2.

Theorem 3. *Let T be a translation $(q + t, t)$-arc of type $(0, 2, t)$ in $PG(2, q)$. If the linear collineation group preserving T acts on the affine t-secants of T as a doubly transitive permutation group, then T*

is projectively equivalent to one of the examples $T(f)$ described in section 1.

Proof. Let A denote the full affine linear collineation group if $PG(2, q)$ preserving T, and let $\overline{A}$ denote the permutation group induced by A on the affine lines through Y_∞. Then $\overline{A}$ is a subgroup of $ASL(1, q)$ regarded, in its natural doubly transitive permutation representation, as the group of all affinities of the one dimensional affine space over $GF(q)$. By our hypothesis, $\overline{A}$ has a doubly transitive orbit. Therefore, we can identify the affine t-secants of T with the elements of a suitable subfield $GF(q')$ in such a way that $\overline{A}$ coincides with the subgroup $ASL(1, q')$. It turns out that $\overline{A}$ contains both the group $\overline{N})$ of all translations $k' = k + u, u \in GF(q')$ and the group $\overline{C}$ of all homotheties $k' = \mu k, \mu \in GF(q')$.

Observe that $A/E = \overline{A}$ where E is a subgroup of order a power of 2 consisting of all elations of A with centre Y_∞. From this we can infer that A contains a normal subgroup N of order a power 2 and of index $q - 1$. By a classical result, due to Zassenhaus (see Huppert 1967; Hauptsatz 18.1), N has a complement C in A. It turns out that A contains a cyclic subgroup C of order $q' - 1$ such that $CE/E = \overline{C}$.

Apart from Y_∞, we find only two points fixed by C. Both belong to T, as t is even: one of them lies on the affine line $[0]$, the other one is an improper point. Choose a frame in such a way that these points become $(0, 0)$ and X_∞, respectively. Then C consists of all collineations $c(\mu)$ with equation $x' = \mu x, y' = \gamma(\mu)y$ such that μ ranges over the non-zero elements of $GF(q')$, and $\gamma(\mu)$ is a homomorphism of the multiplicative group $GF(q')^*$ of $GF(q')$ into itself. Actually, $\gamma(\mu)$ is an isomorphism, as $c(1)$ is the only collineation of C preserving a horizontal line different from the x-axis.

Now let us arrange our frame to have the unit point $(1, 1)$ on T. Then T contains each point $P(\mu) = (\mu, \gamma(\mu))$ where $\mu \in GF(q')^*$. These $q' - 1$ points $P(\mu)$ together with $P(0) = (0, 0)$ are all the affine points of T lying on the subplane $PG(2, q')$, because every affine line through X_∞ meets T in a unique point.

In our frame, T has a canonical form $T(f)$ described in §2. Actually the value set of $f(z)$ consists of all elements of $GF(q')$. In particular, $f(\mu) = \gamma(\mu), \mu \in GF(2, q')^*$, and this remains valid for $\mu = 0$ if we extend γ to $GF(q')$ by putting $\gamma(0) = 0$. We thus deduce that $\gamma(\mu)$ is additive on $GF(q')$. Since we have already seen that $\gamma(\mu)$ for $\mu \neq 0$ is an isomorphism of $GF(q')$, so $\gamma(\mu)$ turns out to be an automorphism of $GF(q')$. Hence $\gamma(\mu) = \mu^{2^n}$.

For every $z \in GF(q)$, the collineation $c(\mu)$ sends $P(z)$ to the point $(\mu f(z), \gamma(\mu)z)$ of T; thus $f(\gamma(\mu)z) = \mu f(z)$ for all $\mu \in GF(q')$, $z \in GF(q)$.

Finally, let $L(z) = \gamma^{-1}(f(z))$. The above discussion shows that $L(z)$ is additive, and $L(\mu z) = \mu L(z)$ holds for all $\mu \in GF(q')$, $z \in GF(q)$. Regarding $GF(q)$ as a vector space over $GF(q')$, the transformation $L(z)$ turns out to be linear from $GF(q)$ to $GF(q')$. Since $L(1) = 1$, from Lidl Niederreiter (1983, Theorem 2.24), it follows that $L(z)$ is the trace function from $GF(q)$ into $GF(q')$. This completes our proof.

Theorem 3 has the following corollary.

Let T be a translation $(2^r + 2^h, 2^h)$-arc of type $(0, 2, 2^h)$ in $PG(2, 2^r)$. If the linear collineation group preserving T acts on the t-secants of T different from l_∞ as a doubly transitive permutation group, then $r - h$ divides r and T can be obtained from a translation oval of $PG(2, 2^{r-h})$ via the trace function over the subfield $GF(2^{r-h})$ of $GF(2^r)$.

Acknowledgement. This paper was supported by "Ministero dell'Università e della Ricerca Scientifica e Tecnologica" and by G.N.S.A.G.A. of C.N.R. (Italy).

References

W. Cherowitzo (1988), Hyperovals in Desarguesian planes of even order, *Ann. Discrete Math.* **37**, 87-94.

C. Hering (1972), On subgroups with trivial normalizer intersection, *J. Algebra* **20**, 622-629.

J.W.P. Hirschfeld (1975), Ovals in Desarguesian planes of even order, *Ann. Mat. Pura Appl.* **102**, 79-88.

J.W.P. Hirschfeld (1979), *Projective Geometries over Finite Fields*, Oxford University Press, Oxford.

B. Huppert (1967), *Endliche Gruppen I*, Springer-Verlag, Heidelberg.

G. Korchmáros and F. Mazzocca (1990), On $(q + t, t)$-arcs of type $(0, 2, t)$ in a Desarguesian plane of order q, *Math. Proc. Cambridge Philos. Soc.* **108**, 445-459.

R. Lidl and H. Niederreiter (1983), *Finite Fields*, Cambridge University Press, Cambridge.

S.E. Payne (1971), A complete determination of translation ovoids in finite Desarguesian plane, *Atti Accad. Naz. Lincei Rend.* **51**, 328-331.

B. Segre (1962), Ovali e curve σ nei piani di Galois di caratteristica due, *Atti Accad. Naz. Lincei Rend.* **34**, 785-790.

J. Tits (1962), Ovoides à translations, *Rend. Mat. e Appl.* **21**, 37-59.

Pseudo-symplectic geometries over finite fields of characteristic two

Y. Liu and Z. Wan

Columbia University and Lund University

1. Introduction

Throughout this paper, let F_q denote a finite field with q elements, where q is a power of 2.

Let S be an $n \times n$ non-singular non-alternating symmetric matrix over F_q. An $n \times n$ matrix T over F_q is said to be *pseudo-symplectic* with respect to S, if

$$TST^T = S. \tag{1}$$

Clearly, $n \times n$ pseudo-symplectic matrices with respect to S are non-singular and they form a group with respect to the matrix multiplication, called the *pseudo-symplectic group* with respect to S over F_q and denoted by $Ps_n(F_q, S)$. The term 'pseudo-symplectic group' is due to Hirschfeld (1979).

Let $V_n(F_q) = \{(x_1, x_2, ..., x_n) \mid x_i \in F_q\}$ be the n-dimensional row vector space over F_q. Define an action of $Ps_n(F_q, S)$ on $V_n(F_q)$ as follows:

$$V_n(F_q) \times Ps_n(F_q, S) \rightarrow V_n(F_q)$$
$$((x_1, x_2, ..., x_n), T)) \mapsto (x_1, x_2, ..., x_n)T.$$

Then $V_n(F_q)$, together with this group action is called the n-dimensional *pseudo-symplectic space* with respect to S over F_q. The main purpose of this paper is to study how the subspaces of $V_n(F_q)$ are divided into transitive sets under $Ps_n(F_q, S)$ and to compute the cardinality of each transitive set.

We shall also introduce the *extended pseudo-symplectic group* and compute the number of $n \times n$ non-alternating symmetric matrices of a given rank r ($0 \le r \le n$) over F_q.

Some known results of symplectic geometries over finite fields will be used below. For the readers' convenience a brief summary of the symplectic geometries is given at the end of this paper.

2. Pseudo-symplectic groups over F_q

Two $n \times n$ non-alternating symmetric matrices S_1 and S_2 over F_q are said to be congruent if there is an $n \times n$ non-singular matrix P such that $S_1 = PS_2P^T$. Assume further that S_1 and S_2 are non-singular; then we have a group isomorphism

$$\psi_P: Ps_n(F_q, S_1) \rightarrow Ps_n(F_q, S_2)$$
$$T \mapsto P^{-1}TP$$

and a ψ_P-invariant vector space isomorphism
$$V_n(F_q) \rightarrow V_n(F_q)$$
$$(x_1, x_2, ..., x_n) \mapsto (x_1, x_2, ..., x_n)P.$$

Therefore it is sufficient to study the pseudo-symplectic groups and pseudo-symplectic spaces with respect to the normal forms of the non-singular non-alternating symmetric matrices under congruences.

The following theorem is well known, see Hirschfeld (1979, Theorem 5.3.3).

Theorem 1. *Let S be an $n \times n$ symmetric matrix over F_q. then S is congruent to one and only one of the following three kinds of normal forms*

$$\left[\begin{array}{cc} 0 & I^{(v)} \\ I^{(v)} & 0 \\ & & 0^{(n-2v)} \end{array}\right], \quad \left[\begin{array}{ccc} 0 & I^{(v)} \\ I^{(v)} & 0 \\ & & 1 \\ & & & 0^{(n-2v-1)} \end{array}\right], \quad \left[\begin{array}{cccc} 0 & I^{(v)} \\ I^{(v)} & 0 \\ & & 0 & 1 \\ & & 1 & 1 \\ & & & & 0^{(n-2v-2)} \end{array}\right] \quad (2)$$

which are of ranks $2v$, $2v + 1$ and $2v + 2$ respectively. ∎

The number v in the normal forms (2) is called the *index* of S. Let

$$K = \left[\begin{array}{cc} 0 & I^{(v)} \\ I^{(v)} & 0 \end{array}\right]. \quad (3)$$

Then we have

Corollary 2. *Let S be an $n \times n$ non-singular non-alternating symmetric matrix. If $n = 2v + 1$ is odd, then S is congruent to*

$$S_1 = \begin{bmatrix} K & \\ & 1 \end{bmatrix},$$ (4)

and if $n = 2\nu + 2$ is even, then S is congruent to

$$S_2 = \begin{bmatrix} K & & \\ & 0 & 1 \\ & 1 & 1 \end{bmatrix}. \quad \square$$ (5)

Thus there are the only two pseudo-symplectic groups $Ps_{2\nu+1}(F_q, S_1)$ and $Ps_{2\nu+2}(F_q, S_2)$ to be considered. In the following we simply write $Ps_{2\nu+\delta}(F_q)$, where $\delta = 1$ or 2, for $Ps_{2\nu+\delta}(F_q, S_\delta)$.

In $V_n(F_q)$ the vector with 1 in the i-th coordinate and 0 elsewhere will be denoted by e_i. Then $e_1, e_2, \ldots, e_n$ form a basis of $V_n(F_q)$. Let S be an $n \times n$ non-singular symmetric matrix. A vector $x \in V_n(F_q)$ is called *isotropic* or *non-isotropic* with respect to S, if $xSx^T = 0$ or $xSx^T \neq 0$ respectively.

Dieudonné (1948) studied the structure of pseudo-symplectic groups over any field of characteristic 2. His main results specialized to the finite field F_q of characteristic 2 follows.

Theorem 3. $Ps_{2\nu+1}(F_q)$ consists of all $(2\nu + 1) \times (2\nu + 1)$ matrices of the form

$$\begin{matrix} \begin{bmatrix} T & \\ & 1 \end{bmatrix} & \begin{matrix} 2\nu \\ 1 \end{matrix} \\ \begin{matrix} 2\nu \ \ 1 \end{matrix} & \end{matrix}$$

where $T \in Sp_{2\nu}(F_q)$. Hence $Ps_{2\nu+1}(F_q) \simeq Sp_{2\nu}(F_q)$ and $|Ps_{2\nu+1}(F_q)| = q^{\nu^2} \prod_{i=1}^{\nu} (q^{2i} - 1)$. $\square$

Corollary 4. With respect to S_1, the vector $(x_1, x_2, \ldots, x_{2\nu+1})$ is non-isotropic if and only if $x_{2\nu+1} \neq 0$; in particular, $e_{2\nu+1}$ is non-isotropic. The dual subspace

$$\langle e_{2\nu+1} \rangle^\perp = \{x \in V_{2\nu+1}(F_q) \mid xS_1 e_{2\nu+1}{}^T = 0\}$$

of $\langle e_{2\nu+1} \rangle$ is the 2ν-dimensional subspace $\langle e_1, e_2, \ldots, e_{2\nu} \rangle$ on which $Ps_{2\nu+1}(F_q)$ acts as the symplectic group $Sp_{2\nu}(F_q)$ does. Also, only $e_{2\nu+1}$ and its multiples are left fixed by every element of $Ps_{2\nu+1}(F_q)$.

Theorem 5. $Ps_{2v+2}(F_q)$ consists of all $(2v + 2) \times (2v + 2)$ matrices of the form

$$
\begin{array}{c}
\begin{bmatrix} T & b^T & 0 \\ 0 & 1 & 0 \\ c & d & 1 \end{bmatrix} & \begin{matrix} 2v \\ 1 \\ 1 \end{matrix} \\
\begin{matrix} 2v & 1 & 1 \end{matrix} &
\end{array}
$$

where $T \in Sp_{2v}(F_q)$, $TKc^T + b^T = 0$, and d is an arbitrary element of F_q. Also, $Ps_{2v+2}(F_q)$ has a normal series

$$Ps_{2v+2}(F_q) \supset \mathcal{K}_1 \supset \mathcal{K}_2 \supset \{I\},$$

where $\mathcal{K}_1$ and $\mathcal{K}_2$ consist of elements of $Ps_{2v+2}(F_q)$ which are of the form

$$
\begin{bmatrix} I & b & 0 \\ 0 & 1 & 0 \\ c & d & 1 \end{bmatrix},
$$

with $Kc^T + b^T = 0$, and of the form

$$
\begin{bmatrix} I & 0 & 0 \\ 0 & 1 & 0 \\ 0 & d & 1 \end{bmatrix},
$$

respectively. Moreover, $Ps_{2v+2}(F_q) / \mathcal{K}_1 \simeq Sp_{2v}(F_q)$, $\mathcal{K}_1 / \mathcal{K}_2$ is abelian of order q^{2v} and $\mathcal{K}_2$ is abelian of order q. Hence $|Ps_{2v+2}(F_q)| = q^{(v+1)^2} \prod_{i=1}^{v} (q^{2i} - 1)$. $\square$

Corollary 6. With respect to S_2, the vector $(x_1, x_2, ..., x_{2v+2})$ is non-isotropic if and only if $e_{2v+2} \neq 0$; in particular, the vector e_{2v+1} is isotropic and the vector e_{2v+2} is non-isotropic. The 2-dimensional subspace $<e_{2v+1}, e_{2v+2}>$ is non-degenerate, that is,

$$
\begin{bmatrix} e_{2v+1} \\ e_{2v+2} \end{bmatrix} S_2 \begin{bmatrix} e_{2v+1} \\ e_{2v+2} \end{bmatrix}^T
$$

is non-singular, and the dual subspace

$$\langle e_{2v+1}, e_{2v+2} \rangle^{\perp} = \{x \in V_{2v+2}(F_q) \mid xS_2 e_{2v+1}^T = 0 \text{ and } xS_2 e_{2v+2}^T = 0\}$$

of $\langle e_{2v+1}, e_{2v+2} \rangle$ is the 2v-dimensional subspace $\langle e_1, e_2, ..., e_{2v} \rangle$ on which the group $Ps_{2v+2}(F_q)$ acts as the symplectic group $Sp_{2v}(F_q)$ does. Moreover, only e_{2v+1} and its multiples are left fixed by every element of $Ps_{2v+2}(F_q)$. □

3. Pseudo-symplectic geometries over F_q

As in §2 we write $n = 2v + \delta$, where $\delta = 1$ or 2. Let P be an m-dimensional subspace of $V_{2v+\delta}(F_q)$. We use the same symbol P to denote any $m \times (2v + \delta)$ matrix, whose m rows form a basis of the subspace P and call it a matrix representation of the subspace P. By Theorem 1, $PS_\delta P^T$ is congruent to one of the following three normal forms:

$$M(m, \ 2s, s) \ = \begin{bmatrix} 0 & I^{(s)} \\ I^{(s)} & 0 \\ & & 0^{(m-2s)} \end{bmatrix} \tag{6}$$

$$M(m, \ 2s + 1, s) = \begin{bmatrix} 0 & I^{(s)} \\ I^{(s)} & 0 \\ & & 1 \\ & & & 0^{(m-2s-1)} \end{bmatrix} \tag{7}$$

and

$$M(m, \ 2s + 2, s) = \begin{bmatrix} 0 & I^{(s)} \\ I^{(s)} & 0 \\ & & 0 \ 1 \\ & & 1 \ 1 \\ & & & 0^{(m-2s-2)} \end{bmatrix} \tag{8}$$

for some s such that $0 \le s \le [m/2]$. We say that P is a subspace of type $(m, 2s + \tau, s, \varepsilon)$ where $\tau = 0, 1,$ or 2 and $\varepsilon = 0$ or 1, if

(i) $PS_1 P^T$ is congruent to $M(m, 2s + \tau, s)$, and

(ii) $e_{2v+1} \notin P$ or $e_{2v+1} \in P$ corresponding to $\varepsilon = 0$ or $\varepsilon = 1$, respectively. In particular, subspaces of types $(m, 0, 0, 0)$ and $(m, 0, 0, 1)$ are called totally isotropic subspaces of dimension m. At first we have

Theorem 7. In the $(2v + \delta)$-dimensional pseudo-symplectic space, subspaces of type $(m, 2s + \tau, s, \varepsilon)$ exist, if and only if

$$(\tau, \varepsilon) = \begin{cases} (0, 0), (1, 0), (1, 1), or (2, 0), & when \ \delta = 1, \\ (0, 0), (0, 1), (1, 0), (2, 0) \ or (2, 1), & when \ \delta = 2, \end{cases} \tag{9}$$

and

$$2s + \max(\tau, \varepsilon) \le m \le v + s + [(\tau + \delta - 1)/2] + \varepsilon. \tag{10}$$

Proof. Consider first the case $\delta = 1$. It is easy to verify that if both (9) and (10) hold, then

$$\left[\begin{array}{ccccccc} I^{(s)} & 0 & 0 & 0 & 0 & 0 & 0 \\ 0 & 0 & 0 & I^{(s)} & 0 & 0 & 0 \\ 0 & I^{(m-2s)} & 0 & 0 & 0 & 0 & 0 \end{array}\right],$$
$$\scriptstyle s \quad\quad m-2s \quad\quad v+s-m \quad\quad s \quad\quad m-2s \quad\quad v+s-m \quad\quad 1$$

$$\left[\begin{array}{ccccccccc} I^{(s)} & 0 & 0 & 0 & 0 & 0 & 0 & 0 & 0 \\ 0 & 0 & 0 & 0 & I^{(s)} & 0 & 0 & 0 & 0 \\ 0 & 1 & 0 & 0 & 0 & 0 & 0 & 0 & 1 \\ 0 & 0 & I^{(m-2s-1)} & 0 & 0 & 0 & 0 & 0 & 0 \end{array}\right]$$
$$\scriptstyle s \quad 1 \quad m-2s-1 \quad v+s-m \quad s \quad 1 \quad m-2s-1 \quad v+s-m \quad 1$$

$$\left[\begin{array}{ccccccc} I^{(s)} & 0 & 0 & 0 & 0 & 0 & 0 \\ 0 & 0 & 0 & I^{(s)} & 0 & 0 & 0 \\ 0 & 0 & 0 & 0 & 0 & 0 & 1 \\ 0 & I^{(m-2s-1)} & 0 & 0 & 0 & 0 & 0 \end{array}\right],$$
$$\scriptstyle s \quad m-2s-1 \quad v+s-m+1 \quad s \quad m-2s-1 \quad v+s-m+1 \quad 1$$

and

$$\left[\begin{array}{ccccccccc} I^{(s)} & 0 & 0 & 0 & 0 & 0 & 0 & 0 & 0 \\ 0 & 0 & 0 & 0 & I^{(s)} & 0 & 0 & 0 & 0 \\ 0 & 1 & 0 & 0 & 0 & 0 & 0 & 0 & 0 \\ 0 & 0 & 0 & 0 & 0 & 1 & 0 & 0 & 1 \\ 0 & 0 & I^{(m-2s-2)} & 0 & 0 & 0 & 0 & 0 & 0 \end{array}\right]$$
$$\scriptstyle s \quad 1 \quad m-2s-2 \quad v+s-m+1 \quad s \quad 1 \quad m-2s-2 \quad v+s-m+1 \quad 1$$

are subspaces of the type $(m, 2s, s, 0)$, $(m, 2s + 1, s, 0)$, $(m, 2s + 1, s, 1)$ and $(m, 2s + 2, s, 0)$, respectively. Hence the sufficiency of condition (9) and (10) is proved.

To prove the necessity of condition (9) it is enough to prove that if $\tau = 0$ or 2, then we have necessarily $\varepsilon = 0$. Suppose first that $\tau = 0$ and let P be an m-dimensional subspace of $V_{2v+1}(F)$ such that $PS_1P^T = M(m, 2s, s)$ (see (6)). Then every vector of P is isotropic, but e_{2v+1} is not. Therefore $e_{2v+1} \notin P$ and hence $\varepsilon = 0$. Next suppose that $\tau = 2$ and let P be an m-dimensional subspace of $V_{2v+1}(F)$ such that $PS_1P^T = M(m, 2s + 2, s)$ (see (8)). If $e_{2v+1} \in P$, then P would contain an isotropic vector not contained in $\langle e_{2v+1}\rangle^{\perp} = \{e_1, e_2, ..., e_{2v}\}$, which is impossible. Hence e_{2v+1} $\notin P$, that is, $\varepsilon = 0$. To prove the necessity of (10) we have to distinguish the four possible cases of (τ, ε).

(a) $(\tau, \varepsilon) = (0, 0)$. Let P be a subspace of the type $(m, 2s, s, 0)$ in $V_{2v+1}(F_q)$. We can assume that $PS_1P^T = M(m, 2s, s)$. Clearly all vectors of P are isotropic. Hence $P \subset \langle e_1, e_2, ..., e_{2v}\rangle$ and we can choose a matrix representation of P of the form

$$P = \begin{bmatrix} P_1 & 0 \end{bmatrix}$$
$$\;2v\;\;\,1$$

where P_1 is a $m \times 2v$ matrix of rank m. From $PS_1P^T = (m, 2s, s)$ we deduce $P_1KP_1^T = M(m, 2s, s)$, that is, P_1 is a subspace of type (m, s) in the $2v$-dimensional symplectic space. Therefore by Theorem 14 (below) we have $2s \le m \le v + s$, which proves the necessity of (10) for the case $(\tau, \varepsilon) = (0, 0)$.

(b) $(\tau, \varepsilon) = (1, 0)$. Let P be a subspace of the type $(m, 2s + 1, s, 0)$ in $V_{2v+1}(F_q)$. Then $e_{2v+1} \notin P$. We can assume that $PS_1P^T = M(m, 2s + 1, s)$, then $2s + 1 \le m$ and P is necessarily of the form

$$P = \begin{bmatrix} P_1 & 0 \\ v & 1 \\ P_2 & 0 \end{bmatrix} \begin{matrix} 2s \\ 1 \\ m-2s-1 \end{matrix} \;\;.$$
$$\;2v\;\;\,1$$

From rank $P = m$ we deduce

$$\mathrm{rank} \begin{bmatrix} P_1 \\ P_2 \end{bmatrix} = m - 1$$

Since $e_{2v+1} \notin P$, v is non-zero and

$$\mathrm{rank} \begin{bmatrix} P_1 \\ v \\ P_2 \end{bmatrix} = m.$$

From $PS_1P^T = M(m, 2s + 1, s)$ we deduce

$$\begin{bmatrix} P_1 \\ v \\ P_2 \end{bmatrix} K \begin{bmatrix} P_1 \\ v \\ P_2 \end{bmatrix}^T = M(m, 2s, s);$$

that is,

$$\begin{bmatrix} P_1 \\ v \\ P_2 \end{bmatrix}$$

is a subspace of type (m, s) in the $2v$-dimensional symplectic space. By Theorem 14 (below) we have $m \le v + s$. Combining the inequalities $2s + 1 \le m$ and $m \le v + s$, we obtain $2s + 1 \le m \le v + s$.

(c) $(\tau, \varepsilon) = (1, 1)$. Let P be a subspace of type $M(m, 2s + 1, s, 1)$ in $V_{2v+1}(F_q)$. Thus $e_{2v+1} \in P$. We can choose a matrix representation of P of the form

$$P = \begin{bmatrix} P_0 & 0 \\ 0 & 1 \end{bmatrix} \begin{matrix} m\text{-}1 \\ 1 \end{matrix} \, ,$$
$$\phantom{P = \begin{bmatrix} P_0 \end{bmatrix}} 2v \quad 1$$

where P_0 is an $(m - 1) \times 2v$ matrix of rank $m - 1$. We have

$$P S_1 P^T = \begin{bmatrix} P_0 K P_0^T & 0 \\ 0 & 1 \end{bmatrix}.$$

Since $P S_1 P^T$ is congruent to $M(m, 2s + 1, s)$, $P_0 K P_0^T$ is congruent to $M(m - 1, 2s, s)$, that is, P_0 is a subspace of type $(m - 1, s)$ in the $2v$-dimensional symplectic space. Therefore $2s \le m - 1 \le v + s$, and hence $2s + 1 \le m \le v + s + 1$.

(d) $(\tau, \varepsilon) = (2, 0)$. Let P be a subspace of type $M(m, 2s + 2, s, 0)$ in $V_{2v+1}(F_q)$. We can assume that $P S_1 P^T = M(m, 2s + 2, s)$. Hence we can further assume that P is of the form

$$P = \begin{bmatrix} P_1 & 0 \\ v & 1 \\ P_2 & 0 \end{bmatrix} \begin{matrix} 2s+1 \\ 1 \\ m\text{-}2s\text{-}2 \end{matrix} \, ,$$
$$\phantom{P = \begin{bmatrix} P_1 \end{bmatrix}} 2v \quad 1$$

where v is a non-zero 2v–dimensional vector. From rank $P = m$ we deduce

$$\operatorname{rank} \begin{bmatrix} P_1 \\ P_2 \end{bmatrix} = m - 1.$$

Since $e_{2v+1} \notin P$, we have also,

$$\operatorname{rank} \begin{bmatrix} P_1 \\ v \\ P_2 \end{bmatrix} = m.$$

From $PS_1P^T = M(m, 2s + 2, s)$ we deduce

$$\begin{bmatrix} P_1 \\ v \\ P_2 \end{bmatrix} K \begin{bmatrix} P_1 \\ v \\ P_2 \end{bmatrix}^T = \begin{bmatrix} 0 & I^{(s)} & & & \\ I^{(s)} & 0 & & & \\ & & 0 & 1 & \\ & & 1 & 0 & \\ & & & & 0^{(m-2s-2)} \end{bmatrix};$$

that is,

$$\begin{bmatrix} P_1 \\ v \\ P_2 \end{bmatrix}$$

is a subspace of type $(m, s + 1)$ in the 2v-dimensional symplectic space. Therefore we have necessarily $2s + 2 \le m \le v + s + 1$.

Thus we have completed the proof of Theorem 7 for the case $\delta = 1$.

Now let us consider the case $\delta = 2$. The sufficiency of conditions (9) and (10) can be proved in the same way and will be omitted.

To prove the necessity of condition (9) it is sufficient to prove that if $\tau = 1$, then $\varepsilon = 0$. Let P be a subspace of type $(m, 2s + 1, s, \varepsilon)$. We can assume that $PS_2P^T = M(m, 2s + 1, s)$, then P must contain a vector whose $(2v + 2)$-th component is 1, from which we deduce easily that $e_{2v+1} \notin P$.

To prove the necessity of condition (10) we distinguish the five possible cases of (τ, ε).

(a) $(\tau, \varepsilon) = (0, 0)$. Let P be a subspace of type $(m, 2s, s, 0)$ in $V_{2v+2}(F_q)$. By Corollary 6, P has a matrix representation of the form

$$P = \begin{bmatrix} P_1 & u^T & 0 \end{bmatrix}.$$
$$\;\; 2v \;\;\; 1 \;\;\; 1$$

Since $e_{2v+1} \notin P$, rank $P_1 = m$. From $PS_2P^T = M(m, 2s, s)$ we deduce $P_1KP_1^T = M(m, 2s, s)$. Thus P_1 is a subspace of type (m, s) in the 2v-dimensional symplectic space. Therefore $2s \le m \le v + s$.

(b) $(\tau, \varepsilon) = (0, 1)$. Let P be a subspace of type $(m, 2s, s, 1)$ in $V_{2v+2}(F_q)$. Then $e_{2v+1} \in P$ and we can assume that P has a matrix representation of the form

$$P = \begin{bmatrix} P_1 & 0 & 0 \\ 0 & 1 & 0 \end{bmatrix} \begin{matrix} m\text{-}1 \\ 1 \end{matrix}$$
$$\; 2v \; 1 \; 1$$

where P_1 is an $(m\text{-}1) \times 2v$ matrix of rank $m\text{-}1$. From $PS_2P^T = M(m, 2s, s)$ we deduce that $P_1KP_1^T = M(m\text{-}1, 2s, s)$. Thus, P_1 is a subspace of type $(m - 1, s)$ in the 2v-dimensional symplectic space. Therefore $2s \le m - 1 \le v + s$ and hence $2s + 1 \le m + 1 \le v + s + 1$.

(c) $(\tau, \varepsilon) = (1, 0)$. Let P be a subspace of type $(m, 2s + 1, s, 0)$ in $V_{2v+2}(F_q)$. We can assume that $PS_2P^T = M(m, 2s + 1, s)$. It follows that the $(2s + 1)$-th row of P is of the form $(x, y, 1)$, where x is a 2v-dimensional row vector over F_q and $y \in F_q$. Let

$$T_1 = \begin{bmatrix} I & Kx^T & 0 \\ 0 & 1 & 0 \\ x & y & 1 \end{bmatrix}. \tag{11}$$

Then $T_1 \in PS_{2v+2}(F_q)$ and $(x, y, 1) T_1 = (0, 0, 1) = e_{2v+2}$.

Since $(x, y, 1) S_2v^T = 0$ for any row v of P different from the $(2s + 1)$-th row, $e_{2v+2}S_2(vT_1)^T = 0$ for any row vT_1 of PT_1 different from the $(2s + 1)$-th row. Hence the $(2v + 1)$-th coordinate of any row of PT_1 different from the $(2s + 1)$-th row is 0. From $(PT_1)S_2(PT_1)^T = M(m, 2s + 1, s)$ we deduce that the $(2v + 2)$-th coordinate of any row of PT_1 different from the $(2s + 1)$-th row is also 0. Therefore PT_1 is of the form

$$PT_1 = \begin{bmatrix} P_1 & 0 & 0 \\ 0 & 0 & 1 \\ P_2 & 0 & 0 \end{bmatrix} \begin{matrix} 2s \\ 1 \\ m\text{-}2s\text{-}1 \end{matrix} .$$
$$ \begin{matrix} 2v & 1 & 1 \end{matrix}$$

Thus

$$\text{rank} \begin{bmatrix} P_1 \\ P_2 \end{bmatrix} = m - 1.$$

From $(PT_1)S_2(PT_1)^T = M(m, 2s + 1, s)$ we deduce also

$$\begin{bmatrix} P_1 \\ P_2 \end{bmatrix} K \begin{bmatrix} P_1 \\ P_2 \end{bmatrix}^T = M(m - 1, 2s, s).$$

That is,

$$\begin{bmatrix} P_1 \\ P_2 \end{bmatrix}$$

is a subspace of type $(m - 1, s)$ in the $2v$-dimensional symplectic space. Therefore $2s \leq m - 1 \leq v + s$ and hence $2s + 1 \leq m \leq v + s + 1$.

(d) $(\tau, \varepsilon) = (2, 0)$. Let P be a subspace of type $(m, 2s + 2, s, 0)$ in $V_{2v+2}(F_q)$. We can assume that $PS_2P^T = M(m, 2s + 2, s)$. Clearly $2s + 2 \leq m$. As in the above case there is a $T_1 \in Ps_{2v+2}(F_q)$ of the form (11), such that the $(2s + 2)$-th row of PT_1 is e_{2v+2}. Then we can assume that PT_1 is of the form

$$PT_1 = \begin{bmatrix} P_1 & 0 & 0 \\ v & 1 & 0 \\ 0 & 0 & 0 \\ P_2 & 0 & 0 \end{bmatrix} \begin{matrix} 2s \\ 1 \\ 1 \\ m\text{-}2s\text{-}2 \end{matrix} .$$
$$ \begin{matrix} 2v & 1 & 1 \end{matrix}$$

Clearly

$$\text{rank} \begin{bmatrix} P_1 \\ P_2 \end{bmatrix} = m - 2.$$

Since $e_{2v+1} \notin P$ and $e_{2v+1} = e_{2v+1} T_1 \notin PT_1$, we must have

$$\text{rank} \begin{bmatrix} P_1 \\ v \\ P_2 \end{bmatrix} = m - 1.$$

From $(PT_1)S_2(PT_1)^T = M(m, 2s + 2, s)$, we deduce

$$\begin{bmatrix} P_1 \\ v \\ P_2 \end{bmatrix} K \begin{bmatrix} P_1 \\ v \\ P_2 \end{bmatrix}^T = M(m - 1, 2s, s).$$

Thus

$$\begin{bmatrix} P_1 \\ v \\ P_2 \end{bmatrix}$$

is a subspace of type $(m - 1, s)$ in the $2v$-dimensional symplectic space. Therefore $2s \leq m - 1 \leq v + s$ and hence $2s + 1 \leq m \leq v + s + 1$. Combining this inequality with the previous one, $2s + 2 \leq m$, we obtain $2s + 2 \leq m \leq v + s + 1$.

(e) $(\tau, \varepsilon) = (2, 1)$. Let P be a subspace of type $(m, 2s + 2, s, 1)$ in $V_{2v+2}(F_q)$. We can assume that $PS_2P^T = M(m, 2s + 2, s)$. As in case (c), there is a $T_1 \in Ps_{2v+2}(F_q)$ of the form (11) such that the $(2s + 2)$-th row of PT_1 is e_{2v+2}. Since $e_{2v+1} \in P$, $e_{2v+1} = e_{2v+1}T_1 \in PT_1$. From $PS_2P^T = M(m, 2s + 2, s)$ we deduce $(PT_1)S_2(PT_1)^T = M(m, 2s + 2, s)$. Hence we can assume that the $(2s + 1)$-th row of PT_1 is e_{2v+1} and that PT_1 is of the form

$$PT_1 = \begin{bmatrix} P_1 & 0 & 0 \\ 0 & 1 & 0 \\ 0 & 0 & 1 \\ P_2 & 0 & 0 \end{bmatrix} \begin{matrix} 2s \\ 1 \\ 1 \\ m-2s-2 \end{matrix} .$$
$$ \begin{matrix} 2v & 1 & 1 \end{matrix}$$

Clearly,

$$\text{rank} \begin{bmatrix} P_1 \\ P_2 \end{bmatrix} = m - 2,$$

and from $(PT_1)S_2(PT_1)^T = M(m, 2s + 2, s)$ we deduce

$$\begin{bmatrix} P_1 \\ P_2 \end{bmatrix} K \begin{bmatrix} P_1 \\ P_2 \end{bmatrix}^T = M(m - 2, 2s, s).$$

Thus

$$\begin{bmatrix} P_1 \\ P_2 \end{bmatrix}$$

is a subspace of type $(m - 2, s)$ in the $2v$-dimensional symplectic space. Consequently, $2s \le m - 2 \le v + s$ and hence $2s + 2 \le m \le v + s + 2$.

Hence the proof of Theorem 7 for the case $\delta = 2$ is also completed. □

Denote the set of subspaces of type $(m, 2s + \tau, s, \varepsilon)$ in $V_{2v+\delta}(F_q)$ by $\mathcal{M}(m, 2s + \tau, s, \varepsilon; 2v + \delta)$, where $\delta = 1$ or 2, $\tau = 0, 1,$ or 2, and $\varepsilon = 0$ or 1. If either (9) or (10) fails to hold, then by Theorem 7 $\mathcal{M}(m, 2s + \tau, s, \varepsilon; 2v + \delta) = \phi$. However, if both (9) and (10) hold, then we have

Theorem 8. *Assume that both (9) and (10) hold. Then $\mathcal{M}(m, 2s + \tau, s, \varepsilon; 2v + \delta)$ is a transitive set of subspaces of $V_{2v+\delta}(F_q)$ under $Ps_{2v+\delta}(F_q)$.*

Proof. Consider first the case $\delta = 1$. The four possible cases of (τ, ε) can be treated in a similar way, so we discuss only the case $(\tau, \varepsilon) = (0, 0)$ as an example.

Let P be a subspace of type $(m, 2s, s, 0)$ in $V_{2v+1}(F_q)$. By the proof of Theorem 7, we can assume that P is of the form

$$P = \begin{bmatrix} P_1 & 0 \end{bmatrix},$$
$$\quad\ \ 2v \ \ 1$$

where P_1 is an $m \times 2v$ matrix of rank m and that $P_1 K P_1^T = M(m, 2s, s)$

Now let Q be another subspace of type $(m, 2s, s, 0)$ in $V_{2v+1}(F_q)$. We can also assume that Q has a matrix representation of the form

$$Q = \begin{bmatrix} Q_1 & 0 \end{bmatrix},$$
$$\quad\ \ 2v \ \ 1$$

where Q_1 is an $m \times 2v$ matrix of rank m such that $Q_1 K Q_1^T = M(m, 2s, s)$. By Theorem 15 (below) the transitivity theorem of the symplectic geometry, there is an element $T \in Sp_{2v}(F_q)$ such that $P_1 = Q_1 T$. Consequently $P = QT_1$, where

$$T_1 = \begin{bmatrix} T & \\ & 1 \end{bmatrix} \in Ps_{2v+1}(F_q).$$

Hence $\mathcal{M}(m, 2s, s, 0; 2v + 1)$ is a transitive set of subspaces under $Ps_{2v+1}(F_q)$.

Then let us consider the case $\delta = 2$.   We distinguish the five possible cases of (τ, ε) as we did in the proof of Theorem 7.

(a) $(\tau, \varepsilon) = (0, 0)$.   Let P be a subspace of type $(m, 2s, s, 0)$ in $V_{2v+2}(F_q)$. By the proof of Theorem 7 we can assume that P has a matrix representation of the form

$$P = \begin{bmatrix} P_1 & u^T & 0 \end{bmatrix},$$
$$\quad\; 2v \quad 1 \quad 1$$

where $P_1 K P_1{}^T = M(m, 2s, s)$.

Similarly, if Q is another subspace of type $(m, 2s, s, 0)$ in $V_{2v+2}(F_q)$, then Q has a matrix representation of the form

$$Q = \begin{bmatrix} Q_1 & v^T & 0 \end{bmatrix},$$
$$\quad\; 2v \quad 1 \quad 1$$

where $Q_1 K Q_1{}^T = M(m, 2s, s)$.   By the transitivity theorem of the symplectic geometry, there is an element $T \in Sp_{2v}(F_q)$ such that $P_1 = Q_1 T$. Since both u and v are m-dimensional row vectors and P_1 is of rank m, we can choose a 2v-dimensional row vector x such that $P_1 K x^T + v^T = u^T$.   Then

$$\begin{bmatrix} P_1 & u^T & 0 \end{bmatrix} = \begin{bmatrix} Q_1 & v^T & 0 \end{bmatrix} \begin{bmatrix} T & & \\ & 1 & \\ & & 1 \end{bmatrix} \begin{bmatrix} I & Kx^T & \\ & 1 & \\ x & 0 & 1 \end{bmatrix},$$

where

$$\begin{bmatrix} T & & \\ & 1 & \\ & & 1 \end{bmatrix} \begin{bmatrix} I & Kx^T & \\ & 1 & \\ x & 0 & 1 \end{bmatrix} \in Ps_{2v+2}(F_q).$$

This proves that $\mathcal{M}(m, 2s, s, 0; 2v + 2)$ is a transitive set.

(b) $(\tau, \varepsilon) = (0, 1)$.   This case can be treated in the same way as the case $\delta = 1$ and will be omitted.

(c) $(\tau, \varepsilon) = (1, 0)$. Let P be a subspace of type $(m, 2s + 1, s, 0)$ in $V_{2v+2}(F_q)$. By the proof of Theorem 7 we can assume that there is an element $T_1 \in Ps_{2v+2}(F_q)$ of the form (11) such that PT_1 is of the form

$$PT_1 = \begin{bmatrix} P_1 & 0 & 0 \\ 0 & 0 & 1 \\ P_2 & 0 & 0 \end{bmatrix} \begin{matrix} 2s \\ 1 \\ m\text{-}2s\text{-}1 \end{matrix}$$
$$\quad\quad\; 2v \;\; 1 \;\; 1$$

where

$$\begin{bmatrix} P_1 \\ P_2 \end{bmatrix} = m - 1,$$

and

$$\begin{bmatrix} P_1 \\ P_2 \end{bmatrix} K \begin{bmatrix} P_1 \\ P_2 \end{bmatrix}^T = M(m - 1, 2s, s).$$

Similarly, if $Q \in \mathcal{M}(m, 2s + 1, s, 0; 2v + 2)$, then we can assume that there is a $T_2 \in Ps_{2v+2}(F_q)$ of the form (11) such that

$$QT_2 = \begin{bmatrix} Q_1 & 0 & 0 \\ 0 & 0 & 1 \\ Q_2 & 0 & 0 \end{bmatrix} \begin{matrix} 2s \\ 1 \\ m\text{-}2s\text{-}1 \end{matrix}$$
$$\quad\quad\; 2v \;\; 1 \;\; 1$$

where

$$\text{rank} \begin{bmatrix} Q_1 \\ Q_2 \end{bmatrix} = m - 1$$

and

$$\begin{bmatrix} Q_1 \\ Q_2 \end{bmatrix} K \begin{bmatrix} Q_1 \\ Q_2 \end{bmatrix}^T = M(m - 1, 2s, s).$$

By the transitivity theorem of the symplectic geometry there is a $T \in Sp_{2v}(F_q)$ such that

$$\begin{bmatrix} P_1 \\ P_2 \end{bmatrix} = \begin{bmatrix} Q_1 \\ Q_2 \end{bmatrix} T.$$

Let

$$T_3 = \begin{bmatrix} T & \\ & I^{(2)} \end{bmatrix},$$

then $T_3 = Ps_{2v+2}(F_q)$ and $P = QT_2T_3T_1^{-1}$.

(d) $(\tau, \varepsilon) = (2, 0)$.

(e) $(\tau, \varepsilon) = (2, 1)$.

These two cases can be treated in the same way as case (c) and the details will be omitted.

Theorem 8 is now completely proved.   □

Let $N(m, 2s + \tau, s, \varepsilon; 2v+ \delta) = |\, \mathcal{M}(m, 2s + \tau, s, \varepsilon; 2v+ \delta)|$.   Our next goal is to compute $N(m, 2s + \tau, s, \varepsilon; 2v+ \delta)$.   It follows from Theorem 7 that $N(m, 2s + \tau, s, \varepsilon; 2v+ \delta) = 0$ if either (9) or (10) is not satisfied. However, if both (9) and (10) are satisfied, then we have the following Anzahl theorem in the $(2v + \delta)$-dimensional pseudo-symplectic geometry over F_q.

Theorem 9. *Assume that both (9) and (10) are satisfied. Then $N(m, 2s + \tau, s, \varepsilon; 2v+ \delta) =$*

$$q^{n_0+2(s+(2-\delta)[\tau/2])(v+s+[(\tau+\delta-1)/2]+\varepsilon-m)} \times$$

$$\frac{\displaystyle\prod_{i=v+s+[(\tau+\delta-1)/2]+\varepsilon-m+1}^{v} (q^{2i} - 1)}{\displaystyle\prod_{i=1}^{s+(2-\delta)[\tau/2]} (q^{2i} - 1) \quad \prod_{i=1}^{m-2s-(3-\delta)[(\tau+\delta-1)/2])-\varepsilon} (q^{i} - 1)} \tag{10}$$

where $n_0 = 0$ if $\delta = 1$, and $n_0 = m, 0, 2(v + s + 1) - m, 2(v + s + 1) - m$, or $2(v + 1) - m$ corresponding to the cases $(\tau, \varepsilon) = (0, 0), (0, 1), (1, 0), (2, 0)$, or $(2, 1)$, respectively, if $\delta = 2$.

Proof. Consider first the case $\delta = 1$.   By the proof of Theorem 7, there is a one-to-one correspondence between $\mathcal{M} (m, 2s, s, 0; 2v + 1)$ in the $(2v + 1)$-dimensional pseudo-symplectic space and $\mathcal{M}(m, s; 2v)$ in the $2v$-dimensional symplectic space over F_q.   Hence $N(m, 2s , s, 0; 2v+ 1) = N(m, s; 2v) =$

$$q^{2s(v+s-m)} \; \frac{\displaystyle\prod_{i=v+s-m+1}^{v} (q^{2i}-1)}{\displaystyle\prod_{i=1}^{s}(q^{2i}-1) \; \prod_{i=1}^{m-2s}(q^i-1)} \;, \quad \text{for } 2s \leq m \leq v+s.$$

Similarly $N(m, 2s+1, s, 0; 2v+1) = N(m, s; 2v) =$

$$q^{2s(v+s-m)} \; \frac{\displaystyle\prod_{i=v+s-m+1}^{v} (q^{2i}-1)}{\displaystyle\prod_{i=1}^{s}(q^{2i}-1) \; \prod_{i=1}^{m-2s}(q^i-1)} \;, \quad \text{for } 2s+1 \leq m \leq v+s,$$

$N(m, 2s+1, s, 1; 2v+1) = N(m-1, s; 2v) =$

$$q^{2s(v+s-m+1)} \; \frac{\displaystyle\prod_{i=v+s-m+2}^{v} (q^{2i}-1)}{\displaystyle\prod_{i=1}^{s}(q^{2i}-1) \; \prod_{i=1}^{m-2s-1}(q^i-1)} \;, \quad \text{for } 2s+1 \leq m \leq v+s+1,$$

and $N(m, 2s+2, s, 0; 2v+1) = N(m, s+1; 2v) =$

$$q^{2(s+1)(v+s-m+1)} \; \frac{\displaystyle\prod_{i=v+s-m+2}^{v} (q^{2i}-1)}{\displaystyle\prod_{i=1}^{s+1}(q^{2i}-1) \; \prod_{i=1}^{m-2s-2}(q^i-1)} \;, \quad \text{for } 2s+2 \leq m \leq v+s+1.$$

These assemble to give Theorem 9 for the case $\delta = 1$.   ☐

Now let us consider the case $\delta = 2$. We compute $N(m, 2s + \tau, s, \varepsilon; 2v+2)$ by considering the five possible cases of (τ, ε) one by one.

(a) $(\tau, \varepsilon) = (0, 0)$. Let $P \in M(m, 2s, s, 0; 2v+2)$. By the proof of Theorem 7 we can assume that

$$P = \begin{bmatrix} P_1 & u^T & 0 \end{bmatrix}$$
$$\phantom{P = \begin{bmatrix}}\; 2v \quad\; 1 \quad\; 1$$

where P_1 is a $m \times 2v$ matrix of rank m such that $P_1 K P_1^T = M(m, 2s, s)$ and u is an m-dimensional row vector. Define a map

$$\mathcal{M}(m, 2s, s, 0; 2v+2) \rightarrow \mathcal{M}(m, s; 2v)$$
$$P = |P_1 \ u^T \ 0| \mapsto P_1.$$

It is easy to verify that it is well-defined, surjective and the inverse image of any element of $\mathcal{M}(m, s; 2v)$ is of cardinality q^m. Therefore $N(m, 2s, s, 0; 2v+2) = q^m N(m, s; 2v)$.

(b) $(\tau, \varepsilon) = (0, 1)$. By the proof of Theorem 7 we know that there is a one-to-one correspondence between $\mathcal{M}(m, 2s, s, 1; 2v+2)$ and $\mathcal{M}(m-1, s; 2v)$. Therefore $N(m, 2s, s, 1; 2v+2) = N(m-1, s; 2v)$.

(c) $(\tau, \varepsilon) = (1, 0)$. By the proof of Theorem 7, for any $P \in \mathcal{M}(m, 2s+1, s, 0; 2v+2)$ there is an element $T_1 \in Ps_{2v+2}(F_q)$ of the form (11) such that PT_1 has a matrix representation of the form

$$PT_1 = \begin{bmatrix} P_1 & 0 & 0 \\ 0 & 0 & 1 \\ P_2 & 0 & 0 \end{bmatrix} \begin{matrix} 2s \\ 1 \\ m-2s-1 \end{matrix} ,$$
$$\qquad\quad 2v \ \ 1 \ \ 1$$

where

$$\begin{bmatrix} P_1 \\ P_2 \end{bmatrix} \tag{13}$$

is a subspace of type $(m-1, s)$ in the $2v$-dimensional symplectic space, and more precisely,

$$\begin{bmatrix} P_1 \\ P_2 \end{bmatrix} K \begin{bmatrix} P_1 \\ P_2 \end{bmatrix}^T = M(m-1, 2s, s). \tag{14}$$

Conversely, given any subspace (13) of type $(m-1, s)$ in the $2v$-dimensional symplectic space such that (14) holds and any element $T_1 \in Ps_{2v+2}(F_q)$ of the form (11),

$$\begin{bmatrix} P_1 & 0 & 0 \\ 0 & 0 & 1 \\ P_2 & 0 & 0 \end{bmatrix} T_1 \tag{15}$$

is a subspace of type $(m, 2s+1, s, 0)$ in $V_{2v+2}(F_q)$ such that

$$\left[\begin{bmatrix} P_1 & 0 & 0 \\ 0 & 0 & 1 \\ P_2 & 0 & 0 \end{bmatrix} T_1 \right] S_2 \left[\begin{bmatrix} P_1 & 0 & 0 \\ 0 & 0 & 1 \\ P_2 & 0 & 0 \end{bmatrix} T_1 \right]^T = M(m, 2s + 1, s).$$

We say that (15) arises from (13) through an element $T_1 \in Ps_{2v+2}(F_q)$ of the form (11). Clearly, subspaces of type $(m, 2s + 1, s, 0)$ in $V_{2v+2}(F_q)$ arising from different subspaces of type $(m - 1, s)$ in the 2v-dimensional symplectic space through elements of the form (11) are different. We want to compute the number of subspaces of type $(m, 2s + 1, s, 0)$ in $V_{2v+2}(F_q)$ arising from a given subspace of type $(m - 1, s)$ in the 2v-dimensional symplectic space through different elements of $Ps_{2v+2}(F_q)$ of the form (11). It is enough to compute the number of elements of $Ps_{2v+2}(F_q)$ of the form (11) through which, from a given subspace of type $(m - 1, s)$ in the 2v-dimensional symplectic space, the same subspace of type $(m, 2s + 1, s, 0)$ in $V_{2v+2}(F_q)$ arises; this number is equal to the number of elements of $Ps_{2v+2}(F_q)$ of the form of (11), which leave a given subspace of the type $(m, 2s + 1, s, 0)$ of the form

$$\begin{bmatrix} P_1 & 0 & 0 \\ 0 & 0 & 1 \\ P_2 & 0 & 0 \end{bmatrix} \tag{16}$$

invariant.

Suppose that

$$T_1 = \begin{bmatrix} I^{(2v)} & Kx^T & \\ & 1 & \\ x & y & 1 \end{bmatrix} \in Ps_{2v+2}(Fq)$$

leaves (16) invariant. Then there is an $m \times m$ non-singular matrix A such that

$$A \begin{bmatrix} P_1 & 0 & 0 \\ 0 & 0 & 1 \\ P_2 & 0 & 0 \end{bmatrix} = \begin{bmatrix} P_1 & 0 & 0 \\ 0 & 0 & 1 \\ P_2 & 0 & 0 \end{bmatrix} T_1;$$

that is

$$A \begin{bmatrix} P_1 & 0 & 0 \\ 0 & 0 & 1 \\ P_2 & 0 & 0 \end{bmatrix} = \begin{bmatrix} P_1 & P_1 Kx^T & 0 \\ x & y & 1 \\ P_2 & P_2 Kx^T & 0 \end{bmatrix}. \tag{17}$$

It is easy to deduce that A is of the form

$$A = \begin{bmatrix} I^{(2s)} & & & \\ a & 1 & b & \\ & & & I^{(m-2s-1)} \end{bmatrix},$$

where a is a $2s$-dimensional row vector and b is an $(m - 2s - 1)$-dimensional row vector. From (17) we deduce also

$$(a\ b) \begin{bmatrix} P_1 \\ P_2 \end{bmatrix} = x,$$

$$\begin{bmatrix} P_1 \\ P_2 \end{bmatrix} Kx^T = 0,$$

and

$$y = 0.$$

Thus

$$0 = \begin{bmatrix} P_1 \\ P_2 \end{bmatrix} Kx^T = \begin{bmatrix} P_1 \\ P_2 \end{bmatrix} K \begin{bmatrix} P_1 \\ P_2 \end{bmatrix}^T \begin{bmatrix} a^T \\ b^T \end{bmatrix} = M(m - 1, 2s, s) \begin{bmatrix} a^T \\ b^T \end{bmatrix}.$$

Therefore $a = 0$ and b is arbitrary. Hence there are q^{m-2s-1} elements of $Ps_{2v+2}(F_q)$ of the form of (11) which leaves a given subspace (16) invariant. Clearly there are q^{2v+1} elements of the form (11) in $Ps_{2v+2}(F_q)$. Consequently

$$N(m, 2s + 1, s, 0; 2v + 2) = q^{2v+1-(m-2s-1)}N(m - 1, s; 2v)$$
$$= q^{2(v+s+1)-m}N(m - 1, s; 2v).$$

(d) $(\tau, \varepsilon) = (2, 0)$. This case can be treated in the same way as case (c) above, and we have
$$N(m, 2s + 2, s, 0; 2v + 2) = q^{2(v+s+1)-m}N(m - 1, s; 2v).$$

(e) $(\tau, \varepsilon) = (2, 1)$. This case can also be treated in the same way as case (c) above and we have
$$N(m, 2s + 2, s, 1; 2v + 2) = q^{2(v+1)-m}N(m - 2, s; 2v).$$

The details of cases (d) and (e) are omitted.   □

Corollary 10   (Pless 1965). *Let* $m \le v + \delta - 1$. *Then the number of totally isotropic subspaces of dimension* m *in the* $(2v + \delta)$-*dimensional pseudo-symplectic space is given by*

$$N(m, 0, 0, 0; 2v + 1) = \frac{\prod\limits_{i=0}^{m-1} (q^{2(v-i)} - 1)}{\prod\limits_{i=1}^{m} (q^i - 1)} \qquad \text{when } \delta = 1,$$

and

$$N(m, 0, 0, 0; 2v + 2) + N(m, 0, 0, 1; 2v + 2)$$

$$= \frac{(q^{2v+2-m} - 1) \prod\limits_{i=1}^{m-1} (q^{2v+2-2i} - 1)}{\prod\limits_{i=1}^{m} (q^i - 1)} , \text{when } \delta = 2. \quad \square$$

Remark.   The second statement of Theorem 8 can also be deduced from the Witt's theorem for non-alternating symmetric bilinear forms over a field of characteristic 2 (Pless 1964); however, the present proof also leads directly to a proof of Theorem 9.

4. Extended pseudo-symplectic groups over F_q

Let

$$S_{\delta,l} = \begin{bmatrix} S_\delta & \\ & O^{(l)} \end{bmatrix},$$

where $\delta = 1$ or 2 and S_δ is the $(2v + \delta) \times (2v + \delta)$ non-singular non-alternating symmetric matrix (4) or (5). The set of all $(2v + \delta + l) \times (2v + \delta + l)$ non-singular matrices T satisfying $TS_{\delta,l}T^T = S_{\delta,l}$ forms a group with respect to matrix multiplication, called the *extended pseudo-symplectic group* over F_q and denoted by $Ps_{2v+\delta+l, \, 2v+\delta}(F_q)$. For $l = 0$, it reduces to the ordinary pseudo-symplectic group $Ps_{2v+\delta}(F_q)$. It can be readily be verified that $Ps_{2v+\delta+l, \, 2v+\delta}(F_q)$ consists of all $(2v + \delta + l) \times (2v + \delta + l)$ non-singular matrices of the form

$$T = \begin{bmatrix} T_{11} & T_{12} \\ 0 & T_{22} \end{bmatrix} \begin{matrix} 2v+\delta, \\ l \end{matrix}$$
$$\begin{matrix} 2v+\delta & l \end{matrix}$$

where $T_{11}S_\delta T_{11}{}^T = S_\delta$ and T_{22} is non-singular.   We have

Theorem 11.

$$|Ps_{2v+\delta+l,\,2v+\delta}(F_q)| = q^{(v+\delta-1)^2+(2v+\delta)l+(l(l-1))/2} \prod_{i=1}^{v} (q^{2i} - 1) \prod_{i=1}^{l} (q^i - 1)$$

Proof. In fact, $|Ps_{2v+\delta+l,\,2v+\delta}(F_q)| = |Ps_{2v+\delta}(F_q)| \, |GL_l(F_q)| \, q^{(2v+\delta)l}$.   It

follows from Theorems 3 and 5 that $|Ps_{2v+\delta}(F_q)| = q^{(v+\delta-1)^2} \prod_{i=1}^{v} (q^{2i} - 1)$

and it is well known that $GL_l(F_q) = q^{(l(l-1))/2} \prod_{i=1}^{l} (q^i - 1)$. $\square$

As an application of Theorem 11, let us compute the number of $n \times n$ non-alternating symmetric matrices of a given rank $r(0 \leq r \leq n)$ over F_q.

Theorem   12. *The number of $n \times n$ non-alternating symmetric matrices of rank $r = 2v + \delta$, where $0 \leq r \leq n$ and $\delta = 1$ or 2, is given by*

$$q^{v(v+1)} \frac{\displaystyle\prod_{i=n-2v-\delta+1}^{n} (q^i - 1)}{\displaystyle\prod_{i=1}^{v} (q^i - 1)}.$$

Proof.   Define an action of $GL_n(F_q)$ on the set $S(n)$ of an $n \times n$ non-alternating symmetric matrices as follows

$$S(n) \times GL_n(F_q) \;\to\; S(n)$$
$$(S, T) \;\mapsto\; TST^T.$$

By Theorem 1, those matrices of $S(n)$ of the same rank r form a transitive set under $GL_n(F_q)$. If r is odd, write $r = 2v + 1$, and if r is even, write $r = 2v + 2$. Then we have $r = 2v + \delta$, where $\delta = 1$ or 2. Let

$$S_{\delta,n-r} = \begin{bmatrix} S_\delta & \\ & O^{(n-r)} \end{bmatrix},$$

where $S_\delta(\delta = 1$ or $2)$ is given by (4) or (5), respectively. Then $S_{\delta,n-r}$ is a non-alternating symmetric matrix of rank $2v + \delta$ and the stabilizer of $S_{\delta,n-r}$ in $GL_n(F_q)$ is $Ps_{2v+\delta+(n-2v-\delta),\,2v+\delta}(F_q)$.   Therefore the number of $n \times n$

non-alternate symmetric matrices of rank $r = 2v + \delta$ is given by the index $GL_n(F_q) : Ps_{2v+\delta+(n-2v-\delta),\ 2v+\delta}(F_q)$.

Hence our Theorem follows from Theorem 11. $\square$

Corollary 13. *The number of* $(2v + \delta) \times (2v + \delta)$ *non-singular non-alternating symmetric matrices, where* $\delta = 1$ *or* 2, *is given by*

$$q^{v(v+1)} \ \frac{\prod\limits_{i=1}^{2v+\delta} (q^i - 1)}{\prod\limits_{i=1}^{v} (q^{2i} - 1)}. \quad \square$$

Remark. The number of $n \times n$ alternating matrices of rank $2v$, where $0 \leq 2v \leq n$, was previously computed by D. Stanton and Y. Wang independently; see Stanton (1980) and Wang (1980).

5. Appendix - A brief summary of the symplectic geometries over F_q

For the reader's convenience, we summarize the main results of symplectic geometries quoted in this paper as follows.   Now, the restriction that q is a power of 2 can be removed.

The *symplectic group* $Sp_{2v}(F_q)$ of degree $2v$ over F_q is defined as $Sp_{2v}(F_q) = \{T\ (2v \times 2v)\ \text{matrix}\ |\ TKT^T = K\}$, where K is given by (3).   A subspace P of $V_{2v}(F_q)$ is said to be of *type* (m, s), if $\dim P = m$ and rank $PKP^T = 2s$.   Denote the set of subspaces of type (m, s) in $V_{2v}(F_q)$ by $M(m, s; 2v)$ and let $N(m, s; 2v) = |M(m, s; 2v)|$.

Theorem 14. *Subspaces of type* (m, s) *exist in* $V_{2v}(F_q)$ *if and only if* $2s \leq m \leq v + s$. $\square$

Theorem 15 (Dieudonné 1948).   *Let* $2s \leq m \leq v + s$. *Then* $M(m, s; 2v)$ *forms a transitive set of subspaces under* $Sp_{2v}(F_q)$. $\square$

Theorem 16. *Let* $2s \leq m \leq v + s$. *Then*

$$N(m, s; 2v) = q^{2s(v+s-m)} \ \frac{\prod\limits_{i=v+s-m+1}^{v} (q^{2i} - 1)}{\prod\limits_{i=1}^{s} (q^{2i} - 1) \ \prod\limits_{i=1}^{m-2s} (q^i - 1)}. \quad \square$$

References

J. Dieudonné (1948), *Sur les Groupes Classiques*, Hermann, Paris.

J.W.P. Hirschfeld (1979), *Projective Geometries over Finite Fields*, Oxford University Press, Oxford.

V. Pless (1964), On Witt's theorem for non-alternating symmetric bilinear forms over a field of characteristic 2, *Proc. Amer. Math. Soc.* **15**, 979-983.

V. Pless (1965), The number of isotropic subspaces in a finite geometry, *Atti Accad. Naz. Lincei Rend.* **39**, 418-421.

D. Stanton (1980), Some q-Krawtchouk polynomials of Chevalley groups, *Amer. J. Math.* **102**, 625-662.

Z. Wan (1964), Notes on finite geometries and the construction of *PBIB* designs. I Some Anzahl theorems in symplectic geometries over finite fields. *Sci. Sinica* **13**, 515-516.

Z. Wan (1991), A note on the invariants of a subspace of the symplectic space, submitted.

Y. Wang (1980), Association schemes with several associate classes, *Acta Math. Appl. Sinica* **3**, 97-105.

Solutions of quadratic equations in a groupring and partial addition sets

S. Löwe
University of Braunschweig

1. Introduction

Let Γ be a strongly regular graph with parameters (v, k, λ, μ) (see Hubaut (1975) for the notation). In this article we are interested in the question whether Γ admits a sharply 1-transitive group G of automorphisms. If so, then Γ can be described (see (3) below) by the set

$$\Delta = \{g \in G \mid \gamma_0{}^g \, I \, \gamma_0\}; \qquad (1)$$

where γ_0 is a previously chosen vertex. Δ is called a *partial addition set* in G with parameters (v, k, λ, μ). If for $S \subseteq G$ we denote the element $\Sigma_{s \in S} s$ of ZG by S^+, where ZG is the integral groupring over G, then Δ fulfils

$$(\Delta^+)^2 = (k - \mu)1 + (\lambda - \mu)\Delta^+ + \mu G^+. \qquad (2)$$

Otherwise , if $\Omega \subseteq G$ is such that Ω^+ fulfils (2), and Ω contains the inverse of each of its elements, then the definition

$$\forall x, y \in G \quad x \, I \, y : \Leftrightarrow xy^{-1} \in \Omega \qquad (3)$$

turns G into a strongly regular graph with parameters (v, k, λ, μ), where v is the cardinality of G and k is the cardinality Ω.

In section 1, we characterise the solutions of the quadratic equation (2) in the group algebra FQ over a field F, both for $\mathrm{char} F \mid v$ and $\mathrm{char} F \nmid v$. The next two paragraphs are devoted to the two separate cases. In section 2, a theorem of Osima is applied to the case $\mathrm{char} F \mid v$ to generalise a theorem of Ma (1984). The case $\mathrm{char} F \nmid v$ is dealt with in section 3. Here ordinary character theory gives a new divisibility condition for the parameters (v, k, λ, μ).

2. Solutions of quadratic equations in a groupring

Let G be a group with v elements, F be a field of $\text{char} F \neq 2$, to be specified when necessary. For $\alpha, \beta, \gamma \in F$, we are looking for solutions of the equation

$$X^2 = \alpha 1 + \beta X + \gamma G^+ \tag{4}$$

in FG.

2.1 *The weight of a solution*

If $X = \sum_{g \in G} x_g g$ is a solution of (4), and let $k_X \in F$ be the *weight* of X. Then k_X is the sum of all coefficients x_g and is characterised by $X \cdot G^+ = k_X G^+$.

Lemma 1. *If X is a solution of (4), then*

$$k_X^2 = \alpha + \beta k_X + \gamma v. \tag{5}$$

Proof. Multiply (4) by G^+:

$$k_X^2 G^+ = \alpha G^+ + \beta k_X G^+ + \gamma v G^+.$$

The assertion follows.   □

In what follows we let $\delta := 4\alpha + \beta^2$. If our quadratic equation is the special one (2) of a partial addition set, then δ is called the *discriminant* of the partial addition set or of the strongly regular graph producing it. In fact, the discriminant of most strongly regular graphs is a square; if not, then $\delta = v$. For more information, the reader is referred to Hubaut (1975).

2.2 *The case* $\text{char} F \nmid v$

The solutions of (4) can be characterised by the idempotents of FG. We shall give the result in full generality, although we shall only need a part of it here. We suppose that both v and $\sqrt{\delta}$ are invertible elements of F.

Lemma 2. *Let X be a solution of (4). Then*

$$E_a(X) := \frac{\sqrt{\delta} - \beta}{2\sqrt{\delta}} 1 + \frac{1}{\sqrt{\delta}} X + \frac{\beta - 2k - \sqrt{\delta}}{2v\sqrt{\delta}} G^+ , \tag{6}$$

$$E_b(X) := \frac{\sqrt{\delta} - \beta}{2\sqrt{\delta}} 1 + \frac{1}{\sqrt{\delta}} X + \frac{\beta - 2k + \sqrt{\delta}}{2v\sqrt{\delta}} G^+ , \qquad (7)$$

is an idempotent in FG.   Furthermore

$$E_a(X)G^+ = 0$$
$$E_b(X)G^+ = G^+.$$

Proof. Check by direct calculation using (5). □
The converse also holds.

Lemma 3. (1) Let E be an idempotent in FG such that $EG^+ = 0$.   Then

$$X_a(E) := \frac{\beta - \sqrt{\delta}}{2} 1 + \sqrt{\delta}E + \frac{2k - \beta + \sqrt{\delta}}{2v} G^+$$

is a solution of (4).

(2) Let E be an idempotent in FG such that $EG^+ = G^+$.   Then

$$X_b(E) := \frac{\beta - \sqrt{\delta}}{2} 1 + \sqrt{\delta}E + \frac{2k - \beta - \sqrt{\delta}}{2v} G^+$$

is a solution of (4).

Proof. By direct calculation. □

2.3 *The case* charF $| v$
Let F be a field of odd characteristic p. We continue to assume that $\sqrt{\delta}$ is not 0 in F.   Then the solutions of (4) can be characterised similarly to the preceding section by the idempotents of FG.   First, however, we have to check that $\beta - 2k_X$ to appear in a denominator is not zero.

Lemma 4. Let X be a solution of (4) in FG.   Then $\beta - 2k_X \neq 0$.

Proof.   Assume that $\beta = 2k_X$.   Let $k := k_X$. By (5), $k^2 = \alpha + \beta k + \gamma v = \alpha + \beta k$, since $v = 0$ in F.   Since by assumption $\beta = 2k$, we have $k^2 = \alpha + 2k^2$.   Therefore

$$\alpha + k^2 \quad = 0$$
$$2\alpha + \beta k \quad = 0, \text{ as } k^2 = \alpha + \beta k$$
$$4\alpha + 2\beta k \quad = 0$$

$$4\alpha + \beta^2 \quad = 0, \text{ as } \beta = 2k.$$

This contradicts the assumption $\delta \neq 0$.   □

Lemma 5.   *Let X be a solution of (4). Then*

$$E := \frac{\sqrt{\delta} - \beta}{2\sqrt{\delta}} 1 + \frac{1}{\sqrt{\delta}} X + \frac{\gamma}{\sqrt{\delta}(\beta - 2k)} G^+$$

is an idempotent of FG.
Proof.   By (4), the denominator of the coefficients of G^+ makes sense.
The rest is, again, proved by direct checking.   □

3. A Theorem of Osima

In this and the following paragraph we want to apply the simple
lemmas of section 1 to the group G and a partial set Δ of G with
parameters (v, k, λ, μ). We know that

$$(\Delta^+)^2 = (k - \mu)1 + (\lambda - \mu)\Delta^+ + \mu G^+$$

in $\mathbf{Z}G$. So, if $\alpha := k - \mu$, $\beta := \lambda - \mu$, $\gamma := \mu$, then Δ^+ is a solution of the
quadratic equation (2):

$$X^2 = \alpha 1 + \beta X + \gamma G^+$$

in FG, where F is any field of characteristic 0.   Reduction of the integer
coefficients modulo a prime p shows that we can consider Δ^+ to be a
solution of (2) for a field of arbitrary characteristic.   Since we want to
apply lemma 5 we assume that F is not of characteristic 2.   We shall
require for the rest of the paper that $\Theta := G\backslash(\Delta \cup \{1\})$ is not empty.

Theorem 6.   *Suppose there is an odd prime p such that $p \mid v$, $p \nmid \delta$. If
Δ is a union of conjugacy classes, then either Δ or Θ is contained in
the set $G_{p'}$ of p'-elements of G.*

The proof is an application of Lemma 5 and the following

Theorem 7 (Osima 1955).   *Let F be a field of characteristic p, G a group
of order divisible by p.   Suppose that $e = \sum_{g \in G} e_g g$ is a central*

idempotent in FG. If $e \neq 0$, then the support $\{g \in G; e_g \neq 0\}$ of e is contained in the set of p'-elements of G.

The proof can be found in Karpilovsky (1987). □

Proof of Theorem 6.   Let $F := Z/pZ$ and reduce the parameters (v, k, λ, μ) of Δ modulo p.   Let E be the idempotent of FG associated with Δ^+ via Lemma 5.

First claim: $\delta \in Z$ is a square.
For otherwise $\delta = v$, which contradicts $p \mid v$ but $p \nmid \delta$.

Second claim: $E \neq 0$.
Let $E = \Sigma_{g \in G} e_g g$.   Take $x \in \Delta, y \in \Theta$.   From Lemma 5 we have that if $E = 0$:

$$e_1 = \frac{\sqrt{\delta} - \beta}{2\sqrt{\delta}} + \frac{\gamma}{\sqrt{\delta}(\beta - 2k)} = 0$$

$$e_x = \frac{1}{\sqrt{\delta}} + \frac{\gamma}{\sqrt{\delta}(\beta - 2k)} = 0$$

$$e_y = \frac{\gamma}{\sqrt{\delta}(\beta - 2k)} = 0 .$$

This implies $1/\sqrt{\delta} = 0$, a contradiction.

End of proof:
We keep the notation of the second claim.   Suppose we can choose x such that $x \notin G_{p'}$.   Then, by Theorem 7, $e_x = 0$.   But e_x is independent of x (as long as $x \in \Delta$).   Hence Δ is not contained in the support of e. Since $1/\sqrt{\delta} = e_x - e_y$, we have $e_y \neq 0$.   Since y was arbitrary in Θ, it follows that Θ is contained in the support of e which, by Theorem 7, is contained in $G_{p'}$. □

Corollary 8. *If $G_{p'}$ is a subgroup of G then either $\Delta \cup \{1\}$ or $\Theta \cup \{1\}$ is a normal subgroup of G.*

Proof.   By the assumption of the theorem, both Δ and Θ are normal subsets of G.   From the definition it is clear that both are invariant under taking the inverse.   Suppose Δ is contained in $G_{p'}$.   Since $G_{p'}$ is properly contained in G, it follows from (2) that $\mu = 0$. Thus $\Delta \cup \{1\}$ is multiplicatively closed. If $\Delta \not\subseteq G_{p'}$, then by the theorem, the only other possibility is $\Theta \subseteq G_{p'}$. We have

$$(\Theta^+)^2 = (k - \lambda - 1)1 + (\mu - \lambda - 2)\Theta^+ + (v - 2k + \lambda)G^+. \tag{8}$$

The same reasoning as above shows that $\Theta \cup \{1\}$ is a subgroup of G. $\square$

Corollary 9. *If G is abelian, then either $\Delta \cup \{1\}$ or $\Theta \cup \{1\}$ is a subgroup of G.*

Corollary 9 was proved by Ma (1984). His proof includes the case $p = 2$.

It is not easy to guess how Theorem 6 would generalise if we dropped the condition that Δ is a union of conjugacy classes.

4. A parameter condition

In this section we prove the following result.

Theorem 10. *Let Δ be a partial addition set in G, with parameters (v, k, λ, μ). Suppose that $\delta = 4(k - \mu) + (\lambda - \mu)^2$ is a square. Let $C \neq \{1\}$ be a conjugacy class in G, $|C| =: c$, $|C \cap \Delta| =: a$. Then*

$$2\sqrt{\delta} \mid (2va/c + v(\beta - \sqrt{\delta})).$$

4.1 *Definitions and notation*
Let F be the v-th cyclotomic field. This is a splitting field for G, which means that we have both orthogonality relations for irreducible characters of FG at our disposal. These can be found in Curtis and Reiner (1962) or Serre (1977).

The left-right regular representation of $FG \otimes FG$ on FG is denoted by $\mathcal{LR}$. The underlying action of $G \times G$ on G is $\mathcal{LR}(g, h)(x) = gxh^{-1}$. It equals $\mathcal{L} \otimes \mathcal{R}$, where $\mathcal{R}$ is the right regular representation of FG and $\mathcal{L}$ is the left regular representation of FG. We have $\mathcal{R}(h)(x) = xh^{-1}$ and $\mathcal{L}(g)(x) = gx$.

For $X = \Sigma x_g g \in FG$, we define $X^{[-1]} := \Sigma x_g g^{-1}$.

As in the statement of Theorem 10, Δ is a partial addition set in G with parameters (v, k, λ, μ). We let $e := E_a(\Delta^+)$, see (6). The character of the FG - module $FG\,e$ is denoted by ϕ.

4.2 *The character of FGe*
In the following two lemmata, *e* could be any idempotent of *FG*.

Lemma 11. *If e is an idempotent of FG, and* ϕ *is the character of FGe,* *then*

$$\forall g \in G: \quad \phi(g) = tr(\mathcal{LR}(g, e^{[-1]})).$$

Proof. We have $\mathcal{LR}(g, e^{[-1]}) = \mathcal{L}(g) \circ \mathcal{R}(e^{[-1]})$. Select a linear basis in *FGe* and *FG*(1 - *e*), and combine them to a basis of *FG*. This is possible since $FG = FGe \oplus FG(1 - e)$ as vectorspaces. With respect to this basis, $\mathcal{R}(e^{[-1]})$ is associated with the matrix

$$r := \begin{pmatrix} 1 & 0 \\ 0 & 0 \end{pmatrix}$$

since $\mathcal{R}(e^{[-1]})$ operates as the identity on *FG e* and annihilates *FG*(1 - *e*). Both *FGe* and *FG*(1 - *e*) remain under $\mathcal{L}(g)$, $g \in G$, invariant. So $\mathcal{L}(g)$ is associated with the matrix

$$l := \begin{pmatrix} l_1 & 0 \\ 0 & l_2 \end{pmatrix}$$

Now, $\phi(g) = tr(l_1) = tr \, (1 \cdot r) = tr \, (\mathcal{LR}(g, e^{[-1]}))$. □

The next lemma gives an expression of the character of $\mathcal{LR}$ in terms of the irreducible characters of *FG*.

Lemma 12. *Let* $\chi_1, \ldots, \chi_s$ *be the irreducible characters of FG.* *Then*

$$tr(\mathcal{LR}(X, Y)) = \sum_{i=1}^{s} \chi_i(X) \, \chi_i(Y^{[-1]})$$

for all $X, Y \in FG$.

Proof. Let $e_1, \ldots, e_s$ be a decomposition of $1 \in FG$ into central primitive idempotents, where e_i carries the character χ_i. The irreducible components of the left-right regular representation of $G \times G$ on *FG* are the blocks e_i, with character $\chi_i \otimes \chi_i$. Here χ_i denotes the contragradient character of χ_i. Since $\chi_i(Y^{[-1]}) = \chi_i(Y)$, the assertion follows. □

In some cases, the sum in Lemma 12 can be easily evaluated.

Lemma 13. *Let* $X, Y \subseteq G$, *where* X *is a union of conjugacy classes.* *Then*

$$|G| \cdot |X \cap Y| = \sum_{i=1}^{s} \chi_i(X^+) \chi_i((Y^+)^{[-1]}).$$

Proof. See Löwe (1987).

4.3 Proof *of Theorem* 10.

Let $g \in C$. In the following calculations we permit $C = \{1\}$. First we apply Lemma 11 and obtain

$$\phi(g) = tr(\mathcal{LR}(g, e^{[-1]})) = tr(\mathcal{LR}(g, \frac{\sqrt{\delta} - \beta}{2\sqrt{\delta}} 1 + \frac{1}{\sqrt{\delta}} \Delta^+ + \frac{\beta - 2k - \sqrt{\delta}}{2v\sqrt{\delta}} G^+))$$

$$= \frac{\sqrt{\delta} - \beta}{2\sqrt{\delta}} tr(\mathcal{LR}(g, 1)) + \frac{1}{\sqrt{\delta}} tr(\mathcal{LR}(g, \Delta^+)) + \frac{\beta - 2k - \sqrt{\delta}}{2v/\sqrt{\delta}} tr(\mathcal{LR}(g, G^+))$$

The first and third summand are easy to deal with:

$$tr(\mathcal{LR}(g, 1)) \quad = \begin{cases} v, & g = 1 \\ 0, & \text{otherwise} \end{cases} \tag{9}$$

$$tr(\mathcal{LR}(g, G^+)) = v. \tag{10}$$

The middle summand yields:

$$\left. \begin{aligned} tr(\mathcal{LR}(g, \Delta^+)) &= \frac{1}{c} \sum_{\chi \in IrrG} \chi(C)\chi((\Delta^+)^{[-1]}) \text{ by Lemma 12} \\ &= \frac{v}{c} a \text{ by Lemma 13.} \end{aligned} \right\} \tag{11}$$

Adding (9), (10) and (11), we obtain

$$\phi(g) = \begin{cases} \dfrac{\sqrt{\delta} - \beta}{2\sqrt{\delta}} v + \dfrac{\beta - 2k - \sqrt{\delta}}{2\sqrt{\delta}}, & g = 1 \\[2mm] \dfrac{v a}{\sqrt{\delta} c} + \dfrac{\beta - 2k - \sqrt{\delta}}{2\sqrt{\delta}}, & g \neq 1. \end{cases}$$

Since $\sqrt{\delta} \in Q$, these values are rational.   But the character values are algebraic integers.   Hence $\phi(g) \in Z$. If $g \neq 1$, we have

$$\phi(g) - \phi(1) = \frac{v a}{c\sqrt{\delta}} - \frac{(\sqrt{\delta} - \beta)v}{2\sqrt{\delta}} \in Z$$

The assertion follows.   □

5. Examples

We do not intend to give serious applications of the theorems proved above.   Our main objective is to show that dropping the assumption that G is abelian and Δ is a union of conjugacy classes is forced by, well, reality.

5.1 *A funny example.*
In section 2 there was an attempt to prove that sometimes the only partial addition sets of a group come from subgroups or their complements.   Now apply Theorem 10 to this situation: $\Delta = H\backslash \{1\}$, where H is a subgroup of order h of G.   It follows that $h \mid (va/c)$, where c is the order of a conjugacy class of G.

5.2 *A serious example.*
The incidence graph of a generalised quadrangle is a strongly regular graph.   We use this to give a class of examples of strongly regular graphs with sharply 1-transitive, nonabelian automorphism group such that for the corresponding partial addition set Δ the following holds:

(1) Δ has parameters $(q^3, (q + 2)(q - 1), q - 2, q + 2)$, where q is an odd prime power.

(2) Δ is not a union of conjugacy classes.

The construction is that of Payne and Thas (1984), in the style of Kantor (1984).   Let q be the power of an odd prime.   On the set $G :=$ $GF(q) \times GF(q) \times GF(q)$ define a multiplication

$$[\alpha, \beta, \gamma][\alpha', \beta', \gamma'] = [\alpha + \alpha', \beta + \beta', \gamma + \gamma' + \beta\alpha' - \alpha\beta'].$$

This turns G into a group with centre $Z = \{[0, 0, \alpha]; \alpha \in GF(q)\}$.   Let ∞ be a new symbol not in $GF(q)$.   For $t \in GF(q) \cup \{\infty\}$ we let

$$A(t) \qquad := \{[-\beta t, \beta, 0]; \beta \in GF(q)\}, t \in GF(q);$$

$$A(\infty) \qquad := \{[\alpha, 0, 0]; \alpha \in GF(q)\}.$$

Then the partial addition set Δ is, by definition;

$$\Delta := \left[\bigcup_{t \in GF(q) \cup \{\infty\}} A(t) \cup Z \right] \setminus \{1\}.$$

This example shows that partial addition sets which are not a union of conjugacy classes turn up in a natural and interesting way.

References

C.W. Curtis and I. Reiner (1962), *Representation Theory of Finite Groups and Associative Algebras*, John Wiley & Sons, New York.

X.L. Hubaut (1975), Strongly regular graphs, *Discrete Math.*, **13**, 357-381.

W.M. Kantor (1984), Generalised polygons, SCABs and GABs, *Buildings and the Geometry of Diagrams*, (ed. L.A. Rosati), Lecture Notes in Mathematics **1181**, 79-158, Springer-Verlag, Berlin.

G. Karpilovsky (1987), *Structure of Blocks of Group Algebras*, Longman, Burnt Mill.

S. Löwe (1987), On multiplier theorems of relative quotient sets, *J. Geom.* **29**, 78-86.

S.L Ma (1984), Partial difference sets, *Discrete Math.* **52**, 75-89.

M. Osima (1955), Notes on blocks of group characters, *Math. J. Okayama Univ.* **4**, 175-188.

S.E. Payne and J. A. Thas (1984), *Finite Generalized Quadrangles*, Pitman, London.

J.P. Serre (1977), *Linear Representations of Finite Groups*, Springer-Verlag, Berlin.

A remark on the derivation of flocks

G. Lunardon
University of Naples

1. Introduction

In this paper q will denote an odd prime power.

Let C_0 be a quadratic cone of $PG(3, q)$ having vertex p_0. A flock $\mathcal{F}_0$ of C_o is a partition of the points of $C_o - \{p_0\}$ into q disjoint irreducible conics. For the connection between flocks of the quadratic cone, translation planes and generalized quadrangles the reader is referred to Thas (1987), or Fisher and Thas (1979).

Embed C_0 into the non-singular quadric $Q(4, q)$ of $PG(4, q)$ as the section of $Q(4, q)$ with the tangent hyperplane H_o to $Q(4, q)$ at the vertex p_0 of C_0. Let $\mathcal{F}_0 = \{C_{01}, C_{02}, ..., C_{0q}\}$ be any flock of C_o and denote by π_{oi} the plane of the conic C_{oi} and by r_{oi} the polar line of π_{oi} with respect to $Q(4, q)$. We have $r_{oi} \cap Q(4, q) = \{p_0, p_i\}$.

Let H_j be the polar hyperplane of the point p_j with respect to $Q(4, q)$ and put $\pi_{ij} = H_i \cap H_j$ for $i, j \in \{0, 1, ..., q\}$ and $i \neq j$. It has been proved in Bader et al. (1990, Theorem 1) that the set $\mathcal{F}_i = \{\pi_{ij} \cap Q(4, q): j = 0, 1, ..., q; i \neq j\}$ is a flock of the quadratic cone $C_i = H_i \cap Q(4, q)$ for any $i \in \{0, 1, ..., q\}$. The flocks $\mathcal{F}_1, \mathcal{F}_2, ..., \mathcal{F}_q$ are derived from $\mathcal{F}_o$. In Payne and Rogers (to appear) it has been proved that the generalized quadrangle associated with $\mathcal{F}_i$ is isomorphic to the one associated with $\mathcal{F}_o$.

Applying the derivation to the known examples of flocks, it has been proved that in many cases the derived flocks are isomorphic to the given one (see Bader to appear; Bader et al. 1990; Payne and Thas to appear), but the derivation applied to the likeable flocks, to the Cohen Ganley flocks for $q > 9$ and to the Kantor flocks with $q \neq \pm 2$ (mod 5) gives new flocks (see Bader et al. 1990 §5; Johnson to appear; Johnson et al. to appear; Payne and Thas to appear).

Using the classification of the simple groups, it has been proved in Johnson et al. (to appear) that if $\mathcal{F}_o$ is either a semifield flock, or a likeable flock such that the flocks $\mathcal{F}_o, \mathcal{F}_1, ..., \mathcal{F}_q$ are mutually isomorphic, then $\mathcal{F}_o$ is a linear flock or $\mathcal{F}_o$ is a Kantor flock constructed via a Knuth semifield (i.e. the flock of type II.2 in Table 1 of Gevaert and Johnson (1988)) or a Fisher-Thas-Walker flock. In this paper we give a new proof of this theorem without using the classification of simple groups. Using the representations of $SL(2, q)$, we obtain some more information on the structure of the set $\{p_0, p_1, ..., p_q\}$.

2. Preliminary results

Let $\mathcal{F}_0 = \{C_{01}, C_{02}, ..., C_{0q}\}$ be a flock of the quadratic cone C_o and let π_{oi} be the plane of the conic C_{oi}. If r_{oi} is the polar line of π_{oi} with respect to $Q(4, q)$ and $r_{oi} \cap Q(4, q) = \{p_o, p_i\}$, let $\mathcal{D} = \{p_o, p_1, ..., p_q\}$.

Embed $Q(4, q)$ into the non-singular hyperbolic quadric $Q^+(5, q)$ as the section of $Q^+(5, q)$ with a non-tangent hyperplane H.

If π_{oi}^* is the polar plane of π_{oi} with respect to $Q^+(5, q)$ and C_{oi}^* is the intersection of π_{oi}^* with $Q^+(5, q)$ then $Q_o = C_{01}^* \cup C_{02}^* \cup ... \cup C_{oq}^*$ is an ovoid of $Q^+(5, q)$ (see Fisher and Thas 1979). Moreover, as r_{oi} is contained in π_{oi}, we have $H \cap Q_o = \mathcal{D}$ (see Bader et al. 1990 §4).

Denote by φ the Klein correspondence mapping the points of $Q^+(5, q)$ onto the lines of $PG(3, q)$ and put $S_o = Q_o\varphi$, $A = p_o\varphi$ and $\mathcal{R}_i = C_{oi}^*\varphi$. Then, S_o is a spread of $PG(3, q)$, $\mathcal{R}_i$ is a regulus of $PG(3, q)$ and $S_o = \mathcal{R}_1 \cup \mathcal{R}_2 \cup ... \cup \mathcal{R}_q$ (see, for instance, Thas (1987)). We notice that if $i \neq j$ then $\mathcal{R}_i \cap \mathcal{R}_j = \{A\}$.

The projective space $PG(3, q)$ can be regarded as the lattice of the subspaces of a 4-dimensional vector space $\mathbf{V}$ over $GF(q)$. Denote by $\mathbf{L}$ the 2-dimensional subspace of $\mathbf{V}$ associated with the line L of $PG(3, q)$ and by $\mathcal{K}_o$ the congruence partition of $\mathbf{V}$ induced by S_o: that is, $\mathcal{K}_o$ is the set of all 2-dimensional subspaces of $\mathbf{V}$ which define the lines of S_o. Let $A(\mathbf{V}, \mathcal{K}_o)$ be the affine translation plane associated with S_o (Lüneburg 1980, Ch. 1).

Let B and C be elements of S_o different from A. There is a unique linear map ' from $\mathbf{A}$ to $\mathbf{B}$ such that $\mathbf{C} = \{a + a': a \in \mathbf{A}\}$. For each element D of S_o different from A, there is a unique linear map J_D from $\mathbf{A}$ to itself such that $\mathbf{D} = \{aJ_D + a': a \in \mathbf{A}\}$ (see Luneburg 1980 Ch. 1). Let $C(A, B, C; S_o) = \{J_D: D \in S_o, D \neq A\}$. We have

Proposition 1 (Gevaert and Johnson 1988, Theorem 3.4). *Let A, B and C be elements of the regulus $\mathcal{R}_1$. For each regulus $\mathcal{R}_i$ of S_o there is a linear map X_i of $C(A, B, C; S_o)$ such that $\{J_D : D \in \mathcal{R}_i; D \neq A\} = \{\lambda I + X_i: \lambda \in GF(q)\}$. For $i = 1$, we have $X_i = 0$.* □

We will say that $\mathcal{F}_0$ is a semifield flock if $C(A, B, C; S_o)$ is closed with respect to the sum of linear maps.

Let $\mathcal{F}_0$ be a semifield flock of C_o. For each element X of $C(A, B, C; S_o)$ let τ_X be the collineation of $PG(3, q)$ associated with the linear map f_X of $\mathbf{V}$ onto itself defined by $f_X(a + b') = a + bX + b'$. Let $\mathcal{E} = \{\tau_X: X \in$

$C(A, B, C; S_o)\}$. If $q = p^s$, then each element of $\mathcal{E}$ has order p, fixes S_o and is the identity on A: i.e. f_X is an elation of $A(\mathbf{V}, \mathcal{K}_o)$ with axis $\mathbf{A}$. We notice that $\mathcal{E}$ is an elementary abelian group of order q^2.

Let $q = p^r$ where p is a prime different from 3. A function J of $GF(q)$ into $GF(q)$ is likeable if and only if

(i) J is additive, and

(ii) if $u^2 = a^2u - a^4/3 + aJ(a)$ then $a = u = 0$.

Let $\{e_1, e_2\}$ be a basis of $\mathbf{A}$. If J is a likeable function of $GF(q)$, let $M(u, a)$ be the linear map defined by the matrix

$$\begin{bmatrix} 1 & 0 & 0 & 0 \\ a & 1 & 0 & 0 \\ u & a & 1 & 0 \\ au - \dfrac{a^3}{3} + J(a) & u & a & 1 \end{bmatrix}$$

with respect to the basis $\{e_1, e_2, e_1', e_2'\}$ of $\mathbf{V}$.

We will say that $\mathcal{F}_o$ is a likeable flock if there is a likeable function J of $GF(q)$ such that $\mathcal{K}_o = \{BM(u, a): u, a \in GF(q)\} \cup \{A\}$. The group $g = \{M(u, a): u, a \in GF(q)\}$ is a collineation group of $A(\mathbf{V}, \mathcal{K}_o)$ of order q^2.

We remark that $g \cap \mathcal{E} = \{\tau_X: X = \lambda I; \lambda \in GF(q)\} = E_1$.

Let $\mathcal{F}_o$ be either a semifield flock, or a likeable flock. We denote by E the group $\mathcal{E}$ when $\mathcal{F}_o$ is a semifield flock and the group g when $\mathcal{F}_o$ is a likeable flock.

If $L = Q(4, q)\varphi$ then L is the linear complex of the isotropic lines of a symplectic polarity of $PG(3, q)$. Let $PSp(4, q)$ be the stabilizer of L in $PSL(4, q)$.

We have already defined $\mathcal{D} = \{p_0, p_1, ..., p_q\}$. Let $A_i = p_i\varphi$ for $i = 0, 1, ..., q$ and $\mathcal{P} = \{A_0, A_1, ..., A_q\} = \mathcal{D}\varphi$. For $i = 0$, we have $A_0 = A$. By construction, $\mathcal{P}$ is a partial spread of $PG(3, q)$ such that $\mathcal{P} = S_0 \cap L$.

Proposition 2 (Bader et al. 1990, page 167). *If τ is a collineation of $PSp(4, q)$ stabilizing the partial spread $\mathcal{P}$ and fixing A_0 then $S_0\tau = S_0$.*

Proposition 3. *If $\mathcal{F}_o$ is either a semifield flock or a likeable flock and E_0 is the stabilizer of L in E, then E_0 has order q, is transitive on the set $\{A_1, A_2, ..., A_q\}$ and fixes A_0*

Proof. As E_0 is a subgroup of E, it is sufficient to prove that E_0 has order q. We can suppose that A, B and C are lines of $\mathcal{R}_1$ such that $A =$

A_0 and $B = A_1$.   By Proposition 1, the subgroup $E_1 = \{\tau_X: X = \lambda I, \lambda \in GF(q)\}$ of E has the following properties:

(a) E_1 fixes all the reguli $\mathcal{R}_1, \mathcal{R}_2, ..., \mathcal{R}_q$ of S_0;

(b) E_1 is transitive on $\mathcal{R}_i - \{A_0\}$ ($i = 1, 2, ..., q$).

We notice that E is a group of order q^2 acting transitively on the set $\{\mathcal{R}_1, \mathcal{R}_2, ..., \mathcal{R}_q\}$.   We denote by $\underline{E}, \underline{E}_0$ and $\underline{E}_1$ the subgroups of $PGO^+(5, q)$ induced via the Klein correspondance by E, E_0 and E_1 respectively.   Let p be the 1-pole of H with respect to $Q^+(5, q)$ where H is the hyperplane of $PG(5, q)$ which intersects $Q^+(5, q)$ at $Q(4, q)$.   If l is the polar line of H_0 with respect to $Q^+(5, q)$, then p and p_0 belong to l.   As E fixes the set $\{\mathcal{R}_1, \mathcal{R}_2, ..., \mathcal{R}_q\}$, the group $\underline{E}$ fixes $\mathcal{F}_0$.   Therefore, $\underline{E}$ fixes H_0 and l.   We notice that any element of $\underline{E}$ fixes p_0.   By Theorem 3(c) of Bader et al. (1990), $\underline{E}_1$ is transitive on the set $l-\{p_0\}$.   This implies that $\underline{E}_0$ has order q because it is the stabilizer of p in $\underline{E}$. $\square$

Proposition 4.   *If $\mathcal{F}_0$ is either a semifield flock or a likeable flock then E_0 is an elementary abelian group.*

Proof.   When $\mathcal{F}_0$ is a semifield flock the group E is elementary abelian. Thus, in this case, E_0 is elementary abelian.   Let $\mathcal{F}_0$ be a likeable flock. If $\{e_1, e_2\}$ is a basis of $\mathbf{A}$, we can suppose that $Q^+(5, q)$ is represented with respect to the basis $\{e_1 \wedge e_2, e_1 \wedge e_1', e_1 \wedge e_2', e_2 \wedge e_1', e_2 \wedge e_2', e_1' \wedge e_2'\}$ of $\wedge^2 V$ by the equation

$$x_1x_6 - x_2x_5 + x_3x_4 = 0$$

and H is the hyperplane $x_3 - x_4 = 0$.   Thus, the 1-pole of H with respect to $Q^+(5, q)$ is $(0, 0, 1, -1, 0, 0)$.   Therefore it is an easy calculation to prove that

$$E_0 = \{M(a^2/2, a): a \in GF(q)\}.$$

As J is an additive function, E_0 is elementary abelian. $\square$

Proposition 5.   *Let $\mathcal{F}_0$ be either a semifield flock or a likeable flock. Denote by G the stabilizer of the partial spread $\mathcal{P}$ in the group $PSp(4, q)$ and by $\mathcal{F}_1, \mathcal{F}_2, ..., \mathcal{F}_q$ the flocks derived from $\mathcal{F}_0$. The flocks $\mathcal{F}_0, \mathcal{F}_1, ..., \mathcal{F}_q$ are mutually isomorphic if and only if G acts 2-transitively on $\mathcal{P}$.*

Proof.   By Theorem 2 of Bader et al. (1990), the flocks $\mathcal{F}_i$ and $\mathcal{F}_j$ are isomorphic if and only if there is a collineation τ of $P\Gamma L(4, q)$ such that

$L\tau = L$, $\mathcal{P}\tau = \mathcal{P}$ and $A_i\tau = A_j$. By Proposition 3, $\mathcal{F}_1$, $\mathcal{F}_2$, ..., $\mathcal{F}_q$ are mutually isomorphic and E_0 is a subgroup of G.   More precisely, if G_0 is the stabilizer of A_0 in G then E_0 is a subgroup of G_0. If τ maps A_0 to A_1 and fixes both L and $\mathcal{P}$, then the group $\tau^{-1} E_0\tau$ is contained in G and acts transitively on $\{A_0, A_2, ..., A_q\}$. Thus, G is 2-transitive on $\mathcal{P}$. □

3. Main theorem

The purpose of this paragraph is to find a characterization of the semifield and likeable flocks, whose derived flocks are isomorphic to the given one.   We will prove the following result.

Theorem 6. *Denote by $\mathcal{F}_1$, $\mathcal{F}_2$, ..., $\mathcal{F}_q$ the flocks derived from $\mathcal{F}_0$. Let $\mathcal{F}_0$ be either a semifield flock or a likeable flock.   If $\mathcal{F}_0$, $\mathcal{F}_1$, ..., $\mathcal{F}_q$ are mutually isomorphic then $\mathcal{F}_0$ is a linear flock or a Kantor flock constructed via a Knuth semifield or a Fisher-Thas-Walker flock.*

Proof. If $\mathcal{F}_0$ is a flock of a quadratic cone of $PG(3, q)$, $q < 8$, then by De Clerck et al. (1988) the flocks $\mathcal{F}_0$, $\mathcal{F}_1$, ..., $\mathcal{F}_q$ are mutually isomorphic.   If $q = 9$ then the semifields of order 81 and kernel of order 9 are classified in Boerner-Lantz (1983);   there only exist the linear flock and the Kantor flock constructed via the Knuth semifield.

We can suppose $q > 9$. Let G be the stabilizer of $\mathcal{P}$ in $PSp(4, q)$. By Proposition 5 the group G acts 2-transitively on $\mathcal{P}$. We denote by G_0 the stabilizer of A_0 in G.

Step 1. Let $q = p^s$. The group E_0 is a Sylow p-subgroup of G_0 and E_0 is normal in G_0.

Proof. Let $\bar{\tau}$ be the rotation of $Q^+(5, q)$ induced by an element τ of G_0 via the Klein correspondence.   Thus, $\bar{\tau}$ fixes $\mathcal{D} = \{p_0, p_1, ..., p_q\}$, p_0 and H_0.   This implies that $\bar{\tau}$ fixes $\mathcal{F}_0$.   Then τ maps S_0 and $\mathcal{P}$ onto themselves.

We will prove that E_0 is a Sylow p-subgroup of G_0.   By way of contradiction, we suppose that E_0 is not a Sylow p-subgroup of G_0. Let P be a Sylow p-subgroup of G_0 which contains E_0. By hypothesis, $E_0 \neq P$. As E_0 is regular on $\mathcal{P} - \{A_0\}$, there is an element τ of P different from the identity which fixes $A_1 = B$.

Let $\mathcal{R}_1'$ be the opposite regulus of $\mathcal{R}_1$. As $\bar{\tau}$ fixes the points p_0 and p_1, τ maps $\mathcal{R}_1'$ onto itself. Thus, $\mathcal{R}_1\tau = \mathcal{R}_1$. As τ induces a p-element on A, τ must fix a point r of A. Therefore, τ fixes the line L of $\mathcal{R}_1'$ incident with r. The collineation τ induces a p-element on L, which fixes the point r and the point $L \cap B$. Then, τ is the identity on L. If $L = \langle e_1, e_1' \rangle$ and $A = \langle e_1, e_2 \rangle$, the collineation τ is induced by the linear map g defined by the matrix

$$\begin{bmatrix} 1 & x & 0 & 0 \\ 0 & 1 & 0 & 0 \\ 0 & 0 & 1 & x \\ 0 & 0 & 0 & 1 \end{bmatrix} (x \neq 0)$$

with respect to the basis $\{e_1, e_2, e_1', e_2'\}$. We notice that g is the identity on L. As $\mathcal{K}_0 g = \mathcal{K}_0$, the map g is a Baer p-element of $A(V, \mathcal{K}_0)$ because L is a Baer subplane of $A(V, \mathcal{K}_0)$. Thus, $A(V, \mathcal{K}_0)$ has a p-group, whose elements are elations and a p-group, whose elements are Baer elements. By Foulser (1974), this is impossible because p is odd. This is the required contradiction. Thus E_0 is a Sylow p-subgroup of G_0.

Now it will be shown that E_0 is normal in G_0. Let $\mathcal{F}_0$ be a semifield flock. Let γ be an element of G_0 and let h be a linear map of $GL(4, q)$, which defines γ. For each τ_X in E_0, the linear map $h^{-1} f_X h$ is an affine elation of $A(V, \mathcal{K}_0)$ with axis A_0. Then, $\gamma^{-1} \tau_X \gamma$ belongs to E. As $\gamma^{-1} \tau_X \gamma$ stabilizes L, we conclude that $\gamma^{-1} \tau_X \gamma$ belongs to E_0; that is, E_0 is normal in G_0. Let $\mathcal{F}_0$ be a likeable flock. Let C be the stabilizer of $\mathcal{K}_0$ in $GL(4, q)$. By Theorem 5.4 of Biliotti et al. (1989), the group $\{M(u, a) : u, a \in GF(q)\}$ is a normal subgroup of C because $q > 9$. Then, E is normal in the stabilizer K of S_0 in $PGL(4, q)$. By Proposition 1, G_0 is contained in K. Thus, for each element τ of G_0, $\tau^{-1} E_0 \tau$ is a subgroup of E which fixes L; that is, $\tau^{-1} E_0 \tau = E_0$.

Step 2. If T is the minimal normal subgroup of G then either q is prime or T is isomorphic to $PSL(2, q)$.

Proof. By Step 1 and Proposition 3, E_0 is a normal subgroup of G_0, which acts regularly on $\mathcal{P} - \{A_0\}$. By Theorem 10.2 of Cameron (1980), one of the following cases holds:
(a) T is an elementary abelian group;
(b) T is isomorphic to $PSL(2, q)$ or to $PSU(3, r)$ with $r^3 = q$ or to $^2G_2(r)$ with $r^3 = q$.

Let T be elementary abelian. As T acts regularly on $\mathcal{P}$, we have $q + 1 = 2^n$ because $q + 1$ is even. If $q = p^s$, we remark that $q^2 - 1$ does not admit a p-primitive divisor because $q + 1 = 2^n$. By Theorem 6.21 of Lüneburg (1980), we have $s = 1$.

Now suppose that T is as in case (b). Then T is 2-transitive on $\mathcal{P}$ and a Sylow p-subgroup P of T must fix an element of $\mathcal{P}$. We can suppose that P fixes A_0. By Step 1, P is contained in E_0. We notice that E_0 is elementary abelian and therefore P is elementary abelian. By Theorem 32.10 of Luneburg (1980), the Sylow p-subgroups of $PSU(3, r)$ are not elementary abelian, and by Theorem 13.2 of Huppert and Blackburn (1982), the Sylow p-subgroups of $^2G_2(r)$ are not elementary abelian. Therefore, T must be isomorphic to $PSL(2, q)$.

Remark. Let q be a prime number. A semifield plane of order q^2 is Desarguesian and a likeable plane of order q^2 is a Walker plane (see Johnson and Wilke 1983-84, §7.1). Therefore the flock $\mathcal{F}_0$ is either a linear flock or a Fisher-Thas-Walker flock.

As the theorem holds when q is a prime number, in the following *we always suppose that q is not a prime number.*

Step 3. Let $Sp(4, q)$ be the subgroup of $GL(4, q)$ of all the linear maps, which define a collineation of $PSp(4, q)$. Let T be the socle of G. If S is the subgroup of $Sp(4, q)$ which defines T, then S is isomorphic to $SL(2, q)$.

Proof. As q is not a prime number, T is isomorphic to $PSL(2, q)$. If $Z = \{\pm I\}$, then Z is the centre of $Sp(4, q)$ and $PSp(4, q) = Sp(4, q)/Z$. By definition, we have that $S/Z = T$.

We first prove that $Sp(4, q)$ has not an elementary abelian subgroup of order 8. We notice that if 2^r is the greatest power of 2 which divides $q^2 - 1$, then 2^{2r+1} is the greatest power of 2 which divides $|Sp(4, q)| = q^4(q^4 - 1)(q^2 - 1)$. Thus, a Sylow 2-subgroup of $Sp(4, q)$ has order 2^{2r+1}.

Let $\{v_1, w_1, v_2, w_2\}$ be a symplectic basis of V and let Σ be the stabilizer of the set of hyperbolic planes $\{\langle v_1, w_1 \rangle, \langle v_2, w_2 \rangle\}$ in $Sp(4, q)$. In the fixed basis, the elements of Σ are represented by a matrix of type $\begin{bmatrix} X & 0 \\ 0 & Y \end{bmatrix}$ or of type $\begin{bmatrix} 0 & X \\ Y & 0 \end{bmatrix}$ with $X, Y \in SL(2, q) = Sp(2, q)$.

If Q is a Sylow 2-subgroup of $SL(2, q)$ then Q is a generalized quaternion group and

$$H = \{g \in \Sigma: g = \begin{bmatrix} X & 0 \\ 0 & Y \end{bmatrix} \text{ or } g = \begin{bmatrix} 0 & X \\ Y & 0 \end{bmatrix} \text{ with } X, Y \in Q\}$$

is a 2-subgroup of order 2^{2r+1}. Thus, H is a Sylow 2-subgroup of $Sp(4, q)$. The involutions of H are the elements of the following types

$$\begin{bmatrix} I \\ & I \end{bmatrix}, \begin{bmatrix} -I \\ & I \end{bmatrix}, \begin{bmatrix} I \\ & -I \end{bmatrix}, \begin{bmatrix} -I \\ & -I \end{bmatrix}, \begin{bmatrix} X & X^{-1} \end{bmatrix}$$

with $X \in Q$.   Therefore, H does not contain an elementary abelian subgroup of order 8.

Let S' be the commutator subgroup of S. By Satz 25.7 of Huppert and Blackburn (1982), if $S' = S$ then S is isomorphic to $SL(2, q)$ because $q > 9$.   Thus, it is sufficient to prove that $S' = S$.   By way of contradiction, we suppose that $S' \neq S$. As S/Z is a non-abelian simple group, we have $S'Z = S$ and therefore, S is the direct product of S' and Z. It follows that S' is isomorphic to $PSL(2, q)$. As in $PSL(2, q)$ there is an elementary abelian subgroup of order 4, S has an elementary abelian subgroup order 8.   This is impossible and we have the required contradiction.

Step 4. If $\mathcal{F}_0$ is a semifield flock, the partial spread $\mathcal{P}$ has at least two distinct transversal lines.

Proof.   The group $\Sigma = \{f_X : \tau_X \in E_0\}$ is a Sylow p-subgroup of S and $\text{Fix}(\Sigma) = \{v \in V : vf_X = v$ for all elements f_X of $\Sigma\} = A_0$. If Λ is a Sylow p-subgroup of S there is an element A_i of $\mathcal{P}$ such that $\text{Fix}(\Lambda) = A_i$. Then $\{\text{Fix}(\Lambda) : \Lambda \in \text{Syl}_p(S)\} = \{A_i, A_i \in \mathcal{P}\}$. We notice that $(f_X - 1)^2 = 0$ and that $\text{Fix}(f_X) = \{v \in V : vf_X = v\} = A_0$.   By Lemma 49.1 of Luneburg (1980), there are at least two distinct transversal lines of $\mathcal{P}$.

Step 5. If $\mathcal{F}_0$ is a semifield flock then $\mathcal{F}_0$ is either a linear flock or a Kantor flock constructed via a Knuth semifield.

Proof. Let T_1 and T_2 be two transversal lines of $\mathcal{P}$. Let t_1 and t_2 be the two points of $Q^+(5, q)$ such that $t_1\varphi = T_1$ and $t_2\varphi = T_2$.

If H is the hyperplane of $PG(5, q)$ such that $Q(4, q) = H \cap Q^+(5, q)$, let p be the 1-pole of H with respect to $Q^+(5, q)$.

If t_1, t_2 and p are not collinear, then $\mathcal{D}$ is contained in the polar plane of the plane $\langle t_1, t_2, p \rangle$ with respect to $Q^+(5, q)$. Thus, $\mathcal{D}$ is a non-singular conic. By §5.a of Bader et al. (1990), $\mathcal{F}_0$ is a linear flock.

If t_1, t_2 and p are collinear, then $\mathcal{D}$ is contained in a 3-dimensional subspace of H. Thus all the planes π_{oi} are incident with a common point r. By Thas (1987, §1.5), $\mathcal{F}_0$ is a Kantor flock constructed with a Knuth semifield.

Step 6. If $\mathcal{F}_0$ is a likeable flock then $\mathcal{F}_0$ is a Fisher-Thas-Walker flock.

Proof. It is sufficient to prove that $A(\mathbf{V}, \mathcal{K}_0)$ is a Walker plane.

For $i = 0, 1, ..., q$ let Σ_i be the Sylow p-subgroup of S which fixes $\mathbf{A}_i$. We notice that $\Sigma_0 = \{M(a^2/2, a): a \in GF(q)\}$ and $\text{Fix}(\Sigma_0) = \langle \mathbf{e}_1 \rangle$. Thus, for $i = 0, 1, ..., q$, $\text{Fix}(\Sigma_i) = \langle \mathbf{v}_i \rangle$ has dimension 1 and is contained in $\mathbf{A}_i$. If $\mathbf{v}_1 = \alpha \mathbf{e}_1' + \beta \mathbf{e}_2'$, then for $j = 1, 2, ..., q$ there is a unique $a \in GF(q)$ such that $\langle \mathbf{v}_j \rangle = \langle \mathbf{v}_1 M(a^2/2, a) \rangle = \langle (\alpha a^2/2 + \beta(a^3/6 + J(a)))\mathbf{e}_1 + (\alpha a + \beta a^2/2)\mathbf{e}_2 + (\alpha + \beta a)\mathbf{e}_1' + \beta \mathbf{e}_2' \rangle$.

Let $\mathcal{H} = \{\langle \mathbf{v}_i \rangle: i = 0, 1, ..., q\}$.

(a) The group T acts 2-transitively on the set $\mathcal{H}$ of points of $PG(3, q)$.

For each h in S and $i = 0, 1, ..., q$ we have $\text{Fix}(h^{-1}\Sigma_i h) = \langle \mathbf{v}_i h \rangle$. Thus if $\mathbf{A}_i h = \mathbf{A}_j$ then $h^{-1}\Sigma_i h = \Sigma_j$ and $\langle \mathbf{v}_i h \rangle = \langle \mathbf{v}_j \rangle$. As S acts 2-transitively on $\{\mathbf{A}_0, \mathbf{A}_1, ..., \mathbf{A}_q\}$ and Σ_i fixes $\mathbf{A}_i$, it is easy to prove that T acts 2-transitively on $\mathcal{H} = \{\langle \text{Fix}(\Sigma_i) \rangle: i = 0, 1, ..., q\}$.

(b) We can suppose that $\beta = 1$.

If $\beta = 0$, then $\mathcal{H}$ is contained in the plane π of $PG(3, q)$ defined by $\pi = \langle \mathbf{e}_1, \mathbf{e}_2, \mathbf{e}_1' \rangle$. We suppose that $\mathcal{H}$ is a line of $PG(3, q)$. If m is the point of $Q^+(5, q)$ such that $m\varphi = \mathcal{H}$, then let L be the line joining the points p and m. Then $\mathcal{D}$ is contained in the polar space of L with respect to $Q^+(5, q)$. Thus all the planes π_{0i} are incident with a common point. By Thas (1987, §1.5) $\mathcal{F}_0$ is a semifield flock. As $\mathcal{F}_0$ is not a semifield flock, $\mathcal{H}$ is not a line of $PG(3, q)$. Therefore T fixes the plane π. As A_0 is contained in π, all the lines $A_1, A_2, ..., A_q$ are contained in π. This is impossible because $\mathcal{P}$ is a partial spread.

(c) $\mathcal{H}$ is a $(q + 1)$-arc of $PG(3, q)$.

We notice that the matrix

$$
\begin{bmatrix}
\dfrac{\alpha a^2}{2} + \dfrac{a^3}{6} + J(a) & \alpha a + \dfrac{a^2}{2} & \alpha + a & 1 \\[2mm]
\dfrac{\alpha b^2}{2} + \dfrac{b^3}{6} + J(b) & \alpha b + \dfrac{b^2}{2} & \alpha + b & 1 \\[2mm]
0 & 0 & \alpha & 1 \\[2mm]
1 & 0 & 0 & 0
\end{bmatrix}
$$

has determinant $ab(b - a)/2$. Thus, the points $\langle \mathbf{v}_1 M(a^2/2, a)\rangle$, $\langle \mathbf{v}_1 M(b^2/2, b)\rangle$, $\langle \mathbf{v}_1\rangle$ and $\langle \mathbf{v}_0\rangle$ are not coplanar when $a \neq b$. As T acts 2-transitively on $\mathcal{H}$ we conclude that $\mathcal{H}$ is a $(q + 1)$-arc.

(d) $A(\mathbf{V}, \mathcal{K}_0)$ is a Walker plane.

As $\mathcal{H}$ is a twisted cubic (Lüneburg (1989), Theorem 43.7), by Theorem 43.3 (op.cit), the stabilizer of $\mathcal{H}$ in $PGL(4, q)$ is isomorphic to $PGL(2, q)$. We recall that $q = p^s$ with $p \neq 3$. By Theorem 43.4 (op.cit), there is a unique symplectic polarity δ of $PG(3, q)$ such that $\langle \mathbf{v}_i\rangle\delta$ is the osculating plane of $\mathcal{H}$ at the point $\langle \mathbf{v}_i\rangle$ for all $i = o, 1, \ldots, q$. We remark that the collineation of $PG(3, q)$ defined by the matrix $M(a^2/2, a)$ $(a \neq 0)$ fixes the point $\langle \mathbf{v}_0\rangle$ and the line A_0, but does not fix any other line through $\langle \mathbf{v}_0\rangle$. As the tangent line to $\mathcal{H}$ at $\langle \mathbf{v}_0\rangle$ is fixed by E_0, the line A_0 is the tangent line to $\mathcal{H}$ at $\langle \mathbf{v}_0\rangle$. Therefore, $\{A_0, A_1, \ldots, A_q\}$ is the set of the tangents to $\mathcal{H}$. As the given flock is not a semifield flock, $\mathcal{D}$ is not contained in a three-dimensional subspace of H (see Thas 1987, §1.5). This implies that L is the set of the isotropic lines of δ. Then, the stabilizer of $\mathcal{H}$ in $PGL(4, q)$ fixes L. Let C be the linear translation complement of $A(\mathbf{V}, \mathcal{K}_0)$. We denote by Γ the subgroup of $PGL(4, q)$, which fixes $\mathcal{H}$ and $\langle \mathbf{e}_1\rangle$. By Corollary 3 of Bader et al. (1990), $E_1\Gamma$ is the collineation group of $PG(3, q)$ induced by C. So C fixes $\langle \mathbf{e}_1\rangle$. By Remark 4 of Bader et al. (1990), C has order $q^2(q - 1)^2$. If C_1 is the subgroup of C, which fixes the vector $\mathbf{e}_1$, then C_1 has order $q^2(q - 1)$ and does not contain kernel-homologies. By Jha et al. (1989), $A(\mathbf{V}, \mathcal{K}_0)$ is a Walker plane.

Remark. Suppose that q is not a prime number. If $\mathcal{F}_0$ is a Fisher-Thas-Walker flock then $\mathcal{P}$ is the set of the tangents to a twisted cubic of $PG(3, q)$. There is another spread S of $PG(3, q)$ not isomorphic to S_0, which contains $\mathcal{P}$, namely the spread of the Hering plane (see Luneburg 1980, §45). We notice that S is invariant under the action of $T = PSL(2, q)$ but S_0 is not invariant under the action of T.

References

L. Bader (to appear), Derivation of Fisher flocks, *J. Geom.* to appear.

L. Bader, G. Lunardon and J.A. Thas (1990), Derivation of flocks of quadratic cones, *Forum Math.* **21**, 163-174.

M. Biliotti, V. Jha, N.L. Johnson and G. Menichetti (1989), A structure theory for two dimensional translation planes of order q^2, *Geom. Dedicata* **29**, 7-43.

V. Boerner-Lantz (1983), A new class of semifields, Ph. D. Thesis, Washington State University.

P.J. Cameron (1980), Finite permutation groups and finite simple groups, *Bull. London Math. Soc.* **13**, 1-22.

F. De Clerck, H. Gevaert and J.A. Thas (1988), Flocks of a quadratic cone in $PG(3, q)$, $q \leq 8$, *Geom. Dedicata* **26**, 215-230.

J.C. Fisher and J.A. Thas (1979), Flocks in $PG(3, q)$, *Math. Z.* **169**, 1-11.

D. Foulser (1974), Baer p-elements in translation planes, *J. Algebra* **31**, 354-366.

H. Gevaert and N.L. Johnson (1988), Flocks of quadratic cones, generalized quadrangles and translation planes, *Geom. Dedicata* **27**, 301-317.

B. Huppert (1979), *Endliche Gruppen I*, Springer-Verlag, Berlin.

B. Huppert and N. Blackburn (1982), *Finite Groups III*, Springer-Verlag, Berlin.

V. Jha, N.L. Johnson and F.W. Wilke (1984), On translation planes of order q^2 that admit a group of order $q^2(q - 1)$; Bartolone's theorem, *Rend. Circ. Mat. Palermo* **33**, 407-424.

N.L. Johnson (to appear), Derivation of partial flocks of quadratic cones, Preprint.

N.L. Johnson and F.W. Wilke (1984), Translation planes of order q^2 that admit a collineation group of order q^2, *Geom. Dedicata* **15**, 293-312.

N.L. Johnson, G. Lunardon and F.W. Wilke (to appear), Semifield skeletons of conical flocks, *J. Geom.*

W.M. Kantor (1982), On point transitive affine planes, *Israel J. Math.* **42**, 227-234.

H. Lüneburg (1980), *Translation Planes*, Springer-Verlag, Berlin.

S.E. Payne and L. Rogers (submitted), Local group actions on generalized quadrangles, Preprint.

S.E. Payne and J.A. Thas (to appear), Conical flocks, partial flocks, derivation and generalized quadrangles, *Geom. Dedicata*.

J.A. Thas (1987), Generalized quadrangles and flocks of cones, *European J. Combin.* **8**, 441-452.

A geometric approach to
semi-cyclic codes

T. Maruta
Okayama University

1. Semi-cyclic codes

Throughout this paper, a code means a linear code over the Galois field $F = GF(q)$ of order q with the minimum distance $d \geq 3$. Let C and C' be (n, k)-codes over F. C and C' are *equivalent* if there exists a monomial matrix M with entries in F such that C' coincides with CM $= \{cM \mid c \in C\}$.

Let α be a fixed non-zero element of F. A code C is α-*cyclic* if $(x_1, x_2, \ldots, x_n) \in C$ implies $(\alpha x_n, x_1, x_2, \ldots, x_{n-1}) \in C$. A code C is called *semi-cyclic* if C is α-cyclic for some $\alpha \in F \setminus \{0\}$. 1-cyclic codes are simply called *cyclic*. We associate the vector $c = (c_0, c_1, \ldots, c_{n-1}) \in F^n$ with the polynomial $c(x) = \sum_{i=0}^{n-1} c_i x^i$. With this association an α-cyclic code C can be identified an ideal of the ring $F[x]/(x^n - \alpha)$. The following theorem states basic properties of semi-cyclic codes which are well-known for $\alpha = 1$.

Theorem 1. *Let C be a non-zero ideal of $F[x]/(x^n - \alpha)$, that is, an α-cyclic code of length n. Then there is a unique monic polynomial $g(x)$ of minimal degree in C, satisfying*

(a) $C = \langle g(x) \rangle$, *that is, $g(x)$ is the generator polynomial of C.*
(b) *$g(x)$ is a factor of $x^n - \alpha$.*
(c) *Any $c(x) \in C$ can be written uniquely as $c(x) = f(x)g(x)$ in $F[x]$, where $f(x) \in F[x]$ has degree $< n - k$, $k = \deg g(x)$. The dimension of C is $n - k$.*

(d) *If $g(x) = \sum_{i=0}^{k} g_i x^i$, then C has the following generator matrix*

$$G = \begin{bmatrix} g_0 & g_1 & g_2 & \cdots & g_k & & & \\ & g_0 & g_1 & g_2 & \cdots & g_k & & \\ & & \cdots & \cdots & \cdots & & & \\ & & & g_0 & g_1 & \cdots & g_k \end{bmatrix} = \begin{bmatrix} g(x) & & & \\ & xg(x) & & \\ & & \cdots & \\ & & & x^{n-k-1}g(x) \end{bmatrix}.$$

(e) The dual code of C is α^{-1}-cyclic and it is equivalent to $\langle h(x) \rangle$, where $x^n - \alpha = g(x)h(x)$.

Proof.   (a)-(e) are easily proved by a similar argument as for cyclic codes. See MacWilliams and Sloane (1977, Chap. 7).

Theorem 2.   *Let C be an $(n, n - k)$-code over F.   Then C is α-cyclic if and only if C has a parity check matrix of the form $[{}^tP, {}^t(PT), {}^t(PT^2), ..., {}^t(PT^{n-1})]$, with $P \in F^k$ and $T \in GL(k, q)$ such that $PT^n = \alpha P$, where ${}^tS$ stands for the transpose of S.*

Proof.   Let C be an α-cyclic $(n, n - k)$-code over F with a parity check matrix $H = [P_1, P_2, ..., P_n]$, where the P_i are column vectors.   Then $H' = [\alpha^{-1}P_n, P_1, P_2, ..., P_{n-1}]$ is also a parity check matrix of C by Theorem 1(e). Hence there exists $T \in GL(k, q)$ with $H = TH'$.   Therefore we have $H = [P_1, TP_1, T^2P_1, ..., T^{n-1}P_1]$ with $T^nP_1 = TP_n = \alpha P_1$.   The converse is trivial.

Theorem 3.   *Let $g(x)$ be a polynomial of degree k in $F[x]$ dividing $x^n - \alpha$. Then C is an α-cyclic $(n, n - k)$-code over F with generator polynomial $g(x)$ if and only if C is a code with parity check matrix $[{}^tP, {}^t(PT), {}^t(PT^2), ..., {}^t(PT^{n-1})]$, where $P = (1, 0, ..., 0)$ and T is the companion matrix of $g(x)$.*

Proof.   We prove only the case $k \geq n/2$. Let C be an α-cyclic $(n, n - k)$-code over F with generator polynomial $g(x) = \sum_{i=0}^{k} a_i x^i, a_k = 1$, and $G = [G_1 \mid G_2]$ be the canonical generator matrix of C (see Theorem 1(d)), where G_1 is square. $g(x) \mid x^n - \alpha$ implies $a_0 \neq 0$.   Put $m = n - k$.   Denote by Δ_i the $(i + 1, 1)$-cofactor of G_1 $(i = 1, 2, ..., m - 1)$, $\Delta_0 = a_0^{m-1}$.   Then

$$G_1^{-1} = (g_{ij}); \quad g_{ij} = \begin{cases} 0 & \text{if } i > j \\ a_0^{-1} & \text{if } i = j \\ a_0^{-m}\Delta_{j-i} & \text{if } i < j \end{cases}$$

Let $S = G_1^{-1}G_2 = (s_{ij})$.   Since $\Delta_s = -a_0^{-1}\sum_{i=0}^{s-1} a_{s-i} \Delta_i, s = 2, 3, ..., m - 1$, we obtain $s_{i+1, j+1} = s_{ij} - s_{ik} a_j$ and $s_{ik} = a_0^{-m} \Delta_{m-i} = -a_0^{-1}s_{i+1,1}$, that is,

$$(s_{i+1, 1}, s_{i+1, 2}, ..., s_{i+1, k}) = (s_{i1}, s_{i2}, ..., s_{ik})T,$$

where T is the companion matrix of $g(x)$.   Since $(s_{11}, s_{12}, ..., s_{1k}) = \alpha^{-1}(a_0, a_1, ..., a_{k-1})$, C has the matrix ${}^t(T^{n-k})[-{}^tS \mid I_k] = [{}^tP, {}^t(PT), {}^t(PT^2), ..., {}^t(PT^{n-1})]$ as a parity check matrix, where $P = (1, 0, ..., 0)$ and I_k is the identity matrix of size k, as desired.   The converse is trivial.

For $2 \leq k \leq n$, $T \in GL(k, q)$ and $P \in F^k$ such that $PT^n = \alpha P$, let $G_{T,P}^{(n)}$ be the $k \times n$ matrix $[{}^t P, {}^t(PT), {}^t(PT^2), \ldots, {}^t(PT^{n-1})]$. Then the code with parity check matrix $G_{T,P}^{(n)}$, which is called the code *defined by* (T, P, n), is an α-cyclic $(n, n - k)$-code if and only if the first k columns of $G_{T,P}^{(n)}$ are linearly independent. Put

$\Gamma_k^\alpha = \{(T, P, n) \in GL(k, q) \times F^k \times \mathbb{Z}\} \mid (T, P, n)$ defines an α-cyclic $(n, n-k)$-code\}.

Theorem 4. *Let* $(T_i, P_i, n) \in \Gamma_k^\alpha$, *and* C_i *be the code defined by* (T_i, P_i, n) $(i = 1, 2)$. *Then the following are equivalent:* (i) $C_1 = C_2$, (ii) T_1 *and* T_2 *are equivalent,* (iii) $\varphi_{T_1} = \varphi_{T_2}$, *where* φ_T *is the characteristic polynomial of* T. *In particular,* (T, P, n) *and* (T, Q, n) *define the same code if they belong to* Γ_k^α.

Proof. "(i) $\Rightarrow$ (ii)": By $C_1 = C_2$, there exists $S \in GL(k, q)$ with ${}^t S G_{T_1, P_1}^{(n)} = G_{T_2, P_2}^{(n)}$. Since C_1 and C_2 have the same dimension and $P_2 T_2^i = P_1 T_1^i {}^i S = P_2(S^{-1} T_1 S)^i$, $i = 1, 2, \ldots, n$, we obtain $S^{-1} T_1 S = T_2$.
"(ii) $\Rightarrow$ (iii)" is trivial. "(iii) $\Rightarrow$ (i)" is easily obtained from the "(i) $\Rightarrow$ (ii)" part and Theorem 3.

Corollary 5. *Let* C *be an* α-cyclic $(n, n - k)$-code *with generator polynomial* $g(x) = \Sigma_{i=0}^k a_i x^i$. *If* $q = p^h$, $r \mid p^h$ *and* $(n, p) = 1$, *then* C *and* $\overline{C}$ *are equivalent, where* $\overline{C}$ *is the* α^r-cyclic code generated by $g_r(x) = \Sigma_{i=0}^k a_i^r x^i$.

Proof. Note that $g(x) = \Sigma_i(x - \lambda_i)$ implies $g_r(x) = \Sigma_i(x - \lambda_i^r)$. Since C is α-cyclic, there exists $(T, P, n) \in \Gamma_k^\alpha$ which defines C. By $r \mid p^h$ and $(n, r) = 1$, (T^r, P, n) belongs to $\Gamma_k^{\alpha^r}$, and it defines a code which is equivalent to C. Hence it follows from $\varphi_T = g(x)$ that $\varphi_{T^r} = g_r(x)$ and from Theorem 4 that C and $\overline{C}$ are equivalent.

Examples.   (1) Let $T \in GL(k, q)$ and $P \in F^k \setminus \{(0, \ldots, 0)\}$.   If φ_T is subprimitive (see Hirschfeld 1979, p74), then $(T, P, (q^k - 1)/(q - 1))$ defines the semi-cyclic code over F which is known as the Hamming code.

(2) Let n, d and b be integers with $n \geq d - 1 \geq 2, b \geq 0$, let n divide $q^m - 1$ for some m, and let α be the n-th root of unity.   For the diagonal matrix $T = a^b (t_{ij})$ with $t_{ii} = a^{i-1}, i = 1, 2 \ldots, d - 1$, and $P = (1, 1, \ldots, 1) \in F^{d-1}$, (T, P, n) defines the 1-cyclic code over F which is known as the BCH code of designed distance d.

2. Semi-cyclic MDS codes

In this section, we are mainly concerned with semi-cyclic MDS codes of length $q + 1$ over $F = GF(q)$.   There is a one to one correspondence between all the equivalence classes of (n, k)-MDS codes over F and all the classes of projectively equivalent n-arcs in $PG(k - 1, q)$.   We call an (n, k)-code over F a *generalized Reed-Solomon code (GRS code)* if it corresponds to a class of n-arcs in $PG(k - 1, q)$, each of which is contained in a normal rational curve.   In particular, a GRS code of length $q + 1$ over F is called a *normal rational code (NR code)*;   it corresponds to the class of normal rational curves.

An $m \times n$ matrix (a_{ij}) is called a *Cauchy matrix* if $a_{ij} = 1/(1 - x_i y_j)$ where each of $\{x_i\}, \{y_j\}$ is a set consisting of distinct elements of F such that $x_i y_j \neq 1$ for any x_i, y_j.   An $m \times n$ matrix A is a *generalized Cauchy matrix (GC matrix)* if it has the form $A = D_1 \overline{A} D_2$, where $\overline{A}$ is an $m \times n$ Cauchy matrix and D_1 and D_2 are non-singular diagonal matrices.   It is well-known that an (n, k)-code C is GRS if and only if C has a generator matrix of the form $[I_k | S]$ such that S is a GC matrix (Roth and Seroussi 1986; Maruta 1990).   Consequently, the dual code of a GRS code is also GRS.

Theorem 6.   *There exist semi-cyclic (n, k)-MDS codes over F with $(n, q) \neq 1, 2 \leq k \leq n - 2$, if and only if $n = p$, where $q = p^h$, with prime p. Moreover, semi-cyclic (p, k)-codes over F are GRS.*

Proof.   Case (1) $n = q, h > 1$.   Let C be an α-cyclic (q, k)-code with generator polynomial $g(x) = (x - \alpha)^r, r = q-k$.   If $r \leq p^{h-1}, (x - \alpha)^{p^{h-1}} = x^{p^{h-1}} - \alpha^{p^{h-1}}$ must be in C and then C is not MDS because of $k \leq n - 2$.   If $p^{h-1} < r < p^h - 1$ and C is MDS, then all coefficients of $g(x)$ must be non-zero.   It follows that $r = sp^{h-1} - 1$ for some $s, 2 \leq s \leq p$.   Then $f(x) = (x - \alpha)^{r+1} = (x^{p^{h-1}} - \alpha^{p^{h-1}})^s \neq 0$ in C.   Hence the minimum distance d of C is $d = r + 1 = sp^{h-1} > s + 1 \geq \mathrm{wt}(f)$, giving a contradiction.

Case (2) $n = p$. Let $\bar{\alpha}^p = \alpha$ and $C = \langle g(x) \rangle$ with $g(x) = (x - \bar{\alpha})^{p-k}$. Let $f_i(x) = \sum_{j=0}^{p-1} 0^j \bar{\alpha}^{p-1-j} x^j$, $0 \le i \le k - 1$. (We define $0^0 \equiv 1$.) Since $g(x)$ divides $(x - \bar{\alpha})^{p-1-i}$ and $(x - \bar{\alpha})^{p-1-i}$ divides $f_i(x)$, $f_i(x)$ must be in C, $0 \le i \le k - 1$. Let G be the matrix consisting of all coefficients of f_i, $0 \le i \le k - 1$. Then

$$G = \begin{bmatrix} f_0 \\ f_1 \\ f_2 \\ \cdots \\ f_{k-1} \end{bmatrix} = \begin{bmatrix} 1 & 1 & 1 & \cdots & 1 \\ 0 & 1 & 2 & \cdots & p-1 \\ 0 & 1^2 & 2^2 & \cdots & (p-1)^2 \\ \vdots & \vdots & \vdots & & \vdots \\ 0 & 1^{k-1} & 2^{k-1} & \cdots & (p-1)^{k-1} \end{bmatrix} \begin{bmatrix} \bar{\alpha}^{p-1} & & & & 0 \\ & \bar{\alpha}^{p-2} & & & \\ & & \bar{\alpha}^{p-3} & & \\ & & & \cdots & \\ 0 & & & & 1 \end{bmatrix}.$$

Clearly, G is a generator matrix of C and hence C is GRS.

Case (3) $n \ne q$, $(n, q) \ne 1$. Let $C = \langle g(x) \rangle$ with $g(x) \mid x^n - \alpha$. We must show the existence of $f(x)$ ($\ne 0$ in $F[x]/(x^n - \alpha)$) in C with weight $\mathrm{wt}(f)$ less than $n - k + 1$. Here we can use a similar argument as in the proof of Zehendner's theorem (Zehendner 1983).

Let K_r be a normal rational curve $\{P(t^r, t^{r-1}, \ldots, t, 1) \in PG(r, q) \mid t \in F \cup \{\infty\}\}$. The projectivities of $PG(r, q)$ fixing K_r are given by

$$P(t^r, t^{r-1}, \ldots, t, 1) \to P((t^r, t^{r-1}, \ldots, t, 1)T_r),$$

where T_r is defined by $((at + b)^r, (at + b)^{r-1}(ct + d), (at + b)^{r-2}(ct + d)^2 \ldots, (ct + d)^r) = (t^r, t^{r-1}, \ldots, t, 1)T_r$, and $M = \begin{bmatrix} a & b \\ c & d \end{bmatrix} \in GL(2, q)$ (note: det $T_r = (ad - bc)^{r(r+1)/2}$). Hence the projectivities given by T_r^i act transitively on K_r if φ_M is subprimitive. (See Hirschfeld 1985 p233, 1979 p74.)

Theorem 7. *For any k, $1 \le k \le q$, there exists a semi-cyclic $(q + 1, k)$-MDS code over F.*

Proof. We may assume $2 \le k \le q - 1$, since the theorem is trivial for $k = 1$ or q. Let K be a normal rational curve $\{P(t^r, t^{r-1}, \ldots, t, 1) \in PG(r, q) \mid t \in F \cup \{\infty\}\}$, where $r = q - k$. Let ξ be a primitive element of $GF(q^2)$ ($\supset F$) and $M = \begin{bmatrix} \xi + \xi^q & \xi^{q+1} \\ -1 & 0 \end{bmatrix}$. Then we have $M \in GL(2, q)$ and φ_M is subprimitive. Hence we obtain $K = \{P(PT^i) \mid i = 1, 2, \ldots, q + 1\}$, where T is the matrix of the form T_r defined by M, and $P = (1, 0, \ldots, 0)$. Therefore $(T, P, q + 1)$ defines a semi-cyclic $(q + 1, k)$-MDS code over F, which is NR.

Let ξ be a primitive element of $GF(q^2)$ ($\supset F$), and put $\zeta = \xi^{q+1}$, $\eta = \xi^{q-1}$, $\alpha = \zeta^s$, $1 \le s \le q - 1$. Further, let $F_0 = \{a \in GF(q^2) \backslash F \mid a^{q+1} = \alpha\} = F_1 \cup F_2$ such that $a \in F_1$ implies $\alpha/a \in F_2$. Then $x^{q+1} - \alpha = \prod_{i=1}^{q+1}(x - \xi^s\eta^i)$ is factorized over F as follows:

$$x^{q+1} - \alpha = \begin{cases} (x + \sqrt{\alpha}) \displaystyle\prod_{a \in F_1}(x^2 + \beta_a x + \alpha) & \text{if } q \text{ is even,} \\[2em] \displaystyle\prod_{a \in F_1}(x^2 - \beta_a x + \alpha) & \text{if } q \text{ is odd and } s \text{ is odd,} \\[2em] (x + \sqrt{\alpha})(x - \sqrt{\alpha}) \displaystyle\prod_{a \in F_1}(x^2 - \beta_a x + \alpha) & \text{if } q \text{ is odd and } s \text{ is even,} \end{cases}$$

where $\beta_a = a + \alpha/a$.

If $g(x) = (x^2 - \alpha) \prod_{i=1}^m, (x^2 - \beta_{a_i} x + \alpha)$, then it is easily verified that the coefficient of x^{m+1} in $g(x)$ is zero. Hence we obtain the following result.

Theorem 8. *Let q be odd and k be even. If α is a quadratic residue of F, then no α-cyclic $(q + 1, k)$-code over F is MDS.*

Note. The above theorem is well-known for $\alpha = 1$ (Georgiades 1982). If q is even or both q and k are odd, cyclic $(q + 1, k)$-codes of BCH type are MDS (MacWilliams and Sloane 1977). In fact, they are NR (see Section 3).

Although many $(q + 2, 3)$-MDS codes over $GF(q)$ exist, it is clear from Theorem 6 that there exist no semi-cyclic ones. There are non-equivalent $(q + 1, k)$-MDS codes for $k = 3$ and 4, and they are all classified for $k = 4$ but not for $k = 3$ (Casse and Glynn 1982). Next, we prove the uniqueness of semi-cyclic $(q + 1, 3)$-MDS codes.

Theorem 9. *Every semi-cyclic $(q + 1, 3)$-MDS code over F is NR.*

Proof. We may assume that q is even since every $(q + 1, 3)$-MDS code over F is NR for odd q (Segre 1955). Let C be an α-cyclic $(q + 1, q - 2)$-MDS code over F. By Theorem 1(e), it is sufficient to verify that C is NR. Let $g(x)$ be the generator polynomial of C, then $g(x) = (x + \sqrt{\alpha})(x^2 + \beta_a x + \alpha)$ for some $a \in F_1$. Let T be the companion matrix of $g(x)$. From Theorem 3, it is sufficient to prove that the matrix $[{}^t(PT^3), {}^t(PT^4), ..., {}^t(PT^q)]$, where $P = (1, 0, 0)$, is GC, but that is easily proved. (See Maruta 1990).

Note. For $q = 2^h, h \geq 3$, 1-cyclic $(q + 1, 4)$-MDS codes corresponding to $(q + 1)$-arcs $K(r) = \{P(t^{r+1}, t^r, t, 1) \in PG(3, q) \mid t \in F \cup \{\infty\}\}, r = 2^n, (n, h) = 1$ exist (Maruta 1990). For odd q, the only known example of a $(q + 1, k)$-MDS code over F which is not NR is a $(10, 5)$-MDS code over $GF(9)$ (Glynn 1986), but it is easily checked that all semi-cyclic $(10, 5)$-MDS codes over $GF(9)$ are NR.

3. Semi-cyclic codes of BCH type

Let z be a primitive element of F. Assume $a = z^s$. Then, we have $x^n - a = \prod_{i=1}^{n}(x - \zeta^s \eta^i)$, where η and ξ are primitive n-th roots of unity and ζ, respectively. C is an α-cyclic $(n, n - k)$-code *of BCH type* if the generator polynomial is $g(x) = \prod_{i=1}^{k}(x - \xi^s \eta^{i+u})$ for some u.

Theorem 10. *Semi-cyclic $(q + 1, k)$-codes of BCH type over F are NR.*

Proof. Let ξ be a primitive element of $GF(q^2)$ $(\supset F)$, and put $\zeta = \xi^{q+1}, \eta = \xi^{q-1}$. Assume $\alpha = \zeta^s$. Note that $s + k$ must be odd if q is odd. We only prove the case where s is even. (We can similarly prove the case of odd s.) We may assume that C is an α-cyclic $(q + 1, q - 2m)$-code over F with generator polynomial $g(x) = (x - \sqrt{\alpha}) \prod_{i=1}^{m}(x - \sqrt{\alpha}\eta^i)(x - \sqrt{\alpha}\eta^{-i})$. (If $x - \sqrt{\alpha}$ is not a factor of $g(x)$, take the dual code.) For the diagonal matrix $T = \sqrt{\alpha}\eta^{-m} (t_{ij})$ with $t_{ii} = \eta^{i-1}, i = 1, 2, \ldots, 2m + 1$, and $P = (1, 1, \ldots, 1)$, $(T, P, q + 1)$ defines C. Put $L = \{P(t^{2m}, t^{2m-1}, \ldots, t, 1) \in PG(2m, q^2) \mid t \in \langle\eta\rangle\} = \{P(PT^i) \in PG(2m, q^2) \mid i = 1, 2, \ldots, q + 1\}$, where $\langle\eta\rangle$ stands for the group generated by η, and $K = \{P(t^{2m}, t^{2m-1}, \ldots, t, 1) \in PG(2m, q^2) \mid t \in F \cup \{\infty\}\}$. Define a map $\psi: GF(q^2) \cup \{\infty\} \to GF(q^2) \cup \{\infty\}$ by $\psi(t) = \{(\xi^{i_0}t - \xi^{qi_0})/(\xi^{i_\infty}t - \xi^{qi_\infty})\}, 1 \leq i_0 \neq i_\infty \leq q + 1$. Then ψ is bijective because we have $i_0 \neq i_\infty \mod q + 1$. It is easily verified that $\psi(t)^q = \psi(t)$ for $t \in F\backslash\{\eta^{i_\infty}\}$ and hence $\psi(\langle\eta\rangle) = F \cup \{\infty\}$. Let T_{2m} be the matrix defined by $M = \begin{bmatrix} \xi^{i_0} & -\xi^{qi_0} \\ \xi^{i_\infty} & -\xi^{qi_\infty} \end{bmatrix}$ (see the section preceding Theorem 7). Using the projectivity given by T_{2m}, we see that L is projectively equivalent to K in $PG(2m, q^2)$. Now we set $i_0 = q + 1$. Then $(\xi^{i_0}\eta^j - \xi^{qi_0})^{2m}\} (\sqrt{\alpha}\eta^{-m})^j = \zeta^{2m} (\beta_j - 2)^m(\sqrt{\alpha})^j \in F$, where $\beta_j = \eta^j + \eta^{-j} (\in F), 1 \leq j \leq q$. Hence all entries of $T_{2m}H$ are in F, where $H = G_{T,P}^{q+1}$. Thus $T_{2m}H$ generates a

$(q + 1, 2m + 1)$-code $\overline{C}$ over F which is NR. Since C is the dual code of $\overline{C}$, C is NR.

References

L.R.A. Casse and D.G. Glynn (1982), The solution to Beniamino Segre's problem $I_{r,q}$, $r = 3$, $q = 2^h$, *Geom. Dedicata* **13**, 157-164.

J. Georgiades (1982), Cyclic $(q + 1, k)$-codes of odd order q and even dimension k are not optimal, *Atti Sem. Mat. Fis. Univ. Modena* **30**, 284-285.

D.G. Glynn (1986), The non-classical 10-arc of $PG(4, 9)$, *Discrete Math.* **59**, 43-51.

J.W.P. Hirschfeld (1979), *Projective Geometries over Finite Fields*, Oxford University Press, Oxford.

J.W.P. Hirschfeld (1985), *Finite Projective Spaces of Three Dimensions*, Oxford University Press, Oxford.

F.J. MacWilliams and N.J.A. Sloane (1977), *The Theory of Error Correcting Codes*, North-Holland, Amsterdam.

T. Maruta (1990), On Singleton arrays and Cauchy matrices, *Discrete Math.* **81**, 33-36.

T. Maruta (1990), On the uniqueness of cyclic MDS codes, *Atti Sem. Mat. Fis. Univ. Modena*, to appear.

R.M. Roth and G. Seroussi (1985), On generator matrices of MDS codes, *IEEE Trans. Inform. Theory* **31**, 826-830.

R.M. Roth and G. Seroussi (1986), On cyclic MDS codes of length q over $GF(q)$, *IEEE Trans. Inform. Theory* **32**, 284-285.

B. Segre (1955), Ovals in a finite projective plane, *Canad. J. Math.* **7**, 414-416.

E. Zehendner (1983), A non-existence theorem for cyclic MDS-codes, *Atti Sem. Mat. Fis. Univ. Modena* **32**, 203-205.

Remarks on non-associative
Galois theory

G. Mason
University of California, Santa Cruz

1. Introduction

Suppose that $k \subseteq K$ is a (finite) Galois extension of (commutative) fields and that G is a group of k-automorphisms of K. Then the normal basis theorem tells us that there is a base of K over k which is permuted semi-regularly by G, ie. there are $x_1, \dots, x_N$ in K such that the elements $\{x_i g \mid 1 \le i \le N, g \in G\}$ form a base of K/k. Another way to say this is that if we think of K as a kG-module then, in fact, K is a free kG-module, that is it is the direct sum of copies of the group algebra kG.

In this paper we discuss the possibility of establishing similar results in more general contexts, mainly concerning translation planes. Here K becomes the (non-associative) co-ordinatizing ring of the plane, k is the kernel, and G a group of automorphisms in the so-called linear translation complement — that is to say G fixes k pointwise. In truth, this philosophy differs only slightly from that of Mason (1987), where the author promoted these ideas from the point of view of representation theory. We believe that it is possible to establish analogues of the "basis theorem," ie. to prove that K is a free kG-module, for "most" groups G and "most" finite co-ordinatizing rings K. Note, however, that it is certainly false in general.

2. An infinite example

Let us first return to the situation of the finite Galois extension $k \subseteq K$ and let V be a 2-dimensional K-vector space. Then our group G of k-automorphisms of K permutes the 1-spaces of V, hence induces a collineation of the corresponding Desarguesian plane. If we think of V as a $2[K : k]$-dimensional vector space over k then we arrive at the following situation:

(a) V is a $2N$-dimensional k-vector space
(b) G is a group of k-linear transformations of V (2.1)
(c) G preserves a spread in V
(d) V is a free kG-module

Main Problems
(A)   Can one show, for "most" groups G, that if k is finite, then
(a)-(c) of (2.1) *imply* (d)?   This is the "basis reduction" problem for
translation planes.

(B)   For a given group G, can one find V, k, N, with k *finite*, so
that (2.1) (a)-(d) hold?   This is the "basis existence" problem for
translation planes.

The condition "k finite" is where all the difficulty lies in (B).
Indeed, infinite fields $k \subseteq K$ with any given group G as k-
automorphisms are in plentiful supply, but if we insist on the
finiteness of k and K we can only get cyclic groups as automorphism
groups!   On the other hand, there is currently no group G for which
the basis existence problem is known to be false for k finite.

3. Results

The remainder of this paper deals with the problem (A). Specifically,
we shall take G to be a symmetric group $G \cong \Sigma_n$, and ask when and if
the basis reduction problem for G has an affirmative solution.
We have dealt previously (Mason and Eastman 1990) with the case
in which $n \geq 4$ and k is a finite field in which $p \mid |k| - 1$ for each
prime $p \leq n$.   Here the answer is "yes".   In this paper we take up the
case char $k = 2$.   One cannot expect an affirmative answer in general
here, for the group Σ_5 induces collineations of the Desarguesian plane
over $GF(q^2)$, q a power of 2, and the vector space V of (2.1) is a
2-dimensional $GF(q^2)A_5$-module.   There are lots more examples; in
each case V is the direct sum of "natural" 2-dimensional modules for
A_5, with coefficients extended to a suitable ground field.   Theorem 1
shows that this is the only exception under suitable conditions on k .
It might be pertinent to recall here that the (absolutely) irreducible
modules for A_5 in characteristic 2 are of dimensions 1, 2, 2, 4, namely
the trivial module 1 and $N, \overline{N}, N \otimes \overline{N}$ where N is the natural
module arising from $A_5 \cong SL_2(4)$ and $\overline{N}$ is the conjugate module.   Note
that $N \otimes \overline{N} = St$ (Steinberg) is a projective A_5-module of dimension 4,
arising from $\Sigma_5 \cong O_4^-(2)$. N admits Σ_5 if we think of it as a 4-
dimensional module over $GF(2)$.

Theorem 1. Let $G \cong \Sigma_5$, and assume that $k = GF(q^4)$ with q a power of
2.   If V is as in (a)-(c) of (2.1) then one of the following holds:
(a) V is the direct sum of copies of N.

(b)   *V is a free kG-module.*

Thus either we get the "basis reduction", or else (a) holds and we are in a situation which is known to exist.   Again, we do not know if there are any examples of (b).

Proof.   The assumption that k is a quartic extension of $GF(q)$ allows us to conclude that $15 \mid |k| - 1$, so the methods of (Mason and Eastman 1990) are applicable to the elements of odd order. Specifically, we may apply Lemma 4.6 (loc. cit) to conclude that if $x \in G$ has odd order then either $C_V(x) = 0$, or x induces a homology on V, or V is a free $k\langle x\rangle$-module and $C_V(x)$ is a subplane of V.

We may assume that part (a) of the theorem does not hold, so by the above some composition-factor of V (as G-module) is either 1 or St. In both cases we get $C_V(x) \neq 0$ for x of order 3.

Next we assert that if $T \cong D_8$ is a 2-Sylow of G then V is a free kT-module.   Here we use the methods of Mason (1987).   Namely, if the assertion is false then there is a four group $U \subseteq T$ such that $C_V(U) = C_V(u)$ for all $u \in U^{\#}$ (cf. section 6 of Mason (1987)).   By 6.9 (loc. cit.) we see that U must centralise every subgroup of odd order in G that it normalizes, whence in fact U is a 2-Sylow subgroup of $G' \cong A_5$.

Now choose the element x of order 3 so that it is inverted by some $u \in U^{\#}$, and let $W = C_V(x) \neq 0$.   Then $W_0 = C_W(u)$ has one-half the dimension of W, and since $G' = \langle U, x \rangle$ then $W_0 = C_V(G) \neq 0$.   Then as an element y of order 5 satisfies $C_V(y) = X \neq 0$, and again if u inverts y, then $G' = \langle U, y \rangle$ is such that $X_0 = C_X(u) = C_V(G')$ has dimension one-half the dimension of W.   So dim W = dim $X \neq 0$; hence as explained above, both x and y are homologies.

Let the axis of x be l, say.   Then $C_V(G') \subseteq l$, so G' acts on l and is trivial on it, so G' acts trivially on l.   But u also acts on the co-axis l' of x, so $C_V(u) = C_V(l') \oplus l$ has dimension greater than one-half dim V, contradiction.   This proves, then, that V is a free kT-module.

Next we show that x of order 3 is not a homology.   If it is, let its axis be l.   Since $C_V(U) \neq 0$ with U a 2-Sylow of G' which we take now to be normalised by x, either $C_V(U) \subseteq l$ or $C_V(U)$ is a subplane and l is a line of $C_V(U)$.   In either case, U fixes l; so $[U, x] = U$ acts trivially on l.   As V is a free kU-module this is impossible.   So x is not a homology and hence it is free; that is, V is a free $k\langle x\rangle$-module, as explained above.   In particular, dim $C_V(x) = 1/3$ dim$V$.

From this we conclude that as G-module, V has composition factors other than St.   On all the others, an element y of order 5 has non-trivial fixed vectors, so this holds on V. So again y is either free on V or y is a homology.

Next, we know that V is a *projective* G'-module since a 2-Sylow U of G' is such that V is a free kU-module by an earlier reduction. So V, as G'-module, is the direct sum of projective indecomposable modules for G'. We work over an algebraic closure for $GF(2)$, where the projective indecomposable modules for A_5 have the following structure (this is well known, cf. Alperin 1979).

$$P\ (1): \qquad 1\begin{vmatrix} \overline{N} & |1| & N \\ \hline N & |1| & \overline{N} \end{vmatrix} 1$$

$$(2.2)$$

$$P\ (N): \qquad N\ \ |1|\ |\overline{N}\ |1|N \quad \text{(and dually)}$$

$St.$

Being a sum of such modules, it is clear that $\dim C_V(y) < 1/2 \dim V$, so V is also free as $k\langle y\rangle$-module. So V is a free $k\langle g\rangle$-module for all g of odd order, and a free kT-module. This is equivalent to case (b) of the theorem, which is thus proved. □

It is possible that the conclusions of the theorem hold even if k is not a quartic extension of some subfield, but a new idea is needed if it is to be proved.

Our second theorem is a step in the direction of the *conjecture* that if $G \cong \Sigma_n$, $n \geq 6$, and if (a)-(c) of (2.1) hold (char = 2) then V is a projective KG-module, ie. V is a projective kT-module for $T \in \mathrm{Syl}_2(G)$.

Theorem 2. *If $G \cong \Sigma_6$ then V is a free kT-module.*

Remark. We get the result *gratis* for Σ_7 since Σ_6 and Σ_7 share a 2-Sylow.

Corollary. *If $G \cong \Sigma_6$ and $k = GF(q^4)$ where q is a power of 2 then V is a free kG-module.*

So for Σ_6 there is no analogue of part (a) of Theorem 1, and thus if k is a quartic extension then we always have "basis reduction".

To prove the corollary on the basis of Theorem 2 it is enough to show that V is a free $k\langle g\rangle$-module for each $g \in G$ of odd order. But any such $g \in G \cong \Sigma_6$ lies in a subgroup $H \cong \Sigma_5$ and we can appeal to Theorem 1.

The proof of Theorem 2 thus is similar to Theorem 1, but makes greater use of the 2-modular representation theory of Σ_6. To prove Theorem 2 we again apply Section 6 of Mason and Eastman (1990) to deduce that if it is false then there is an elementary abelian 2-subgroup $E \leq G$ with the property that $C_V(E) = C_V(E_0)$ for each subgroup $E_0 \leq E$ of index 2. Obviously $4 \leq |E| \leq 8$.

Case 1: $|E| = 4$: Suppose that $E \not\leq G' \cong A_6$. Then if $e \in E \backslash G$ we have $C_G(e) \cong Z_2 \times \Sigma_4$ and so $\langle C_G(e) \mid e \in E^\# \rangle = G$ acts on $C_V(E)$. Then $[C_V(E), G] = 0$ and by the $P \times Q$ lemma applied to a subgroup $R \leq C_G(e)$ of order 3, the condition $[C_V(e), R] = 0$ yields $[V, R] = 0$, contradiction. So $E \leq G'$ and we may take E to be one of $A = \langle (12)(34), (13)(24) \rangle$, or $B = \langle (12)(34), (12)(56) \rangle$. If $C_V(D) = C_V(d)$ for all $d \in D^\#$ where D is *either* of A or B then again $G = \langle C_G(d) \mid d \in A^\# \cup B^\# \rangle$ acts on $C_V(E)$ and we get $[C_V(E), G'] = 0$. But then if R is as above and $x \in G'$ an involution inverting R we get dim ≤ 2 dim $C_V(R) \Rightarrow$ dim $V \leq 4/3$ dim $C_V(x)$, an impossibility.

Hence one of A, B (call it F) is free on V (that is, V is a free kF-module) and the other — which is E — is not. We can build a subgroup $H \cong \Sigma_4$ of G' in which $O_2(H) = F$, H has Sylow 3-subgroup R, and $H = FRE$. Setting $W = C_V(F)$ we have dim $V = 4$ dim W as F is free, and as $C_V(E) = C_V(E \cap F)$ we can conclude that $W = C_V(H)$. Let $J \cong A_5$ be a subgroup of G of the indicated isomorphism type and with 2-Sylow normalizer FR. As F is free then F is a projective as J-module and hence a sum of modules of the type (2.2). The only such module for which $C_V(F) = C_V(FR)$ is St, so $V \cong k \cdot St$ as J-module.

Next, the irreducible modules for Σ_6 in characteristic 2 are the following: $1, 4_1, 4_2, St_1$ where St_1 is the (projective) 16-dimensional Steinberg $4_1 \otimes 4_2$ (recall $\Sigma_6 \cong Sp(4, 2)$). Since V is dispersive but not projective as G-module (cf. Mason (1987), St_1 cannot be a composition-factor of V, and neither is 1 since $V \cong k \cdot St$ as J-module. But only one of 4_1, 4_2 is isomorphic to St as J-module, so in fact V has only 4_1 (say) as composition-factor (as G-module). As is well known, this makes V a completely reducible G-module. But now we see that a suitable involution t, acting as a transvection on 4_1, satisfies dim $V = 4/3$ dim $C_V(t)$. This impossibility establishes that Case 1 cannot occur.

Case 2: $|E| = 8.$ After Case 1 we know that each four-group of G is free, in particular V is projective as G'-module. The characteristic 2 irreducibles for A_6 are $1, 4_1, 4_2, 8_1, 8_2$ where the first three restrict to irreducibles of A_5 in a notation consistent with that used above, and where $8_1 \oplus 8_2 \cong St_1$ as G-module. As V is not projective as G-module, it has no summands of type 8_1 or 8_2 as G'-module. So as G'-module, V is the direct sum of a certain number of the projective covers of $1, 4_1, 4_2$.

We claim that V has no summand which is a projective cover of 1. If it does, call such a summand P. Then P has a unique G'-submodule $K \cong 1$, and since $[K, E \cap G'] = 0$ then E also fixes K, ie. K is G-invariant.

Next, the socle of $P/K \cong 4_1 \oplus 4_2$. Let us choose notation so that the four-group $E_0 = E \cap G'$ satisfies $|C_{4_i} (E_0)| = 2^i$, $i = 1, 2$; this is always possible. Now, if M_i/K are the two modules in question, that is, $M_i/K \cong 4_i, i = 1, 2$, then one calculates that we have $|C_{M_i} (E_0)| = 2^{i+1}$, $i = 1, 2$. In particular as $[E, C_V (E_0)] = 0$ then both M_1 and M_2 are G-invariant.

Next, we have $E = C_G(E_0) \trianglelefteq ER$ where $R \cong Z_3$, so $E = E_0 \times E_1$ where $E_1 = C_E(R) \cong Z_2$. Now we have $[E_1 \times R, C_{M_1}(E_0)] = 0$, and as $[R, C_{M_1}(E_1)]$ $\neq 0$ by the $P \times Q$-lemma, then $C_{M_1}(E_1)$ is a hyperplane of M_1. Choose $e \in E_0^{\#}$. As dim $M_1 = 5$ then dim $C_{M_1}(e) = 3$. Since $C_{M_1}/K (e) \leq C_{M_1}/K (E_1)$ it follows that $C_{M_1}(e) \geq C_{M_1}(\langle E_1, e \rangle)$. But then $[C_{M_1}(e), E]$ $= 0$, ie. $C_{M_1}(E)$ is a 3-space, and hence $C_{M_1}/K(E_0)$ is a 2-space. This is a contradiction which proves that indeed V has no summand isomorphic to P.

Finally, let P be a summand of V (as G-module) isomorphic to the projective cover $P(4_i)$ of 4_i, $i = 1$ or 2. As before G leaves the socle M of P invariant, where of course $M \cong 4_i$. The socle of P/M is just 1, so we can find a G-invariant submodule W of V such that $M \subseteq W$, $W/M \cong 1$ and W is a *non-split* extension as G'-module. Now $|W| = 2^5$ and the orbits of G on $W - M$ are of lengths 6, 10 with isotropy groups Σ_5, $(Z_3 \times Z_3) \cdot D_8$ respectively. We deduce that E fixes no vector in $W - M$. On the other hand, some hyperplane E_1 of E is (conjugate to) a subgroup $(Z_3 \times Z_3) \cdot D_8$, so $C_W(E) \underset{\neq}{\subseteq} C_W(E_1)$. This contradiction completes the proof. $\square$

Acknowledgement. This work is supported by the National Science Foundation, U.S.A.

References

J. Alperin (1979), Projective modules for $SL(2, 2^n)$, *J. Pure Appl. Algebra* **15**, 219-234.

G. Mason (1987), Representation theory and translation planes, *Proc. Sympos. Pure Math.* **47**, 211-222, American Mathematical Society, Providence.

G. Mason and M. Eastman (1990), Spread-invariant representations of symmetric groups, *J. Algebra* **128**, 86-100.

Codes of nets with translations

G.E. Moorhouse
University of Wyoming

1. Introduction

A k-net of order n is an incidence structure consisting of a set $\mathcal{P}$ of n^2 points, together with nk distinguished subsets $l_{ij} \subset \mathcal{P}$ called lines ($1 \leq i \leq k, 1 \leq j \leq n$), such that

$$|l_{ij} \cap l_{i'j'}| = \begin{cases} 1 & \text{if } i \neq i'; \\ 0 & \text{if } i = i', j \neq j'; \\ n & \text{if } i = i', j = j'. \end{cases}$$

Thus the lines are partitioned into k parallel classes $\{l_{ij} : 1 \leq j \leq n\}$, $1 \leq i \leq k$, and each line has n points. Note that an $(n + 1)$-net is the same thing as an affine plane of order n. For any k-net $\mathcal{N}$ and any k'-subset of the parallel classes of $\mathcal{N}$, we obtain a k'-subnet $\mathcal{N}' \subseteq \mathcal{N}$ of order n, also with point set $\mathcal{P}$. The description of nets found here is more amenable to our specialised rank computations than to a broader presentation; more general information concerning nets is found in Baer (1939); Beth et al. (1985); Bruck (1951), (1963); Jungnickel (1990); and Pickert (1975).

We denote the p-rank of a net $\mathcal{N}$ (the p-rank of its incidence matrix) by $\mathrm{rank}_p \mathcal{N}$. In Moorhouse (1990) we proposed the following conjecture, supported by numerous computational examples:

Conjecture 1. Let $\mathcal{N}_k$ be any k-net of order n, and let $\mathcal{N}_{k-1}$ be any $(k - 1)$-subnet thereof. If p is any prime such that $p^2 \nmid n$, then

$$\mathrm{rank}_p \mathcal{N}_k - \mathrm{rank}_p \mathcal{N}_{k-1} \geq n - k + 1.$$

The only interesting primes are those dividing n, since as shown in Moorhouse (1990), for $k \geq 1$ we have $\mathrm{rank}_p \mathcal{N}_k \leq (n - 1)k + 1$, in which equality holds if $p \nmid n$. Furthermore Conjecture 1 holds trivially for $k \leq 2$, and its validity for $k = 3$ has been shown using the theory of loop characters (see Theorem 5 below).

In sections 2 and 3 we prove Theorems 2, 3, and Corollary 4.

Theorem 2. *Conjecture 1 holds for 4-nets constructed from $3 \times p$ difference matrices over cyclic groups of prime order p, that is, for 4-nets of prime order p admitting a central translation of order p.*

Theorem 3. *Let $\mathcal{N}_k$ be a translation k-net with abelian translation group $T = G \times G$, $k \geq 2$, and let $\mathcal{N}_{k-1}$ be any $(k-1)$-subnet thereof. Let $\mathbf{F}_p$ be a field of prime order p, and let $\mathcal{A}$ be the augmentation ideal of the group algebra $\mathbf{F}_p[G]$. Then*

$$rank_p\ \mathcal{N}_k - rank_p\ \mathcal{N}_{k-1} \geq dim\mathcal{A}^{k-1}.$$

Corollary 4. *Conjecture 1 holds for translation nets with abelian translation groups.*

The values $dim\mathcal{A}^{k-1}$ are determined by Jennings (1941); see Theorem 11 below. In case $k = 3$, the lower bound of Theorem 3 holds with equality since the subgroup $G_2 = \langle [G, G], G^p \rangle$ of Theorem 11 coincides with the subloop K of Theorem 5. It is hoped that these results may generalise to other nets without the assumption of translations, after replacing group algebras by loop algebras; see Bruck (1944). We are encouraged in this direction by the fact that many of the relevant facts concerning group algebras, including parts of Theorem 11 below, would seem to extend to loop algebras.

The following five additional results are shown in Moorhouse (1990); our reference for loop theory is Bruck (1958).

Theorem 5. *Let G be a loop of order n, and let $\mathcal{N}$ be the corresponding 3-net. Let p be a prime such that $p^e \parallel n$ (meaning that p^e is the highest power of p dividing n). Then*

$$rank_p\ \mathcal{N} = 3n - 2 - s \geq 3n - 2 - e$$

where $p^s = [G:K]$ and K is the unique minimal normal subloop of G such that G/K is an elementary abelian p-group.

Corollary 6. *Let $\mathcal{N}$ be a 3-net of order $n \equiv 2\ mod\ 4$. If $\mathcal{N}$ is extendable to a 4-net, then $rank_2\ \mathcal{N} = 3n - 2$.*

A 3-net is *cyclic* if it is coördinatised by a cyclic group, or equivalently, if it is obtainable from a Latin square in which all rows are cyclic shifts of the first row.

Theorem 7. *Let G be a loop of order n, with corresponding 3-net $\mathcal{N}$.*
(i) *Suppose that n is squarefree (that is, n is a product of distinct primes), or that G is a nilpotent group. Then $\mathcal{N}$ is cyclic if and only if $\text{rank}_p \mathcal{N} = 3n - 3$ for every prime $p \mid n$.*
(ii) *Suppose that G is a nilpotent loop. Then G is generated by a single element, if and only if $\text{rank}_p \mathcal{N} \geq 3n - 3$ for every prime $p \mid n$.*

Theorem 8. *An explicit basis for the $\mathbf{F}_p$-code of $AG(2, p)$ is obtained by choosing all p lines of some parallel class, followed by any p - 1 lines from any other parallel class, plus any p - 2 lines from yet another parallel class, and so on, finally taking 0 lines from the last remaining parallel class; this gives $p(p + 1)/2$ lines in all. In particular, Conjecture 1 holds with equality in the case of subnets of $AG(2, p)$.*

Actually, the final assertion of Theorem 8 is obtainable from Corollary 4 without appeal to the proof in Moorhouse (1990).

Our main interest in Conjecture 1 is due to the following:

Theorem 9. *Suppose that Π is a projective plane of order n, where n is squarefree or $n \equiv 2 \bmod 4$. If Conjecture 1 holds for n, then n is prime and Π is Desarguesian.*

2. 4-nets of prime order with translations

A *central* (or "strict") *translation* of a net $\mathcal{N}$ is an automorphism of $\mathcal{N}$ preserving each parallel class of $\mathcal{N}$, and fixing every line in some parallel class (called the *direction* of the translation). We show how nets with such strict translations may be constructed; see also Beth et al. (1985); Jungnickel (1990).

Suppose that G is a group of order n , and that $\sigma_3, \sigma_4, ..., \sigma_k : G \to G$ are bijections such that whenever $i \neq j$, the map $g \mapsto \sigma_i(g)\sigma_j(g)^{-1}$ is also a bijection $G \to G$. Then the sets

$$l_{1g} = \{(g, y) : y \in G\}, \; g \in G,$$
$$l_{2g} = \{(x, g) : x \in G\}, \; g \in G,$$
$$l_{ig} = \{(x, g\sigma_i(x)) : x \in G\}, \; 3 \leq i \leq k, \; g \in G$$

form a k-net $\mathcal{N}$ of order n on the point set $\mathcal{P} = G \times G$. For $h \in G$ define $\tau_h : \mathcal{P} \to \mathcal{P}$ by $\tau_h(x, y) = (x, hy)$. Then τ_h maps $l_{1g} \mapsto l_{1g}$, $l_{ig} \mapsto l_{i, hg}$ for $2 \leq i \leq k$, and thus $\mathcal{N}$ admits a group of central translations whose common direction is the first parallel class $\{l_{1g} : g \in G\}$. Moreover, every net with the latter property is constructible in this way. It is customary to specify $\mathcal{N}$ by the *difference matrix*

$$\begin{bmatrix} 1 & 1 & \dots & 1 \\ \sigma_3(g_1) & \sigma_3(g_2) & \dots & \sigma_3(g_n) \\ \sigma_4(g_1) & \sigma_4(g_2) & \dots & \sigma_4(g_n) \\ \vdots & \vdots & & \vdots \\ \sigma_k(g_1) & \sigma_k(g_2) & \dots & \sigma_k(g_n) \end{bmatrix}$$

where $g_1, g_2, \dots, g_n$ are the elements of G is some order (see Beth et al. (1985); Jungnickel (1990)); however, the language of difference matrices will not be required for our presentation. We also remark that there is no loss of generality in assuming that $\sigma_3(g) = g$ for all $g \in G$, and $\sigma_i(g_1) = g_1 = 1$ for all i.

We develop some notation which will be useful in handling codes of such nets. Let $F_p{}^{\mathcal{P}}$ be the n^2-dimensional vector space over F_p consisting of all functions $\mathcal{P} \to F_p$. To each line l_{ig} of $\mathcal{N}$ there corresponds the characteristic function

$$\chi_{ig} : \mathcal{P} \to F_p, \quad \chi_{ig}(x, y) = \begin{cases} 1, & (x, y) \in l_{ig}, \\ 0, & (x, y) \notin l_{ig}. \end{cases}$$

We may define the F_p-code of $\mathcal{N}$ as

$$C_p(\mathcal{N}) = \sum_{i=1}^{k} \sum_{g \in G} F_p \chi_{ig} ,$$

which is the subspace of $F_p{}^{\mathcal{P}}$ spanned by the characteristic functions of the line sets. The p-rank of $\mathcal{N}$ is then

$$\operatorname{rank}_p \mathcal{N} = \dim C_p(\mathcal{N}).$$

If $1 \le i_0 \le k$, then a $(k-1)$-subnet $\mathcal{N}_{k-1} \subset \mathcal{N}$ is formed by excluding the i_0-th parallel class of $\mathcal{N}$ and the F_p-code of $\mathcal{N}_{k-1}$ is

$$C_p(\mathcal{N}_{k-1}) = \sum_{\substack{1 \le i \le k \\ i \ne i_0}} \sum_{g \in G} F_p \chi_{ig} .$$

Clearly we have:

Lemma 10. *With the above notation, we have* $rank_p \, \mathcal{N} - rank_p \, \mathcal{N}_{k-1} =$ $n - \dim \, \mathcal{V}$, *where* $\mathcal{V}$ *is the vector space consisting of all sequences* $(a_g :$ $g \in G)$, *where* $a_g \in F_p$, *such that* $\sum_{g \in G} a_g \chi_{i_0 g} \in C_p(\mathcal{N}_{k-1})$.

For a group G, recall that the *augmentation ideal* of the group algebra $F_p[G]$ is the ideal defined by

$$\mathcal{A} = \left\{ \sum_{g \in G} a_g g \in F_p[G] : \sum_{g \in G} a_g = 0 \right\}.$$

The j-th power of $\mathcal{A}$ is the ideal spanned over F_p by

$$\{(1 - x_1)(1 - x_2) \ldots (1 - x_j) : x_1, x_2, \ldots, x_j \in G\}.$$

In case G is cyclic of order p, it is well known that $\mathcal{A}$ is nilpotent, and that the *only* ideals of $F_p[G]$ are those in the chain

$$F_p[G] \supset \mathcal{A} \supset \mathcal{A}^2 \supset \ldots \supset \mathcal{A}^{p-1} \supset \mathcal{A}^p = 0,$$

so $\mathcal{A}^{p-j}$ is the unique ideal of $F_p[G]$ of dimension j. It follows that $\text{Ann} \, \mathcal{A}^j = \mathcal{A}^{p-j}$ where $\text{Ann} \, \mathcal{B} = \{\alpha \in F_p[G] : \alpha\beta = 0 \text{ for all } \beta \in \mathcal{B}\}$ is the *annihilator* of an arbitrary ideal $\mathcal{B} \subseteq F_p[G]$.

Proof of Theorem 2. Let G be a cyclic group of prime order p, and let $\mathcal{N}$ be a 4-net of order p with a central translation of order p. Thus we may assume that $\mathcal{N}$ is constructed as above with $k = 4$, $\sigma_3 = $ identity, $\sigma_4 = \sigma$, $\sigma(1) = 1$. Corresponding to the lines l_{ig} defined as above, we have the characteristic functions

$$\chi_{1g}(x, y) = \delta_{x,g}, \; \chi_{2g}(x, y) = \delta_{y,g}, \; \chi_{3g}(x, y) = \delta_{gx,y}, \; \chi_{4g}(x, y) = \delta_{g\sigma(x),y}$$

for $g \in G$. With Lemma 10 in mind, for given scalars $a_{ig} \in F_p$ we define $\chi = \sum_{i=1}^{4} \sum_{g \in G} a_{ig} \chi_{ig} \in F_p{}^p$ and suppose that $\chi = 0$. Then for arbitrary $x, y \in G$ we have $0 = \chi(1, 1) - \chi(x, 1) - \chi(1, y) + \chi(x, y)$, or

$$0 = a_{3,1} - a_{3,x^{-1}} - a_{3,y} + a_{3,x^{-1}y} + a_{4,1} - a_{4,\sigma(x)^{-1}} - a_{4,y} + a_{4,\sigma(x)^{-1}y}.$$

Or writing $\alpha_i = \sum_{y \in G} a_{i,y} \, y \in F_p[G]$ for $i = 3, 4$, we obtain

$$\alpha_3(1-x) + \alpha_4(1-\sigma(x)) = \sum_{y \in G} (a_{3,y} - a_{3,\,x^{-1}y} + a_{4y,y} - a_{4,\sigma(x)^{-1}y})y$$

$$= (a_{3,1} - a_{3,x^{-1}} - a_{4,1} + a_{4,\sigma(x)^{-1}})\sum_{y \in G} y \in F_p\gamma$$

for all $x \in G$, where $F_p\gamma = \mathcal{A}^{p-1}$ is the one dimensional ideal of F_p spanned by $\gamma = \sum_{g \in G} g \in F_p[G]$. Since $x \in G$ is arbitrary, we may replace x by z or by xz, where $x, z \in G$ are both arbitrary, so that

$$[\alpha_3(1-x) + \alpha_4(1-\sigma(x))]x^{-1} \in F_p\gamma,$$

$$\alpha_3(1-z) + \alpha_4(1-\sigma(z)) \in F_p\gamma,$$

$$[\alpha_3(1-xz) + \alpha_4(1-\sigma(xz))](-x^{-1}) \in F_p\gamma.$$

Adding the above relations yields

$$\alpha_4(1 - \sigma(x)x^{-1} - \sigma(z) + \sigma(xz)x^{-1}) \in F_p\gamma$$

for all $x, z \in G$. This means that $\alpha_4 \mathcal{B} \subseteq F_p\gamma$ where $\mathcal{B} \subset F_p[G]$ is the ideal generated by all expressions of the form

$$1 - \sigma(x)x^{-1} - \sigma(z) + \sigma(xz)x^{-1} = (1 - \sigma(x)x^{-1})(1 - \sigma(z)) + (\sigma(xz) - \sigma(x)\sigma(z))x^{-1}$$

for all $x, z \in G$. We may assume that $\sigma(xz) \neq \sigma(x)\sigma(z)$ for some $x, z \in G$; otherwise $\sigma \in \text{Aut}G$, in which case, as will appear in section 3, $\mathcal{N}$ is a translation 4-net of order p and hence is a 4-subnet of $AG(2, p)$, and the result follows by Corollary 4 (which will be proved in section 3), or by Theorem 8, (proved in Moorhouse (1990)).

Thus $\mathcal{B} \not\subseteq \mathcal{A}^2$ and $\mathcal{B} \subseteq \mathcal{A}$, and so by the preceding remarks, we obtain $\mathcal{B} = \mathcal{A}$. Now $\alpha_4 \mathcal{A} \subseteq F_p\gamma$ is equivalent to $\alpha_4 \in \text{Ann } \mathcal{A}^2 = \mathcal{A}^{p-2}$, so by Lemma 10,

$$\text{rank}_p \mathcal{N} - \text{rank}_p \mathcal{N}_3 \geq p - \dim \mathcal{A}^{p-2} = p - 2$$

where $\mathcal{N}_3$ is the 3-subnet of $\mathcal{N}$ formed by omitting the fourth parallel class. But $\text{rank}_p \mathcal{N}_3 = 3p - 3$ by Theorem 5, and so $\text{rank}_p \mathcal{N} \geq 4p - 5$. Now for *any* 3-subnet $\mathcal{N}' \subset \mathcal{N}$ we have $\text{rank}_p \mathcal{N}' \leq 3p - 2$ by the remarks following Conjecture 1, and so

$$\text{rank}_p \mathcal{N} - \text{rank}_p \mathcal{N}' \geq (4p - 5) - (3p - 2) = p - 3$$

as required. □

3. Translation nets

A net $\mathcal{N}$ is a *translation net* if it admits an automorphism group T (called a *translation group* of $\mathcal{N}$) which acts regularly (=sharply transitively) on the points of $\mathcal{N}$, and preserving each parallel class of $\mathcal{N}$. For a given translation net $\mathcal{N}$, the translation group T need not be abelian, or even unique. We describe how the most general translation nets with abelian translation groups are constructed, then proceed to prove Theorem 3 and Corollary 4. For more on translation nets, see Beth et al. (1985); Jungnickel (1990).

Suppose that G is a group of order n, with bijections $\sigma_3, \sigma_4, \ldots, \sigma_k$ as in section 2, with corresponding k-net $\mathcal{N}$ of order n defined as before. But now assume additionally that $\sigma_3, \sigma_4, \ldots, \sigma_k$ are homomorphisms, so that $\sigma_3, \sigma_4, \ldots, \sigma_k \in \text{Aut} G$. For $u, v \in G$ define $\tau_{uv} : \mathcal{P} \to \mathcal{P}$ by $\tau_{uv}(x, y) = (ux, vy)$. Then τ_{uv} maps $l_{1,g} \mapsto l_{1,ug}, l_{2g} \mapsto l_{2,vg}, l_{ig} \mapsto l_{i, vg\sigma_i(u)^{-1}}$ for $3 \le i \le k$, so that $\mathcal{N}$ is a translation net with a translation group $T = (\tau_{uv} : u, v \in G) \cong G \times G$. This construction does not yield the most general translation net; however every translation net with an *abelian* translation group is constructible in this way. Indeed, in proving Theorem 3 we shall assume that G is abelian, though this hypothesis is not required in the foregoing construction of $\mathcal{N}$, and so perhaps might be avoidable.

In order for Theorem 3 to be useful, for a given group algebra $F_p[G]$ we shall require a knowledge of the dimensions of various powers of its augmentation ideal. Fortunately these dimensions may be derived from G by purely group-theoretic methods, as shown by Jennings (1941). Although the following result is stated in Jennings (1941) and Passman (1977) only for p-groups (when $\mathcal{A}$ is nilpotent, and coincides with the radical of $F_p[G]$), in fact it holds for general n, by the same proof.

Theorem 11 (Jennings 1941). *For $i \ge 1$ define $G_i = \{g \in G: g - 1 \in \mathcal{A}^i\}$. Then the following statements hold:*

(i) $G = G_1 \trianglerighteq G_2 \trianglerighteq G_3 \trianglerighteq \ldots$ *is a sequence of characteristic subgroups of G, such that G_i / G_{i+1} is an elementary abelian p-group.*

(ii) *The subgroups G_i may be recursively determined by $G_1 = G$ and $G_i = \langle [G, G_{i-1}], G_{\lceil i/p \rceil}^p \rangle$ for $i \ge 2$, where $\lceil i/p \rceil$ is the least integer $\ge i/p$,* and $G^p_{\lceil i/p \rceil}$ consists all p-th powers of the elements of $G_{\lceil i/p \rceil}$.

(iii) $dim\ (\mathcal{A}^j / \mathcal{A}^{j+1})$ equals the coefficient of X^j in the expansion of

$$\prod_{i=1}^{\infty} (1 + X^i + X^{2i} + \ldots + X^{(p-1)i})\ e_i$$

where $p^{e_i} = |G_i / G_{i+1}|$.

Observe that the above product is finite since $e_i = 0$ for i sufficiently large.

We shall also require the following, in which Ann $\mathcal{B}$ is the annihilator of $\mathcal{B}$:

Lemma 12.   *If $\mathcal{B}$ is any ideal of $F_p[G]$, then $dim\ \mathcal{B} + dim\ Ann\ \mathcal{B} = n$.*

Proof.   Ann $\mathcal{B}$ is the orthogonal complement of $\mathcal{B}$ with respect to the non-degenerate symmetric bilinear form $\left(\sum_{g \in G} a_g g\ ,\ \sum_{g \in G} b_g g \right) = \sum_{g \in G} a_g b_{g^{-1}}$ defined on $F_p[G]$. $\square$

Actually, Lemma 12 will be required only for $\mathcal{B}$ a power of the augmentation ideal $\mathcal{A}$, in which case an explicit basis may be obtained for Ann $\mathcal{A}$ if desired; see Hill (1970).

Proof of Theorem 3.   The proof is trivial for $k = 2$;   hence assume that $k \geq 3$.   We begin with the same steps as the proof of Theorem 2 in section 2.   The characteristic functions of the line sets are

$$\chi_{1g}(x, y) = \delta_{x,g},\ \chi_{2g}(x, y) = \delta_{y,g},\ \chi_{ig}(x, y) = \delta_{g\ \sigma_i(x),y}$$

for $3 \leq i \leq k$; $g \in G$.   We may assume that $\mathcal{N}_{k-1}$ consists of all but the k-th parallel class $\{l_{kg} : g \in G\}$; this is more evident from Beth et al. (1985) and Jungnickel (1990) than from the latter expressions for the χ_{ig}'s, which unfortunately make the first two parallel classes appear different by construction.   For given scalars $a_{ig} \in F_p$ we define $\chi = \sum_{i=1}^{4} \sum_{g \in G} a_g \chi_{ig} \in F_p^{\mathcal{P}})$, and suppose that $\chi = 0$.   Then for arbitrary x, y

$\in G$ we have

$$0 = \chi(1, 1) - \chi(x, 1) - \chi(1, y) + \chi(x, y)$$

$$= \sum_{i=3}^{k} (a_{i,1} - a_{i,\sigma_i(x)^{-1}} - a_{i,y} + a_{i,\sigma_i(x)^{-1}y}).$$

Letting $\alpha_i = \sum_{y \in G} a_{iy}y \in F_p[G]$, we obtain

$$\sum_{i=3}^{k} \alpha_i(1 - \sigma_i(x)) = \sum_{i=3}^{k} \sum_{y \in G} (a_{i,y} - a_{i,\sigma_i(x)^{-1}y})y = \left(\sum_{i=3}^{k}(a_{i,1} - a_{i,\sigma_i(x)^{-1}})\right) \sum_{y \in G} y \in F_p\gamma$$

for all $x \in G$. Replacing x by x_1, x_2, or $x_1x_2 \in G$ gives

$$\sum_{i=3}^{k} \alpha_i(1 - \sigma_i(x_1))\sigma_3(x_1)^{-1} \in F_p\gamma,$$

$$\sum_{i=3}^{k} \alpha_i(1 - \sigma_i(x_2)) \in F_p\gamma,$$

$$\sum_{i=3}^{k} \alpha_i(1 - \sigma_i(x_1x_2)(-\sigma_3(x_1)^{-1})) \in F_p\gamma.$$

Adding these relations yields

$$\sum_{i=4}^{k} \alpha_i(1 - \sigma_3(x_1)^{-1}\sigma_i(x_1))(1 - \sigma_i(x_2)) \in F_p\gamma$$

for all $x_1, x_2 \in G$. Again replacing x_2 by x_2, x_3, $x_2x_3 \in G$ and combining yields

$$\sum_{i=5}^{k} \alpha_i(1 - \sigma_3(x_1)^{-1}\sigma_i(x_1))(1 - \sigma_4(x_2)^{-1}\sigma_i(x_2))(1 - \sigma_i(x_3)) \in F_p\gamma$$

for all $x_1, x_2, x_3 \in G$. Continuing in this way, we eventually obtain

$$\alpha_k(1 - \sigma_3(x_1)^{-1}\sigma_k(x_1))(1 - \sigma_4(x_2)^{-1}\sigma_k(x_2))\ldots$$

$$\ldots(1 - \sigma_{k-1}(x_{k-3})^{-1}\sigma_k(x_{k-3}))(1 - \sigma_k(x_{k-2})) \in F_p\gamma$$

for all $x_1, x_2, ..., x_{k-2} \in G$. Since the maps $G \to G$, $x \mapsto \sigma_i(x)^{-1}\sigma_k(x)$ are bijective for $3 \leq i < k$, we have $\alpha_k \mathcal{A}^{k-2} \subseteq F_p\gamma$, or equivalently, $\alpha_k \in$ Ann $\mathcal{A}^{k-1}$. By Lemma 12 we have dim Ann $\mathcal{A}^{k-1} = n -$ dim $\mathcal{A}^{k-1}$, and so the result follow by Lemma 10. $\square$

Proof of Corollary 4. By assumption $p^2 \nmid n$. By Theorem 11, we have dim $\mathcal{A}^{k-1} = \max\{n - k + 1, n - p^s\}$ where $p^s = p$ if G has a normal subgroup of index p; $p^s = 1$ otherwise. The result follows from Theorem 3. $\square$

References

R. Baer (1939), Nets and groups, *Trans. Amer. Math. Soc.* **46**, 110-141.

T. Beth, D. Jungnickel and H. Lenz (1985), *Design Theory*, Bibliographisches Institut, Mannheim.

R.H. Bruck (1944), Some results in the theory of linear non-associative algebras, *Trans. Amer. Math. Soc.* **56**, 141-199.

R.H. Bruck (1951), Finite nets I. Numerical invariants, *Canad. J. Math.* **3**, 94-107.

R.H. Bruck (1958), *A Survey of Binary Systems*, Springer-Verlag, Berlin.

R.H. Bruck (1963), Finite nets II. Uniqueness and embedding, *Pacific J. Math.* **13**, 421-457.

E.T. Hill (1970), The annihilator of radical powers in the modular group ring of a p-group, *Proc. Amer. Math. Soc.* **25**, 811-815.

S.A. Jennings (1941), The structure of the group ring of a p-group over a modular field, *Trans. Amer. Math. Soc.* **50**, 175-185.

D. Jungnickel (1990), Latin squares, their geometries and their groups. A survey, *Coding Theory and Design Theory part II: Design Theory* (ed. D. Ray-Chaudhuri), 166-225, Springer-Verlag, Berlin.

G.E. Moorhouse (1990), Bruck nets, codes and characters of loops, *Designs, Codes and Cryptography*, to appear.

D.S. Passman (1977), *The Algebraic Structure of Group Rings*, Wiley, New York.

G. Pickert (1975), *Projektive Ebenen*, 2nd edition, Springer-Verlag, Berlin.

Stabilisers of hyperovals
in $PG(2, 32)$

Christine M. O'Keefe, Tim Penttila and Cheryl E. Praeger
University of Western Australia

1. Introduction

A *hyperoval* $\mathcal{H}$ of the Desarguesian projective plane $PG(2, q)$ of even order $q = 2^e$, $e > 1$, is a set of $q + 2$ points, no three of which are collinear. Such a set $\mathcal{H}$ can be written, with a particular choice of coordinates for $PG(2, q)$, as

$$\mathcal{H} = \mathcal{D}(f) = \{(1, t, f(t)) : t \in GF(q)\} \cup \{(0, 1, 0), (0, 0, 1)\}$$

for some function f on $GF(q)$ satisfying $f(0) = 0$ and $f(1) = 1$, see Hirschfeld (1979). $\mathcal{H}$ is said to be *monomial* if, for some choice of coordinates, $\mathcal{H} = \mathcal{D}(f)$ where f is monomial, that is, $f(t) = t^n$ for some integer n. One well known class of monomial hyperovals is the *regular* hyperovals. A hyperoval is *regular* if it contains $q + 1$ points which are the points of a non-degenerate conic, that is, their coordinates satisfy an absolutely irreducible quadratic equation. Thus, with a suitable choice of coordinates $\mathcal{H} = \mathcal{D}(f)$ with $f(t) = t^2$.

We shall be concerned with the collineation stabiliser G and the homography stabiliser H of a hyperoval $\mathcal{H}$ in $PG(2, q)$, that is, the stabilisers of $\mathcal{H}$ in the collineation group $P\Gamma L(3, q)$ and homography group $PGL(3, q)$ of $PG(2, q)$ respectively.

If $\sigma : x \mapsto x^\sigma$ is an automorphism of $GF(q)$ then σ induces a collineation of $PG(2, q)$, called an *automorphic collineation*, as follows: if P is the point (x, y, z) then P^σ is $(x^\sigma, y^\sigma, z^\sigma)$. Let A denote the group of automorphic collineations of $PG(2, q)$, so that $|A| = e$.

Cherowitzo (1988) has discovered three hyperovals, conjectured to belong to an infinite sequence. These are

$$C = \{(1, t, t^\sigma + t^{\sigma+2} + t^{3\sigma+4}) : t \in GF(q)\} \cup \{(0, 1, 0), (0, 0, 1)\},$$

where $q = 2^h$ with $h = 5, 7$, or 9 and σ is an automorphism of $GF(q)$ with $\sigma^2 \equiv 2 \pmod{q - 1}$. The collineation stabiliser of a Cherowitzo hyperoval has been conjectured to be the group A of automorphic collineations (Cherowitzo 1988). We prove this for $q = 32$:

Theorem 1. *The collineation stabiliser of the Cherowitzo hyperoval C in PG(2, 32) is A, the group of automorphic collineations of PG(2, 32).*

The Payne hyperovals (1985) are

$$\mathcal{P} = \{(1, t, t^{1/6} + t^{3/6} + t^{5/6}) : t \in GF(q)\} \cup \{(0, 1, 0), (0, 0, 1)\}$$

for $q = 2^h$ where $h \geq 5$ is odd and the exponents are read modulo $q - 1$. The collineation stabilisers of the Payne hyperovals have been determined (Thas et al. 1988), but the case of $PG(2, 32)$ was treated separately and included the use of a computer.

Theorem 2. *The collineation stabiliser of the Payne hyperoval $\mathcal{P}$ in PG(2, 32) is $\langle A, \theta \rangle$, where A is the group of automorphic collineations of PG(2, 32) and θ is the homography $\theta : (x, y, z) \mapsto (y, x, z)$.*

Section 2 gives some preliminaries, then in Section 3 we study the homography stabiliser of a hyperoval $\mathcal{H}$ in $PG(2, 32)$, under the assumption that $\mathcal{H}$ is fixed by the group of automorphic collineations. We consider the special cases of the Cherowitzo and the Payne hyperovals separately in Sections 4 and 5.

2.   Preliminaries

In the following, we use $|X|$ to denote the number of elements in the set, or group X, and $\langle x, Y \rangle$ denotes the subgroup generated by the element x together with the elements of the subset, or subgroup Y. We write $Y \leq X$ to denote that Y is a subgroup of X and $Y \lhd X$ if Y is normal in X. $\mathrm{Aut}(X)$ denotes the automorphism group of the structure X. If a group X acts on a set Z with b orbits of length B, c orbits of length C and so on, then we say that X has orbit structure $B^b C^c \ldots$ on Z. We denote the pointwise stabiliser of a subset $W \subseteq Z$ in X by $X_{(W)}$. If Y is a subgroup of the group X then $N_{(X)}(Y)$ denotes the normalizer of Y in X, and $C_{(X)}(Y)$ denotes the centraliser of Y in X. For more details on these and other permutation group definitions and results, see Wielandt (1964).

Theorem 3 (The Frattini Argument, Gorenstein 1980). *Let Y be a normal subgroup of the group X and let P be a Sylow p-subgroup of Y. Then $X = YN_X(P)$. Further, $|X:Y| = |N_X(P):N_Y(P)|$.*

Theorem 4 (Gorenstein 1980).   *Let K be a group and let X and Y be subgroups of K.   Then the quotient $N_Y(X)/C_Y(X)$ is isomorphic to a subgroup of Aut(X).*

In the following, we use $\mathcal{H}$ to denote a hyperoval of $PG(2, q)$, G to denote its collineation stabiliser and H to denote its homography stabiliser.

Since $P\Gamma L(3, q)$ is transitive on quadrangles of $PG(2, q)$ we can assume that a hyperoval $\mathcal{H}$ contains the points of a fundamental quadrangle $(1, 0, 0)$, $(0, 1, 0)$, $(0, 0, 1)$ and $(1, 1, 1)$.   Let Q denote the homography stabiliser of the fundamental quadrangle, a group of order 24.

We shall suppose that A stabilises $\mathcal{H}$ and hence $|G| = e|H|$. Note that A stabilises $\mathcal{D}(f)$ if and only if all the coefficients of f are in $GF(2)$.

Theorem 5. *Let H be the homography stabiliser of a hyperoval $\mathcal{H}$ in $PG(2, q)$ which contains the fundamental quadrangle.   The normaliser $N_H(A)$ in H of A is a subgroup of Q, where A and Q are defined as above.*

Proof.   Each element of $N_H(A)$ permutes the fixed points of A.   These are precisely the points of the subplane

$$PG(2, 2) = \{(x, y, z) : x, y, z \in GF(2), \text{not all zero}\}.$$

The points of $PG(2, 2)$ on $\mathcal{H}$ are exactly the points of the fundamental quadrangle, so each element of $N_H(A)$ is a homography permuting those points and the result follows.

Recall that, as in Dembowski (1968), for an *arc* $\mathcal{K}$ (that is, a set of points no three collinear), a line l is a *secant* or an *external line* to $\mathcal{K}$ if $|l \cup \mathcal{K}|$ is 2 or 0 respectively.

Theorem 6. *Let H be the homography stabiliser of a hyperoval $\mathcal{H}$ in $PG(2, q)$ which contains the fundamental quadrangle.   Suppose that $h \in H$ has odd order m and has a unique fixed line l.   Suppose also that A is a subgroup of the normaliser $N_G(\langle h \rangle)$ in G of $\langle h \rangle$.   Then l is external to $\mathcal{H}$ and in fact it is the line $x + y + z = 0$.   Further, if e is prime then all orbits of $\langle h, A \rangle$ on points not contained in $\mathcal{H} \cup l$ have length divisible by e.*

Proof. First we show that l is external to $\mathcal{H}$. If not, then because a hyperoval has only secants and external lines, it is a secant of $\mathcal{H}$. Then h must either fix, or interchange the two points of $\mathcal{H}$ on l. In the first case we have a contradiction to the the fact that the number of fixed points of a collineation is equal to the number of its fixed lines (see Theorem 7, below) and the second possibility is contrary to the fact that the order of h is odd.

Since A is a subgroup of $N_H(\langle h \rangle)$, elements of A permute the fixed lines of $\langle h \rangle$, and as the fixed line is unique, all elements of A fix l. Also, l is external to $\mathcal{H}$, so the line $x + y + z = 0$ is the only possibility for l (the fixed lines of A are the seven lines of $PG(2, 2)$, six of which meet $\mathcal{H}$).

Now $|A| = e$ so if e is prime, then the orbits of A on the points $PG(2, q)$ have length 1, or e, and the orbits of length 1 (the fixed points) all lie on $\mathcal{H} \cup l$. An orbit of $\langle h, A \rangle$ is a union of orbits of A and if such an orbit contains no point of $\mathcal{H} \cup l$ then it contains no fixed point of A and hence is a union of orbits of A of length e.

We now recall some results on collineations of $PG(2,q)$, all of which are found in Dembowski (1968). A point P of $PG(2, q)$ is a *centre* of a collineation σ of $PG(2, q)$ if all lines through P are fixed by σ. A line l is an *axis* of σ if every point of l is fixed by σ. A collineation of $PG(2, q)$ has a centre if and only if it has an axis. An axial collineation is either an *elation* or a *homology* dependent upon having an incident, or non-incident centre-axis pair. An elation has prime order p, where q is a power of p, and a homology has order dividing $q - 1$.

Theorem 7 (Dembowski 1968, 4.1.2). *Let σ be a collineation of $PG(2, q)$. Then σ has the same number of fixed lines as fixed points.*

3. Stabilisers of hyperovals in $PG(2, 32)$

From now on, we have $q = 32$, so that $e = 5$. We use the notation introduced in Section 2, in particular we assume that A stabilises the hyperoval $\mathcal{H}$ and hence $|G| = e |H|$.

By O'Keefe and Penttila (to appear, Theorem 1.8) it follows that $|G|$ divides $5.34.33.32.31 = 2^6.3.5.11.17.31$. Since $|G| = 5|H|$, $|H|$ divides $2^6.3.11.17.31$. Also $|H|$ divides $|PGL(3, 32)| = 2^{15}.3.7.11.31^2.151$ which implies that $|H|$ divides $2^6.3.11.31$.

If 31 divides $|H|$ then $\mathcal{H}$ is a monomial hyperoval (see O'Keefe and Penttila 1989).   From now on we assume that $\mathcal{H}$ is not monomial so that $|H|$ divides $2^6.3.11$.

Theorem 8.   11 *does not divide* $|H|$.

Proof.   Suppose, to the contrary, that 11 divides $|H|$ and choose $h \in H$ with order 11.   The plane $PG(2, 32)$ has $32^2 + 32 + 1$ points and the orbits of h have length 1 or 11, so the number of fixed points of h is congruent to 1 modulo 11.   The fixed point configuration of h is an arc, for if h has three collinear fixed points then it is axial, but axial homographies of $PG(2, 32)$ have order 2 or 31.   Since the homography stabiliser of a 4-arc is trivial, h must have less than 4 fixed points and therefore has a unique fixed point.   Thus h has a unique fixed line l and we can apply Theorem 6 provided we can show that $A \leq N_G(\langle h \rangle)$.

Since $\langle h \rangle$ is a Sylow 11-subgroup of H, the Frattini argument (Theorem 3) implies that $|G : H| = 5 = |N_G(\langle h \rangle):N_H(\langle h \rangle)|$ and hence 5 divides $|N_G(\langle h \rangle)|$.   Let A_1 be a Sylow 5-subgroup of $N_G(\langle h \rangle)$; so it is a Sylow 5-subgroup of G.   Since A is also a Sylow 5-subgroup of G and all Sylow 5-subgroups are conjugate in G there exists an element $g \in G$ such that $A = g^{-1}A_1g$.   Now $A_1 {\leq} N_G(\langle h \rangle)$; so $g^{-1}A_1g \leq g^{-1}N_G(\langle h \rangle)g = N_G(g^{-1}\langle h \rangle g)$.   If we replace h in the above argument by $h' = g^{-1}hg$ then h' is an element of G of order 11 stabilising $\mathcal{H}$ with $A \leq N_G(\langle h' \rangle)$ as required.   So without loss of generality we assume that $A \leq N_H(\langle h \rangle)$. By Theorem 6, orbits of $\langle h, A \rangle$ on points not contained in $\mathcal{H} \cup l$ have length divisible by 5.   Now $\langle h, A \rangle$ has order 55; so its orbits have length 1, 5, 11, or 55.   An orbit $\langle h, A \rangle$ has length 1 or 5 if and only if it contains a fixed point of $\langle h \rangle$.   But $\langle h \rangle$ has a unique fixed point, which must also be fixed by A $(N_G(\langle h \rangle))$.   Thus $\langle h, A \rangle$ has a unique orbit of length 1 on $\mathcal{H}$ and no orbit of length 5.   Orbits of $\langle h, A \rangle$ not containing any point of $\mathcal{H} \cup l$ have length divisible by 5, and hence all have length 55.   Apart from these orbits of length 1 and 55, the group $\langle h, A \rangle$ has three orbits of length 11 on $\mathcal{H}$, and has three orbits of length 11 on l.

These results imply that $\langle h, A \rangle$ fixes a unique hyperoval, namely $\mathcal{H}$.   For if $\langle h, A \rangle$ fixes a hyperoval then that hyperoval is a union of orbits of $\langle h, A \rangle$.   It cannot contain an orbit of $\langle h, A \rangle$ on l as they have triples of collinear points, and orbits of length 55 are too large to be contained in a hyperoval.

We proceed to show that this hyperoval must be regular, contrary to hypothesis. First we show that the stabiliser of a regular hyperoval, which is isomorphic to $P\Gamma L(2,32)$, has a subgroup of order 55. Let S be a Sylow 11-subgroup of $P\Gamma L(2,32)$, noting that $|P\Gamma L(2,32)| = 5.33.32.31$. By the Frattini argument (Theorem 3),

$$|N_{P\Gamma L(2,32)} : N_{PGL(2,32)}(S)| = |P\Gamma L(2,32) : PGL(2,32)| = 5$$

so that 5 divides $|N_{P\Gamma L(2,32)}(S)|$. Let $n \in N_{P\Gamma L(2,32)}(S)$ have order 5. We show that $\langle n, S \rangle$ has order 55. First $S \vartriangleleft \langle n, S \rangle$; so by an isomorphism theorem (Gorenstein 1980)

$$\frac{\langle n \rangle S}{S} \cong \frac{\langle n \rangle}{\langle n \rangle \cap S} = \langle n \rangle.$$

Thus S has index 5 in $\langle n, S \rangle$, which therefore has order 55 as required.

Next we show that all subgroups of $P\Gamma L(3,32)$ of order 55 are conjugate. Let J and K be subgroups of $P\Gamma L(3,32)$ of order 55. Then J and K have unique Sylow 11-subgroups, say J_1 and K_1. Now J_1 and K_1 are Sylow 11-subgroups of $P\Gamma L(3,32)$, and hence they are conjugate in $P\Gamma L(3,32)$. Thus we can assume that $J_1 = K_1$, and hence that J and K lie in $N = N_{P\Gamma L(3,32)}(J_1)$. Then as all subgroups of N of order 5 are conjugate in N it follows that J and K are conjugate in N.

We have just proved that all subgroups of $P\Gamma L(3,32)$ of order 55 are conjugate and that a regular hyperoval is stabilised by a subgroup of $P\Gamma L(3,32)$ of order 55. Thus the unique hyperoval fixed by the group $\langle h, A \rangle$ of order 55 is the image under an element of $P\Gamma L(3,32)$ of a regular hyperoval, and therefore is regular. Thus $\mathcal{H}$ is regular, contrary to the hypothesis that it is not monomial.

Theorem 9. *If $\mathcal{H}$ is a non-monomial hyperoval which is not stabilised by an element of Q of order 3, then 3 does not divide $|H|$.*

Proof. Suppose to the contrary that 3 divides $|H|$. Choose $h \in H$ such that h has order 3; then $\langle h \rangle$ is a Sylow 3-subgroup of H. Recall also that 5 divides $|N_G(\langle h \rangle)|$ (by the Frattini argument, Theorem 3 above).

By similar arguments to those used in the previous theorem, we can assume that $A \leq N_G(\langle h \rangle)$. Also by Theorem 6, since $Aut(\langle h \rangle) = Aut(C_3)$ which is isomorphic to C_2, so $N_G(\langle h \rangle)/C_G(\langle h \rangle)$ is a subgroup of

C_2. Since 5 divides $|N_G(\langle h \rangle)|$, it divides $|C_G(\langle h \rangle)|$.   Thus $A \leq C_G(\langle h \rangle)$, which means that $\langle h \rangle \leq C_G(\langle h \rangle) \leq N_G(\langle h \rangle)$.   So $h \in N_G(\langle A \rangle)$ and therefore h permutes the fixed points of A. By definition, $h \in Q$, contrary to hypothesis.

So far, we have proved that if $\mathcal{H}$ is a non-monomial hyperoval which is not stabilised by an element of Q of order 3 then $|H|$ divides 2^6.   We can assume that $A \leq N_G(H)$ by similar arguments to those used in Theorems 8 and 9, noting that a Sylow 2-subgroup of H is H itself.   Also since $C_H(A) \leq N_G(A) \leq Q$, it follows that $C_H(A) \leq Q$.

For further analysis, we look at the cases of the Cherowitzo and the Payne hyperovals separately in sections 4 and 5.   But first we give an interesting characterisation of the 5 known classes of hyperovals in $PG(2, 32)$.   These are the regular $\mathcal{D}(x^2)$, the irregular translation $\mathcal{D}(x^4)$, the Segre $\mathcal{D}(x^6)$, the Payne $\mathcal{D}(x^6 + x^{16} + x^{26})$ and the Cherowitzo $\mathcal{D}(x^8 + x^{10} + x^{28})$ hyperovals (see O'Keefe 1990).

Theorem 10. *Let $\mathcal{H}$ be a hyperoval of $PG(2, 32)$ such that 5 divides the order of the collineation stabiliser G of $\mathcal{H}$. Then $\mathcal{H}$ belongs to one of the known classes of hyperovals, and conversely.*

Proof. Since 5 divides $|G|$, it follows that G has a subgroup K of order 5.   The orbits of K on $\mathcal{H}$ have length 1 or 5 and since $|\mathcal{H}| = 34$ it follows that K has at least 4 fixed points on $\mathcal{H}$.   We can assume that these points are the points of the fundamental quadrangle and so K is A, the group of automorphic collineations of $PG(2, 32)$.   Now $\mathcal{H}$ is fixed by A and hence all the coefficents of its o-polynomial belong to $GF(2)$.   But all such o-polynomials represent known hyperovals (see Cherowitzo 1988).   For the converse, see, for example, O'Keefe and Penttila (1989). □

4.   The Cherowitzo hyperoval in $PG(2, 32)$

The Cherowitzo hyperoval in $PG(2,32)$ is

$$C = \{(1, t, t^8 + t^{10} + t^{28}) : t \in GF(32)\} \cup \{(0, 1, 0), (0, 0, 1)\}.$$

Certainly the collineation stabiliser G of C contains the group A of automorphism collineations of order 5 generated by $(x, y, z) \mapsto (x^2, y^2, z^2)$.

Lemma 11 (Payne 1985, V.1). *Let $\mathcal{H} = \mathcal{D}(f)$ be a hyperoval of $PG(2, q)$, q even, containing the fundamental quadrangle.   The pointwise stabiliser of $(1, 0, 0)$, $(0, 1, 0)$ and $(0, 0, 1)$ in the collineation stabiliser of $\mathcal{H}$ is transitive on $\mathcal{H}\backslash\{(1, 0, 0), (0, 1, 0), (0, 0, 1)\}$ if and only if f is multiplicative.*

Lemma 12. *The pointwise stabiliser of $(0, 1, 0)$, $(0, 0, 1)$ and $(1, 1, 1)$ in G is not transitive on $C\backslash\{(0, 1, 0), (0, 0, 1), (1, 1, 1)\}$.*

Proof.   First consider the homography $\phi : (x, y, z) \mapsto (x, x + y, x + z)$. This fixes $(0, 1, 0)$ and $(0, 0, 1)$ and interchanges $(1, 0, 0)$ and $(1, 1, 1)$. Let $t \in GF(q)\backslash\{0\}$.   Then ϕ maps the point $(1, t, f(t))$ to $(1, 1 + t, 1 + f(t)) = (1, u, 1 + f(u + 1))$, so that the hyperoval $C = \phi C$ is equal to $\mathcal{D}(g)$ where $g(x) = 1 + f(x + 1)$.   Let τ be a collineation as described in the statement of the lemma.   The collineation $\phi\tau\phi^{-1}$ stabilises C, fixes $(1, 0, 0)$, $(0, 1, 0)$ and $(0, 0, 1)$ and maps $(1, 1, 1)$ to the point $(1, t, g(t))$ of C.   By Lemma 11, g is multiplicative;   hence $g(x) = x^k$ for some integer k.   Thus $1 + f(x + 1) = x^k$ so that $1 + f(x) = (x + 1)^k$ and $f(x) = 1 + (x + 1)^k$.   This is impossible as $f(x) = x^8 + x^{10} + x^{28}$.

Lemma 13. *The pointwise stabiliser of $(1\ 0, 0)$, $(0, 0, 1)$ and $(1, 1, 1)$ in G is not transitive on $C\backslash\{(1\ 0\ 0), (0, 0, 1), (1, 1, 1)\}$.*

Proof.   Let ϕ be the homography $\phi : (x, y, z) \mapsto (x + y, y, y + z)$.   This fixes $(1, 0, 0)$ and $(0, 0, 1)$ and interchanges $(0, 1, 0)$ and $(1, 1, 1)$.   Let $t \in GF(q)\backslash\{0\}$.   Then ϕ maps the point $(1, t, f(t))$ to $(1 + t, t, t + f(t)) = (1, u, u + (u + 1)f(u/(u + 1)))$, so that the hyperoval $C = \phi C$ is equal to $\mathcal{D}(g)$ where $g(x) = x + (x + 1)f(x/(x + 1))$.   Let τ be a collineation as described in the statement of lemma.   The collineation $\phi\tau\phi^{-1}$ stabilises C, fixes $(1, 0, 0)$, $(0, 1, 0)$ and $(0, 0, 1)$ and maps $(1, 1, 1)$ to the point $(1, t, g(t))$ of C.   By Lemma 11, g is multiplicative;   hence $g(x) = x^k$ for some integer k.   Thus $x + (x + 1)f(x/(x + 1)) = x^k$ so that $1 + f(x) = x + (x + 1)(x/(x + 1))^k$ and $f(x) = 1 + (x + 1)^k$.   This is impossible as $f(x) = x^8 + x^{10} + x^{28}$.

Lemma 14. *C is not monomial.*

Proof.   Suppose that C is monomial.   Then its homography stabiliser H has an orbit of length at least 31 (O'Keefe and Penttila, 1989).   Now H cannot have an orbit of length 34 as this implies that $q = 2, 4,$ or 16 (O'Keefe and Penttila, 1989).   Suppose H has an orbit of length 33 on

C. Then C is regular and f must be either monomial, or have 15 terms (O'Keefe and Penttila, to appear, Theorem 1.11). This is not the case.

Suppose now, that H has an orbit of length 32 on C. Then C is a translation hyperoval (O'Keefe and Penttila 1989), that is, C is a homographic image of a hyperoval

$$\{(1, t, t^\alpha) : t \in GF(32) \} \cup \{(0, 1, 0), (0, 0, 1)\}$$

for some $\alpha \in Aut \, (GF(32))$. The collineation group of such a hyperoval fixes $(0, 1, 0)$ and $(0, 0, 1)$ and is doubly transitive on the remaining points. If C were a translation hyperoval, it would have fixed points two of the fundamental points since A stabilises C. By Lemma 11 these cannot be two of $(1, 0, 0)$, $(0, 1, 0)$ and $(0, 0, 1)$, by Lemma 12 they cannot be two of $(0, 1, 0)$, $(0, 0, 1)$ and $(1, 1, 1)$ and by Lemma 13 they cannot be two of $(1, 0, 0)$, $(0, 0, 1)$ and $(1, 1, 1)$. But these are all the possibilities, so C is not a translation hyperoval.

Finally, suppose that H has an orbit of length 31 on C. This orbit must contain a unique fixed point of A. We apply an element h of Q (a homography stabilising the fundamental quadrangle) so that this point is $(1, 1, 1)$. We choose h so that the hyperoval $C = \mathcal{D}(f) = \{(1, t, f(t)) : t \in GF(32) \} \cup \{(0, 1, 0), (0, 0, 1)\}$, where $f(x) = x^8 + x^{10} + x^{28}$, is replaced by the homographic image $\mathcal{D}(g)$ where g is one of the following functions (see Cherowitzo 1988):

(1) $g(x) = f(x)$,
(2) $g(x) = 1 + f(x + 1)$, $g(0) = 0$,
(3) $g(x) = x + (x + 1)f(x/(x + 1))$, $g(1) = 0$,
(4) $g(x) = 1 + (x + 1)f^{-1}(1/(x + 1))$, $g(1) = 0$.

The stabiliser H_1 of $\mathcal{D}(g)$ is a conjugate of H in $PGL(3, 32)$, so has orbit structure $1^3 31^1$ or $3^1 31^1$ on $\mathcal{D}(g)$. An element $h \in H$ of order 31 fixing $(1, 0, 0)$, $(0, 1, 0)$ and $(0, 0, 1)$ setwise must fix them pointwise (as 6 and 31 are coprime), so generates a subgroup transitive on $\mathcal{D}(g) \backslash \{(1, 0, 0), (0, 1, 0), (0, 0, 1)\}$. Thus g is multiplicative, by Lemma 11. Hence $g(x) = x^r$ for some integer r. But none of the functions (1)-(4) is monomial, a contradiction.

Lemma 15. *C is not stabilised by any non-identity element of Q.*

Proof. The result follows by direct calculation.

By the results of Section 3, $|H|$ divides 2^6. We will prove that $|H| = 1$.

Lemma 16. *If $H \neq \{1\}$ then $C_H(A) = \{1\} = C_A(H)$.*

Proof. Firstly $C_H(A) \leq H$ and $C_H(A) \leq Q$ (see section 3), but the intersection $H \cap Q = \{1\}$ (by Lemma 12) so that $C_H(A) = \{1\}$. Finally, the centralizer $C_A(H)$ must be trivial, for if $C_A(H)$ contains a non-identity element a then $A = \langle a \rangle$ is centralized by H and $C_H(A) = H \neq \{1\}$.

Lemma 17. *Let M be a subgroup of H and suppose that $A \leq N_G(M)$. Then $|M| \equiv 1 \pmod{5}$.*

Proof. Since $A \leq N_G(M)$, A acts on M by conjugation. As $|A| = 5$, the orbits of A on M have length 1, or 5. An orbit of length 1 is fixed by a point under conjugation by elements of A, so is an element of $C_M(A)$ $\leq C_H(A) = \{1\}$ (Lemma 16). Thus there is a unique orbit of length 1 and $|M| \equiv 1 \pmod 5$.

Applying Lemma 17 with $M = H$ we find that $|H| \equiv 1 \pmod 5$ and, as $|H| = 2^i$ for some $0 \leq i \leq 6$, we have $|H| = 1$, or 16.

Theorem 18. $|H| = 1$.

Proof. Suppose that $|H| = 16$.

Orbits of H on $\mathcal{H}$ have lengths dividing 16; so the orbits have length 1 or a power of 2. As $|\mathcal{H}| = 34 = 2 \pmod 4$, H has at least one orbit of length 2, or at least two fixed points.

Suppose that H has more than one orbit of length 2 on $\mathcal{H}$. The pointwise stabiliser of the four points belonging to two of these orbits is trivial; so $|H| \leq 4$, a contradiction.

If H has exactly one orbit X of length 2 on $\mathcal{H}$, let $H_{(X)}$ be its pointwise stabiliser in H, of order 8. Since $A \leq N_G(H_{(X)})$, so A permutes the orbits of H of each fixed length. Since X is the unique orbit of length 2, A fixes X. Also, $A \leq N_G(H_{(X)})$ (for if $a \in A$ and $h \in H_{(X)}$ acts on the left then $a^{-1}ha \in H$ and $h(ax) = ax$ so that $a^{-1}hax = x$ for all $x \in X$). By Lemma 16, $|H_{(X)}| \equiv 1 \pmod 5$, a contradiction.

Finally, suppose that H has at least two fixed points U and V. Consider the action of an element $h \in H, h \neq 1$, on the line UV. Recall that h has even order. As the line UV has an odd number of points and U and V are fixed by h, so h must fix at least one further point of UV. Hence UV is an axis of h and h is an elation.

Thus H is a group of elations with axis the line through the two fixed points of H, so any $h \in H \setminus \{1\}$ fixes no further point of $\mathcal{H}$. Therefore the orbit structure of H on $\mathcal{H}$ is $1^2 16^2$. The group $\langle A, H \rangle$ has order $5.16 = 80$ and its orbits are unions of orbits of H with orders dividing 80, hence $\langle A, H \rangle$ also has orbit structure $1^2 16^2$ on $\mathcal{H}$. (The orbit structure $2^1 16^2$ is impossible since this implies the existence of an element of order 2 in A.) Since $16 \equiv 1 \pmod{5}$ there is a unique fixed point of A in each order if H of length 16.

Without loss of generality, we can re-label the fundamental quadrangle so that the fixed points of $\langle A, H \rangle$ are $(1, 0, 0)$ and $(0, 1, 0)$ and the fixed points of A in the orbits of $\langle A, H \rangle$ of length 16 are $(0, 0, 1)$ and $(1, 1, 1)$. This re-labelling amounts to taking an image of $\mathcal{H}$ under an element τ of the homography stabiliser Q of the fundamental quadrangle. The resulting hyperoval is a Cherowitzo hyperoval with collineation stabiliser $\tau G \tau^{-1}$ containing no non-trivial element of Q.

An element h of H can be represented as a matrix

$$h = h_{(a, b)} = \begin{bmatrix} 1 & 0 & a \\ 0 & 1 & b \\ 0 & 0 & 1 \end{bmatrix}$$

for some $a, b \in GF(32)$. Now $h_{(a,b)}(0, 0, 1) = (a, b, 1) \in \mathcal{H}$. Since the points $(1, 0, 0)$, $(a, b, 1)$ and $(a, c, 1)$ are collinear for all $c \in GF(32)$, only $(1, 0, 0)$ and one point $(a, b, 1)$ can belong to $\mathcal{H}$ for each value of a. Thus, if $h_{(a,b)} \in H$ then $h_{(a,c)} \notin H$ for all $c \neq b$. So there exists a subset $E \subseteq GF(32)$ and a function f on $GF(32)$ such that $H = \{h_{(a,f(a))} : a \in E\}$. Since $(0, 0, 1)$ and $(1, 1, 1)$ are in different orbits of $\langle A, H \rangle$ on $\mathcal{H}$,

$\mathcal{H} = \{(1, 0, 0), (0, 1, 0)\} \cup H(0, 0, 1) \cup H(1, 1, 1)$
$= \{(1, 0, 0), (0, 1, 0)\} \cup \{(a + 1, f(a), 1) : a \in E\} \cup \{(a + 1, f(a) + 1, 1) : a \in E\}$.

Now $h_{(1,1)}(a, f(a), 1) = (a + 1, f(a) + 1, 1)$ and $h_{(1,1)}(a + 1, f(a) + 1, 1) = (a, f(a), 1)$; so $h_{(1,1)} \in H$. But $h_{(1,1)}(0, 0, 1) = (1, 1, 1)$ contradicting the fact that these points are in different orbits of H on $\mathcal{H}$.

Theorem 1. *The collineation stabiliser G of the Cherowitzo hyperoval C in PG(2, 32) is A, the group of automorphic collineations of PG(2, 32).*

Proof. By Theorem 18, $|H| = 1$. Since $|G:H| = 5$ we have $|G| = 5$ and so $G = A$.

In fact, we have proved more than this. The proof of Theorem 18 shows that if $\mathcal{H}$ is a non-monomial hyperoval in $PG(2, 32)$ not fixed by any non-identity element of Q and stabilised by the group A of automorphic collineations, then the collineation stabiliser of $\mathcal{H}$ is precisely A. The Cherowitzo hyperoval C is the only known example of such a hyperoval.

These techniques can be used to show that the homography stabiliser of the Cherowitzo hyperoval in $PG(2, 32)$ has order 1, or 8.

5. The Payne Hyperoval in $PG(2, 32)$

The Payne hyperoval (1985) in $PG(2, 32)$ is

$$\mathcal{P} = \{(1, t, t^6 + t^{16} + t^{26}) : t \in GF(32)\} \cup \{(0, 1, 0), (0, 0, 1)\}.$$

The collineation stabiliser of the Payne hyperoval contains the group A of automorphic collineations.

Lemma 19 (Payne 1985, V-5). *$\mathcal{P}$ is not monomial.*

Lemma 20. *The subgroup Θ of Q which stabilises $\mathcal{P}$ is the group of order two generated by the homography θ where*

$$\theta : (x, y, z) \mapsto (y, x, z).$$

Proof. The result follows by direct calculation, noting that since

$$(x^{-1})^6 + (x^{-1})^{16} + (x^{-1})^{26} = (x^6 + x^{16} + x^{26})/x$$

the map θ fixes $(0, 0, 1)$ and $(1, 1, 1)$, interchanges $(1, 0, 0)$ and $(0, 1, 0)$ and interchanges $(1, x, x^6 + x^{16} + x^{26})$ and $(1, x^{-1}, (x^{-1})^6 + (x^{-1})^{16} + (x^{-1})^{26})$.

From Section 3, $|H|$ divides 2^6. We will prove that $|H| = 2$, so that $H = \theta$. As in Section 4, we find a contradiction in the case that

$|H| \neq 2$. First we prove two lemmas, analogous to Lemmas 16 and 17.

Lemma 21. If $|H| > 2$ then $C_H(A) = \Theta$ and $C_A(H) = 1$.

Proof. Firstly $C_H(A) \leq H \cap Q = \Theta$ and $C_H(A) = \Theta$ since θ commutes with every automorphic collineation. If $C_H(A) \neq \{1\}$, then $C_A(H) = A$ whence $H = C_H(A)$ which is not the case.

Lemma 22. Let M be a subgroup of H and suppose that $A \leq N_G(M)$. Then $|M| \equiv |M \cap Q|$ (mod 5), noting that $|M \cap Q|$ is 2, or 1 according as $\theta \in M$ or not.

Proof. Since $A \leq N_G(M)$, A acts on M by conjugation. As $|A| = 5$, the orbits of A on M have lengths 1 or 5. An orbit of length 1 is a fixed point on M under conjugation by elements of A, so is an element of $C_M(A) \leq C_H(A) = \Theta$. The result follows.

Applying Lemma 22 with $M = H$ we find that $|H| \equiv 2$ (mod 5), and as $|H| = 2^i$ for some $0 \leq i \leq 6$, we have $|H| = 2$ or 32.

Theorem 23. $|H| = 2$ so that $H = \Theta$.

Proof. Suppose that $|H| = 32$.

Orbits of H on $\mathcal{H}$ have lengths dividing 16; so the orbits have length 1 or a power of 2. As $|\mathcal{H}| = 34 \equiv 2$ (mod. 4), H has at least one orbit of length 2 or at least two fixed points.

Suppose that H has at least two fixed points U and V. Consider the action of an element $h \in H, h \neq 1$, on the line UV. Recall that h has even order. As the line UV has an odd number of points and U and V are fixed by h, so h must fix at least one further point of UV. Hence UV is an axis of h and h is an elation. Thus H is a group of elations with axis the line through the two fixed points of H, so fixes no further point of $\mathcal{H}$. Therefore the orbit structure of H on $\mathcal{H}$ is $1^2 32^1$ and $\mathcal{H}$ is monomial (O'Keefe and Penttila 1989), a contradiction.

Suppose that H has more than one orbit of length 2 on $\mathcal{H}$. The pointwise stabiliser of the 4 points belonging to two of these orbits is trivial; so $|H| \leq 4$, a contradiction.

If H has exactly one orbit X of length 2 on $\mathcal{H}$. The pointwise stabiliser $H_{(X)}$ has order 16 and is an elation group. Thus as above $H_{(X)}$ has orbit structure $1^2 16^2$ on $\mathcal{H}$.

The group $\langle A, H_{(X)} \rangle$ has order $5.16 = 80$ and its orbits are unions of orbits of $H_{(X)}$ with orders dividing 80, hence $\langle A, H_{(X)} \rangle$ also has orbit structure $1^2 16^2$ on $\mathcal{H}$. (The orbit structure $2^1 16^2$ is impossible since this implies the existence of an element of order 2 in A.) Since $16 \equiv 1$ (mod 5) there is a unique fixed point of A in each orbit of $H_{(X)}$ of length 16.

The fixed points A are $(1, 0, 0)$, $(0, 0, 1)$ and $(1, 1, 1)$. The element $\theta \in H$ switches the points $(1, 0, 0)$ and $(0, 1, 0)$ and fixes $(0, 0, 1)$ and $(1, 1, 1)$. Since X is an orbit of H it must consist of either of the points $(1, 0, 0)$ and $(0, 1, 0)$, or of the points $(0, 0, 1)$ and $(1, 1, 1)$. But X cannot consist of the points $(0, 0, 1)$ and $(1, 1, 1)$ for then θ is an element of $\langle A, H_{(X)} \rangle$ switching the points $(1, 0, 0)$ and $(0, 1, 0)$, a contradiction since these lie in different orbits of $\langle A, H_{(X)} \rangle$.

Thus X consists of the points $(1, 0, 0)$ and $(0, 0, 1)$.

We argue exactly as in the case of the Cherowitzo hyperoval in Section 4, noting that the element of H used there is also an element of $H_{(X)}$. Thus this case also leads to contradiction.

Theorem 24. *The collineation stabiliser G of the Payne hyperoval $\mathcal{P}$ in $PG(2, 32)$ is $\langle A, \theta \rangle$, where A is the group of automorphic collineations of $PG(2, 32)$ and θ is the homography $\theta: (x, y, z) \mapsto (y, x, z)$.*

Proof. By Theorem 23, $|H| = 2$. Since $|G:H| = 5$ we have $|G| = 10$ and so $G = \langle A, \theta \rangle$.

The proof of Theorem 23 in fact shows that if $\mathcal{H}$ is a non-monomial hyperoval in $PG(2, 32)$ fixed by the element θ of Q, and no other non-trivial element of Q, and stabilised by the group A of automorphic collineations, then the collineation stabiliser of $\mathcal{H}$ is precisely $\langle A, \theta \rangle$. The Payne hyperoval $\mathcal{P}$ is the only known example of such a hyperoval.

Acknowledgement. The first author acknowledges the support of an Australian Research Council Research Fellowship.

References

W. Cherowitzo (1988), Hyperovals in Desarguesian planes of even order, *Ann. Discrete Math.* **37**, 87-94.

P. Dembowski (1968), *Finite Geometries*, Springer-Verlag, Berlin.

D. Gorenstein (1980), *Finite Groups*, 2nd ed., Chelsea, New York.

J.W.P. Hirschfeld (1979), *Projective Geometries over Finite Fields*, Oxford University Press, Oxford.

C.M. O'Keefe (1990), Ovals in Desarguesian planes, *Australas. J. Combin.* **1**, 149-159.

C.M. O'Keefe and T. Penttila (1989), Symmetries of arcs, University of Western Australia Research Report 62, (Submitted).

C.M. O'Keefe and T. Penttila (to appear), Polynomials for hyperovals of Desarguesian planes, *J. Austral. Math. Soc. Ser. A.*

S.E. Payne (1985), A new infinite family of generalized quadrangles, *Congr. Numer.* **49**, 115-128.

J.A. Thas, S.E. Payne and H. Gevaert (1988), A family of ovals with few collineations, *European J. Combin.* **9**, 353-362.

H. Wielandt (1964), *Finite Permutation Groups*, Academic Press, New York.

On a problem of Wielandt and a question by Dembowski

J. Siemons and B. Webb
University of East Anglia

1. Introduction

Let G be a group with finite permutation representations on two sets Ω_1 and Ω_2. Let $\pi(g)$ and $\beta(g)$ be the number of points fixed by g in Ω_1 and Ω_2 respectively. Our principal assumption is $\pi(g) = \beta(g)$ for all g in G. Then clearly $|\Omega_1| = \pi(1) = \beta(1) = |\Omega_2|$ and both representations have the same kernel. Hence assume that they are both faithful. Furthermore, as the number of orbits on Ω_1 or Ω_2 is equal to the average number of points fixed by the group elements, G has the same number of orbits in both representations. Hence assume that they are transitive. Similar reasoning shows that G has the same permutation rank in both representations.

In the Kourovka Notebook (1983) Wielandt poses the following problem: If G acts primitively on Ω_2 does G always act primitively on Ω_1? There are many examples of inequivalent permutation representations with $\pi = \beta$, but none are known to contradict Wielandt's conjecture[†]. Generally it seems to be a hard problem and only little progress has been made towards a proof. The main result to date (see §2) is due to Förster and Kovács (1990) who show that Wielandt's problem can be reduced to the consideration of almost simple groups.

In Dembowski's book (page 212) a similar question appears in a more restricted situation: If G is a group of automorphisms of a finite projective plane, acting primitively on lines, is G then necessarily primitive on points? The answer to this question is now known, due to the classification of projective planes with primitive collineation group by Kantor (1987).

First we observe that Dembowski's problem is a subcase of Wielandt's as automorphisms of finite projective planes fix an equal number of points and lines. (In passing we note that this is *not* the case in infinite planes, see the paper of Mäurer 1988). As a proof one can quote Baer (1947) or alternatively an earlier result in Brauer (1941). There the "permutation lemma" states that if two permutation matrices A and B satisfy $AS = SB$ for some non-singular matrix S, then A and B represent similar permutations. In the case of a projective plane one can take S to be the incidence matrix of the plane.

By an incidence structure $S = (\mathbb{P}, \mathbb{B}; \mathbb{I})$ we mean two disjoint and finite sets $\mathbb{P}$ and $\mathbb{B}$, the *point* set and *block* set of S, with some incidence relation $\mathbb{I} \leq \mathbb{P} \times \mathbb{B}$. Dembowski's question now has an obvious generalisation: Let S belong to some class of incidence structures and let G be a group of automorphisms of S. Does primitivity of G on $\mathbb{B}$ always imply primitivity of G on $\mathbb{P}$?

We need of course some general conditions on S. The remark above concerning Brauer's permutation lemma suggests we might assume that the linear rank of an incidence matrix for S is equal to $|\mathbb{P}|$. This has the consequence that the associated permutation characters $\pi(g) = \mathrm{fix}_P(g)$ and $\beta(g) = \mathrm{fix}_B(g)$ satisfy $\pi \leq \beta$ so that $\beta = \pi + \chi$ where χ is some character of the automorphism group of S. For details see theorem 3.2 in Camina and Siemons (1989).

Returning to Wielandt's problems one might therefore ask about groups G with faithful representations on sets Ω_1 and Ω_2, with (G, Ω_2) primitive and (G, Ω_1) imprimitive, such that $\pi \leq \beta$ for the corresponding permutation characters. In this situation we say that G is an *exception* with $\pi \leq \beta$. One can view Ω_1 and Ω_2 as the point and block sets of an incidence structure in which the incidence relation is given as some union of G-orbits on $\Omega_1 \times \Omega_2$. In section 3 we give examples of exceptions arising from the groups $PSL(2, p)$ and discuss the associated incidence structures.

2. Some observations and known results

Suppose that G acting faithfully on Ω_1 and Ω_2 is an exception with $\pi \leq \beta$. Let $P_1, \ldots, P_i, \ldots, P_t$ be a non trivial system of imprimitivity for the action of G on Ω_1. Let H be the subgroup fixing all the P_i sets. If $H \neq 1$ then it is transitive on Ω_2 as it is normal in G and it is also transitive on Ω_1 as $\pi \leq \beta$. Hence the action of fegree t is faithful and by choice we may assume it is primitive. Let π_0 denote the corresponding character. It is a simple matter to show that then $\pi_0 < \pi \leq \beta$. This gives first of all

Lemma 1. *Let (G, Ω_1, Ω_2) be an exception with $\pi \leq \beta$. Then every abelian subgroup acts intransitively on Ω_2.*

Proof. Let A be an abelian subgroup and consider the restriction of $\pi_0 < \pi \leq \beta$ to A. If A is transitive on Ω_2 then it must be transitive and faithful in all three representations. But then A is regular contradicting $\pi_0 < \pi$.

For exceptions with $\pi = \beta$ this observation is mentioned already in the Kourovka Notebook. The lemma implies in particular that exceptions can not have a solvable normal subgroup.

Theorem 2 (Förster and Kovács 1990). *An exception G with $\pi = \beta$ and minimal normal subgroup $T \times T \ldots \times T$, where T is a non-abelian simple group, gives rise to an exception $\underline{G}$ with $\underline{\pi} = \beta$ and $T \leq \underline{G} \leq Aut(T)$.*

Their proof shows a little more, namely that every almost simple exception conversely extends to exceptions with several simple factors. We expect that Theorem 2 carries through to the general situation $\pi \leq \beta$ but we have not yet completed a proof.

We now come to the combinatorial and geometric aspect of the problem and assume that additional requirements on the type of incidence structure is made. Delandtsheer and Doyen (1989) give an elegant and elementary proof for the following

Theorem 3. *If a 2-(v, k, λ) design with $v > (\binom{k}{2} - 1)^2$ has a block-transitive automorphism group then the group acts primitively on points.*

This paper is also a good source of reference to conditions on 2-designs which imply point-primitivity. It contains the conjecture that block-primitivity and $\lambda = 1$ should imply point-primitivity. This has been settled in Delandtsheer (1988, 1989) for small values of k and small permutation rank in the block representation and in Kantor (1973) for projective spaces of dimension at least 3.

Kantor (1987) characterizes finite projective planes with point-primitive collineation group. This work is based on a list of primitive permutation groups of odd degree and ultimately on the classification of finite simple groups. As a by-product, Dembowski's original question finally has an affirmative answer:

Theorem 4 (Kantor 1987). *Let Π be a projective plane of finite order q and let G be a group of automorphisms permuting the points primitively. Then either* (i): *Π is desarguesian and $G \geq PSL(3, q)$*

or (ii): *G is regular or a Frobenius group and the number of points of* Π *is a prime.*

In this context one should also mention the classification of affine planes with primitive collineation group by Hirame (1990).

Further evidence for the conjecture that block-primitivity implies point-primitivity in 2-(v, k, 1) designs can be derived from a result in Camina and Siemons (1989) which is independent of the classification theorem.

Theorem 5. *Let D be a* 2-(v, k, 1) *design with automorphism group G acting primitively on the blocks of D. Suppose that the stabilizer in G of two points* $p \neq p'$ *fixes all points in the block through p and p'. Then G acts primitively on the points of D.*

Proof. In Theorem 2 of Camina and Siemons (1989) we have shown that a block-*transitive* group as above either is flag-transitive or has odd order. In the first case *G* is primitive by a theorem of Higman and McLaughlin (1961). In the second case *G* is solvable and hence point primitive by Lemma 1.

3. Splitting designs

We construct examples of groups *G* with two faithful representations $(G, \mathbb{P})$ and $(G, \mathbb{B})$ such that the associated permutation characters satisfy $\pi \leq \beta$. The motivation is, of course, to try and find examples with $|\mathbb{P}| = |\mathbb{B}|$, contradicting Wielandt's conjecture. There has been no success so far nor have we been able to construct 2-designs with this property.

The construction can be described as follows: In $G = PSL(2, p)$ let G_0 be the one point stabilizer in the usual representation on the projective line $\mathbb{P}_0$ of $p + 1$ points with character $\pi_0 = 1 + \psi$ and let B be some maximal subgroup of *G* with induced character β. When $\mathbb{B}$ are the cosets of *B* in *G*, it is clear that ($\mathbb{P}_0$, $\mathbb{B}$) either is a 2-design and $\pi_0 \leq \beta$ or ($\mathbb{P}_0$, $\mathbb{B}$) is a trivial structure and π_0 and β have only the principle character in common. Now let *P* be a subgroup of G_0 with associated character π and let $\mathbb{P}$ be the cosets of *P* in *G*. This corresponds to splitting each point of the projective line into $|G_0 : P|$ *new points.* On the sets $\mathbb{P}$ and $\mathbb{B}$ incidence can be defined in various ways. Of course, it remains to be seen if $\pi \leq \beta$. For the characters of *PSL* we follow Dornhoff's (1971) notation.

Example 6. $B \cong A_4$. For p at least 7, the subgroup B is maximal if and only if $p \equiv 3$ or $7 \bmod 10$ and $p^2 \equiv 9 \bmod 80$. The cosets of B can be identified with sets of 4 harmonic points on the projective line and the resulting structure $(\mathbb{P}_0, \mathbb{B})$ is a 3-$(p + 1, 4, 1)$ design. The character β can be calculated as $\beta = 1 + [(p + t_p)/12] \psi + [(p + t_p)/24] (\xi_1 + \xi_2) + \Sigma_k[(p + t_{pk})/12]\theta_k + \Sigma_l[(p + t_{pl})/12] \chi_l$ where the values for t_p, t_{pl} and t_{pk} are given in terms of various congruences on p, k and l.

Taking P as the subgroup of index 2 in G_0, hence $p \equiv 1 \bmod 4$, we find that $\pi = 1 + \psi + (\xi_1 + \xi_2)$ so that $\pi \leq \beta$ in all cases. Defining incidence between $\mathbb{P}$ and $\mathbb{B}$ suitably one obtains $2 - (2(p + 1), 4, \Lambda)$ designs with $\Lambda = \{0, (p - 1)/4\}$.

The smallest exception overall, in terms of group order, arises in this way when G is $PSL(2, 13)$ and $(\mathbb{P}, \mathbb{B})$ is a 2-$(28, 4, \{0, 3\})$ design. If S denotes its 28×91 incidence matrix, then SS^T has eigenvalues 52^1, 10^{13}, and $(13 \pm 3\sqrt{13})^7$, from which one can see that the multiplicities are the degrees of the characters $1, \psi, \xi_1$ and ξ_2.

Example 7. $B \cong D_{p-1}$. In this case B is the stabilizer of the 2-set $\{0, \infty\}$ on the projective line whose remaining two orbits are the squares and non-squares in the field. Choosing either, one obtains a 2-$(p + 1, (p - 1)/2, (p - 1)(p - 3)/8)$ design. The corresponding character β can be calculated as $\beta = 1 + 2 \psi + t_p(\xi_1 + \xi_2) + \Sigma_k \theta_k + 2 \Sigma_{1 \leq i \leq (p-5)/8} \chi_i$ where $t_p = 1$ if $p \equiv 1 \bmod 8$ and 0 otherwise. As above, we let P be a subgroup of index 2 in G_0. Hence $\pi \leq \beta$ if $p \equiv 1 \bmod 8$.

Defining incidence between $\mathbb{P}$ and $\mathbb{B}$ suitably we now obtain 2-$(2(p + 1), (p - 1)/2, \Lambda)$ designs with $\Lambda = \{0, (p - 1)(p - 5)/32, (p - 1)^2/32\}$.

† *Note added in proof*: We mentioned Wielandt's problem to R Guralnick at the Durham Symposium. Later at the meeting he produced a counter example: it involves the triple cover of M_{22} in $PSL(45, 43)$. See his preprint note *Primitive Permutation Characters*.

References

R. Baer (1947), Projectivities in finite projective planes, *Amer. J. Math.* **69**, 653 - 684.

R. Brauer (1941), On the connections between the ordinary and the modular characters of groups of finite order, *Ann. of Math.* **42**, 926 - 935.

A.R. Camina and J Siemons (1989), Block-transitive automorphism groups of 2-(v, k, 1) block designs, *J. Combin. Theory Ser. A* **51**, 268 - 276.

A.R. Camina and J. Siemons (1989), Intertwining automorphisms in finite incidence structures, *Linear Algebra Appl.* **117**, 25-34.

A. Delandtsheer and J. Doyen (1989), Most block-transitive t-designs are point-primitive, *Geom. Dedicata* **29**, 307 - 310.

A. Delandtsheer (1988), Line-primitive groups of small rank, *Discrete Math.* **68**, 103 - 106.

A. Delandtsheer (1989), Line-primitive automorphism groups of finite linear spaces, *European J. Combin.* **10**, 161 - 169.

P. Dembowski (1968), *Finite Geometries*, Springer-Verlag, Berlin.

L. Dornhoff (1971), *Group Representation Theory*, Dekker, New York.

P. Förster and L.G. Kovács (1990), A problem of Wielandt on finite permutation groups, *J. London Math. Soc.* **41**, 231-243.

R. Guralnick (1983), Subgroups of prime power index in a simple group, *J. Algebra* **81**, 304 - 311.

R. Guralnick (1983), Subgroups inducing the same permutation representation, *J. Algebra*, **81**, 312 - 319.

D.G. Higman and J.E. McLaughlin (1961), Geometric ABA groups, *Illinois J. Math.* **5**, 382 - 397.

Y. Hirame (1990), Affine planes with primitive collineation group, *J. Algebra* **128**, 366 - 383.

W.M. Kantor (1973), Line-transitive collineation groups of finite projective spaces, *Israel J. Math.* **14**, 229 - 235.

W. M. Kantor (1987), Primitive permutation groups of odd degree and an application to finite projective planes, *J. Algebra* **106**, 15 - 45.

The Kourovka Notebook (1978), translated in *American Mathematical Society Translations* 2 (1983) 121, Providence.

H. Mäurer (1988), Collineations of projective planes with different numbers of fixed points and fixed lines, *European J. Combin.* **9**, 375 - 378.

A Wagner (1986), Orbits on finite incidence structures, *Combinatorica*, Symposia Mathematica **28** , 219- 229, Academic Press, London.

(40, 13, 4)-designs derived from strongly regular graphs

E. Spence
University of Glasgow

1. Introduction

A graph G is said to be *strongly regular* with parameters (v, k, λ) if it has v vertices, is regular of degree k and each pair of distinct vertices has λ or μ common neighbours depending on whether or not the vertices are adjacent. Its adjacency matrix C then satisfies

$$C^2 = (k - \mu)I + \mu J + (\lambda - \mu)C, \tag{1}$$

where, as usual, I is the identity matrix and J is the all-one matrix, both of size v. The complement of G is also strongly regular with parameters $(v, v - k - 1, v - 2k + \mu - 2, v - 2k + \lambda)$. For a survey of recent results on the existence and construction of such graphs see Brouwer and van Lint (1984). Here we shall be interested in the particular case $\lambda = \mu$ when the strongly regular graphs are called (v, k, λ) graphs. A (v, k, λ) design is a collection of v k-subsets (blocks) of a v-set such that any pair of distinct blocks have λ elements in common. Thus a (v, k, λ) graph is a (v, k, λ) design, each vertex of the graph yielding a block comprising its neighbours. We denote by $D(G)$ the (v, k, λ) design corresponding to a (v, k, λ) graph G. Then, since the incidence matrix of $D(G)$ is symmetric and has zero diagonal, D has a polarity with no absolute points. In this case the complementary design has a polarity with every point absolute. The article by Rudvalis (1971) gives some necessary conditions for the existence of (v, k, λ) graphs and more recently Bussemaker et al. (1989) and Haemers and Spence (to appear) have explored the relationship between the automorphism group of a (v, k, λ) graph and the automorphism group of the corresponding design $D(G)$.

In this article we consider (40, 27, 18) graphs, or more precisely, their complements, (40, 12, 2, 4) graphs, and show how they may be used to construct at least 389 non-isomorphic (40, 13, 4) designs. It turns out that many of these designs possess ovals, an oval being a set of four points, no three of which belong to the same block. We shall also describe, using maximal arcs, how these 389 designs themselves may be used to construct many 2-(10, 4, 4) designs.

2. Designs from graphs

2.1 Symmetric Designs

Let D be a (v, k, λ) design with point set Π and block set Δ. A *polarity* of D is a one-one mapping $\pi: \Pi \leftrightarrow \Delta$, of order 2, such that for $p \in \Pi$ and $B \in \Delta, p \in B \Rightarrow B\pi \in p\pi$. If the points of a design with a polarity π are $p_1, p_2, ..., p_v$ and the blocks are $p_1\pi, p_2\pi, ..., p_v\pi$ then the incidence matrix C, defined by

$$C_{ij} = \begin{cases} 1, & \text{if } p_i \in p_j\pi \\ 0, & \text{otherwise,} \end{cases}$$

is symmetric and satisfies

$$C^2 = (k - \lambda)I + \lambda J. \tag{2}$$

A point p is *absolute* with respect to π if $p \in p\pi$. The number of absolute points of π is simply the trace of C. As was mentioned in the introduction, a (v, k, λ) graph is a (v, k, λ) design possessing a polarity with no absolute points, and the complement of such a graph (as a design) is a design having a polarity with every point absolute.

The following theorem is couched in more general terms than are needed for this article as we shall be interested here in applying it to the case of (40, 13, 4) designs with a polarity having every point absolute.

Theorem 1. *Let D be a (v, k, λ) design having a polarity π. By complementation, if necessary, we may suppose that π has at least one absolute point. The incidence matrix for D can then be written in the form*

$$\begin{bmatrix} 1 & j^t & 0 \\ j & A & N \\ 0 & N^t & B \end{bmatrix}, \tag{3}$$

where A and B are square (symmetric) matrices of order $k - 1, v - k$, respectively, and j is the all-one vector. Clearly A and B have row and column sums $\lambda - 1$ and $k - \lambda$, respectively. If every eigenvalue of A other than $\lambda - 1$ is zero or $\pm\sqrt{(k - \lambda)}$ then for every permutation matrix P that commutes with A, the matrix C_p defined by

$$C_P = \begin{bmatrix} 1 & j^t & 0 \\ j & PA & N \\ 0 & N^t & B \end{bmatrix}, \tag{4}$$

is the incidence matrix of a (v, k, λ) design.

Proof. From (2) and (3) we have

$$A^2 + NN^t = (k - \lambda)I + (\lambda - 1)J, \tag{5}$$
$$AN + NB = \lambda J, \tag{6}$$
$$N^tN + B^2 = (k - \lambda)I + \lambda J. \tag{7}$$

Since P commutes with A it is clear that

$$(PA)(PA)^t + NN^t = (k - \lambda)I + (\lambda - 1)J.$$

Thus, to verify that C_P is the incidence matrix of a (v, k, λ) design we need only check that $PAN + NB = \lambda J$, which, on account of (6), is equivalent to $(P - I)AN = 0$. Let $Z = (P - I)AN$. We prove $Z = 0$ by showing that $Zx = 0$ for all eigenvectors x of B. First, since B has constant row sums it has an eigenvector j, and since A and N also have constant row sums, ANj is a multiple of j, and hence $(P - I)AN$j = 0. Now let x be any other eigenvector corresponding to the eigenvalue α, say, so that $Bx = \alpha x$, $x^t \cdot j = 0$. Then from (6)

$$ANx = -\alpha Nx. \tag{8}$$

Also, from (7) we deduce that $\|Nx\|^2 = (k - \lambda - \alpha^2)\|x\|^2$. Hence if $\alpha^2 = k - \lambda$, then $Nx = 0$ and consequently $Zx = 0$. On the other hand, if $\alpha^2 \neq k - \lambda$, it follows that $Nx \neq 0$ so that $-\alpha$ is an eigenvalue of A, whence $-\alpha = \lambda - 1$ or 0. If $-\alpha = \lambda - 1$, then Nx is a multiple of j, and again $Zx = 0$. Finally, if $\alpha = 0$, then $ANx = 0$.

Remark 2. The permutation matrices that commute with A form a group $Z(A)$ say. Let $N(A)$ denote the subgroup of $Z(A)$ defined by $N(A) = \{P \in Z(A) : PA = A\}$. Then in determining the non-isomorphic (v, k, λ) designs constructed by the theorem we need only consider those permutation matrices P that lie in distinct cosets of $N(A)$ in $Z(A)$.

Let G be a (40, 12, 2, 4) graph and let G_x be the subgraph of G induced by the neighbours of a vertex x (the *neighbour* graph of x).

Then clearly G_x is regular of degree 2 and so is a disjoint sum of cycles. The next result was proved by Haemers (1980) in his thesis.

Lemma 3. G_x *is one of the following graphs:*
(1) *a 12-cycle,*
(2) *the disjoint sum of a 9-cycle and a triangle,*
(3) *the disjoint sum of two 6-cycles,*
(4) *the disjoint sum of a 6-cycle and two triangles,*
(5) *the disjoint sum of four triangles.*

He associates with each G a 5-tuple $(a_1, a_2, a_3, a_4, a_5)$ of non-negative integers a_i, where a_i denotes the number of vertices x of the graph for which G_x has the form (i). Using tactical decompositions, Haemers constructed three of these graphs, two having the 5-tuple (0, 0, 0, 0, 40) and the other (0, 0, 4, 24, 12). Using a computer program based on an algorithm of Weisfeiler (1976) Haemers also discovered a further 21 graphs which were shown to be non-isomorphic by comparing their 5-tuples. Independently of Haemers, also using a (different) computer program, the author has established that there are precisely 23 such graphs that have 5-tuples with $a_5 \neq 0$. Moreover, we have found 4 that have $a_5 = 0$ and $a_4 \neq 0$. An exhaustive search for all (40, 12, 2, 4) graphs has not yet run its course, but we could guess that the 27 graphs found so far comprise the complete set. In particular, it seems that every such graph must have a clique size 4. However, all attempts to prove this conjecture without a computer have failed. For the sake of completeness, we give in Table 1 the three graphs found that are not listed in Haemers (1980). The vertices of the graphs are 1, 2, ..., 40 and the neighbours of the i-th vertex are the vertices of the i-th column $(i = 1, 2, ..., 40)$.

(1)

1	2	3	4	5	6	7	8	9	10	11	12	13	14	15	16	17	18	19	20
2	1	1	1	1	1	1	1	1	1	1	1	1	2	2	2	2	2	2	2
3	3	2	2	6	5	5	9	8	8	12	11	11	5	5	5	6	6	6	7
4	4	4	3	7	7	6	10	10	9	13	13	12	8	8	8	9	9	9	10
5	14	23	32	14	17	20	14	17	20	14	17	20	11	11	11	12	12	12	13
6	15	24	33	15	18	21	15	18	21	15	18	21	17	18	19	14	15	16	14
7	16	25	34	16	19	22	16	19	22	16	19	22	20	21	22	20	22	17	19
8	17	26	35	23	26	29	26	29	23	29	23	26	23	24	25	24	25	23	25
9	18	27	36	24	27	30	27	30	24	30	24	27	26	27	28	27	26	28	28
10	19	28	37	25	28	31	28	31	25	31	25	28	29	30	31	31	29	30	30
11	20	29	38	32	35	38	38	32	35	35	38	32	32	33	34	33	34	32	34
12	21	30	39	33	36	39	39	33	36	36	39	33	35	36	37	37	35	36	36
13	22	31	40	34	37	40	40	34	37	37	40	34	38	39	40	40	39	38	39

21	22	23	24	25	26	27	28	29	30	31	32	33	34	35	36	37	38	39	40
2	2	3	3	3	3	3	3	3	3	3	4	4	4	4	4	4	4	4	4
7	7	5	5	5	6	6	6	7	7	7	5	5	5	6	6	6	7	7	7
10	10	10	10	10	8	8	8	9	9	9	9	9	9	10	10	10	8	8	8
13	13	12	12	12	13	13	13	11	11	11	13	13	13	11	11	11	12	12	12
15	16	14	15	16	14	15	16	14	15	16	14	15	16	14	15	16	14	15	16
19	18	19	17	18	18	17	19	18	19	17	19	17	18	18	19	17	19	18	17
24	23	22	21	20	21	22	20	22	20	21	21	22	20	21	20	22	22	20	21
26	27	26	28	27	23	25	24	24	25	23	25	23	24	25	23	24	24	23	25
31	29	31	29	30	30	31	29	28	28	27	27	28	26	28	27	26	27	28	26
32	33	33	34	32	34	32	33	32	33	34	29	30	31	31	29	30	30	31	29
35	37	36	37	35	37	36	35	36	37	35	37	35	36	33	34	32	34	32	33
40	38	39	38	40	40	38	39	40	38	39	39	40	38	38	40	39	35	37	36

(2)

1	2	3	4	5	6	7	8	9	10	11	12	13	14	15	16	17	18	19	20
2	1	1	1	1	1	1	1	1	1	1	1	1	2	2	2	2	2	2	2
3	3	2	2	6	5	5	9	8	8	9	10	11	5	5	5	6	6	6	7
4	4	4	3	7	7	6	10	11	12	13	13	12	8	10	12	8	9	11	8
5	14	23	32	14	17	20	14	14	15	15	16	16	9	11	13	10	12	13	13
6	15	24	33	15	18	21	17	18	17	19	18	19	19	18	17	16	15	14	15
7	16	25	34	16	19	22	20	21	21	22	22	20	22	20	21	22	20	21	18
8	17	26	35	23	26	29	23	24	24	25	25	23	25	23	24	23	24	23	26
9	18	27	36	24	27	30	26	27	28	28	26	27	30	29	29	25	25	24	27
10	19	28	37	25	28	31	29	30	31	29	30	31	31	31	30	27	28	26	30
11	20	29	38	32	25	28	32	33	34	32	33	34	34	33	32	36	35	37	32
12	21	30	39	33	36	39	35	36	37	36	37	35	35	36	35	39	38	38	34
13	22	31	40	34	37	40	38	39	40	40	38	39	37	37	36	40	39	40	40

21	22	23	24	25	26	27	28	29	30	31	32	33	34	35	36	37	38	39	40
2	2	3	3	3	3	3	3	3	3	3	4	4	4	4	4	4	4	4	4
7	7	5	5	5	6	6	6	7	7	7	5	5	5	6	6	6	7	7	7
9	11	8	9	11	8	9	10	8	9	10	8	9	10	8	9	10	8	9	10
10	12	13	10	12	12	13	11	11	12	11	12	13	13	11	12	12	13	12	11
16	14	15	16	14	19	17	18	15	14	14	16	15	14	14	15	14	18	17	17
19	17	17	18	17	20	20	21	16	16	15	20	21	20	16	16	15	19	18	19
26	27	19	19	18	21	22	22	22	20	21	21	22	22	18	17	19	22	21	20
28	28	28	27	26	25	24	23	24	23	25	25	23	24	28	26	27	23	23	24
31	29	30	29	31	29	31	30	26	28	27	27	26	26	29	30	29	24	25	25
32	33	33	34	32	33	32	32	35	36	35	28	27	28	31	31	30	31	29	30
33	34	38	38	39	34	33	34	37	37	36	37	35	36	33	34	32	32	34	33
39	38	39	40	40	36	37	35	39	40	38	38	40	39	40	38	39	36	37	35

(3)

1	2	3	4	5	6	7	8	9	10	11	12	13	14	15	16	17	18	19	20
2	1	1	1	1	1	1	1	1	1	1	1	1	2	2	2	2	2	2	2
3	3	2	2	6	5	5	9	8	8	9	10	11	5	5	5	6	6	6	7
4	4	4	3	7	7	6	10	11	12	13	13	12	8	10	12	8	9	11	8
5	14	23	32	14	17	20	14	14	15	15	16	16	9	11	13	13	10	12	11
6	15	24	33	15	18	21	17	18	18	19	19	17	19	17	18	15	16	14	16
7	16	25	34	16	19	22	20	21	22	20	21	22	22	21	20	21	22	20	19
8	17	26	35	23	26	29	23	24	24	25	25	23	25	23	24	27	28	26	23
9	18	27	36	24	27	30	26	27	26	28	27	28	27	26	26	29	29	30	24
10	19	28	37	25	28	31	29	30	31	31	29	30	28	28	27	30	31	31	29
11	20	29	38	32	35	38	32	33	34	32	33	34	34	33	32	32	32	33	36
12	21	30	39	33	36	39	35	36	37	37	35	36	38	39	38	34	33	34	37
13	22	31	40	34	37	40	38	39	40	39	40	38	40	40	39	37	36	35	40

21	22	23	24	25	26	27	28	29	30	31	32	33	34	35	36	37	38	39	40
2	2	3	3	3	3	3	3	3	3	4	4	4	4	4	4	4	4	4	4
7	7	5	5	5	6	6	6	7	7	7	5	5	6	6	6	7	7	7	7
9	10	8	9	11	8	9	11	8	9	10	8	9	10	8	9	10	8	9	10
12	13	13	10	12	10	12	13	12	13	11	11	12	13	12	13	11	13	11	12
15	14	15	16	14	15	14	14	17	17	18	16	15	14	19	18	17	14	15	14
17	18	20	20	21	16	16	15	18	19	19	17	18	17	21	20	20	16	16	15
24	23	22	21	22	19	17	18	20	22	21	18	19	19	22	21	22	21	22	20
25	25	27	28	26	25	23	24	25	24	23	25	23	24	23	23	24	26	26	27
31	30	31	30	29	30	31	29	28	26	27	30	29	29	29	24	25	28	27	28
35	35	33	34	32	36	37	35	33	32	32	31	30	31	28	26	27	31	29	30
36	37	35	35	36	38	39	38	34	33	34	35	37	36	32	34	33	33	34	32
38	39	36	37	37	39	40	40	39	40	38	40	38	39	39	40	38	37	35	36

Table 1

Now let G be a $(40, 12, 2, 4)$ graph with $a_5 \neq 0$. Take a vertex x whose neighbour graph is a disjoint sum of four triangles and partition the vertices of G into three constituents according to the vertex x, the vertices of G_x and the remaining vertices. With respect to these constituents let the adjacency matrix be C_x say, so that $C_x + I$ is the incidence matrix of a $(40, 13, 4)$ design with an obvious polarity having all points absolute. Write $C_x^* = C_x + I$, so that C_x^* is of the form (3). In this case, A of (3) is a direct sum of four copies of the all-one matrix of order 3 and so has eigenvalues 0, 3 of multiplicities 8 and 4, respectively. Now the permutation matrices P that commute with A form a group $Z(A)$ of order $6^4.4!$ while $N(A)$ has order 6^4. The distinct coset representatives of $N(A)$ in $Z(A)$ are of the form $I_3 \otimes Q$ where Q is a permutation matrix of order 4. It is clear that if y is in the same orbit as x under the action of Aut(G) then the designs constructed via the matrices C_x^* and C_y^* are isomorphic. Thus in determining the non-isomorphic designs that can be manufactured from the graph G using

Theorem 1 it is only necessary to find vertices x not in the same orbit under Aut(G) and to let P run through all permutation matrices of the form $I_3 \otimes Q$.

Remark 3.   Without going into details we mention that it is possible to construct further (40, 13, 4) designs in the case the vertex x has neighbour graph of type (4).   Here C^* has a submatrix of the form $\begin{bmatrix} J & 0 \\ 0 & J \end{bmatrix}$.   Replacing it with $\begin{bmatrix} 0 & J \\ J & 0 \end{bmatrix}$ gives a design with a polarity having 34 absolute points.

2.2   BIBD's from (40, 13, 4) designs.
Let us call a set of points in a (40, 13, 4) design an *arc* if it meets every block of the design in 1 or 4 points.   It is easy to see that an arc can have at most 10 points.   Arcs having this number of points are called *maximal* arcs.   Again it is a simple matter to verify that there are precisely 10 blocks that meet a maximal arc in one point.   If the points of this design are partitioned into two sets, one comprising the points of a maximal arc M and the other the remaining points, and the blocks are similarly partitioned into two sets, one containing the blocks that meet M in one point and the other remaining blocks, we obtain the incidence matrix of the design in the form

$$\begin{bmatrix} I_{10} & X \\ Y & * \end{bmatrix},$$

where, as is easily checked, X and Y^t are the incidence matrices of 2-(10, 4, 4) designs.   Many of the (40, 13, 4) designs constructed by means of Theorem 2 have maximal arcs and consequently give rise to such bibd's.   The details are give in the next section.

3. **Results of computer analysis**

In Assmus and van Lint (1977) an oval was defined in (v, k, λ) designs.   Its meaning differed according to the parameters of the design, but in the case of (40, 13, 4) designs it is a set of four points no three of which lie on the same block.   It turns out that of all the designs manufactured by the methods above, only those from graphs 18, 19, ..., 27 of Table 2 have ovals.

No.	Autorder	5-tuple	Orbit structure	Orbits of G_x
1	72	(0, 12, 0, 18, 10)	$2^2, 6^1, 12^1, 18^1$	(0, 1, 0, 1, 3)
2	432	(0, 0, 0, 36, 4)	$4^1, 36^1$	(0, 0, 0, 1, 1)
3	51840	(0, 0, 0, 0, 40)	40^1	(0, 0, 0, 0, 1)
4	648	(0, 36, 0, 0, 4)	$4^1, 36^1$	(0, 1, 0, 0, 1)
5	2	(8, 18, 2, 9, 3)	2^{16}	(4, 11, 1, 5, 3)
6	6	(18, 18, 0, 3, 1)	$3^7, 6^3$	(3, 6, 0, 0, 1)
7	4	(12, 12, 0, 14, 2)	$2^6, 4^6$	(3, 4, 0, 7, 2)
8	6	(6, 22, 6, 3, 3)	$2^4, 3^3, 6^3$	(1, 9, 1, 1, 3)
9	4	(4, 20, 4, 10, 2)	$2^6, 4^6$	(1, 6, 1, 6, 2)
10	9	(9, 18, 0, 9, 4)	$3^4, 9^3$	(1, 4, 0, 1, 2)
11	48	(0, 32, 0, 0, 8)	$3^1, 4^1, 8^1, 24^1$	(0, 2, 0, 0, 3)
12	6	(18, 20, 0, 0, 2)	$2^4, 3^3, 6^3$	(3, 10, 0, 0, 3)
13	1	(16, 14, 4, 5, 1)		(16, 14, 4, 5, 1)
14	3	(12, 18, 6, 3, 1)	3^{13}	(4, 6, 2, 1, 1)
15	18	(18, 18, 0, 3, 1)	$3^1, 9^2, 18^1$	(1, 2, 0, 1, 1)
16	9	(18, 12, 9, 0, 1)	$3^4, 9^3$	(2, 4, 1, 0, 1)
17	27	(27, 12, 0, 0, 1)	$3^1, 9^1, 27^1$	(1, 2, 0, 0, 1)
18	144	(0, 24, 0, 12, 4)	$4^1, 12^1, 24^1$	(0, 1, 0, 1, 1)
19	16	(8, 8, 0, 20, 4)	$4^2, 8^4$	(1, 1, 0, 3, 1)
20	48	(0, 16, 6, 12, 6)	$2^1, 4^1, 6^1, 12^1, 16^1$	(0, 1, 1, 1, 2)
21	8	(2, 8, 12, 16, 2)	$2^8, 4^2, 8^2$	(1, 1, 5, 4, 1)
22	192	(0, 0, 4, 24, 12)	$4^1, 12^1, 24^1$	(0, 0, 1, 1, 1)
23	51840	(0, 0, 0, 0, 40)	40^1	(0, 0, 0, 0, 1)
24	64	(16, 0, 8, 16, 0)	$8^1, 16^2$	(1, 0, 1, 1, 0)
25	12	(6, 12, 12, 10, 0)	$3^3, 6^3, 12^1$	(1, 1, 3, 3, 0)
26	8	(10, 8, 18, 4, 0)	$2^8, 4^2, 8^2$	(2, 1, 7, 2, 0)
27	384	(0, 0, 24, 16, 0)	$16^1, 24^1$	(0, 0, 1, 1, 0)

Table 2

We also mention the so-called Hamada conjecture concerning the p rank of the incidence matrix of (v, k, λ) designs with the parameters of a projective geometry. Here the p-rank of the incidence matrix is its rank over $GF(p)$, where p is a prime divisor of $k - \lambda$. As a result of the paper by Hamada (1973) it was conjectured that the p-rank was minimum when the design arose from a projective geometry. We computed the 3-rank of all the 389 (40, 13, 4) designs manufactured and found the minimum of 11 to be achieved only in the case of Graph Number 3, which is indeed a projective geometry. The remaining 3-ranks ranged from 12 to 18, so the conjecture remains open.

The tables need a word of explanation. Table 2 contains data about the 27 (40, 12, 2, 4) graphs that were found by computer search. Numbers 15, 24 and 27 are those not found by Haemers (1980). Column 4 describes the non-trivial orbits, a^b meaning b orbits of length a, while column 5 contains the numbers of orbits of points with neighbour graphs of types (1), (2), ..., (5) respectively. A representative was chosen from each of the orbits corresponding to types (4) and (5) and used to manufacture (40, 13, 4) designs by the methods previously described.

No	Autorder	Total(4)	3-rank(4)	Total(5)	3-rank(5)
1	144	1	14	19	13
2	1728	1	14	7	13
3	12130560	0		5	11
4	1994	0		4	13
5	2	5	16	40	15
6	6	1	16	7	15
7	4	7	16	27	15
8	6	1	16	26	15
9	4	6	16	27	15
10	9	1	16	19	15
11	96	0		21	15
12	6	0		20	15
13	1	5	16	24	15
14	3	1	16	10	15
15	18	1	16	7	15
16	9	0		10	15
17	27	0		4	15
18	144	1	16	14	15
19	16	3	16	14	15
20	48	1	16	15	15
21	8	4	16	12	15
22	192	1	16	8	15
23	51840	0		5	15
24	64	1	16		
25	12	3	16		
26	8	2	16		
27	384	1	16		

Table 3

Table 3 contains the details about these manufactured designs. In column 2 we give the order of the automorphism group of the corresponding design $D(G)$ say, with a polarity having all points absolute.   Of course, Aut(G) is a subgroup of Aut($D(G)$) but it is interesting to note that Aut(G) = Aut($D(G)$) in all cases except graphs 1, 2, 3, 4 and 11.   When Aut(G) is odd this has already been proved in Haemers and Spence (to appear).   Columns 3, 4 contain the number of designs manufactured from neighbour graphs of type (4) and the minimum value of the 3-rank, respectively, while columns 5 and 6 contain the corresponding data for neighbour graphs of type (5).   The designs in column 3 are all non-isomorphic (47 in total) and indeed non-isomorphic to any in column 5.   However, exactly three designs appear twice in column 5, so the total number of non-isomorphic designs from columns 3 and 5 is 389.   Of these, there are 91 with trivial automorphism group.

A large number of the designs obtained have maximal arcs, and using these we found a total of 807 non-isomorphic 2-(10, 4, 4) designs, while the number of non-isomorphic residual and derived designs, with parameters 2-(27, 9, 4) and 2-(13, 4, 3) amounted to 8071 and 3702, respectively.

Acknowledgement.   The author acknowledges with gratitude the debt he owes to F.C. Bussemaker for the use of two of his computer programs in the above analysis.

References

A. Rudvalis (1971), (v, k, λ)-graphs and polarities of (v, k, λ)-designs, *Math. Z.* **120**, 224-230.

N. Hamada (1973), On the p-rank of the incidence matrix of a balanced or partially balanced incomplete block design and its application to error correcting codes, *Hiroshima Math. J.* **3**, 154-226.

B. Weisfeiler (ed) (1976), On construction and identification of graphs, *Lecture Notes in Math.* **558**, Springer-Verlag, Berlin.

E.F. Assmus and J.H. van Lint (1977), Ovals in projective designs, *J. Combin. Theory Ser. A* **27**, 307-324.

W.H. Haemers (1980), *Eigenvalue Techniques in Design and Graph Theory*, Tract **121**, Mathematische Centrum, Amsterdam.

A.E. Brouwer and J.H. van Lint (1984), Strongly regular graphs and partial geometries, *Proc. Silver jubilee Conf. Waterloo*, 1982.

F.C. Bussemaker, W.H. Haemers, J.J. Seidel and E. Spence (1989), On (v, k, λ) graphs and designs with trivial automorphism group, *J. Combin. Theory Ser. A* **50**, 33-46.

W.H. Haemers and E. Spence (to appear), On (v, k, λ) graphs and designs without involutions, *Proc. Combinatorics '88*, Ravello.

Generalized Reed-Solomon codes and normal rational curves: an improvement of results by Seroussi and Roth

L. Storme and J.A. Thas
State University of Ghent

1.   Introduction

Let $\Sigma = PG(n, q)$ denote the n-dimensional projective space over the field $GF(q)$.   A k-arc of points in Σ (with $k \geq n + 1$) is a set K of k points such that no $n + 1$ points of K belong to a hyperplane.   A k-arc is *complete* if it is not contained in a $(k + 1)$-arc.

    A linear $[n, k, d]$ code C satisfies the condition $d \leq n - k + 1$ (Bruen et al. 1988).   When d equals $n - k + 1$, C is called an M.D.S. (Maximum Distance Separable) code.

    It is well known that k-arcs of $PG(n, q)$ and linear M.D.S. codes of dimension $n + 1$ and length k over $GF(q)$ are equivalent objects (Bruen et al. 1988).

    Segre (1955) posed the following problems:

(1)   For given n and q, what is the maximum value of k for which there exist k-arcs in $PG(n, q)$?

(2)   For what values of n and q, with $q > n + 1$, is every $(q + 1)$-arc of $PG(n, q)$ the point set of a normal rational curve?

(3)   For given n and q, with $q > n + 1$, what are the values of k for which every k-arc of $PG(n, q)$ is contained in a normal rational curve of this space?

In particular, we will consider k-arcs which are subsets of a normal rational curve, that is, the k-arcs which correspond to generalized Reed-Solomon codes and generalized doubly-extended Reed-Solomon codes (Seroussi and Roth 1986).

2.   Definitions

A normal rational curve in $PG(n, q)$ is any $(q + 1)$-arc projectively equivalent to the $(q + 1)$-arc $\{(1, t, ..., t^n) \,|\, t \in GF(q)\} \cup \{(0, ..., 0, 1)\} = \{(1, t, ..., t^n) \,|\, t \in GF(q)^+\}$ $(GF(q)^+ = GF(q) \cup \{\infty\}$; ∞ corresponds to $(0, ..., 0, 1))$.

A linear $[n, k]$ code C is called a generalized Reed-Solomon (G.R.S.) code if and only if it has a generator matrix $G = [g_{ij}]$ with $g_{ij} = v_j \alpha_j^{i\text{-}1}$, $1 \le i \le k$, $1 \le j \le n$, $\alpha_1, ..., \alpha_n$ are distinct elements of $GF(q)$

(with $0^0 = 1$), and $v_1, ..., v_n$ are non-zero (not necessarily distinct). The elements $\alpha_1, ..., \alpha_n$ are called the column generators of C.

Thus, G.R.S. codes are defined for any length $n \le q$. The columns of G can be identified with n points of the q-arc $\{(1, t, ..., t^n) \mid t \in GF(q)\}$.

G.R.S. codes can be extended while preserving the M.D.S. property, by extending $G = [v_j \alpha_j^{i\text{-}1}]$ by an extra column $\begin{bmatrix} 0 \\ \vdots \\ 0 \\ v \end{bmatrix}$, $v \ne 0$. We

say that this column has column generator ∞. The resulting codes, but also all G.R.S. codes are called generalized doubly-extended Reed-Solomon (G.D.R.S.) codes.

The matrix $G = [v_j \alpha_j^{i\text{-}1}]$, $1 \le i \le k$, $1 \le j \le n$, with $\alpha_1, ..., \alpha_n$ distinct elements of $GF(q)^+$ and $v_1, ..., v_n$ non-zero, is called a canonical generator matrix of the corresponding G.D.R.S. code.

3.   Known results

We shall investigate how these G.D.R.S. $[n, k]$ codes can be extended to linear codes with larger length, while preserving the M.D.S. property. This means, we will determine how the corresponding n-arc on a normal rational curve of $PG(k - 1, q)$ can be extended to a $(n + 1)$-arc of $PG(k - 1, q)$.

A solution to this problem is known in the following cases. Some theorems hold for arcs which do not necessarily belong to a normal rational curve. These results can all be found in Bruen et al. (1988), and Thas (to appear).

3.1 q odd

Theorem 1 (Segre 1961). *For any k-arc in $PG(n, q)$, $q > 3$, with $n = 2, 3$ or 4, we have $k \le q + 1$. For any n, the point set of a normal rational curve is a $(q + 1)$-arc. For $n = 2, 3$, the converse is true; a $(q + 1)$-arc of $PG(2, q)$ is an irreducible conic and a $(q + 1)$-arc of $PG(3, q)$ is a twisted cubic.*

Theorem 2 (Thas 1968, 1987).
(a)   *For any k-arc of $PG(n, q)$, $q > (4n - 23/4)^2$, we have $k \le q + 1$.*

(b)   *In $PG(n, q)$, with $q > (4n - 23/4)^2$, every $(q + 1)$-arc is the point set of a normal rational curve.*

(c)   *In $PG(n, q)$, every k-arc with $k > q - (\sqrt{q})/4 + n - 7/16$ is contained in one and only one normal rational curve of this space.*

Remark 3. By a recent note, due to   Kaneta and Maruta (1989), the bound of Theorem 2(a) can be improved to $q > (4n - 39/4)^2$.

Remark 4.   By results of Voloch (1990, this volume), combined with the note mentioned in Remark 3, the bounds in (a), (b) and (c) of Theorem 2 can be improved respectively to $q > 45n - 140$, $q > 45n - 95$ and $k > 44q/45 + n - 10/9$ when q is prime, and to $q > ((29/4)p + 4n - 12)^2/p$, $q > ((29/4)p + 4n - 8)^2/p$, and $k > q - \sqrt{(pq)}/4 + (29/16)p + n - 1$ when $q = p^{2h+1}$, p an odd prime.

Theorem 5 (Seroussi and Roth 1986). *In $PG(n, q)$, a k-arc K consisting of k points of a normal rational curve L, such that $2 \le n \le k - (q + 1)/2$, can be extended by a point p to a $(k + 1)$-arc if and only if $p \in L \backslash K$.*

3.2   *q* even

In $PG(2, q)$, any $(q + 1)$-arc is incomplete. It can be extended to a $(q + 2)$-arc by its nucleus.

Theorem 6 (Segre: see Hirschfeld 1979, §10.3). *A k-arc of $PG(2, q)$ with $k > q - \sqrt{q} + 1$ is contained in a $(q + 2)$-arc; for $q > 2$ and $q = 2, k = 3$, this $(q + 2)$-arc is unique and for $q = 2$, $k = 2$, there are two such 4-arcs.*

Remark 7. By a recent result of Voloch (this volume), the bound in Theorem 6 can be improved to $k > q - \sqrt{(2q)} + 2$ when $q = 2^{2h+1}$.

Theorem 8 (Casse: see Hirschfeld 1985, §21.3). *For any k-arc of $PG(3, q)$, $q > 2$, we have $k \le q + 1$; for any k-arc of $PG(3, 2)$, we have $k \le 5$. For any k-arc of $PG(4, q)$, $q > 4$, we have $k \le q + 1$; for any k-arc of $PG(4, 2)$ or $PG(4, 4)$, we have $k \le 6$.*

Theorem 9 (Kaneta and Maruta 1989). *For any k-arc of $PG(5, q)$, $q > 4$, we have $k \le q + 1$; for any k-arc of $PG(5, 2)$, or $PG(5, 4)$, we have $k \le 7$.*

Theorem 10 (Thas 1969b). *In $PG(q - 2, q)$, any k-arc satisfies $k \le q + 2$ and a $(q + 2)$-arc can be constructed by adjoining to a normal rational curve of $PG(q - 2, q)$, a certain point, its nucleus.*

Theorem 11 (Storme and Thas, to appear).

(a) Let K be a k-arc of $PG(3, q), q \neq 2$. If $k > q - (\sqrt{q})/2 + 9/4$, then K can be completed to a $(q + 1)$-arc which is uniquely determined by K.

(b) In $PG(n, q), n \geq 4$ and $q > (2n - 11/2)^2$, any k-arc K satisfies $k \leq q + 1$.

(c) In $PG(n, q), n \geq 4$ and $q > (2n - 7/2)^2$, every $(q + 1)$-arc is a normal rational curve.

(d) Let K be a k-arc in $PG(n, q), n \geq 4, q = 2^h, h \geq 2$ and $k > q - (\sqrt{q})/2 + n - 3/4$.   Then K lies in a normal rational curve L of $PG(n, q)$, and L is completely determined by K.

Theorem 12 (Seroussi and Roth 1986). In $PG(n, q)$, a k-arc K contained in a normal rational curve L, such that $2 \leq n \leq k - q/2$, can be extended to a $(k + 1)$-arc by a point p if and only if either

(a) $p \in L \backslash K$, or

(b) $n = 2$ and p is the nucleus of the conic L.

4. Alternative proof for the result of Seroussi and Roth

Lemma 13.   In $PG(n, q), n > 2$, a k-arc K consisting of k points of the q-arc $\{(1, t, ..., t^n) | t \in GF(q)\}$ such that $2 < n \leq k - \lfloor(q - 1)/2\rfloor$ cannot be extended to a $(k + 1)$-arc by the point $(0, ..., 0, 1, \alpha), \alpha \in GF(q)$.

Proof.   This proof is a combination of arguments of Seroussi and Roth (1986).

$$\begin{vmatrix} 0 & 1 & ... & 1 \\ 0 & x_1 & ... & x_n \\ \vdots & \vdots & & \vdots \\ 0 & x_1^{n-2} & .. & x_n^{n-2} \\ 1 & x_1^{n-1} & .. & x_n^{n-1} \\ \alpha & x_1^n & ... & x_n^n \end{vmatrix} = \begin{vmatrix} 0 \\ 0 \\ \vdots & A_n \\ 0 \\ 1 \\ \alpha\, x_1^n\, ...\, x_n^n \end{vmatrix} = \left[\alpha - \sum_{j=1}^{n} x_j\right] (-1)^n \det A_n = 0$$

$$\Leftrightarrow \alpha = \sum_{j=1}^{n} x_j \quad (x_i \in GF(q), \text{ all } x_i \text{ different}).$$

We suppose that K consists of the k points with parameters in $S = \{x_1, \ldots, x_k\}$. We select $n - 2$ parameters x_i and we define

$\beta = \alpha - \Sigma_{i=1}^{n-2} x_i$. We require that β is not equal to 0 when q is even.

In $GF(q)$, there are $\lfloor q/2 \rfloor$ unordered pairs $\{a_i, b_i\}$ such that $\beta = a_i + b_i$. Since $k \geq n + \lfloor (q - 1)/2 \rfloor$, we have $k - (n - 2) \geq \lfloor (q + 3)/2 \rfloor$. Hence there must be a pair $\{x_{n-1}, x_n\} \subseteq S$ such that $x_{n-1} + x_n = \beta$ and x_{n-1}, $x_n \notin \{x_1, \ldots, x_{n-2}\}$. This means $\alpha = \Sigma_{i=1}^{n} x_i$, all x_i different, which proves that p does not extend K.

The main result of Seroussi and Roth (1986) is the following theorem.

Theorem 14. *Let C be a G.D.R.S. $[n, k, n-k+1]$ code over $GF(q)$ such that $2 \leq k \leq n - \lfloor (q - 1)/2 \rfloor$. Let G be a canonical generator matrix of C, and let g be a k-dimensional column vector over $GF(q)$. The extension of C generated by the matrix $[G \mid g]$ is M.D.S. if and only if either*

(a) $\quad g = v \begin{bmatrix} 1 \\ \beta \\ \vdots \\ \beta^{k-1} \end{bmatrix}$, *where $v \in GF(q) \setminus \{0\}$, $\beta \in GF(q) \cup \{\infty\}$ and β is not a*

column generator of C, or

(b) $\quad q$ *is even, $k = 3$ and $g = v \begin{bmatrix} 0 \\ 1 \\ 0 \end{bmatrix}$ for some $v \in GF(q) \setminus \{0\}$.*

This gives the following equivalent theorem about k-arcs on a normal rational curve.

Theorem 15. *In $PG(n, q)$, a k-arc belonging to the normal rational curve $L = \{(1, t, \ldots, t^n) \mid t \in GF(q)^+\}$, such that $2 \leq n \leq k - \lfloor (q + 1)/2 \rfloor$, can be extended to a $(k + 1)$-arc by a point p if and only if either*
(a) $p \in L \setminus K$, *or*
(b) q *is even, $n = 2$ and $p = (0, 1, 0)$, the nucleus of the conic L.*

Proof. We first consider the case $n = 2$. Then K contains at least $\lfloor (q + 5)/2 \rfloor$ points.

(a) q odd

If p is an internal point of the conic L, then p belongs to $(q + 1)/2$ bisecants and 0 tangents of L.   When p is external to L, then p belongs to $(q - 1)/2$ bisecants and 2 tangents of L.   In both cases, p belongs to a bisecant of K.   Thus, the only points which extend K are the points of $L \backslash K$.

(b) q even

If p does not belong to L and p is not the nucleus of L, then p belongs to one tangent and $q/2$ bisecants of L. Thus, p is on a bisecant of K. This proves the theorem when $n = 2$.

We now assume $n > 2$.   We choose the projective reference system in such a way that $e_0(1, 0, ..., 0), ..., e_n(0, ..., 0, 1), e_{n+1}(1, ..., 1)$ belong to K.   The points $e_0, ..., e_{n-2}$ generate an $(n - 2)$-dimensional subspace β of $PG(n, q)$.   Consider the $(n - 3)$-dimensional subspace α_i of β generated by $e_0, ..., e_{i-1}, e_{i+1}, ..., e_{n-2}$ $(0 \leq i \leq n - 2)$.   We shall project from α_i onto a plane β_i skew to it.   Let β_i have equations $X_j = 0$ for all $j \neq i, n - 1, n$.   All points of K, not belonging to α_i, are projected onto points of a $(k - (n - 2))$-arc K_i on a conic L_i in β_i.   The arc K_i contains at least $\lfloor (q + 5)/2 \rfloor$ points.   A point p of $PG(n, q)$ which extends K is projected onto a point p_i of β_i which extends K_i.

(a) q odd

From the proof for $n = 2, q$ odd, we conclude that $p_i \in L_i$.   The points e_i, e_{n-1}, e_n belong to L_i.   The equations of L_i are

$$\begin{cases} X_i a_i(X_{n-1}, X_n) = \beta_i X_{n-1} X_n, \ \beta_i \neq 0, \ \deg_{X_{n-1}} a_i = \deg_{X_n} a_i = 1 \\ \qquad\quad X_j = 0, \qquad\qquad \text{for all } j \neq i, n - 1, n. \end{cases}$$

Since $a_i(X_{n-1}, X_n) = 0$ contains β and the tangent of L at e_i, all $n - 1$ hyperplanes $a_i(X_{n-1}, X_n) = 0$ are distinct.

Let $(V^2_{n-1})_i$ be the quadratic hypercone of $PG(n, q)$ with vertex α_i and with base L_i.   This hypercone contains β, L and p.   We determine the intersection of all $(V^2_{n-1})_i$.   This intersection is β together with the set L^* of all points

$$\left(\frac{\beta_0 X_{n-1} X_n}{a_0(X_{n-1}, X_n)}, \dots, \frac{\beta_{n-2} X_{n-1} X_n}{a_{n-2}(X_{n-1}, X_n)}, X_{n-1}, X_n\right),$$

with $(X_{n-1}, X_n) \neq (0, 0)$. Clearly L^* is a set of $q + 1$ points containing e_0, ..., e_{n-2}. Hence this set must be the normal rational curve L. Thus p belongs to L.

(b) q even

From the proof for $n = 2$, q even, we conclude that either $p_i \in L_i$ or p_i is the nucleus of L_i. We have $\alpha_i e_{n+1} \cap \beta_i = (z_0, \dots, z_{i-1}, z_i, z_{i+1}, \dots, z_n) = (0, \dots, 0, 1, 0, \dots, 0, 1, 1) \in L_i$. Thus, in β_i, the equation of the conic L_i is $X_i(X_{n-1} + \gamma_i X_n) = (1 + \gamma_i) X_{n-1} X_n$ $(\gamma_i \neq 1, 0)$. As in the odd case, all γ_i are distinct. The nucleus of the conic L_i is $(y_0, \dots, y_{i-1}, y_i, y_{i+1}, \dots, y_n) = (0, \dots, 0, 1 + \gamma_i, 0, \dots, 0, \gamma_i, 1)$. Since K contains $e_0, \dots, e_{n+1}$, all coordinates of $p(a_0, \dots, a_n)$ are different. This implies that p_i is at most once the nucleus of a conic L_i.

When p is never projected onto the nucleus of L_i, we can use the same proof as for q odd.

However, if p is projected onto the nucleus of a conic L_i, choose $i = n - 2$, and determine the intersection of the other hypercones $(V_{n-1}^2)_i$ $(0 \leq i \leq n - 3)$. This intersection is β union the surface of points

$$\left(\frac{(1 + \gamma_0) X_{n-1} X_n}{X_{n-1} + \gamma_0 X_n}, \dots, \frac{(1 + \gamma_{n-3}) X_{n-1} X_n}{X_{n-1} + \gamma_{n-3} X_n}, X_{n-2}, X_{n-1}, X_n\right)$$

with $(X_{n-2}, X_{n-1}, X_n) \neq (0, 0, 0)$. This surface is a cone V_2^{n-1} with vertex $(0, \dots, 0, 1, 0, 0)$ and with base a normal rational curve in $X_{n-2} = 0$. In fact V_2^{n-1} is a cone which projects L from $e_{n-2} \in L$. Since p is projected from α_{n-2} onto the nucleus of L_{n-2}, the point p is on the tangent of L at e_{n-2}.

There is a projectivity θ of $PG(n, q)$ which maps V_2^{n-1} onto the cone with vertex $(0, \dots, 0, 1)$ and base $\{(1, t, \dots, t^{n-1}, 0) \mid t \in GF(q)^+\}$, with $\theta(p)$ belonging to the line $e_n e_{n-1}$. Thus $\theta(p) = (0, \dots, 0, 1, \alpha)$ $(\alpha \in GF(q))$. These points were treated in Lemma 13: $\theta(p)$ cannot extend $\theta(K)$, a contradiction.

We conclude that p belongs to L.

Corollary 16. *Each normal rational curve of $PG(n, q)$, $2 < n \leq \lfloor (q + 2)/2 \rfloor$ is complete.*

Remark 17. If q is odd and $k = n + (q - 1)/2$, then the projection K_i, in the preceding theorem, is a $((q + 3)/2)$-arc in β_i. A point p of β_i, not belonging to the conic L_i, which extends K_i to a $((q + 5)/2)$-arc, must be external to L_i and the 2 tangents of L_i through p must intersect L_i in points of K_i.

Relying on the proof of Theorem 15, in combination with an argument of Seroussi and Roth, we are able to improve Corollary 16 when q is odd.

5. Normal rational curves in $PG((q + 3)/2, q)$, q odd

The Method of Seroussi and Roth (1986)

We will only consider q odd.

Seroussi and Roth proved Theorem 15 as follows by induction on n. For $n = 2$ we refer to the first part of the proof of Theorem 15. Assume now that $n > 2$ and that the theorem holds in $PG(n - 1, q)$. We are considering a k-arc K on the normal rational curve $L = \{(1, t, ..., t^{n-1}, t^n) \mid t \in GF(q)^+\}$ such that $k \geq n + (q + 1)/2$. We also assume that K contains $e_n = (0, ..., 0, 1)$ and the points $(1, \alpha_i, ..., \alpha_i^n)$, $\alpha_i \in S$, $|S| = k - 1$. Suppose $p(a_0, ..., a_n)$ extends K. We project $K \backslash \{e_n\}$ from e_n on $X_n = 0$. We obtain a $(k - 1)$-arc K_1, on the normal rational curve $L_1 = \{(1, t, ..., t^{n-1}) \mid t \in GF(q)^+\}$ in $X_n = 0$, which can be extended by $p_1(a_0, ..., a_{n-1})$ to a k-arc. We know that Theorem 15 holds in $n - 1$ dimensions; thus $p_1 \in L_1 \backslash K_1$, so $p_1 = (1, \lambda, ..., \lambda^{n-1})$.

If $(a_0, ..., a_{n-1}) = (0, ..., 0, b)$ $(b \neq 0)$, then $p(0, ..., 0, 1, \alpha)$. These points are treated in Lemma 13; p doesn't extend K. Thus $p_1 (1, \lambda, ..., \lambda^{n-1})$, $\lambda \in GF(q)$. If $p = (1, \lambda, ..., \lambda^n)$, the theorem holds. We assume the contrary; $p = (1, \lambda, ..., \lambda^n) + (0, ..., 0, b)$, $b \neq 0$.

Seroussi and Roth (1986) prove as follows that there exist n distinct points $(1, \alpha_i, ..., \alpha_i^n)$ of $K \backslash \{e_n\}$ such that $p = \Sigma_{i=1}^n d_i (1, \alpha_i, ..., \alpha_i^n)$, $\alpha_i \in S$.

We have

$$p = \sum_{i=1}^{n} d_i (1, \alpha_i, ..., \alpha_i^n) \Leftrightarrow \prod_{i=1}^{n} (\lambda - \alpha_i) + b = 0. \tag{1}$$

Select $n - 2$ different values α_i in S and replace α_{n-1} and α_n by x.

We obtain the quadratic equation

$$\prod_{i=1}^{n-2}(\lambda - \alpha_i)(\lambda - x)^2 + b = 0. \tag{2}$$

This equation has at most two roots in $GF(q)$, which will be denoted by β_1 and β_2.

We will select α_{n-1} and α_n in S in such a way that $\prod_{i=1}^{n}(\lambda - \alpha_i) +$

$b = 0$. In order to find α_{n-1} and α_n , select α_{n-1} such that

(a) $\alpha_{n-1} \neq \alpha_i$ ($1 \leq i \leq n - 2$)
(b) $\alpha_{n-1} \neq \beta_1, \beta_2$ and
(c) $\alpha_{n-1} \in S$.

This gives us at least $k - n - 1$ elements of S to choose α_{n-1} from.
For a given choice of α_{n-1}, (1) holds if and only if

$$\alpha_n = \lambda + b/\{(\lambda - \alpha_{n-1}) \prod_{i=1}^{n-2}(\lambda - \alpha_i)\} = f(\alpha_{n-1}).$$

For $\alpha_{n-1} \in GF(q)\backslash\{\lambda\}$, f is a bijective function on $GF(q)\backslash\{\lambda\}$. Clearly α_{n-1} $\neq f(\alpha_{n-1}) = \alpha_n$ and α_n must belong to $S\backslash\{\alpha_1, ..., \alpha_{n-2}\}$. Remark that $\lambda \notin S$, since otherwise p is on the line joining two points of K. There are exactly $q - 1 - (k - 1 - (n - 2))$, that is, at most $(q - 5)/2$ elements $f(\alpha_{n-1})$, $\alpha_{n-1} \in GF(q)\backslash\{\lambda\}$, which do not satisfy this condition. We have at least $k - n - 1 \geq (q - 1)/2$ elements to choose a suitable α_{n-1} from. That is why there must be a pair of distinct elements $\{\alpha_{n-1}, \alpha_n\}$ in S which satisfy the imposed conditions. This implies $p = \sum_{i=1}^{n} d_i(1, \alpha_i, ..., \alpha_i^n)$; so p is

linearly dependent on n elements of K. Hence p does not extend K.

Lemma 18. *Consider a normal rational curve K in $PG(n, q)$, $q > n + 1$. Let $PG(m - 1, q)$ be the space defined by the points $p_1, ..., p_m$ of K. Project $K\backslash\{p_1, ..., p_m\}$ from $PG(m - 1, q)$ onto a projective space α of dimension $n - m$ skew to $PG(m - 1, q)$. This projection K_1 is a $(q + 1 - m)$-arc contained in a normal rational curve L of α. The remaining points of $L\backslash K_1$ are the intersections of α with the m-dimensional subspaces generated by $p_1, ..., p_m$ and the respective tangents to K in these points.*

Proof. Let $K = \{(1, t, ..., t^n) \mid t \in GF(q)^+\}$. We use induction on m.

(a) $m = 1$

Select $p_1(1, 0, ..., 0)$ and project $K \backslash \{p_1\}$ from p_1 onto $X_0 = 0$. The points of $K \backslash \{p_1\}$ are projected onto the points $(0, 1, t, ..., t^{n-1})$, $t \neq 0$. The last point $r_1(0, 1, 0, ..., 0)$ of the normal rational curve $\{(0, 1, t, ..., t^{n-1})$ $\mid t \in GF(q)^+\}$ in $X_0 = 0$ is the intersection of the line $p_1 r_1$ with $X_0 = 0$, where $p_1 r_1$ is the tangent to K at p_1.

(b) $m > 1$

We now project from the $PG(m - 1, q)$ defined by the m points p_1, ..., p_m. We assume that the lemma holds for any $m - 1$ points. We first project $K \backslash \{p_1\}$ from $p_1(1, 0, ..., 0)$ onto $X_0 = 0$. This gives a normal rational curve K_1 in $X_0 = 0$. The points $p_2, ..., p_m$ are projected onto the points $p'_2, ..., p'_m$ of K_1. The point $p'_1(0, 1, 0, ..., 0)$ is not the projection of any point of $K \backslash \{p_1\}$.

We then project $K_1 \backslash \{p'_1, ..., p'_m\}$ from the $PG(m - 2, q)$ defined by $p'_2, ..., p'_m$, onto an $(n - m)$-dimensional subspace α of $X_0 = 0$ skew to $PG(m - 2, q)$, giving a normal rational curve K_2 of α. The point $(0, 1, 0, ..., 0)$ is projected onto a point r of K_2, where r is the intersection of α with the projective space generated by $p'_2, ..., p'_m$ and $(0, 1, 0, ..., 0)$. It is also the intersection of α with the m-dimensional subspace generated by $p_1, ..., p_m$ and $e_1(0, 1, 0, ..., 0)$ which contains the tangent $p_1 e_1$ to K in p_1.

Each point r_i of K_2 which is not the projection of a point of K_1 is the intersection of α with the space generated by $p'_2, ..., p'_m$ and the tangent L'_i to K_1 at a point p'_i, $2 \leq i \leq m$. The point r_i also belongs to the space generated by $p_1, p'_2, ..., p'_m$ and L'_i. Clearly L'_i is the projection from p_1 onto $X_0 = 0$ of the tangent L_i to K at p_i, $2 \leq i \leq m$. Thus r is the intersection of α with the m-dimensional space generated by $p_1, ..., p_m$ and L_i, $2 \leq i \leq m$.

Lemma 19. *In $PG((q + 3)/2, q)$, q odd and $q > 5$, a point p which extends a normal rational curve K to a $(q + 2)$-arc, belongs to a $(n - 2)$-dimensional subspace generated by $n - 2$ points of K and the tangent to K at one of these points.*

Proof. As in the proof of Theorem 15, we select $n - 1$ points r_i of K. These points generate an $(n - 2)$-dimensional subspace α. We then project from each $(n - 3)$-dimensional subspace α_i generated by $n - 2$ points $r_1, ..., r_{i-1}, r_{i+1}, ..., r_{n-1}$ onto a plane β_i skew to it.

The remaining points of K are projected onto points of a conic L_i in β_i. If for some i, p is projected onto a point of L_i, then by Lemma 18, Lemma 19 holds.

We now assume that for all i, p is projected onto a point p_i not belonging to L_i. Exactly $(q + 3)/2$ points of L_i are the projections of the points of K. Following Remark 17, p_i must be an external point of L_i such that the 2 tangents to L_i through p_i intersect L_i in points t_i and s_i which are the projections of points t'_i and s'_i of K. So p belongs to the hyperplanes generated by α_i and the respective tangents to K at t'_i and s'_i.

As i varies, it takes $(q + 1)/2$ values. This gives a total of $(q + 1)$ tangents to the conics L_i through the points p_i.

At most one of these $q + 1$ tangents intersects L_i in the projection of the unique point r_i of K in $\alpha \setminus \alpha_i$. For when this occurs, p belongs to the hyperplane γ_1 generated by $r_1, \ldots, r_{n-1}$ and the tangent to K at one of those points r_i. If this happens twice, we obtain hyperplanes γ_1 and γ_2. These must be equal since they are generated by n points $r_1, \ldots, r_{n-1}$, p of a $(q + 2)$-arc. Thus γ_1 contains $n - 1$ points of K and 2 tangents to K. This is impossible.

Therefore, at least q tangents to L_i through p_i ($i = 1, \ldots, n - 1$) have their tangent point in the set of projections of the $(q + 1)/2$ points of K, not belonging to α.

There must be 2 projections, say from α_1 and α_2, and a point s' of K, not belonging to α, such that p_i belongs to the tangent at s_i to L_i, where s' is projected from α_i onto s_i, $i = 1, 2$. The point p then belongs to the hyperplanes generated by α_i, $i = 1, 2$, and the tangent to K at s'. These hyperplanes intersect in a $(n - 2)$-dimensional subspace containing p, $n - 3$ points of K in $\alpha_1 \cap \alpha_2$ and the tangent to K at s'.

Lemma 20. *For q odd, $2 \leq n \leq q - 2$, $a \in GF(q)$, there exist n different elements t_i in $GF(q)$ such that $a = \sum_{i=1}^{n} t_i$.*

Proof. We have $\sum_{\delta \in GF(q)} \delta = 0$. Thus, if $a = \sum_{i=1}^{l} t_i$, then $a = \sum_{i=l+1}^{q} (-t_i)$, with $GF(q) = \{t_1, \ldots, t_q\}$. So, if $q = 2r + 1$, we need only consider $n \leq r$.

We shall use induction on n.

(a) $n = 2$

$a = \beta + \gamma$ for two distinct elements in $GF(q)$.

(b) $n > 2$

Suppose $a = \Sigma_{i=1}^{n} t_i$, $r > n \geq 2$, all t_i different.

Select t_1; there are r pairs $\{\beta, \gamma\}$ of distinct elements such that $t_1 = \beta + \gamma$. The elements β and γ must be different from $t_2, \ldots, t_n$, that is, from at most $r - 2$ elements. Hence there must be a pair $\{\beta, \gamma\}$ not containing $t_2, \ldots, t_n$. For any such pair, we have $a = \Sigma_{i=2}^{n} t_i + \beta + \gamma$, whence a is the sum of $n + 1$ distinct elements.

Lemma 21. *No point* $(0, \ldots, 0, 1, \alpha)$, $\alpha \in GF(q)$, *extends the normal rational curve* $\{(1, t, \ldots, t^n) \mid t \in GF(q)^+\}$.

Proof.

$$\begin{vmatrix} 0 & 1 & \ldots & 1 \\ 0 & t_1 & \ldots & t_n \\ \vdots & \vdots & & \vdots \\ 0 & t_1^{n-2} & \ldots & t_n^{n-2} \\ 1 & t_1^{n-1} & \ldots & t_n^{n-1} \\ \alpha & t_1^{n} & \ldots & t_n^{n} \end{vmatrix} = 0 \Leftrightarrow \alpha = \sum_{i=1}^{n} t_i, \text{ all } t_i \in GF(q) \text{ distinct,}$$

(Seroussi and Roth 1986). From Lemma 20, such values t_i exist.

Theorem 22. *In* $PG((q + 3)/2, q)$, *q odd and* $q > 3$, *every normal rational curve is complete.*

Proof. By 3.1 we may assume that $q > 5$.

Suppose there exists a point p which extends the normal rational curve $K = \{(1, t, \ldots, t^n) \mid t \in GF(q)^+\}$ to a $(q + 2)$-arc.

From Lemma 19, p belongs to an $(n - 2)$-dimensional subspace α generated by $n - 2$ points $p_0, \ldots, p_{n-3}$ of K and the tangent to K at one of these points. Choose the reference system in such a way that $p_0 = e_0(1, 0, \ldots, 0)$, $p_{n-3} = e_n(0, \ldots, 0, 1)$ and such that the tangent $e_{n-1}e_n$ through p_{n-3} belongs to α.

Then $(a_0, \ldots, a_n)$ with $p(a_0, \ldots, a_n)$, is a linear combination of the rows of the $(n - 1) \times (n + 1)$ matrix

$$
\begin{bmatrix}
1 & 0 & \dots & 0 & 0 & 0 \\
1 & \alpha_1 & \dots & \alpha_1^{n-2} & \alpha_1^{n-1} & \alpha_1^{n} \\
\vdots & \vdots & & \vdots & \vdots & \vdots \\
1 & \alpha_{n-4} & \dots & \alpha_{n-4}^{n-2} & \alpha_{n-4}^{n-1} & \alpha_{n-4}^{n} \\
0 & 0 & \dots & 0 & 1 & 0 \\
0 & 0 & \dots & 0 & 0 & 1
\end{bmatrix}
=
\begin{bmatrix}
1 & 0 & \dots & 0 & 0 \\
1 & & & & \alpha_1^{n} \\
\vdots & & H & & \vdots \\
1 & & & & \alpha_{n-4}^{n} \\
0 & & & & 0 \\
0 & 0 & \dots & 0 & 1
\end{bmatrix}
\qquad (\alpha_i \neq 0, \infty).
$$

Hence $(a_1, \dots, a_{n-1})$ is a linear combination of the $n - 3$ rows of H. These rows define $n - 3$ points of the normal rational curve $K_1 = \{(1, t, \dots, t^{n-2}) \mid t \in GF(q)^+\}$ in $PG(n - 2, q)$; these rows also include the point $(0, \dots, 0, 1)$ in $PG(n - 2, q)$. Consequently $(a_1, \dots, a_{n-1}) \neq l(1, 0, \dots, 0)$, $l \neq 0$. Since $p \notin p_0 p_{n-3}$, we have $(a_1, \dots, a_{n-1}) \neq (0, \dots, 0)$. Further, $(a_1, \dots, a_{n-1}) \neq l(1, t, \dots, t^{n-2})$, $t \neq 0, \infty$, $l \neq 0$, since otherwise p is incident with the plane through $(1, 0, \dots, 0)$, $(0, \dots, 0, 1)$, $(1, t, \dots, t^n)$.

We first assume $(a_1, \dots, a_{n-1}) \neq l(0, \dots, 0, 1)$, $l \neq 0$. Let $(a_1, \dots, a_{n-1}) = \sum_{i=1}^{n-4} b_i (1, \alpha_i, \dots, \alpha_i^{n-2}) + b_{n-3}(0, \dots, 0, 1)$. Then not all b_i, $1 \leq i \leq n - 4$, are zero; for instance $b_1 \neq 0$. Clearly $b_{n-3} \neq 0$ as otherwise p is in the space generated by the points $(1, \alpha_1, \dots, \alpha_1^{n})$, $\dots$, $(1, \alpha_{n-4}, \dots, \alpha_{n-4}^{n})$, $(0, \dots, 0, 1)$, $(1, 0, \dots, 0)$.

The element $(1, \alpha_1, \dots, \alpha_1^{n-2}) + b(0, \dots, 0, 1)$, with $b = b_{n-3}/b_1$, is a linear combination of $(1, \alpha_i, \dots, \alpha_1^{n-2})$ $(2 \leq i \leq n - 4)$ and 3 other elements $(1, \alpha_{n-3}, \dots, \alpha_{n-3}^{n-2})$, $(1, \alpha_{n-2}, \dots, \alpha_{n-2}^{n-2})$, $(1, \alpha_{n-1}, \dots, \alpha_{n-1}^{n-2})$ of $K_1 \setminus \{(0, \dots, 0, 1)\}$ if and only if $\left[\Pi_{i=2}^{n-4}(\alpha_1 - \alpha_i) \right](\alpha_1 - \alpha_{n-3})(\alpha_1 - \alpha_{n-2})(\alpha_1 - \alpha_{n-1}) + b = 0$ (see Seroussi and Roth 1986). We will choose $\alpha_{n-3}, \alpha_{n-2}, \alpha_{n-1}$ in such a way that all α_i, $1 \leq i \leq n - 1$, are distinct and non-zero. We have exactly $(q + 3)/2$ choices for α_{n-3} ($\alpha_{n-3} \neq 0$; $\alpha_{n-3} \neq \alpha_i$, $i = 1, \dots, n - 4$).

When we apply the method of Seroussi and Roth (1986) §5, we first consider the quadratic equation

$$
\left[\prod_{i=2}^{n-3}(\alpha_1 - \alpha_i) \right] (\alpha_1 - x)^2 + b = 0. \tag{3}
$$

We then need to find an element α_{n-2} such that (a) $\alpha_{n-2} \neq \alpha_r$, $1 \leq r \leq n-3$, (b) α_{n-2} different from the roots of (3) above, (c) $\alpha_{n-2} \neq 0$. The equation (3) has at most 2 roots in $GF(q)$. We have exactly $(q+3)/2$ choices for α_{n-3}. There must be a certain α_{n-3} such that (3) contains no roots in $GF(q)$. Select such an α_{n-3}. We therefore have $(q+1)/2$ possibilities for α_{n-2}.

The last element $\alpha_{n-1} = \alpha_1 + b/\{(\alpha_1 - \alpha_{n-2})\Pi_{i=2}^{n-3}(\alpha_1 - \alpha_i)\}$ must be

different from 0 and α_r, $2 \leq r \leq n-3$. Hence, we must avoid $(q-3)/2$ values. We therefore have at least 2 values of α_{n-2} which give a satisfactory value for α_{n-1}. Select such a α_{n-2}. We then know from §5 that $(1, \alpha_1, ..., \alpha_1^{n-2}) + (0, ..., 0, b) = \Sigma_{i=2}^{n-1} c_i(1, \alpha_i, ..., \alpha_1^{n-2})$ with all α_i distinct

and $\alpha_i \neq 0, \infty$. Thus $(a_1, ..., a_{n-1}) = \Sigma_{i=1}^{n-4} b_i(1, \alpha_i, ..., \alpha_i^{n-2}) + (0, ..., 0, b_{n-3})$

$= \Sigma_{i=2}^{n-4} b_i(1, \alpha_i, ..., \alpha_i^{n-2}) + b_1\Sigma_{i=2}^{n-1} c_i(1, \alpha_i, ..., \alpha_i^{n-2})$. So, p is linearly

dependent on $(1, 0, ..., 0)$, $(0, ..., 0, 1)$, $(1, \alpha_i, ..., \alpha_i^n)$, $2 \leq i \leq n-1$.

Consequently, p does not extend K.

We still need to consider the points $(a_0, ..., a_n) = (a_0, 0, ..., 0, a_{n-1}, a_n)$ $(a_{n-1} \neq 0)$. We may assume $a_0 \neq 0$, since the points $(0, ..., 0, a_{n-1}, a_n)$ were treated in Lemma 21. We use a transformation f that leaves K and $(0, ..., 0, 1)$ invariant, but such that the image of $p_0 = (1, 0, ..., 0)$ is the point p_1 $(p_1 \in \alpha)$. We then apply the previous method to $f(p)$. This completes the proof.

6. k-arcs on a conic C in PG(2, q)

Introduction.

Consider the conic $C = \{(t, t^2, 1) | t \in GF(q)^+\}$ and a set K of k points of C, which contains $(0, 0, 1)$ (parameter 0) and $(0, 1, 0)$ (parameter ∞). We assume the existence of a point p, not belonging to C and different from the nucleus of C when q is even, which extends the k-arc K. Let ϕ be the involution of $PGL(2, q)$ on C which fixes the tangent points of the tangents to C through p (these may lie in an extension of $GF(q)$) and such that $\phi(r_1) = r_2$, $r_1 \neq r_2$, if and only if $p \in r_1 r_2$. Thus $\phi : x$ a $(ax + b)/(cx - a)$, with $a^2 + bc \neq 0$. Since p extends the k-arc K, no two different points of K are the image of one another under ϕ.

We will now look for k-arcs K on C which are only extendable to a $(k + 1)$-arc by means of the other points of C, and the nucleus of C if q is even. We will therefore select the points of K in such a way that, for each involution ϕ (of $PGL(2, q)$) on C, there exist two different points p_1, p_2 of K such that $\phi(p_1) = p_2$.

We will now determine when this holds.

Conditions on the set K.

We will always assume that the points with parameters 0 and ∞ belong to K. Identify each point of K with its parameter.

(a) The involutions ϕ which fix ∞: $\phi : y = -x + d$ ($d \neq 0$, if q is even). There must be two distinct elements x_1, x_2 of $K \backslash \{\infty\}$ such that

$$x_1 + x_2 = d. \tag{4}$$

(b) The involutions ϕ which do not fix ∞: $\phi : y = (ax + b)/(x - a)$ ($a^2 + b \neq 0$), $\phi(\infty) = a$. If $(a, a^2, 1) \in K$, we have a solution to the condition. We now assume the contrary; this implies $a \neq 0$. Two points x, y of $K \backslash \{\infty\}$, $x, y \in GF(q)$, are the image of each other if and only if

$$y = \frac{ax + b}{x - a} \Leftrightarrow xy - ay - ax - b = 0. \tag{5}$$

Conclusion.

Let $0, \infty \in K$. For any d, with $d \neq 0$ if q is even, there must be two distinct $x_1, x_2 \in K \backslash \{\infty\}$ such that $x_1 + x_2 = d$. For any two elements a, b $\in GF(q)$ with $a^2 + b \neq 0$ we have either $a \in K \backslash \{\infty\}$, or there exist two distinct elements $x, y \in K \backslash \{\infty\}$ satisfying $xy - ay - ax - b = 0$.

Segre's irreducibility criterion.

In Bartocci and Segre (1971) it is proved that a curve Γ in $PG(2, q)$ is absolutely irreducible; that is, it is irreducible over the algebraic closure $\overline{GF(q)}$ of $GF(q)$, if there exists a point p (over $GF(q)$) of Γ and a tangent L (over $\overline{GF(q)}$) in p such that

(a) L has multiplicity 1 in the set of all tangents of Γ at p;

(b) the intersection multiplicity $i(L, \Gamma; p)$ of L and Γ in p is $\deg(\Gamma)$;

(c) over $\overline{GF(q)}$, the curve Γ does not contain a linear component through p.

Theorem 23 (The Hasse-Weil bound, see Hirschfeld 1979). *Let Γ be an absolutely irreducible curve of degree n defined over $GF(q)$. Suppose that N is the number of non-singular points of Γ (over $GF(q)$) plus the number of singular points p_i of Γ (over $GF(q)$) counted according to their real index (this is the number of tangents (over $GF(q)$) of Γ at p_i, where each tangent is counted according to its multiplicity in the set of tangents of Γ at p_i).*
  Then $|N - (q + 1)| \leq 2g \sqrt{q}$ with g the genus of Γ.

Remark 24. Korchmáros (1983) proved: For every divisor s of $q - 1$ such that $q > \left[(s-1)^2 + \sqrt{s^4 - 4s^3 + 8s^2 + 1} \ \right]^2$, let $R_s = \{g^s, g^{2s}, ..., g^{ds}\}$ $(d = (q - 1)/s;$ g a primitive element of $GF(q)^*)$ and consider the set $K = \{(1, g^{2is}, g^{is}) | i = 1, ..., (q-1)/s\}$ of the conic $C = \{(1, t^2, t) | t \in GF(q)^+\}$.
  Then a point p extends K to a $(((q - 1)/s) + 1)$-arc if and only if p lies on $C \backslash K$, or $p = (1, t, 0)$ with $-t \notin R_s$, or $p = (0, 0, 1)$ and $2s$ does not divide $q - 1$.
  We will modify this result in Theorem 25 in order to obtain a set K on C which can only be extended to a larger arc by the remaining points of C and the nucleus of C when q is even.

Theorem 25. *In $PG(2, q)$, consider the set $K = \{(a^r, a^{2r}, 1) | a \in GF(q)\} \cup \{(0, 1, 0)\}, r > 1$, of points of the conic $C = \{(t, t^2, 1) | t \in GF(q)\} \cup \{(0, 1, 0)\}$.*
  Suppose that:
(a) $2r | (q - 1)$ *when q is odd and r is even, and $r | (q - 1)$ in all other cases.*
(b) $q + 1 - 2r^2 - 2r - 2(r - 1)^2 \sqrt{q} > 0$ *when q is odd, and $q + 1 - r^2 - 2r - 2(r - 1)^2 \sqrt{q} > 0$ when q is even.*

  Then K can only be extended to a $(k + 1)$-arc by the remaining points of C and the nucleus of C when q is even.

Proof. We will show that the above conditions on the set K are satisfied. We must find two elements ξ^r, η^r $(\xi^r \neq \eta^r)$ such that $\xi^r \eta^r - a(\xi^r + \eta^r) - b = 0$ (a, b any two elements in $GF(q)$ with a no r-th power, $a^2 + b \neq 0$) and two elements ζ^r, δ^r $(\zeta^r \neq \delta^r)$ such that $\zeta^r + \delta^r = d$ (d any element in $GF(q)$, with $d \neq 0$, if q is even).

We therefore consider the curve Γ: $X^r Y^r - aZ^r(X^r + Y^r) - bZ^{2r} = 0$, with $a \neq 0$, $a^2 + b \neq 0$.

Korchmáros (1983) has proved that:

(1)   (1, 0, 0) and (0, 1, 0) are ordinary singular points of Γ with multiplicity r.

(2)   The tangents of Γ in (1, 0, 0) (resp. (0, 1, 0)) are F_i:$Z = \lambda_i Y$ (resp. H_i:$Z = \lambda_i X$) with λ_1, ..., λ_r the solutions of $\lambda^r = 1/a$ over the algebraic closure of $GF(q)$.

(3) $i(F_i, \Gamma; (1, 0, 0)) = i(H_i, \Gamma; (0, 1, 0)) = 2r$.

(4) Γ has no linear component through (1, 0, 0) as well as through (0, 1, 0).

Hence Γ satisfies the conditions of Segre's irreducibility criterion. This is to say that Γ is absolutely irreducible.   We will now apply the Hasse-Weil bound on Γ.

When $b \neq 0$, (1, 0, 0) and (0, 1, 0) are the only singular points on Γ. This implies that the genus of Γ is $(r - 1)^2$.   We then know from the Hasse-Weil bound that $R + 2r \geq q + 1 - 2(r - 1)^2\sqrt{q}$ where R is the number of non-singular points of Γ in $PG(2, q)$.

When $b = 0$, Γ contains one other singular point (0, 0, 1) of multiplicity r.   Moreover, (0, 0, 1) is also an ordinary singular point. In this case, the genus of Γ is $g = (r - 1)^2 - r(r - 1)/2$.   The Hasse-Weil bound implies that $R + 3r \geq q + 1 - 2g\sqrt{q}$, with R the number of non-singular points of Γ in $PG(2, q)$.   Therefore

$$R \geq q + 1 - 2(r - 1)^2\sqrt{q} + r(r - 1)\sqrt{q} - 3r$$
$$> q + 1 - 2(r - 1)^2\sqrt{q} - 2r.$$

We conclude that Γ contains at least $q + 1 - 2(r - 1)^2\sqrt{q} - 2r$ non-singular points in $PG(2, q)$.

If q is even (resp. odd), at most r^2 (resp. $2r^2$) points $(x, y, 1)$ of Γ satisfy the condition $x^r = y^r$.   The intersection of Γ and the line $Z = 0$ are the singular points (1, 0, 0) and (0, 1, 0).   So, we will find a point $(x, y, 1)$ $(x^r \neq y^r)$ of Γ if $q + 1 - 2(r - 1)^2\sqrt{q} - 2r - r^2 > 0$ for q even, and if $q + 1 - 2(r - 1)^2\sqrt{q} - 2r - 2r^2 > 0$ for q odd.   These inequalities hold.

We must also find two elements ξ^r and η^r $(\xi^r \neq \eta^r)$ such that $\xi^r + \eta^r = d$ $(d \neq 0$, if q is even).   If $d \neq 0$, consider Δ: $X^r + Y^r - dZ^r = 0$.

The curve Δ is absolutely irreducible and contains no singular points. Hence Δ contains at least $q + 1 - 2g\sqrt{q}$ points, $g = ((r - 1)(r - 2))/2$. We do not look at the points of Δ on $Z = 0$ and we need to avoid the points $(x, y, 1)$ of Δ for which $x^r = y^r$. If q is odd, at most r^2 points $(x, y, 1)$ of Δ satisfy $x^r = y^r$. If q is even, no point $(x, y, 1)$ satisfies $x^r = y^r$. On $Z = 0$, the curve Δ has exactly r points. When q is odd and $d \neq 0$, Δ contains a point $(x, y, 1)$ $(x^r \neq y^r)$ when $q + 1 - r^2 - r - (r - 1)(r - 2)\sqrt{q} > 0$. If q is even and $d \neq 0$, Δ contains such a point if $q + 1 - r - (r - 1)(r - 2)\sqrt{q} > 0$. These inequalities hold.

If q is odd and $d = 0$, we need to find x and y for which $x^r = -y^r$. This means $(x/y)^r = -1$. Such elements x and y exist when r is odd. If r is even, $2r | (q - 1)$, thus x and y exist in $GF(q)$.

7. Completeness of normal rational curves

Theorem 26. *Let $q = p^h$, p prime. Suppose r, $r > 1$, exists such that*
(a) $2r | (q - 1)$ *when q is odd and r even, and $r | (q - 1)$ in all other cases;*

(b) $q + 1 - 2r^2 - 2r - 2(r - 1)^2\sqrt{q} > 0$, *$q$ odd, and $q + 1 - r^2 - 2r - 2(r - 1)^2$ $\sqrt{q} > 0$, q even.*

Then every normal rational curve of $PG(n,q)$ is complete for
(a) *q even and $3 \leq n \leq (r - 1)(q/r) + 1/r$,*
(b) *q odd and $2 \leq n \leq (r - 1)(q/r) + (1/r)$.*

Proof. From Theorem 15, we may assume $n > (q /2) + 1$. Choose the reference system in such a way that $e_0(1, 0, ..., 0)$, ..., $e_n(0, ..., 0, 1)$, $e_{n+1}(1, ..., 1)$ belong to the normal rational curve K.

Then K is the set of points $\left\{ \left(\dfrac{a_0}{(a_0 - 1)t + 1}, ..., \dfrac{a_n}{(a_n - 1)t + 1} \right) \mid t \in GF(q)^+ \right\}$, where all elements a_i are different and non-zero (Hirschfeld 1985).

We consider the set of parameters $S = \{a^r \mid a \in GF(q)\} \cup \{\infty\}$. Then $|S| = ((q - 1)/r) + 2$. We select a_i so that $-(1/(a_i - 1)) \notin S$, $i = 0, ..., n - 2$. This is possible if $n - 1 \leq q + 1 - (((q - 1)/r) + 2) \Leftrightarrow n \leq (r - 1)(q/r) + 1/r$.

We now proceed as in the proof of Theorem 15. The points $e_0, ..., e_{n-2}$ determine a $(n - 2)$-dimensional subspace α: $X_{n-1} = X_n = 0$. We project from $\alpha_i : X_i = X_{n-1} = X_n = 0$ $(0 \leq i \leq n - 2)$ onto the plane β_i:

$X_j = 0$ for all $j \neq i, n - 1, n$. The points of K which do not belong to α_i are projected onto points of the conic $C_i = \left\{ \left(\dfrac{a_i}{(a_i - 1)t + 1}, \dfrac{a_{n-1}}{(a_{n-1} - 1)t + 1}, \right. \right.$ $\left. \left. \dfrac{a_n}{(a_n - 1)t + 1} \right) \mid t \in GF(q)^+ \right\}$ in β_i. The points of K, not belonging to α_i, are projected onto the points of a $(q + 3 - n)$-arc K_i on C_i. K_i contains the points with parameters in S. We know from Theorem 25 that K_i can only be extended to a $(q + 4 - n)$-arc in β_i by the remaining points of C_i and also by the nucleus of C_i if q is even.

If there exists a point p of $PG(n, q)$, which extends K to a $(q + 2)$-arc, then p is projected from α_i onto a point p_i of β_i which extends K_i to a $(q + 4 - n)$-arc. Thus p is projected onto C_i or possibly to the nucleus of C_i if q is even. We may now use the arguments in the proof of Theorem 15 combined with Lemma 21. It follows that p belongs to K, a contradiction. We conclude that K is complete.

Remark 27. From the proof of the preceding theorem, we may also conclude the following. In $PG(n, q)$, q odd, $n > 2$, any k-arc L of points on a normal rational curve $K = \left\{ \left(\dfrac{a_0}{(a_0 - 1)t + 1}, \cdots, \dfrac{a_n}{(a_n - 1)t + 1} \right) \mid t \in GF(q)^+ \right\}$, all a_i different and non-zero, where L contains $e_0, \ldots, e_{n-2}$ and the points of K with parameters in $S = \{a^r \mid a \in GF(q)\} \cup \{\infty\}$, with $-(1/(a_i - 1)) \notin S$, $i = 0, \ldots, n - 2$, and r satisfying the conditions of the preceding theorem, can only be extended to a $(k + 1)$-arc by the remaining points of K.

Remark 28. We will determine an upper bound for r in Theorem 26:

$$q + 1 - 2r^2 - 2r - 2(r - 1)^2\sqrt{q} > 0 \text{ when } r < \frac{-1 + 2\sqrt{q} + \sqrt{(2q\sqrt{q} + 2q - 6\sqrt{q} + 3)}}{2 + 2\sqrt{q}},$$

and,

$$q + 1 - r^2 - 2r - 2(r - 1)^2\sqrt{q} > 0 \text{ when } r < \frac{-1 + 2\sqrt{q} + \sqrt{(2q\sqrt{q} + q - 4\sqrt{q} + 2)}}{1 + 2\sqrt{q}}.$$

In both cases, it is sufficient that $r \leq (\sqrt[4]{q})/\sqrt{2}$.

Remark 29. Some examples where the conditions of Theorem 26 are satisfied:

$q = 13^6$	$r = 28$;
$q = 3^{10}$	$r = 11$;
$q = 2^{5h}$	$r = 2^h - 1$ $(h \geq 2)$;
$q = p^{4h}$ $(p > 2)$	$r = (p^h - 1)/2$ $(h \geq 2$ when $p = 3)$.

8. The special case: q prime

Lemma 30 (Dirichlet's Theorem). *If a and m (a, m $\geq$ 1) are relatively prime integers, there exist infinitely many prime numbers p such that $p \equiv a$ (mod m)* (Serre 1973).

Theorem 31. *There exist infinitely many prime numbers p, such that for each n, $2 \leq n \leq p - 1$, every normal rational curve in PG(n, p) is complete.*

Proof. Assume $m \geq 45$, m odd. From Lemma 30, we know that there is an infinite number of primes p for which $p \equiv 1$ (mod m). If $p \equiv 1$ (mod m) and $p + 1 - 2m^2 - 2m - 2(m - 1)^2\sqrt{p} > 0$, $r = m$ satisfies the conditions of Theorem 26. Thus, in $PG(n, p)$, with $2 \leq n \leq (m - 1)(p/m) + (1/m)$, every normal rational curve is complete.

Voloch (1990) proved that for $2 \leq n < (p + 140)/45$ and $p \neq 2$, we have $k \leq p + 1$ for every k-arc in $PG(n, p)$. By dualizing this result (Thas 1969a, to appear), we then obtain: $p - 2 \geq n > (44p - 140)/45$, with $p \neq 2$, implies $k \leq p + 1$ for any k-arc in $PG(n, p)$. Moreover it is clear that any $(p + 1)$-arc of $PG(p - 1, p)$ is complete.

We conclude: in $PG(n, p)$, p prime, $p \equiv 1$ (mod m), m odd, $m \geq 45$ and $p + 1 - 2m^2 - 2m - 2(m - 1)^2\sqrt{p} > 0$, $2 \leq n \leq p - 1$, any normal rational curve is complete.

We may also consider m even. When $p \equiv 1$ (mod $2m$), $m \geq 45$, $p + 1 - 2m^2 - 2m - 2(m - 1)^2\sqrt{p} > 0$, then, by using a similar proof, we conclude: in $PG(n, p)$, $2 \leq n \leq p - 1$, every normal rational curve is complete.

Acknowledgement. The first author is a Research Assistant of the National Fund for Scientific Research (Belgium).

References

U. Bartocci and B. Segre (1971), Ovali ed altre curve nei piani di Galois di caratteristica due, *Acta Arith.* **18**, 423-449.

A.A. Bruen, J.A. Thas and A. Blokhuis (1988), On *M.D.S.* codes, arcs in $PG(n, q)$ with q even, and a solution of three fundamental problems of B. Segre, *Invent. Math.* **92**, 441-459.

J.W.P. Hirschfeld (1979), *Projective Geometries over Finite Fields*, Oxford University Press, Oxford.

J.W.P. Hirschfeld (1985), *Finite Projective Spaces of Three Dimensions,* Oxford University Press, Oxford.

H. Kaneta and T. Maruta (1989), An elementary proof and an extension of Thas' theorem on k-arcs, *Math. Proc. Cambridge Philos. Soc.* **105**, 459-462.

G. Korchmáros (1983), New examples of complete k-arcs in $PG(2, q)$, *European J. Combin.* **4**, 329-334.

B. Segre (1955), Curve razionali normali e k-archi negli spazi finiti, *Ann. Mat. Pura Appl.* **39**, 357-379.

B. Segre (1961), *Lectures on Modern Geometry,* Cremonese, Rome.

B. Segre (1962), Ovali e curve σ nei piani di Galois di caratteristica due, *Atti Accad. Naz. Lincei Rend.* **(8)** 32, 785-790.

G. Seroussi and R.M. Roth (1986), On *M.D.S.* extensions of generalized Reed-Solomon codes, *IEEE Trans. Information Theory* **32**, 349-354.

J.-P. Serre (1973), *A Course in Arithmetic,* Springer-Verlag, Berlin.

L. Storme and J.A. Thas (to appear), *M.D.S.* codes and arcs in $PG(n, q)$ with q even: An improvement of the bounds of Bruen, Thas and Blokhuis.

J.A. Thas (1968), Normal rational curves and k-arcs in Galois spaces, *Rend. Mat.* **1**, 331-334.

J.A. Thas (1969a), Connection between the Grassmannian $G_{k-1;n}$ and the set of the k-arcs of the Galois space $S_{n,q}$, *Rend. Mat.* **2**, 121-134.

J.A. Thas (1969b), Normal rational curves and $(q + 2)$-arcs in a Galois space $S_{q-2,q}$ $(q = 2^h)$, *Atti Accad. Naz. Lincei Rend.* **47**, 249-252.

J.A. Thas (1987), Complete arcs and algebraic curves in $PG(2, q)$, *J. Algebra* **106**, 451-464.

J.A. Thas (to appear), Projective geometry over a finite field, in *Handbook of Incidence Geometry,* (ed. F. Buekenhout), North-Holland, Amsterdam.

J.F. Voloch (1990), Arcs in projective planes over prime fields, *J. Geom.* **38**, 198-200.

J.F. Voloch (1991), Complete arcs in Galois planes of non-square order, this volume.

Common characterizations of the finite
Moufang polygons

H. Van Maldeghem
State University of Ghent

1. Introduction and notation

The notion of a generalised polygon was introduced by Tits (1959). The main idea was to give a geometric interpretation of the Chevalley groups and the twisted groups of rank 2. A *generalised polygon* is a generalised n-gon for some positive integer $n \neq 1$. A *generalised n-gon of order* (s, t), $s, t, \in N^* \cup \{\infty\}$ is a point-line incidence geometry $S = (P, B, I)$ satisfying (GP1) up to (GP4).

(GP1)   *There are $s + 1$ points incident with each line.*
(GP2)   *There are $t + 1$ lines incident with each point.*
(GP3)   *Every two incident pairs of varieties (a variety is a point or a line) lie in a common circuit consisting of $2n$ varieties.*
(GP4)   *Every (proper) circuit in S contains at least $2n$ varieties.*

In this paper, we will deal mainly with *thick* generalised polygons, ie. $s, t, > 1$. If both s and t are finite, then S is also finite (by a simple counting argument). Thick generalised n-gons exist for every $n > 1$ (eg. via free construction). A generalised 2-gon is a trivial geometry: every point is incident with every line. Although this structure is important in the theory of diagrams (see Buekenhout (1979)), we will not consider it in this paper. A thick generalised 3-gon is an ordinary projective plane. Generalised 4-gons, 6-gons and 8-gons are also called respectively *generalised quadrangles, hexagons* and *octagons*. An important result of Feit and Higman (1964) says that finite thick generalised n-gons exist only if $n = 2, 3, 4, 6,$ or 8 and there are examples in each of these cases. An important subclass of the class of generalised polygons are the so-called *Moufang polygons*. They satisfy the *Moufang condition* (condition (M) below). Before stating (M), we need some definitions. Let $S = (P, B, I)$ be a generalised n-gon, $n > 2$. A (proper) circuit consisting of $2n$ varieties is called an *apartment*. Let A be an apartment and x a variety of S. We denote the set of all varieties incident with x but distinct from the $2n$ varieties of A by $A^*(x)$. A chain of $n + 1$ distinct consecutively incident varieties is called a *root*. If n is even, then every root has a unique middle element. If this is a point, then we call the root *short*, otherwise *long*. Let $\Re = (x_0 I x_1 I \ldots I x_n)$ be a root. If α is an automorphism of S fixing all elements of $A^*(x_1)$, $A^*(x_2)$,

$A^*(x_{n-1})$, then we call α an $\Re$-*elation* or, in general, a *root-elation*. If the group of $\Re$-elations act transitively on the set of apartments containing $\Re$ (for fixed $\Re$), then we call $\Re$ a *transitive root*. In that case the action just mentioned is *regular*.

We now state the Moufang Condition:

(*M*)   *Every root in S is transitive.*

It is easily seen that the condition (M) is equivalent to asking that every root in a certain fixed apartment of S is transitive.   A celebrated result of Tits (1976) and Weiss (1979) states that infinite thick Moufang n-gons can only exist for n = 3, 4, 6, 8 (disregarding the generalised 2-gons) and there are finite and infinite examples in each case.   A theorem of Fong and Seitz (1973) implies that a finite thick generalised polygon S is Moufang if and only if S or its point-line dual arises naturally from the Chevalley groups or twisted groups of rank 2. Moreover it was proved by Tits (1979, personal communication, to appear) that also all infinite Moufang generalised n-gons are known. So a characterization of a class of Moufang polygons is very often at the same time a characterization of a class of geometries associated with Chevalley groups or twisted groups, partly motivating the results in this paper.

In §2 - 4 we collect characterization theorems about big classes of (mostly finite) Moufang polygons.   By *big class* we understand, as a general rule, a class of Moufang polygons containing all finite n-gons for at least one $n \in \{4, 6, 8\}$, thus leaving out the numerous characterizations of classical projective planes and all the characterizations of certain classical quadrangles gathered in Payne and Thas (1984).   Also, almost all characterizations we will mention include projective planes and they mostly follow from Wagner's Theorem (1965).   In that case, we will not mention references.

2. Characterizations by elations

2.1 Definitions

Let S = (P, B, I) be a generalised n-gon.   A *flag* in S is an incident point-line pair.   Let $\{p, L\}$ be a flag of S, $p \in P$, $L \in B$.   If an automorphism α of S fixes every point incident with L and every line incident with p, then we call α an *elation* (*with source* $\{p,L\}$).   For instance, a root-elation is an elation.   A *central root-elation* (*with centre p*) is an automorphism of S fixing all varieties at distance $\le$

$n/2$ from p.   If n is even, it is a root-elation for every short root having p as middle element.   Dually, one defines an *axial root-elation*.   Now let $C = (x_0 I x_1 I \ldots I x_k)$ be a chain of length k, $3 \le k \le n$. Then a *C-elation* is an elation with sources (x_{i-1}, x_i), $i = 2, 3, \ldots, k - 1$. If every chain C of fixed length k, $3 \le k \le n$, the group of C-elations acts transitively (and hence regularly) on the set of apartments containing C, then we call S *k-Moufang*.   Finally, we call S *half-Moufang* provided n is even and every short root of S is transitive, or every long root of S is transitive.

2.2 Characterizations

Theorem 1 (Van Maldeghem et al. to appear, $n = 4$; Van Maldeghem 1991a, $n = 6, 8$). *Let S be a finite thick generalised n-gon, $n = 3, 4, 6,$ or 8 and suppose $3 \le k \le n$.   Then S is Moufang if and only if it is k-Moufang.*

Theorem 2 (Walker).   *Let S be a finite thick generalised n-gon, $n = 3$, 6, or 8.   Then S is Moufang if and only if it admits an automorphism group acting on the set of ordered pairs (F, A), where F is a flag contained in an appartment A, and for every root $\mathfrak{R}$ in S, there exists a non-trivial $\mathfrak{R}$-elation.*

Theorem 3 (Walker).   *Let S be a finite thick generalised n-gon, $n = 6$, or 8.   Then S is Moufang if and only if every point of S is the centre of a non-trivial central root-elation, or every line of S is the axis of a non-trivial axial root-elation.*

Theorem 4 (Thas et al. to appear).   *A finite thick generalised quadrangle is Moufang if and only if it is half-Moufang .*

2.3 Remarks

(a)   Theorem 1 is trivial for $k = n$ and it is almost trivial for $k > 3$.
(b)   Suppose a generalised polygon S admits an automorphism group G acting transitively on the ordered pairs (F, A) as in Theorem 2. Then G is a group with a (saturated) (B, N)-pair and it is conjectured by Tits (1974) that every such finite thick generalised polygon is Moufang.
(c)   Similarly, as one can reduce the Moufang condition to the condition that every root in a fixed appartment is transitive, one can reduce the hypotheses of the above theorems (but not always that far).

(d) Theorem 3 has a partial analogue for $n = 4$, namely the result of Ealy (thesis) characterizing the finite Moufang quadrangles of *even characteristic.*

3. Characterizations by homologies

3.1 Definitions

Let S again be a generalised n-gon. There is a natural distance map d inherited from the incidence graph of S. We call the two varieties in S *opposite* if they are at a distance n (maximal distance) from each other. Let x, y be two opposite varieties in S and suppose σ is an automorphism of S fixing every variety incident with x or y. Then we call σ a *generalised homology*, or an (x, y)-*homology*. Now, let A be an apartment containing both x and y and let z be incident with x and lie in A. If the group of all (x, y)-homologies acts transitively on the set $A^*(z)$, then we call S (x, y)-*transitive*. This definition does not depend on the choice of A. If u is in A and if it is incident with z (so u is at a distance 2 from x), then we call S (x, y)-*quasi-transitive* provided the group of all (x, y)-homologies acts transitively on the set $A^*(u)$. Note that S is (x, y)-(quasi)-transitive if and only if S is (y, x)-(quasi)-(transitive).

3.2 Characterizations

Theorem 5. *Let S be a finite thick generalised n-gon of order (s, t), $n = 3, 4, 6,$ or 8 and if $n = 8$, then $s, t > 2$. Then S is Moufang if and only if at least one of the following occurs:*
(a) *$n = 3$ or 6 and S is (x,y)-transitive for every pair (x, y) of opposite varieties.*
(b) *$n = 4$ and S is (x, y)-transitive for every pair (x,y) of opposite points, or S is (X, Y)-transitive for every pair (X, Y) of opposite lines.*
(c) *$n = 8$ and S, or its point-line dual, is (X, Y)-transitive for every pair (X,Y) of opposite lines and (x, y)-quasi-transitive for every pair (x, y) of opposite points.*

Proof: For the case $n = 4$, see Thas (1986); for $n = 6$, see Van Maldeghem (1990). Let $n = 8$. Note first that all finite thick Moufang octagons have the desired property by the commutator relations in Tits (1983) and the calculations at the end of Tits (1983). We will show the converse in detail in §5 below.

4. Geometric characterizations

4.1 Definitions

Let S be a thick generalised polygon (not necessarily finite).   Suppose $d(x, y) = l < n$ for two varieties x, y in S, then for every $k, 1 \le k \le l$, there exists a unique variety x_k at a distance k from x and at a distance $l - k$ from y. We denote $x_k = \Pi_x^k (y)$.   Now let $C = (x_1 I x_2 I \ldots I x_{k-1})$ be a chain of length $k - 2$, $3 \le k \le n$. Let $A = (y_1 I y_2 I \ldots I y_{2n} I y_1)$ and $A' = (y'_1 I y'_2 I \ldots I y'_{2n} I y'_1)$ be two ordered apartments in S. We call A and A' in perspective from C if for every $i \in \{1, \ldots, k - 1\}$ and every $j \in \{1, \ldots, 2n\}$, we have $d(x_i, y_j) = d(x_i, y'_j)$ and $\Pi_{x_i}^1 (y_j) = \Pi_{x_i}^1(y'_j)$ if $d(x_i, y_j) < n$.   The configuration (C, A, A') is called a generalised k-Desargues configuration.   For $n = 3$, we have $k = 3$ and we get a usual (little) Desargues configuration, (see Hughes and Piper (1973)).   Now S is called C-Desarguesian if for every ordered apartment $A = (y_1 I y_2 I \ldots I y_{2n} I y_1)$ with $d(x_1, y_1) < n$, and every variety $y \ne x_1$, $y I \Pi_{x_1}^1 (y_1)$, there exists an ordered apartment $A' = (y'_1 I y'_2 I \ldots I y'_{2n} I y'_1)$ in perspective with A from C such that $\Pi_{x_i}^2 (y'_1) = y$.   In that case C is called a Desarguesian chain.   If every chain of length k is Desarguesian, then S is called k-Desarguesian and if $k = 3$, then we simply say that S is Desarguesian.

4.2   Characterizations

Theorem 6 (Hughes and Piper 1973, $n = 3$), (Thas and Van Maldeghem to appear, $n = 4$) (Van Maldeghem 1991a, $n > 4$).   A thick generalised n-gon S, $n > 2$, is Moufang if and only if S is n-Desarguesian.

Theorem 7 (Hughes and Piper 1973, $n = 3$), (Van Maldeghem et al. to appear, $n = 4$).   A finite thick generalised n-gon S, $n = 3$ or 4, is Moufang if and only if S is Desarguesian.

4.3   Remarks
(a)   For projective planes, the definition above of Desarguesian is in classical terms in fact little Desarguesian.
(b)   Theorem 7 is probably also true for $n > 4$, but it seems that one will not be able to avoid a very long, tiresome, dull and messy proof.
(c)   There are some other beautiful geometric characterizations of Moufang n-gons due to Thas (1983) $n = 4$ and Ronan (1980, 1981) $n = 6$

not included here, because the conditions in these theorems depend strongly on n.

(d)   As a general concluding remark, we could say that the generalization of all theorems above, which do not include all Moufang polygons to all Moufang polygons is an open question.   It is also an open question as to which characterizations of finite Moufang polygons above also hold in the infinite case.

5. Proof of Theorem 5, case $n = 8$

Throughout this section we suppose that S is a finite generalised octogon of order (s, t), $s, t > 2$ and S is (X, Y)-transitive for every pair of opposite lines X, Y and S is (x, y)-quasi-transitive for every pair of mutually opposite points x, y.   All there is left to show is that in this case S is Moufang.

5.1 Step 1

We will use the following property frequently.   Suppose an automorphism σ fixes an apartment A, all varieties incident with a certain element x of A (suppose eg. without loss of generality that x is a line) and also at least three lines through a certain point of A.   Then σ fixes a suboctogon S' of order (s, t'), $t' > 1$ (cf. Walker thesis) and hence $t' = t$ by Thas (1979), (cf. also Van Maldeghem (1991b)).   So σ is the identity.   Also, if $s > t$, then S does not contain suboctogons of order $(s, 1)$, see Thas (1979).

5.2 Step 2

Now suppose $A = p_1 I L_1 I p_2 I L_3 I p_4 I L_5 I p_6 I L_7 I p_8 I L_8 I p_7 I L_6 I p_5 I L_4 I p_3 I L_2 I p_1)$ is an apartment in S.   We denote the group of automorphisms of S fixing every variety incident with at least one of the varieties belonging to a certain set D and fixing in addition all varieties in E by $\mathcal{H}_E(D)$, (omitting the curl brackets of D and E for short).   Now by Step 1, the group $\mathcal{H}(p_1, p_8)$ acts semi-regularly on $A^*(L_1)$ and since S is (p_1, p_8)-quasi-transitive, we have $s > t$.

Suppose $\sigma \in \mathcal{H}(L_1, L_8)$ fixes an element $p \in A^*(L_3)$.   Then σ fixes the flag complex S' of a generalised quadrangle of order (s, s') with $1 < s' < s$ (indeed, $s' = s$ implies that S' is a suboctogon of order $(s, 1)$, contradicting $s > t$).   Let $L \in A^*(p_2)$ and put $L' = L^\sigma$.   Let $\sigma' \in \mathcal{H}(L_3, L_6)$ be such that $L'^{\sigma'} = L$; then $\sigma\sigma'$ fixes a suboctogon S'' of order (s', t') with $1 < t' < t$. But $S' \cap S''$ is a suboctogon of order $(s', 1)$ of S'', hence $s' < t'$ and so $s' < t$. Now consider $\alpha \in \mathcal{H}(p_2, p_7)$ not preserving S'' (α exists since $t > s'$ and $\mathcal{H}(p_2, p_7)$ acts semi-regularly on $A^*(L_1)$). Then $S'' \cap S''^\alpha$ is a suboctogon of order (s'', t') of S'' with $s'' < s$. Hence $s'' = 1$ and $t' < s'$, a contradiction. So $\mathcal{H}(L_1, L_8)$ acts semi-regularly on $A^*(L_3)$.

5.3 Step 3

Choose a fixed $L \in A^*(p_2)$ and let $L' \in A^*(p_2)$ and $p \in A^*(L_3)$. Denote by $\sigma(p, L')$ the unique (L_1, L'_8)-homology mapping L to L', where L'_8 is at distance 2 from L_6 and at distance 5 from p. If $p \neq p' \in A^*(L_3)$, then by Step 1, $p_4\sigma(p, L') \neq p_4\sigma(p', L')$. Hence, by a simple counting argument, for every point $p'_4 I L_3$, $p'_4 \neq p_2$, there exists a unique $p\, IL_3$, $p \neq p_2$, such that $p_4\sigma(p, L') = p'_4$. Varying L', we find $t - 2$ distinct non-trivial generalised homologies of the form $\sigma(p, L')$ mapping p_4 to p'_4. In particular, it is now easy to see that the automorphism group of S has a (B, N)-pair, (cf. Van Maldeghem (1991b), Lemma 2).

5.4 Step 4

In this step we show that, if an automorphism β of S fixes A, all lines through p_1 and at least three lines through p_3, then it must fix all lines through p_3. Suppose the contrary. Then β fixes the flag complex S of a generalised quadrangle of order (t', t) with $1 < t' < t$. But by transitivity, there exists a (p_1, p_8)-homology σ not preserving S', hence $S' \cap S'^\sigma$ is the flag complex of a generalised subquadrangle of S' of order (t'', t) with $t'' < t'$, contradicting the result of Thas (1982). Hence, $t' = t$ and β fixes all lines through p_3.

5.5 Step 5

The following results are proved in Van Maldeghem (1991c) for generalised hexagons, but it is easy to adjust the proof for generalised octogons.

Lemma 8. Let $X \subseteq \{p_1, p_3, p_5, L_2, L_4, L_6\}$. If $\mathcal{H}_{(p_2, p_7)}(X) \neq 1$, then $|\mathcal{H}_{(p_2, p_7)}(X)| = t$.

Lemma 9.   (1) *We have* $\mathcal{H}_{p_7}(p_1, L_1) \neq 1$.

(2) *Let* $X \subseteq \{p_3, p_5, L_2, L_4, L_6\}$ *and suppose for every* $x \in X$, $\mathcal{H}(x, y)$ *acts non-trivially on* $A^*(p_2)$, *where* y *is opposite to* x *in* A.   *Then,* $\mathcal{H}_{p_7}(X, p_1, L_1) \neq 1$

These two lemmas, together with Step 1, imply readily that

$$\mid \mathcal{H}(L_1, p_1, L_2, L_4, p_5, L_6) \mid = t.$$

Now, let $\sigma \in \mathcal{H}(L_1, p_1, L_2, L_4, p_4, L_6)$.   Suppose σ does not fix all lines through p_3 and let $L \text{ I } p_3$ be such that $L^\sigma = L' \neq L$. Set $L_3{}^\sigma = L'_3$ and let M , $M' \in A^*(p_2)$ with $M \neq M' \neq L'_3 \neq M$.   Let p'_8 resp. p''_8 be at distance two from p_7 and at distance five from M resp. M' and consider the (p_1, p'_8)-homology α_M , resp. (p_1, p''_8)-homology α_M ', mapping L_3 to L'_3. Then by Step 4, $L^{\alpha_M} \neq L^{\alpha_{M'}}$.   Since there are t - 2 choices for M, we can choose M such that it maps L to L'. Considering $\sigma \alpha_M{}^{-1}$ we see, by Step 4, that this is impossible, hence $L = L'$ and $\sigma \in \mathcal{H}(L_1, p_1, L_2, p_3, L_4, p_5, L_6)$, and consequently all short roots are transitive.

5.6 Step 6

In view of Theorem 2, all we need to show is that there exists a non-trivial "long-root-elation".   So suppose $\mathcal{H}(p_5, L_4, p_3, L_2, p_1, L_1, p_2)$ is trivial.   Then the commutator

$$\left[\ \mathcal{H}(L_6, p_5, L_4, p_3, L_2, p_1, L_1), \ \mathcal{H}(L_4, p_8, L_2, p_1, L_1 p_2, L_3) \ \right]$$

is also trivial and this means geometrically that every element of $\mathcal{H}(L_6, p_5, L_4, p_3, L_2, p_1, L_1)$ fixes every every point collinear with p_1 or p_5. Now let $q_4 \in A^*(L_3)$ and suppose M_8 is at a distance two from L_6 and at a distance five from q_4.   Let $L \in A^*(p_2)$ and denote by σ any non-trivial (L_1, L_8)-homology.   Let σ' be the unique (L_1, M_8)-homology mapping L^σ back to L. Since $\sigma \sigma'$ is non-trivial, it must act semi-regularly on the set of points incident with L_3 and distinct from p_2 (by Step 1).   Similarly, as in Step 5, using the results of Step 3, we see that $\sigma \sigma'$ fixes every line incident with p_1, p_2, p_3 or p_5.   Note that it also fixes at least three points incident with L_4.   Now, by varying q_4, one can see that, in fact, $\mid \mathcal{H}(p_2, L_1, p_1, p_3, p_5) \mid = s$.   So, by symmetry, there exists

$\alpha \in \mathcal{H}(p_2,p_1, p_3,L_4,p_5)$ such that $\alpha\sigma\sigma'$ fixes A. By Step 1, this must be the identity, hence $|\mathcal{H}(p_2, L_1, p_1, p_3,L_4,p_5)| = s$.

5.7 Step 7

Let $1 \neq \sigma \in \mathcal{H}(p_5, L_4, p_3, p_1,L_1, p_2)$ and $1 \neq \tau \in \mathcal{H}(L_2, p_1, L_1, p_2, L_3,p_4,L_5)$. Then $\sigma\tau\sigma^{-1}\tau^{-1} \in \mathcal{H}(L_4, p_3, L_2, p_1,L_1, p_2, L_3)$, otherwise $\mathcal{H}(p_3, L_2,p_1,L_1, p_2, L_3, p_4) \neq 0$. Geometrically, this implies that σ fixes every point collinear with p_3 or p_1. Now define the chain $p_1IM_2Iq_3IM_4Iq_5IM_6I$ $q_7IM_8Ip_8$; then clearly, $\sigma \in \mathcal{H}(M_2, p_1, L_1, p_2)$. We can now choose elations $\alpha'' \in \mathcal{H}(M_2, p_1, L_1, p_2, L_3, p_4, L_5)$, $\alpha' \in \mathcal{H}(q_3,M_2, p_1, p_2, L_3, p_4)$, $\alpha'' \in \mathcal{H}(M_4, q_3,M_2, p_1, L_1, p_2, L_3)$, and $\alpha''' \in \mathcal{H}(q_5,M_4, q_3,M_2, p_1, L_1, p_2)$ such that $\sigma' = \sigma\alpha\alpha'\alpha''\alpha'''$ fixes the apartment $p_1IM_2Iq_3IM_4Iq_5IM_6Iq_7IM_8Ip_8$ $IL_7Ip_6IL_5Ip_4IL_3Ip_2IL_1Ip_1$. But σ' fixes all lines through p_1 and at least three points incident with M_2, (because the order of α''' divides s and α''' fixes already two points on M_2, namely p_1 and q_3). By Step 1, $\sigma' = 1$. Hence $\alpha''' = (\sigma\alpha\alpha'\alpha'')^{-1}$. But $\sigma\alpha\alpha'\alpha''$ fixes every point incident with M_2, hence so does α'''. But since $\sigma \neq 1$, we have $1 \neq \alpha''' \in \mathcal{H}(q_5,M_4, q_3,M_2, p_1, L_1, p_2)$. This completes the proof of the Theorem.

Acknowledgement. The author is a Research Associate at the National Fund for Scientific Research (Belgium).

References

F. Buekenhout (1979), Diagrams for geometries and groups, *J. Combin. Theory Ser. A* **27**, 121-151.

W. Feit and G. Higman (1964), The nonexistence of certain generalised polygons, *J. Algebra* **1**, 114-131.

P. Fong and G. Seitz (1973), Groups with a (B,N)-pair of rank 2, I and II, *Invent. Math.* **21**, 1-57 and *Invent. Math.* **24** (1974), 191-239.

D.R. Hughes and F.C. Piper (1973), *Projective Planes*, Springer-Verlag, Berlin.

M.A. Ronan (1980), A geometric characterization of the Moufang hexagons, *Invent. Math.*, **57**, 227-262.

M.A. Ronan (1981), A combinatorial characterization of the dual Moufang hexagons, *Geom. Dedicata* **11**, 61-67.

C.E. Ealy, Jr., Generalised quadrangles and odd transpositions, *Thesis*.

S.E. Payne and J.A. Thas (1984), *Finite Generalized Quadrangles*, Pitman, London.

J.A. Thas (1972), 4-gonal subconfigurations of a given 4-gonal configuration, *Atti Accad. Naz. Lincei Rend.* **53**, 520-530.

J.A. Thas (1979), A restriction on the parameters of a suboctagon, *J. Combin. Theory Ser. A* , **40**, 385-387.

J.A. Thas (1981), New combinatorial characterizations of generalized quadrangles, *European J. Combin.* **2**, 299-303.

J.A. Thas (1986), The classification of all (x, y)-transitive generalized quadrangles, *J. Combin. Theory Ser. A* **42**, 154-157.

J. Tits (1959), Sur la trialité et certain groupes qui s'en déduisant, *Inst. Hautes Etudes Sci. Publ. Math.* **2**, 14-60.

J. Tits (1974), *Buildings of spherical type and finite BN-pairs* , Lecture Notes in Mathematics 386, Springer-Verlag, Berlin.

J. Tits (1976), Non-existence de certains polygons généralisès, I and II, *Invent. Math.* **36**, 275-284 and *Invent. Math.* **51** (1979) 267-269.

J. Tits (1979), Classification of buildings of spherical type and Moufang polygons: a survey, *Teorie Combinatorie, Volume* 1, 229-246, Accademia Nazionale dei Lincei, Rome.

J. Tits (1983), Moufang octagons and the Ree groups of type $^2F_4$, *Amer. J. Math.* **105**, 539-594.

J. Tits (1990), Quadrangles de Moufang I, *preprint*.

J.A. Thas, S.E. Payne and H. Van Maldeghem (1991), Half Moufang implies Moufang for finite generalised quadrangles, *submitted*.

J.A. Thas and H. Van Maldeghem (to appear). Generalised Desargues configurations in generalised quadrangles. *Bull. Soc. Math. Belg.*.

H. Van Maldeghem (1991a), A configurational characterization of the Moufang generalised polygons, *European. J. Combin.*, to appear.

H. Van Maldeghem (1991b), On finite Moufang polygons, *submitted*.

H. Van Maldeghem (1991c), A characterization of the finite Moufang hexagons by generalized homologies, *submitted*.

H. Van Maldeghem, S.E. Payne and J.A. Thas (1991), Desarguesian generalized quadrangles are classical or dual classical, *submitted*.

A. Wagner (1965), On finite affine line transitive planes, *Math. Z.* **87**, 1-11.

M. Walker, On central root automorphisms of finite generalised polygons, *Ph.D. Thesis*, University of London.

R. Weiss (1979), The nonexistence of certain Moufang polygons, *Invent. Math.* **51**, 261-266.

Complete arcs in Galois
planes of non-square order

J. F. Voloch
I.M.P.A., Rio de Janeiro

1.   Introduction

Let $PG(2, q)$ be the projective plane over the field $GF(q)$ of q elements. A k-arc $K \subseteq PG(2, q)$ is a set of k points, no three of which are collinear. A k-arc is called complete if it is not contained in a $(k + 1)$-arc.   It was first shown by Bose (see Hirschfeld (1979) Theorem 8.13) that a k-arc satisfies $k \leq q + 2$ if q is even and $k \leq q + 1$ if q is odd. Both bounds are sharp and arcs attaining these bounds are called ovals.   A celebrated theorem of Segre (see Hirschfeld (1979) Theorem 8.24) states that, for q odd, an oval is a conic.

A fundamental quantity in the geometry of projective spaces over finite fields is the cardinality of the second largest complete arc in $PG(2, q)$ denoted by $m'(2, q)$.   Equivalently one can define $m'(2, q)$ as being the smallest k_0 such that, if $k > k_0$, a k-arc is contained in an oval.   The value of $m'(2, q)$ is not only important in the geometry of the plane, but also is a basic quantity in bounds for the sizes of arcs and caps in higher dimensional spaces (see Hirschfeld (1983, 1985); Hirschfeld and Thas (1987, 1991); Thas (1968)).

The basic result on $m'(2, q)$ is the following bounds, due to Segre:

$$m'(2, q) \leq q - \sqrt{(q)} + 1, \qquad q \text{ even}; \tag{1}$$
$$m'(2, q) \leq q - \sqrt{(q)}/4 + 7/4, \quad q \text{ odd}. \tag{2}$$

(See Hirschfeld (1979), Theorem 10.3.3 for (1) and Theorem 10.4.4 for (2)).   Segre's method of proof of these inequalities consists of associating to an arc an algebraic curve over $GF(q)$ with many rational points and then using Weil's Theorem (1948) to get an upper bound on the number of points of the algebraic curve.   Thas (1983, 1987) has given an alternative approach to (1) and (2), bounding the points on the algebraic curve by using Bézout's theorem only, for q even, and the Plücker formulæ, for q odd.   Thas sometimes obtains improvements to (1) or (2) but his bounds differ from (1) or (2) at most by two.

In the opposite direction, when q is a square, there exist complete k-arcs for $k = q - \sqrt{(q)} + 1$, Fisher et. al. (1986).   Thus (1) is sharp when q is a square.   When q is not a square, the situation is completely different.   The author has shown (1990) that, for q an odd prime, $m'(2, q) \leq 44q/45 + 2$. It is the purpose of this paper to substantially improve the bounds (1) and (2) for an arbitrary non-square q, in

Theorems 1 and 2 below. Note that, for q not a square, the best lower bound for $m'(2, q)$, is at present $(q + 1)/2 + \sqrt{(q)}$, which follows from the results of Voloch (1987).

The author would like to thank James Hirschfeld for his interest in this work.

2. Bounds for the number of rational points on curves

To bound the number of rational points of algebraic curves defined over finite fields we will use the following result of Stöhr and Voloch (1986).

Theorem 1. If $X \subseteq PG(n, q)$ is an irreducible curve of degree d, genus g, not contained in any hyperplane and with Frobenius orders $v_0, \ldots, v_{n-1}$, then the number N of $GF(q)$-rational points of X satisfies

$$N \leq [(v_0 + \ldots + v_{n-1})(2g - 2) + d(q + n)]/n .$$

The Frobenius orders $0 = v_0 < \ldots < v_{n-1}$ is a certain well defined subset of the order sequence $0 = \varepsilon_0 < \ldots < \varepsilon_n$ of X, defined as the possible intersection multiplicities of X with hyperplanes at a generic point of X; see Stöhr and Voloch (1986) for more details.

Let us consider a plane curve X of degree d. Then

$$N \leq [v_1(2g - 2) + d(q + 2)]/2. \tag{3}$$

Note also that $v_1 = \varepsilon_1 = 1$ or $v_1 = \varepsilon_2$. We have that ε_2 is the order of contact of X with its tangent at a generic point and so $\varepsilon_2 \leq d$. Also, $\varepsilon_2 = 2$ or a power of the characteristic p (as follows from Garcia and Voloch (1987) Proposition 2). When $p = 2$ we conclude that v_1 is a power of 2 and $v_1 \leq d$. This will be used in § 3.

The plane curve X of degree d can be embedded in $PG(5, q)$ as a curve of degree $2d$ not contained in a hyperplane by the Veronese embedding $PG(2, q) \rightarrow PG(5, q)$, given by

$$(x_0 : x_1 : x_2) \mapsto (x_0 x_1 : x_0 x_2 : x_1 x_2 : x_0^2 : x_1^2 : x_2^2).$$

It follows that

$$N \leq [(v_1 + v_2 + v_3 + v_4)(2g - 2) + 2d(q + 5)]/5. \tag{4}$$

Suppose that X, as a plane curve, is classical (that is, has order sequence 0, 1, 2). Then the orders of X in 5-space are 0, 1, 2, 3, 4, ε_5 for some $\varepsilon_5 \leq 2d$ (see Garcia and Voloch (1987) p.464). If $p \neq 2, 3$ then, by

Garcia and Voloch (1987), Proposition 2, $\varepsilon_5 = 5$ or ε_5 is a power of p. When $p = 2, 3$ then $\varepsilon_5 = 6$ is the only other possibility. Suppose now that X has a rational point P so that X meets its tangent at P with multiplicity exactly 2 at P. Then the orders at P of X in 5-space begin with $0, 1, 2, 3, 4$ and it follows from Stöhr and Voloch (1986) Corollary 2.6 that $v_i = i$, $i \leq 3$. Thus $v_4 = 4$ or ε_5.

3. q even

Let q be even.

To prove his result on $m'(2, q)$, Segre showed that, with $t = q + 2 - k$, the kt unisecants to a k-arc $K \subseteq PG(2, q)$ lie, in the dual projective plane, on an algebraic curve C of degree t. (See Hirschfeld (1979) Theorem 10.3.1). Note that, given P in K, the line P^v in the dual plane will intersect C in the t unisecants of K through P. It follows readily from this that the unisecants of K are simple points of C and that an irreducible component X of C of degree d contains at least kd unisecants to K. Segre then applied Weil's Theorem (1948) to X to conclude that $m'(2, q) \leq q - \sqrt{(q)} + 1$. To improve on Segre's result when q is not a square we will apply instead the results of §2.

Theorem 2. *If $q \neq 2$ is even and not a square, then*

$$m'(2, q) \leq q - \sqrt{(2q)} + 2.$$

Proof. As above, we obtain for a k-arc K, an irreducible algebraic curve X of degree $d \leq t = q + 2 - k$ containing kd unisecants to K. If $k = m'(2, q)$ and K is complete, then $d \geq 2$ (as in the proof of Theorem 10.3.3 of Hirschfeld (1979)). If X is not defined over $GF(q)$, then by Hirschfeld (1979) Lemma 10.1.1, $kd \leq d^2$. So $k \leq d \leq t = q + 2 - k$; that is, $k \leq (q + 2)/2 \leq q - \sqrt{(2q)} + 2$.

If X is of degree $d \geq 2$ and is defined over $GF(q)$, then (3) of §2 is valid. Therefore

$$kd \leq [(v_1(2g - 2) + d(q + 2)]/2 \leq d [v_1(d - 3) + q + 2]/2.$$

Hence

$$2k \leq v_1(d - 3) + q + 2 \leq v_1(t - 3) + q + 2$$

and, since $k = q + 2 - t$,

$$t \geq \frac{q + 2 + 3v_1}{v_1 + 2}.$$

If $v_1 \leq \sqrt{(q/2)}$, then, for $q \neq 8$,

$$t \geq \frac{2(q + 2) + 3\sqrt{(2q)}}{\sqrt{(2q)} + 4} > \left[\frac{2(q + 2) + 3\sqrt{(2q)}}{\sqrt{(2q)} + 4} \right] = \sqrt{(2q)} - 1.$$

Therefore,

$$k = q + 2 - t < q + 3 - \sqrt{(2q)}.$$

As $\sqrt{(2q)}$ is an integer, we get the theorem in this case.

Recall that, as remarked in §2, $v_1 \leq d \leq t$ and is a power of 2. Thus if $v_1 > \sqrt{(q/2)}$ then $\sqrt{(2q)} \leq v_1 \leq t$, whence $k \leq q + 2 - \sqrt{(2q)}$, as was to be proved.

For $q = 8$, the bound in the theorem is sharp, as the only complete arcs other than ovals are 6-arcs (Hirschfeld (1979), Theorem 9.2.5).

4. q odd

In this section q is odd.

Similarly to the case q even, a k-arc K in $PG(2, q)$, q odd, has its kt unisecants ($t = q + 2 - k$) lying, in the dual plane, on an algebraic curve C. Differently from the case q even, this time C has degree $2t$ and for each P in K, the line P^v meets C with a multiplicity two at each of the t unisecants of K through P (Hirschfeld (1979), Theorem 10.4.1). Again we will use the results of Stöhr and Voloch (1986) presented in §2, instead of Weil's results, to improve on Segre's bound. When q is prime, this was already done in Voloch (1990).

Theorem 3. *Let q be odd, not a square or a prime. Then, if q is a power of the prime p,*

$$m'(2, q) \leq q - \sqrt{(pq)}/4 + 29p/16 + 1.$$

Proof. Let K be a complete k-arc in $PG(2, q)$, q odd, not a square, $k = m'(2, q)$. Let X be an irreducible component of the curve C constructed above. It follows from the argument in Voloch (1990) that

$$k \leq 44q/45 + 2 \leq q - \sqrt{(pq)}/4 + 29p/16 + 1,$$

unless X is degree $d \geq 3$ defined over $GF(q)$ and with the Veronese embedding of X in $PG(5, q)$ having $v_4 > 4$. Suppose that we are in this last case. If the $kd/2$ unisecants of K which are in X are double points of X, then

$$kd/2 \leq (d-1)(d-2)/2 \leq d^2/2;$$

so $k \leq d \leq 2t$, and hence $k \leq 2(q+2)/3 \leq 44q/45 + 2$, as desired. Otherwise, X has a point, corresponding to a unisecant ℓ to K through P in K, which is simple, and therefore, P^v meets X at ℓ with multiplicity two. It follows from the discussion in §2 that $X \subseteq PG(5, q)$ has order sequence $0, 1, 2, 3, 4, \varepsilon_5 \leq 2d$ and Frobenius orders $0, 1, 2, 3, \varepsilon_5$. Finally ε_5 is a power of p unless $p = 3$ and $\varepsilon_5 = 6$. From (4) and the argument in Voloch (1990) it follows that

$$kd/2 \leq [(\varepsilon_5 + 6)d(d-3) + 2d(q+5)]/5 .$$

Hence

$$k \leq 2/5[(\varepsilon_5 + 6)(d-3) + 2(q+5)] \leq 2/5[(\varepsilon_5 + 6)(2t-3) + 2(q+5)].$$

As $k = q + 2 - t$, it follows that

$$t \geq \frac{q}{4\varepsilon_5 + 29} + \frac{6\varepsilon_5 + 26}{4\varepsilon_5 + 29} \geq \frac{q}{4\varepsilon_5 + 29} + 1. \tag{5}$$

Write $q = p^{2h+1}, h \geq 1$. If $\varepsilon_5 \geq p^{h+1}$ then, as $\varepsilon_5 \leq 2d \leq 4t$, it follows that $k \leq q + 2 - p^{h+1}/4$ and the theorem is proved. Suppose now that $\varepsilon_5 < p^{h+1}$. If $p \neq 3$ then ε_5 is a power of p; so $\varepsilon_5 \leq p^h$ and, from (5), $t \geq q/(4p^h + 29) + 1$, which proves the theorem. If $p = 3$, the same argument gives the result unless $\varepsilon_5 = 6$, which also satisfies $\varepsilon_5 \leq p^h$ unless $h = 1$, that is, $q = 27$. However, the result is trivially true for $q = 27$.

References

J. C. Fisher, J.W.P. Hirschfeld and J. A. Thas (1986), Complete arcs in planes of square order, *Ann. Discrete Math.* **30**, 243-250.

A. Garcia and J. F. Voloch (1987), Wronskians and linear independence in fields of prime characteristic, *Manuscripta Math.* **59**, 457-469.

J. W. P. Hirschfeld (1979), *Projective Geometries over Finite Fields*, Oxford University Press, Oxford.

J. W. P. Hirschfeld (1983), Caps in elliptic quadrics, *Ann. Discrete Math.* **18**, 449-466.

J. W. P. Hirschfeld (1985), *Finite Projective Spaces of Three Dimensions* , Oxford University Press, Oxford.

J. W. P. Hirschfeld and J. A. Thas (1987), Linear independence in finite spaces, *Geom. Dedicata* **23**, 15-31.

J. W. P. Hirschfeld and J. A. Thas (1991), *General Galois Geometries*, Oxford University Press, Oxford, to appear.

K. O. Stöhr and J. F. Voloch (1986), Weierstrass points and curves over finite fields, *Proc. London Math. Soc.* **52**, 1-19.

J. A. Thas (1968), Normal rational curves and k-arcs in Galois spaces, *Rend. Mat.* **1**, 331-334.

J. A. Thas (1983), Elementary proofs of two fundamental theorems of B. Segre without using the Hasse-Weil theorem, *J. Combin. Theory Ser. A* **34**, 381-384.

J. A. Thas (1987), Complete arcs and algebraic curves in $PG(2, q)$, *J. Algebra* **106**, 457-464.

J. F. Voloch (1987), On the completeness of certain plane arcs, *European J. Combin.* **8**, 453-456.

J. F. Voloch (1990), Arcs in projective planes over prime fields, *J. Geom.* **38**, 198-200.

A. Weil (1948), *Sur les Courbes Algébriques et les Variétés qui s'en Déduisent*, Hermann, Paris.

Internal nuclei of k-sets in finite projective spaces of three dimensions

F. Wettl
Technical University, Budapest

1. Introduction

It is well known that an oval, that is a $(q + 1)$-arc in an even order plane has a *nucleus*, that is a point such that every line through this point intersects the oval in exactly one point. Every point of a $(q + 2)$-arc (hyperoval) is a nucleus of the remaining points. If we drop the condition that the pointset is an arc, we may get two generalisations of the notion of nucleus depending on whether the nucleus is an element of the pointset (in the case of the hyperoval), or not (in the case of oval):

Definition 1. (Bichara and Korchmáros 1982). Let K be a set of $q + 2$ points in a finite projective plane π. A point $P \in K$ is called a *nucleus* if every line through P intersects K in exactly two points. The set of nuclei of K is denoted by $I(K)$.

Definition 2. (Mazzocca 1985). Let K be a set of $q + 1$ points in a finite projective plane π. A point $P \notin K$ is called a *nucleus* if every line through P intersects K in exactly one point. The set of nuclei of K is denoted by $N(K)$.

To distinguish these two different notions Szönyi (1989) used the expression *internal nucleus* in the case of Definition 1. The next result is about this type of nuclei.

Result 3. (Bichara and Korchmáros 1982). *If K is a $(q + 2)$-set in $PG(2, q)$ and $|I(K)| \geq 3$ then q is even.*

This result implies that a further weakening of conditions in the Definition 1 is necessary when q is odd. One weakening led to the notion of a semi-nucleus. Let K be a set of $q + 2$ points in a finite projective plane π. A point $P \in K$ is called a *semi-nucleus* if there is exactly one tangent line of K at P. This implies that there is exactly one 3-secant at P, and the other $q - 1$ lines are 2-secants. Further details are in Bruen (1988), Blokhuis and Bruen (1989) and Blokhuis (to appear). Another weakening led to the next definition.

Definition 4. (Wettl 1987).   Let K be a set of k points in a finite projective plane π. A point $P \in K$ is called an *internal nucleus* if every line through P intersects K in at most two points.   The set of nuclei of K is denoted by $I(K)$.

This definition and Definition 1 are the same if $k = q + 2$.   Bichara and Korchmáros (1982) proved that if q is even, K is a $(q + 2)$-set in $PG(2, q)$ and $|I(K)| > q/2$, then $I(K) = K$, that is $|I(K)| = q + 2$.   A similar result was proved for $(q + 1)$-sets when q is even or odd (Wettl 1987).   If K is a $(q + 1)$-set in $PG(2, q)$ and $|I(K)| > (q+ 1)/2$ then $I(K) = K$, that is $I(K)$ is an oval.   Szönyi (1988) described the extreme cases of these two results and proved that if $k \geq q+ 1$ and $|I(K)| = \lfloor (q + 1)/2 \rfloor$ then the points of $K \backslash I(K)$ are collinear and $I(K)$ is an affinely regular $(q + 1)/2$-gon when q is odd.   Also, he generalised these two results.

Result 5. (Szönyi 1988).   *If a k-set of $PG(2, q)$, q odd and $q > q_0$, has $k > q - \sqrt(q)/8 + 2$, then $|I(K)| \leq (q + 1)/2$ or $I(K) = K$.*

The nuclei of non-desarguesian planes was studied by Biscarini and Conti (1982) and Szönyi (1989).   The next result is about nuclei of $(q + 1)$-sets.

Result 6.   (Wettl 1987).   *If K is a $(q + 1)$-set in $PG(2, q)$, q odd, then $I(K)$ is a subset of a conic.*

We will use a proposition of Szönyi (1988) on complete k-arcs.   A similar result for higher dimensional spaces can be found in Bruen (1988) and Blokhuis et al. (1990).

Result 7. (Szönyi 1988).   *Let K be a k-arc in $PG(2, q)$ with $k > (q + 2)/2$ if q is even and with $k > (2q + 5)/3$ if q is odd.   Then there exists a unique complete arc containing K.*

The other type of nuclei is discussed, for example, in Blokhuis and Willbrink (1987), Bruen (1988), Korchmáros and Mazzocca (1990) and Blokhuis and Mazzocca (this volume).   We recall the main result of nuclei.

Result 8. (Blokhuis and Wilbrink 1987).   *In $PG(2, q)$ the lines are the only $(q + 1)$-sets admitting at least q nuclei.*

In this paper, we define and study the notion of internal nuclei of k-sets in projective spaces of higher dimensions, mainly of three

dimensions.   In section 2 we give a general definition of nucleus and
nuclear set of k-sets and describe their connections with caps and arcs.
In section 3 we study the nuclei of point-sets of $PG(3, q)$, give a
generalisation of the theorem on ovaloids of Barlotti (1955) and
Panella (1955), generalise Results 5 and 7 for complete caps.   The
results of §4 are connected to the arcs of $PG(n, q)$.

2. The notion of m-nucleus

Definition 9.   Let K be a set of k points in an n-dimensional finite
projective (or affine) space.   A point P is called the m-nucleus of K if
every m-dimensional subspace through P intersects $K\backslash\{P\}$ in at most
m points.   If $P \in K$ then P is called an *internal m-nucleus*.   If $P \notin K$
then P is called an *external m-nucleus*.   The set of internal m-nuclei
of K is denoted by $I_m(K)$ and called the *internal m-nuclear set* of K.
$|I_m(K)|$ is denoted by $i_m(K)$, or simply i_m.   Instead of 1-nucleus, $I_1(K)$
and i_1 we may write nucleus, $I(K)$ and i.

If $n = 2$ and $m = 1$ then we obtain Definition 1 when $k = q + 2$, $P \in K$,
Definition 2 when $k = q + 1$, $P \notin K$, and Definition 4 when $k \leq q + 2$,
$P \in K$.   If $n \geq 2$, $m = 1$, $k = q^{n-1} + ... + q + 1$ and $P \notin K$ then we get the
notion studied by Bruen and Mazzocca (1990).
    A k-arc K in the projective space $PG(n, q)$ is a k-set such that every
point of K is an internal l-nucleus for $l = 1, 2, ..., n - 1$.   A k-cap of
$PG(n, q)$ is a k-set such that every point of K is an internal 1-nucleus.
It is clear that in $PG(n, q)$ the pointset $I_1(K)$ is a cap and $\cap_{l=1}^{n-1} I_l(K)$ is an
arc.
    If $k > n$ then $I_{n-1}(K)$ is also an arc;   also K has no $l + 2$ points in an l-
dimensional subspace and so $I_l(K) = K$ for $l = 1, 2, ..., n - 2$.   If P_1, P_2 are
two different points of $K\backslash I_1(K)$ then $I_1(K) \cup \{P_1, P_2\}$ is an $(i_1 + 2)$-cap.
Similarly, if $k > n$ and $P_1, P_2, ..., P_n$ are different points of $K\backslash I_{n-1}(K)$ then
$I_{n-1}(K) \cup \{P_1, P_2, ..., P_n\}$ is an $(i_{n-1} + n)$-arc.   Of course we also get an arc if
we choose less than n points from $K\backslash I_{n-1}(K)$.   In $PG(3, q)$ this means
that if $k \geq 4$ and K is a k-set then $I_2(K) \cup \{P_1, P_2, P_3\}$ is an arc for every
three points P_1, P_2, P_3 of $K\backslash I_2(K)$.   In general, we get the next
proposition.

Proposition 10.   *If $k \geq m + 2$, $P_1, P_2, ..., P_{n+1}$ are different points of
$K\backslash I_m(K)$ and $K' = I_m(K) \cup \{P_1, P_2, ..., P_{m+1}\}$ then $K' \subseteq I_l(k)$ for $l = 1, 2, ...,
m$.*

Using the notion of caps of kind s (see Tallini 1961) this means that K' is an $(i_m + m + 1)$-cap of kind m.

Proposition 11. *Let $k_{n,\,m}$ denote the maximum value of k for which there exists a k-set in $PG(n, q)$ with an internal m-nucleus. Then $k_{n,\,m} \leq q^{n-m} + q^{n-m-1} + \ldots + q + 1 + m$.*

Proof. It is clear that $k_{n,\,1} \leq q^{n-m} + q^{n-m-1} + \ldots + q + 2$. Let us project the points of K from an internal m-nucleus to a hyperplane. We get $k_{n,\,m} - 1 \leq k_{n-1,\,m-1}$ and this completes the proof.

In $PG(3, q)$ this means that $k_{3,\,1} \leq q^2 + q + 2$ and $k_{3,\,2} \leq q + 3$. Trivial examples show that equality holds in these two formulae.

3. 1-nuclei of k-sets in $PG(3, q)$

First of all we describe some trivial examples. Let us denote $PG(3, q)$ by Σ.

Let α be the pointset of a plane of Σ, and P_1, P_2, P_3 be three non-collinear points not in α. Let A_i be that point of α for which A_i, P_j, P_k are collinear. Let $K_1 = \alpha \cup \{P_1\}$, $K_2 = \alpha \cup \{P_1, P_2\} \backslash \{A_3\}$, $K_3 = \alpha \cup \{P_1\} \backslash \{A_3\}$, $K_4 = \alpha \cup \{P_1, P_2\} \backslash \{A_3, A_1\}$, $K_5 = \alpha \cup \{P_1, P_2, P_3\} \backslash \{A_1, A_2, A_3\}$. If k_i denotes $|K_i|$ then $k_1 = k_2 = q^2 + q + 2$, $k_3 = k_4 = k_5 = q^2 + q + 1$, $I(K_1) = I(K_3) = \{P_1\}$, $I(K_2) = I(K_4) = \{P_1, P_2\}$, $I(K_5) = \{P_1, P_2, P_3\}$.

Let α and β be two different planes of Σ and P a point not on these planes. Let Ω_α be an oval on the plane α. Let us project it from P to β and let us denote this oval by Ω_β. Let $K_6 = \alpha \backslash \beta \cup \Omega_\beta$, $K_7 = \alpha \backslash \beta \backslash \Omega_\alpha \cup \Omega_\beta \cup \{P\}$. Then $k_6 = q^2 + q + 1$, $I(K_6) = q + 1$, q or $q - 1$; $k_7 = q^2 + 1$, $q^2 + 2$ or $q^2 + 3$, $I(K_7) = q + 1$, q or $q - 1$ depending on whether $|\Omega_\beta \cap \alpha| = 0, 1$ or 2.

An ovoid O is a $(q^2 + 1)$-set and $I(O) = O$ so $i = q^2 + 1$.

For the rest of this section q is odd. Let K be a k-set in Σ and $P \in I(K)$ be a point of it. Let us project the points of $I(K) \backslash \{P\}$ from P into a plane α which is not through P. The set of projected points is denoted by K_P. Every tangent line at P intersects α in a point. The set of these points is denoted by T_P.

Lemma 12. *If q is odd then the secant lines of K_P are blocked by the points of T_P.*

Proof. Let P_1 and P_2 be two points of K_P. Let us denote the plane of P, P_1, P_2 by β. It is clear that every point of $I(K) \cap \beta$ is an internal nucleus of $K \cap \beta$ in β and so P, P_1, P_2 are nuclei of it. From Result 1 it follows that $k < q + 2$, that is P has at least one tangent in α. This means that the line $P_1 P_2$ is blocked by a point T_P.

Corollary 13. *Let q be odd.* (a) *If $k = q^2 + q + 2$ then $i \leq 2$;* (b) *if $k = q^2 + q + 1$ or $k = q^2 + q$ then $I(K)$ is a subset of a plane and $i \leq q + 1$;* (c) *if $k = q^2 + q - 1$ then $i \leq 5$.*

Proof. If $k = q^2 + q + 2$ then K has no tangent, so $T_P = 0$, that is K_P has no secant, so $i = |I(K)| \leq 2$ (see K_1 and K_1). If $k = q^2 + q + 1$ or $k = q^2 + q$ then $|T_P| = 1$ or 2; so the points of K_P are on a line, that is the points of $I(K)$ are in the plane (see K_3, K_4, K_5 and K_6). If $k = q^2 + q - 1$ then $|T_P| = 3$ and three points cannot block the secant of 5 points but can block the secants of 4 points; so $i \leq 5$.

Corollary 14. *If q is odd and $i > q\sqrt{q} + 2$ then $k \leq q^2 + 1$.*

Proof. If K_P has a point through which every line is a secant of K_P then $|T_P| \geq q + 1$. If K_P has no such point then it has a tangent at every point of it. These types of sets are called strong representative systems. The upper bound of these sets is $q\sqrt{q} + 1$ (see Illés et. al. (1989)). But if $i > q\sqrt{q} + 2$ then $|K_P| > q\sqrt{q} + 1$.

Corollary 15. *If q is odd, $k = q^2 + 1$ and $i \geq q\sqrt{q} + q + 2$, then every point of $I(K)$ has a tangent plane.*

Proof. Similarly to the previous proof, K_P has at most $q\sqrt{q} + 1$ points through which we have a tangent. So there are at least $(q\sqrt{q} + q + 2) - 1 - (q\sqrt{q} + 1) = q$ points through which every line is a secant to K_P. But $|T_P| = q + 1$, $|K_P| \geq q$; so from Result 8 it follows that T_P is a line. This means that all tangent lines at P are in a plane. We do not know whether the inequality $i \geq q\sqrt{q} + q + 2$ is sharp or not, but as the set K_7 shows $i = q + 2$ is not enough.

Theorem 16. *Let q be odd, K be a $(q^2 + 1)$-set and let J be the set of those points of $I(K)$ at which K has a tangent plane. If J has 4 non-coplanar points then J is a subset of a quadric. The tangent planes of J are the tangent planes of this quadric.*

Proof. If $P \in J$ then on the one side the tangent lines at P are in a plane; so there is a tangent line in every plane through P. On the other hand, P is a nucleus; so every plane through P has $q + 1$ points of K except the tangent plane.   If such a plane has three points of J then from Result 3 it follows that the points of J in this plane are in a conic.   We follow the method of the presentation of Segre's theorem (1955) by Kárteszi (1976).   It is proved that the triangle inscribed in an oval in $PG(2, q)$ (q odd) and the triangle which consists of the tangents at its vertices are in perspective.   This property is said to be the π-property.   Let P_1, P_2, P_3 be three points of J, α be the plane of them, τ_i be the tangent plane at P_i, and $t_{\alpha,i}$ denote the tangent line in α at P_i. J satisfies the π-property, because the triangles $\{P_1, P_2, P_3\}$ and $\{t_{\alpha,1}, t_{\alpha,2}, t_{\alpha,3}\}$ are perspective non-degenerate triangles.   Now, our theorem follows from the next lemma.

Lemma 17.   *Let q be odd, J be a cap of Σ and let a tangent plane of J be given at every point of it.   If J satisfies π-property then J is a subcap of a quadric.   The given tangent planes of J are the tangent planes of this quadric.*

Proof.   Let A_1, A_2, A_3, A_4 be four non-coplanar points of J. Let D be the intersection of the tangent planes at A_1, A_2 and A_3.   Take the system of coordinates as follows:

A_1: $(1, 0, 0, 0)$, A_2: $(0, 1, 0, 0)$, A_3: $(0, 0, 1, 0)$, A_4: $(0, 0, 0, 1)$, D: $(1, 1, 1, -2)$.

The coordinates of the tangent plane τ_i at A_i ($i = 1, 2, 3$) depend on a parameter because two points of each are given; so

$$\tau_1: [0, 1, \lambda, (1 + \lambda)/2], \quad \tau_2: [\mu, 0, 1, (1 + \mu)/2], \quad \tau_3: [1, \rho, 0, (1 + \rho)/2].$$

If α_4 denotes the plane of A_1, A_2 and A_3, and B_i denotes the intersection of α_4, τ_j and τ_k where (i, j, k) is a permutation of $(1, 2, 3)$, then

$$B_1: (1, \lambda\mu, -\mu, 0), \quad B_2: (-\rho, 1, \mu\rho, 0), \quad B_3: (\lambda\rho, -\lambda, 1, 0).$$

A_1B_1, A_2B_2, A_3B_3 are concurrent lines.   The intersection is $(x, y, z, 0)$ where x, y, z satisfy the equations

$$\lambda\mu x - y = 0, \quad \mu\rho y - z = 0, \quad \rho\lambda z - x = 0.$$

If this system of equations has a non-trivial solution then $\lambda^2\mu^2\rho^2 = 1$. B_1, B_2, B_3 are different points; so $\lambda\mu\rho \neq -1$, that is $\lambda\mu\rho = 1$. Let τ_4: $[a, b, c, 0]$ be a tangent plane at A_4 and α_3: $[0, 0, 1, 0]$ be the plane of A_1, A_2, A_4. The intersection of the planes α_3, τ_4, τ_1 is the point

$$(-b(1 + \lambda)/2, a(1 + \lambda)/2, 0, -a),$$

the intersection of α_3, τ_4, τ_2 is

$$(-b(1 + \mu)/2, a(1 + \mu)/2, 0, b\mu),$$

and the intersection of α_3, τ_1, τ_2 is

$$(-(1 + \mu)/2, -\mu(1 + \lambda)/2, 0, \mu).$$

The triangle of these three points and A_1, A_2, A_4 are in perspective. The coordinates of the centre $(x, y, 0, z)$ of the perspectivity satisfy the equations:

$$bz(1 + \lambda)/2 - ax = 0, \quad az(1 + \mu)/2 - \mu by = 0, \quad y(1 + \mu)/2 - x\mu(1 + \lambda)/2 = 0.$$

This system of equations has a solution if $b^2\mu^2(1 + \lambda)^2 - a^2(1 + \mu)^2 = 0$. Now, $b\mu(1 + \lambda) + a(1 + \mu) \neq 0$ because the point $(-(1 + \mu)/2, -\mu(1 + \lambda)/2, 0, \mu)$ is not on the plane τ_4: $[a, b, c, 0]$; so $b\mu(1 + \lambda) - a(1 + \mu) = 0$. After similar calculations we get $a\lambda (1 + \rho) - c(1 + \lambda) = 0$ and $c\rho(1 + \mu) - b(1 + \rho) = 0$. From the solution of these equations, we get the coordinates of τ_4:

$$\tau_4: [1 + \lambda, 1 + \lambda\rho, \lambda + \lambda\rho, 0].$$

Now let P: (x, y, z, q) be an arbitrary point of J and β: $[u, v, w, t]$ be the tangent plane at P. The plane $[0, 0, q, -z]$ through A_1, A_2, P is denoted by π_{12}. The intersection of π_{12}, β, τ_1 is

$$(tq - vz\lambda + wz - vq(1 + \lambda)/2, uq(1 + \lambda)/2 + uz\lambda, -uz, -uq),$$

the intersection of π_{12}, β, τ_2 is

$$(-vz - vq(1 + \mu)/2, uz - tq\mu - wz\mu + uq(1 + \mu)/2, vz\mu, vq\mu),$$

and the intersection of π_{12}, τ_1, τ_2 is

$$(-z -q(1 + \mu)/2, -qu(1 + \lambda)/2 - z\lambda\mu, \mu z, \mu q).$$

The triangle of these three points and the triangle A_2, A_1, P are in perspective from which it follows that $\mu^2 v^2 (y + z\lambda + q(1 + \lambda)/2)^2 - u^2(z + x\mu + q(1 + \mu)/2)^2 = 0$. The point $(-z - q(1 + \mu)/2, -qu(1 + \lambda)/2 - z\lambda\mu, \mu z, \mu q)$ is not on the plane β: $[u, v, w, t]$. So $\mu v(y + z\lambda + q(1 + \lambda)/2 + u(z + x\mu + q (1 + \mu)/2) \neq 0$; that is $\mu v(y + z\lambda + q(1 + \lambda)/2) = u(z + x\mu + q(1 + \mu)/2)$. After similar calculations with the planes π_{13} and π_{23} we get similar equations; so u, v, w, t and x, y, z, q satisfy the equations:

$$\mu v(y + z\lambda + q(1 + \lambda)/2) = u(z + x\mu + q(1 + \mu)/2);$$
$$\rho w(z + x\mu + q(1 + \mu)/2) = v(x + y\rho + q(1 + \rho)/2); \qquad (1)$$
$$\lambda u(x + y\rho + q(1 + \rho)/2) = w(y + z\lambda + q(1 + \lambda)/2).$$

Calculating with the plane π_{14}: $[0, z, -y, 0]$ the intersection of π_{14}, τ_1, τ_4 is

$$((1+\lambda)(\lambda z(1+\rho)+y(1+\lambda\rho))/2, -y(1+\lambda)^2/2, -z(1 + \lambda)^2/2, (1+\lambda)(z\lambda+y)),$$

the intersection of π_{14}, β, τ_1 is

$$(wz(1+\lambda)/2-ty+vy(1+\lambda)/2-z\lambda, -uy(1\ \lambda)/2, -uz(1+\lambda)/2, u(z\lambda+y)),$$

and the intersection of π_{14}, β, τ_4 is

$$(-ty(1+\lambda\rho)-tz(\lambda+\lambda\rho),ty(1+\lambda),tz(1+\lambda),-wz+vy)(1+\lambda)+uy(1+\lambda\rho)+uz(\lambda+\lambda\rho)).$$

The triangle of these three points and the triangle P, A_4, A_1 are in perspective, from which it follows that

$$u(x(1 + \lambda) + y(1 + \lambda\rho) + z(\lambda + \lambda\rho)) = t(q(1 + \lambda) + 2(y + z\lambda)).$$

After similar calculations with the planes π_{24} and π_{34} we get similar equations, whence u, v, w, t and x, y, z, q satisfy the equations:

$$u(x(1 + \lambda) + y(1 + \lambda\rho) + z(\lambda + \lambda\rho)) = t(q(1 + \lambda) + 2(y + z\lambda));$$
$$\mu v(x(1 + \lambda) + y(1 + \lambda\rho) + z(\lambda + \lambda\rho)) = t(q(1 + \mu) + 2(z + x\mu)); \qquad (2)$$
$$\mu\rho w(x(1 + \lambda) + y(1 + \lambda\rho) + z(\lambda + \lambda\rho)) = t(q(1 + \rho) + 2(x + y\rho)).$$

The system of linear equations of (1) and (2) has a non-trivial solution:

$$c \begin{bmatrix} v \\ u \\ w \\ t \end{bmatrix} = \begin{bmatrix} 0 & 1 & \lambda & (1+\lambda)/2 \\ 1 & 0 & \rho & (1+\lambda\rho)/2 \\ \lambda & \lambda\rho & 0 & (\lambda+\lambda\rho)/2 \\ (1+\lambda)/2 & (1+\lambda\rho)/2 & (\lambda+\lambda\rho)/2 & 0 \end{bmatrix} \begin{bmatrix} x \\ y \\ z \\ q \end{bmatrix}, \quad (3)$$

where c is a non-zero constant. Eliminating u, v, w, t we get that (x, y, z, q) is on the quadric

$$2xy + 2\lambda xz + 2\lambda\rho yz + (1 + \lambda)xq + (1 + \lambda\rho)yq + (\lambda + \lambda\rho)zq = 0.$$

The determinant of the coefficient matrix of (3) is $-\lambda^2\rho(1 + \lambda + \lambda\rho)$; so the matrix is singular if $1 + \lambda + \lambda\rho = 0$. In this case, every tangent plane is incident with the point D:$(1, 1, 1, -2)$; so the quadric is a cone. In every other case the linear transformation of (3) is non-singular and defines a polarity established by the quadric. This completes the proof.

As a hyperbolic quadric or a cone is the union of $q + 1$ lines, there is no $(2q + 3)$-cap on them. From this fact and the previous results follow the next corollaries.

Corollary 18. *If the cap J given in Lemma 17 has more than $2q + 2$ points then J is the subcap of an elliptic quadric.*

Corollary 19. *If q is odd, K is a $(q^2 + 1)$-set in Σ and $i \geq q\sqrt{q} + q + 2$ then the points of $I(K)$ are on an elliptic quadric.*

This corollary is a generalisation of the result of Barlotti (1955) and Panella (1955), because using the notion of nuclei it says that if $|K| = |I(K)| = q^2 + 1$ then the points of $K = I(K)$ are on an elliptic quadric.

In the next result we study the unique extendability of caps and give a bound for $i = |I(K)|$ when $K \neq I(K)$.

Lemma 20. *Let J be a j-cap in Σ with $j > (q + 1)(q + 2)/2$ if q is even and with $j > (q + 1)(2q + 5)/3$ if q is odd. Then there exists a unique complete cap containing J.*

Proof. Let us suppose that q is even. Let K be the union of all those caps which contain J. If J has a unique extension then K is a cap. If not, then K has three collinear points and at most one of them is from J. Let us consider all the planes through the line of the three collinear points. The set of points of J in every plane is an arc which has more than one extension, so the number of points of J in every plane is not greater than $(q + 2)/2$, and so $j \le (q + 1)(q + 2)/2$; this is a contradiction. If q is odd then after similar calculations we get $j \le (q + 1)(2q + 5)/3$.

Theorem 21. *If K is a k-set in Σ with $i > (q + 1)(q + 2)/2$ if q is even and with $i > (q + 1)(2q + 5)/3$ if odd then K is a cap, that is $K = I(K)$.*

Proof. From Proposition 11 it follows that we can extend $I(K)$ by every two points of K; so we may repeat the proof of Lemma 20 with $J = I(K)$.

4. $(n - 1)$-nuclei of k-sets

Observation 22. Let K be a k-set in $PG(n, q)$, $P \in I_m(K)$, $Q \in K \backslash I_m(K)$ and let π be a hyperplane not containing P. If we project the points of K from P to π then the image will be a $(k - 1)$-cap of kind $m - 1$, that is, no $m + 1$ points will be in an $(m - 1)$-dimensional subspace. If we project the points of K from Q then the image of an internal m-nucleus will be an internal $(m - 1)$-nucleus.

Theorem 23.   *Let K be a k-set in $PG(3, q)$:*
(i)   *if $i_2 \ge 1$ then $k \le q + 3$;*
(ii)   *if q is odd and $i_2 \ge 1$ then $k \le q + 2$;*
(iii)   *if q is odd and $i_2 \ge 2$ then $k \le q + 1$;*
(iv)   *q is odd, $i_2 \ge 2$, and $q - \sqrt{q}/4 + 11/4 < k \le q + 1$ then K is a subarc of a normal rational curve.*

Proof.   (i) See Proposition 11.   The points of a hyperoval in a plane and a point out of the plane is a $(q + 3)$-set with an internal 2-nucleus.
(ii)   If $k = q + 3$ and we project the points of K from an internal 2-nucleus to a plane not through this point then we get a $(q + 2)$-arc; so q is even.
(iii)   Let P and Q be two internal 2-nuclei of K. If $k = q + 2$ and we project the points of K from P to a plane then we get a $(q + 1)$-arc; so the points of K are on a cone.   Doing the same with Q we get another cone.   The intersection of these two cones is the line PQ and a twisted cubic curve.   This cubic curve contains the points of K, whence $k \le q + 1$.   This is a contradiction.   If q is even then there exists a k-set such that $k = q + 3$ and $i_2 = 2$.   Take a $(q + 1)$-arc in a plane and take two points from this plane.   If the line of these points intersects the plane in the nucleus of the $(q + 1)$-arc then the previous conditions are satisfied.
(iv)   A theorem of Segre (1967) says that a k-arc in $PG(2, q)$, q odd, $k > q - \sqrt{q}/4 + 7/4$ is a subarc of a conic.   Using this theorem and repeating

the projections given in (iii), we get that the points of K are on a twisted cubic.

Using Result 7 we give a new and simple proof of the theorem of Blokhuis, Bruen and Thas (1990).

Theorem 24. (Szönyi (1988) for $n = 2$; Blokhuis et al. (1990, Theorem 4 and 6) for $n \geq 3$). *Let K be a k-arc in $\Sigma = PG(n, q)$. If $k > (q/2) + n - 1$ for even q and $k > 2(q + 1)/3 + n - 1$ for odd q then K is contained in a unique complete arc of Σ.*

Proof. Let k_n denote the bound for k given in the Theorem. Let L denote the set of points of Σ that can be added to K, that is, let $L = \{P \in \Sigma : K \cup \{P\}$ is an arc$\}$. If K is contained in a unique complete arc C then $C = K \cup L$, otherwise $K \cup L$ is not an arc. If K has more than one completion then there exists a hyperplane π such that $(K \cup L) \cap \pi$ has more than n points. Let $K_\pi = K \cap \pi$ and $L_\pi = L \cap \pi$.

First assume that K_π has a point Q such that there are no two points $P_1, P_2 \in L_\pi$ which are collinear with Q. Let α be a hyperplane not through Q. Let us project the points of $K \cup L_\pi$ from Q to α. The images of K and L_π are denoted by K' and L'_π. It is clear that $|K'| = k - 1$ and L'_π has n points. So K' is an arc in α having more than one completion. Suppose that the theorem is true for $n - 1$. For K' this means that $k - 1 \leq k_{n-1}$. Suppose to the contrary that $k > k_n$. But $k_{n-1} + 1 = k_n$; so there is a contradiction. For $n = 2$ we use Result 7.

If there are three collinear points P_1, P_2, P_3 such that $P_1, P_2 \in L$, $P_3 \in K$, then there exists a hyperplane π containing $n - 1$ points of K and $P_1, P_2 \in L$ ($k > k_n$ is assumed). Let $Q \in K \cap \pi$, $Q \neq P_1$ and let α be a hyperplane not through Q. Then we may repeat the projection and induction as above.

Finally suppose that there is no plane π having more than n points of $K \cup L$ and intersecting K. Let us add a point $P \in L$ to K. Then we may extend $K \cup \{P\}$ by any point of $L \setminus \{P\}$. So we may repeat the previous steps with the set $K \cup \{P\}$.

Theorem 25. *Let K be a k-set in $\Sigma = PG(n, q)$. If $i_2 > q/2 + n - 1$ for even q and $i_2 > 2(q + 1)/3 + n - 1$ for odd q then K is an arc, that is $K = I(K)$.*

Proof. This follows directly from Theorem 24.

Acknowledgement. The author thanks T. Szönyi for fruitful discussions.

References

A. Barlotti (1955), Un'estensione del teorema di Segre - Kustaanheimo, *Boll. Un. Mat. Ital.* **10**, 498-506.

A. Bichara and G. Korchmáros (1982), Note on (q + 2)-sets in a Galois plane of order q, *Ann. Discrete Math.* **14**, 117-122.

A. Blokhuis (to appear), The characterisation of seminuclear sets in a finite projective plane, *J. Geom.*

A. Blokhuis and A.A. Bruen (1989), The minimal number of lines intersected by a set of q + 2 points, blocking sets and intersecting circles, *J. Combin. Theory Ser. A* **50**, 308-315.

A. Blokhuis and H. Wilbrink (1987), Characterisation of exterior lines of certain sets of points in $PG(2, q)$, *Geom. Dedicata* **23**, 253-254.

A. Blokhuis, A. A. Bruen and J.A. Thas (1990), Arcs in $PG(n, q)$, MDS codes and three fundamental problems of B. Segre - some extensions, *Geom. Dedicata* **35**, 1-11.

A.A. Bruen (1988), Finite geometries - a miscellany, Lecture given at the International Conference "Combinatorics '88", Ravello 23-28 May 1988.

A.A. Bruen (1990), Nuclei of sets of q + 1 points in $PG(2, q)$ and blocking sets of Rédei type, *J. Combin. Theory Ser. A* **55**, 130-132.

A.A. Bruen and F. Mazzocca (1990), Nuclei of sets in finite projective and affine spaces, submitted to *Combinatorica*.

A.A. Bruen, J.A. Thas and A. Blokhuis (1988), On MDS codes, arcs in $PG(n, q)$ with q even, and a solution to three fundamental problems of B. Segre, *Invent. Math.* **92**, 441-459.

T. Illés, T. Szönyi and F. Wettl (1989), Blocking sets and maximal strong representative systems in finite projective planes, Proc. "Conference on Blocking Sets", Giessen 1989, to appear.

F. Kárteszi (1976), *Introduction to Finite Geometries*, Akadémiai Kiadó, Budapest.

G. Korchmáros and F. Mazzocca (1990), Nuclei of point sets of size q + 1 contained in the union of two lines in $PG(2, q)$, Preprint.

F. Mazzocca (1985), unpublished.

G. Panella (1955), Caratterizzazione delle quadriche di uno spazio (tridimensionale) lineare sopra un corpo finito, *Boll. Un. Mat. Ital.* **10**, 507-513.

B. Segre (1955), Ovals in a finite projective plane, *Canad. J. Math.* **7**, 414-416.

B. Segre (1967), Introduction to Galois geometries (ed. J.W.P. Hirschfeld), *Atti Accad. Naz. Lincei Mem.* **8**, 133-236.

T. Szönyi (1988), k-sets in $PG(2, q)$ having a large set of internal nuclei, *Proc. "Combinatorics '88"*, to appear.

T. Szönyi (1989), Arcs and k-sets with large nucleus-set in Hall planes, *J. Geom.* **34**, 187-194.

G. Tallini (1961), On caps of kind s in a Galois r-dimensional space, *Acta Arith.* **7**, 19-28.

F. Wettl (1987), On the nuclei of a finite projective plane, *J. Geom.* **30**, 157-163.

Menon difference sets and relative difference sets

P.R. Wild
Royal Holloway and Bedford New College, London

1. Introduction

A (v, k, λ)-difference set is a set D of k elements in a group G of order v such that for every $g \in G \setminus \{1\}$ there are exactly λ pairs $d_1, d_2 \in D$ such that $d_1 d_2^{-1} = g$.   The incidence structure whose points are the elements of G and whose blocks are all translates Dg for $g \in G$ is a symmetric 2-(v, k, λ) design.   The complement $\overline{D} = G \setminus D$ of D is a $(v, v - k, v - 2k + \lambda)$-difference set.

A Menon difference set is one whose parameters satisfy $v = 4(k - \lambda)$. Menon (1962) has shown that if D_1, D_2 are difference sets in groups G_1, G_2 respectively then $D_1 \times D_2 \cup \overline{D_1} \times \overline{D_2}$ is a difference set in $G_1 \times G_2$ if and only if D_1 and D_2 are both Menon difference sets (and then D is also a Menon difference set).   The parameters of a Menon difference set satisfy $v = 4l^2, k = 2l^2 - l, \lambda = l^2 - l$ for some integer l.   The complement of a Menon difference set with parameter l is a Menon difference set with parameter $-l$.

Turyn (1965) discusses connections between Menon difference sets and Hadamard matrices and Barker sequences.   If A is the incidence matrix of a symmetric 2-(v, k, λ)-design then $M = 2A - J$ (where J is the matrix with every entry 1) is a Hadamard matrix if and only if $v = 4(k - \lambda)$.   Thus a Menon difference set determines a Hadamard matrix. A sequence $x_1, x_2, ..., x_v$, where $x_i \in \{1, -1\}$ for $i = 1, ..., v$, is a Barker sequence if the aperiodic autocorrelation coefficients $c_j = \sum_{i=1}^{v-j} x_i x_{i+j}$ satisfy $|c_j| \leq 1$ for all $j > 0$.   Turyn and Storer (1961) have shown that if $v > 13$ then necessarily v is even and the periodic autocorrelation coefficients $a_j = c_j + c_{v-j}$ satisfy $a_j = 0$ for $j = 1, ..., v - 1$.   A sequence whose periodic autocorrelation coefficients have this property is called perfect and then $D = \{i \mid x_i = 1\}$ is a Menon difference set in the cyclic group Z_v of integers modulo v.

The only known cyclic Menon difference set is the trivial difference set in Z_4 corresponding to the Barker sequence 111-1.   It has been conjectured that no other cyclic Menon difference set exists. Turyn (1968) has shown that if there exists a Menon difference set in

Z_v, $v > 4$ then $v \geq 12,100$.   Several authors (Calabro and Wolf 1968; Bomer and Antweiller 1987) have considered binary arrays with perfect periodic autocorrelation coefficients.   These correspond to abelian Menon difference sets (Chan et al. 1979).   In particular 2-dimensional perfect binary arrays correspond to Menon difference sets in groups which are the direct product of two cyclic groups.   Jedwab et al. (1990) have constructed infinite families of 2-dimensional perfect binary arrays.   These together with the construction of Menon (1962) may be used to obtain infinite families of Menon difference sets in abelian groups which are direct products of more that two cyclic groups.

The construction of Jedwab et al. (1990) relies on two techniques. The first is the use of supplementary difference sets and the second is the use of relative difference sets to provide these supplementary difference sets.   It turns out that the relative difference sets may be constructed by application of the same techniques and the construction may be iterated to give the infinite families.

2. Supplementary difference sets

Let G be a group of order v.   Let $D_1, D_2 \subseteq G$ be such that for every $g \in G \backslash \{1\}$ there are exactly λ pairs d_1, d_2 such that $d_1, d_2^{-1} = g$ where either both $d_1, d_2 \in D_1$ or both $d_1, d_2 \in D_2$.   Then D_1, D_2 are called (a pair of) supplementary difference sets (for λ).   If $|D_1| = |D_2| = k$ then the collection of all translates D_1g, D_2g for $g \in G$ forms the set of blocks of a 2-(v, k, λ)-design.   If the subsets D_1 and D_2 have different size then the collection of translates determines a 2-structure with two block sizes. It is clear how this concept may be generalized to a collection $D_1, ..., D_t$ of t-supplementary difference sets.

In some circumstances supplementary difference sets may be used to construct a difference set in a larger group.   Let G be an abelian group of order v and H a subgroup of G of index 2.   Suppose that D_1, D_2 are a pair of supplementary difference sets for λ in H.   Suppose there is an element $x \in G \backslash H$ such that for every $h \in H$ there are exactly λ pairs d_1, d_2 with $d_1 \in D_1, d_2 \in D_2$ such that $d_1d_2^{-1} = h$ or $d_1d_2^{-1} = h^{-1}x^2$.   Then we say that D_1, D_2 are compatible supplementary difference sets and the following holds.

Lemma 1. *Let G, H, D_1, D_2 and x be as above.   Then $D = D_1 \cup D_2x$ is a (v, k, λ)-difference set in G where $k = |D_1| + |D_2|$.*

Proof. An element $h \in H \backslash \{1\}$ can be expressed as $h = g_1 g_2^{-1}$ with $g_1, g_2 \in D$ exactly when $h = g_1 g_2^{-1}$ with $g_1, g_2 \in D_1$ or $h = (g_1 x^{-1})(g_2 x^{-1})^{-1} = g_1 g_2^{-1}$ with $g_1 x^{-1}, g_2 x^{-1} \in D_2$. Since D_1 and D_2 are supplementary difference sets there are exactly λ pairs g_1, g_2. An element $g \in G \backslash H$ can be expressed as $g_1 g_2^{-1}$ with $g_1, g_2 \in D$ exactly when $hx^{-1} = g = g_1(g_2 x^{-1})^{-1} x^{-1}$ with $g_1 \in D_1$, $g_2 x^{-1} \in D_2, h \in H$ or $hx^{-1} = g = g_1 x^{-1}(g_2 x^{-1})^{-1} x$ with $g_2 \in D_1, g_1 x^{-1} \in D_2, h \in H$. But since D_1 and D_2 are compatible there are λ such pairs $g_1, g_2 x^{-1}$ and $g_2, g_1 x^{-1}$. $\square$

The problem remains to construct compatible supplementary difference sets. The supplementary difference sets used by Jedwab et al. (1990) correspond to relative difference sets.

3. Relative difference sets

Let H be a group and K a normal subgroup of H. A subset S of H is called a relative λ-difference set of H with respect to K if for every $h \in H \backslash K$ there are exactly λ pairs of elements $s_1, s_2 \in S$ such that $h = s_1 s_2^{-1}$ and there are exactly λ pairs of elements $s_3, s_4 \in S$ such that $h = s_3^{-1} s_4$. (Note that these two conditions are equivalent if H is ableian.)

The following is one way in which relative difference sets may arise. Let H be an abelian group and K a subgroup of H. Let D be a (v, k, λ)-difference set of the factor group H/K. Let $\phi : H \to H/K$ be a homomorphism with kernel K. Put $S = \phi^{-1}(D)$.

Lemma 2. *Let H, K and S be as above. Then S is a relative $\lambda |K| -$ difference set of H with respect to K.*

Proof. Let $g \in H \backslash \{1\}$. Then there are λ pairs $Kg_1, Kg_2 \in H/K$ with $\phi(g) = Kg_1(Kg_2)^{-1} = Kg_1 g_2^{-1}$. Moreover, there are $|K|$ pairs $h_1 \in Kg_1$ and $h_2 \in Kg_2$ with $g = h_1 h_2^{-1}$. Thus there are $\lambda |K|$ pairs $h_1, h_2 \in S$ with $g = h_1 h_2^{-1}$. $\square$

Note that for every $k \in K \backslash \{1\}$ there are exactly $|D||K|$ pairs $s_1, s_2 \in S$ with $k = s_1 s_2^{-1}$. Any subset $S' \subseteq H$ such that S and S' are supplementary difference sets in H will be a relative μ-difference set of H with respect to K. Moreover for each element $k \in K \backslash \{1\}$ there will be exactly $\mu - |K|(|D| - \lambda)$ pairs of elements $s_1, s_2 \in S'$ such that $k = s_1 s_2^{-1}$.

Example (Jedwab et al. 1990). Suppose $H = Z_{2s} \times Z_t$ is the direct product of two cyclic groups and $K = \langle (s, 0) \rangle$. Suppose D is a Menon difference set with parameters $v = st = 4l^2$, $k = l^2 - l$, $\lambda = l^2 - l$. Thus S is a relative 2λ-difference set with respect to K and $|S| = 4l^2 - 2l$. Counting differences between distinct elements of S' we have $(8l^2 - 2)\mu + w = |S'|(|S' - 1)$ where w is the number of times $(s, 0)$ occurs as a difference of two elements of S'. For S and S' to be supplementary difference sets we must have $2\lambda + \mu = |S| + w$ so that $\mu = 2l^2 + w$ and it follows that $|S'| \geq 4l^2$. Now since $|S'| \geq |H|/2$ it follows that $w \geq 2(|S'| - 4l^2)$. Substituting for $|S'|$ we obtain $w^2 - (16l^2 - 2)w \geq 0$. Hence $w = 0$ and $|S'| = 4l^2$. Thus S' is a relative $2l^2$-difference set of H with respect to K such that no difference $s_1 s_2^{-1}$, where $s_1, s_2 \in S'$, equals $(s, 0)$.

Now let R be a set of representatives of the cosets of K in H. Then $S = (S \cap R) \cup (S \cap R)x$ and $S' = (S' \cap R) \cup (H \backslash (S \cap R))x$, and it is easy to see that every element $h \in H$ may be expressed as $g = s_1 s_2^{-1}$ with $s_1 \in S$, $s_2 \in S'$ in exactly $|S \cap R| = k$ ways. It follows that if H is a subgroup of index 2 in the abelian group $G = \langle H, y \rangle$ then S and S' are compatible supplementary difference sets for $2\lambda + \mu = 2|S \cap R|$ and $S \cup S'$ is a Menon difference set of G.

Theorem 3. *If there exists a Menon difference set in $Z_s \times Z_t$ and a relative $st/2$-difference set of $Z_{2s} \times Z_t$ relative to $\langle (s, 0) \rangle$ then there exists a Menon difference set in $Z_{4s} \times Z_t$ and one in $Z_{2s} \times Z_{2t}$.* $\square$

4. Supplementary relative difference sets

The construction of the previous section relies on the existence of the relative difference set S'. We shall now see that such relative difference sets may be constructed by the same method given above for constructing difference sets.

Let G be an abelian group. Let H be a subgroup of G of index 2 and let $x \in G \backslash H$. Further let K be a subgroup of H. Let D_1, D_2 be relative difference sets of H with respect if K. We define D_1 and D_2 to be compatible supplementary relative difference sets in a way completely analagous to the difference set case above. Then D_1 and D_2 are compatible supplementary relative difference sets exactly when $D_1 \cup D_2 x$ is a relative difference set of G with respect to K.

Again we consider the possibility that $D_1 = \phi^{-1}(D)$ where D is a relative difference set of $H \backslash N$ with respect to $\langle K, N \rangle / N$ for some subgroup N of H and ϕ is the appropriate homomorphism.

It is easy to see that D_1 is a relative difference set of H with respect to $\langle K, N \rangle$ and it follows that D_2 must also be a relative difference set of H with respect to $\langle K, N \rangle$.

Example (Jedwab et al. 1990).   Suppose $H = Z_{2s} \times Z_{2t}$, $K = \langle (s, 0) \rangle$, $N = \langle (0, t) \rangle$ and suppose D is a relative $st/2$-difference set of H/N with respect to $\langle K, N \rangle / N$. Then D_1 is a relative st-difference set of H relative to $\langle (s, 0), (0, t) \rangle$.   The elements $(s, 0)$ and (s, t) never occur as a difference $d_1 d_2^{-1}$ with $d_1, d_2 \in D_1$, while $(0, t)$ occurs as such a difference $2st$ times.

If D_1 and D_2 are supplementary relative difference sets of H with respect to $\langle (s, 0), (0, t) \rangle$ then a similar calculation as before shows that D_2 is a relative st-difference set of H with respect to $\langle (s, 0), (0, t) \rangle$ with $|D_2| = 2st$ such that the elements $(s, 0)$ and $(0, t)$ never occur as a difference and the element (s, t) occurs $2st$ times as a difference. Furthermore $D_1 \cup D_2 x$ is a relative $2st$-difference set of G with respect to $\langle (s, 0), (0, t) \rangle$.

5. Menon difference sets

The following observation shows that under certain conditions the relative difference set D_2 of the previous section exists if and only if D_1 (that is D) exists.

Lemma 4. *Let* $G = Z_s \times Z_t$. *Suppose that* $t/(s, t)$ *is odd.   Then the mapping* $\alpha: G \rightarrow G$ *given by* $(a, b)\alpha = (a + bs/(s, t), b)$ *is an isomorphism fixing* $(s, 0)$ *and interchanging* $(0, t)$ *and* (s, t).

Proof.   The mapping α maps $(1, 0), (0, 1)$ to $(1, 0), (s/(s, t), 1))$ respectively.   Both these pairs of elements generate G. □

The isomorphism α maps D_1 of the example above onto D_2. Thus the following result holds.

Lemma 5. *If there exists a relative* $st/2$-*difference set in* $Z_{2s} \times Z_t$ *with respect to* $\langle (s, 0) \rangle$ *and* $t/(s, t)$ *is odd then there exist a relative* $2st$-*difference set in* $Z_{4s} \times Z_{2t}$ *with respect to* $\langle (2s, 0) \rangle$. □

Combining Theorem 3 and Lemma 5 we have the following theorem (Jedwab et al. 1990):

Theorem 6. *If there exists a Menon difference set in $Z_s \times Z_t$ and a relative $st/2$-difference set in $Z_{2s} \times Z_t$* with respect to $\langle (s, 0) \rangle$ and $t/(s, t)$ is *odd, then there exist Menon difference sets in $Z_{2s} \times Z_t$ and $Z_{4s} \times Z_{2t}$ and a relative $2st$-difference set in $Z_{4s} \times Z_{2t}$ with respect to $\langle (2s, 0) \rangle$.* $\square$

Clearly Theorem 6 may be applied repeatedly. Since there exist trivial examples when $s = t = 1$ and examples when $s = t = 6$ (Bomer and Antweiler 1987; Jedwab et al. 1990) infinite families of Menon difference sets result.

We remark that the condition $t/(s, t)$ is odd is not restrictive as we can state a similar result with s and t interchanged and at least one of $t/(s, t)$ and $s/(s, t)$ is odd while the conditions on a Menon difference set are symmetric in s and t.

References

R.H. Barker (1953), Group synchronizing of binary digital systems, *Communication Theory* (ed. W. Jackson), 273-287, Academic Press, New York.

L.D. Baumert (1971), *Cyclic Difference Sets*, Lecture Notes in Math. **182**, Springer-Verlag, New York.

L. Bomer and M. Antweiler (1987), Perfect binary arrays with 36 elements, *Electronics Letters* **23**, 730-732.

C. Calabro and J.K. Wolf (1968), On the synthesis of two dimensional arrays with desirable correlation properties, *Information and Control* **11**, 537-560.

Y.K. Chan, M.K. Siu and P. Tong (1979), Two-dimensional binary arrays with good autocorrelation, *Information and Control* **42**, 125-130.

J. Jedwab, C. Mitchell, F.C. Piper and P.R. Wild (1990), Perfect binary arrays and difference sets, submitted.

P.K. Menon (1962), On difference sets whose parameters satisfy a certain relation, *Proc. Amer. Math. Soc.* **13**, 739-745.

R.J. Turyn (1965), Character sums and difference sets, *Pacific J. Math.* **15**, 319-346.

R.J Turyn (1968), Sequences with small correlation, *Error Correcting Codes* (ed. H.B. Mann), 195-228, Wiley, New York.

R.J. Turyn and J. Storer (1961), On binary sequences, *Proc. Amer. Math. Soc.* **12**, 394-399.

List of Talks

T. Maruta	A geometric approach to linear cyclic codes
G. Mason	Is there a Galois theory for translation generalized n-gons?
V.C. Mavron	Designs and conics
G.E. Moorhouse	Codes of Bruck nets
C.M. O'Keefe	Some recent results on hyperovals in Desarguesian planes
A. Pasini	A common characterization of projective geometries, truncated projective geometries, polar spaces and their truncations, affine polar spaces and their standard quotients, attenuated spaces (including dual affine spaces), their truncations and their duals, affine attenuated spaces and something else
S.E. Payne	Fisher flocks and partial flocks of quadratic cones
T. Penttila	Some recent results on ovoids in $PG(3, q)$
J.J. Seidel	Groups presented by two-graphs
E.E. Shult	Geometric hyperplanes of non-embeddeble Grassmannians
J. Siemons	On two problems of Wielandt and Dembowski
L. Storme	k-arcs on a normal rational curve in $PG(n, q)$
T. Szönyi	A problem about squares of $GF(p^h)$ and its application to geometry
V.D. Tonchev	Intersection numbers of quasi-multiples of symmetric designs
H. Van Maldeghem	Common characterizations of the finite Moufang polygons
F. Wettl	Internal nuclei of k-sets in finite spaces of three dimensions
Z. Wan	Lattices generated by transitive sets of subspaces under finite classical groups
P. Wild	Menon difference sets and relative difference sets

List of Participants

A. Ali, Sussex
L. Bader, Rome
A. Barlotti, Florence
L.M. Batten, Manitoba
A. Beutelspacher, Giessen
A. Bichara, Rome
J. Bierbrauer, Heidelberg
A. Blokhuis, Eindhoven
A.E. Brouwer, Eindhoven
J.M.N. Brown, York (Toronto)
F. Buekenhout, Brussels
A.R. Calderbank, Bell Labs.
A. Cohen, C.W.I. Amsterdam
H. Cuypers, Michigan State
F. De Clerck, Ghent
M. de Finis, Rome
A. Delandtsheer, Brussels
M.J. de Resmini, Rome
V. De Smet, Ghent
P. de Vito, Naples
G.L. Ebert, Delaware
J.C. Fisher, Regina
P. Fisher, Wyoming
D. Ghinelli, Rome
W. Haemers, Tilburg
J.I. Hall, Michigan State
R. Hill, Salford
J.W.P. Hirschfeld, Sussex
C.Y. Ho, Florida
S.A. Hobart, Wyoming
D.R. Hughes, QMW
H. Kaneta, Okayama
W.M. Kantor, Oregon
J.D. Key, Birmingham

G. Korchmáros, Basilicata
C. Lefèvre-Percsy, Brussels
S. Löwe, Braunschweig
G. Lunardon, Naples
M. Marchi, Brescia
T. Maruta, Okayama
G. Mason, Calif. (Santa Cruz)
V.C. Mavron, Wales (Aberyst.)
F. Mazzocca, Naples
G.E. Moorhouse, Wyoming
C.M. O'Keefe, Western Aust.
D. Olanda, Naples
U. Ott, Braunschweig
S.E. Payne, Colorado
A. Pasini, Naples
T. Penttila, Western Aust.
N. Percsy, Mons
S. Rees, Warwick
J. Saxl, Cambridge
J.J. Seidel, Eindhoven
E.E. Shult, Kansas State
J. Siemons, East Anglia
E. Spence, Glasgow
L. Storme, Ghent
T. Szönyi, Budapest
J.A. Thas, Ghent
V.D. Tonchev, Sofia
J. van Bon, Tufts
H. Van Maldeghem, Ghent
Z. Wan, Beijing
F. Wettl, Budapest
P. Wild, RHBNC
B.J. Wilson, RHBNC